Predicting Future Oceans

Predicting Future Oceans

Sustainability of Ocean and Human Systems Amidst Global Environmental Change

Edited by

Andrés M. Cisneros-Montemayor

Nippon Foundation Nereus Program, Institute for the Oceans and Fisheries, The University of British Columbia, Vancouver, BC, Canada

William W.L. Cheung

Changing Ocean Research Unit & Nippon Foundation Nereus Program, Institute for the Oceans and Fisheries, The University of British Columbia, Vancouver, BC, Canada

Yoshitaka Ota

Nippon Foundation Nereus Program, Vancouver, BC, Canada; School of Marine and Environmental Affairs, University of Washington, Seattle, WA, United States

Elsevier
Radarweg 29, PO Box 211, 1000 AE Amsterdam, Netherlands
The Boulevard, Langford Lane, Kidlington, Oxford OX5 1GB, United Kingdom
50 Hampshire Street, 5th Floor, Cambridge, MA 02139, United States

British Library Cataloguing-in-Publication Data
A catalogue record for this book is available from the British Library

Library of Congress Cataloging-in-Publication Data
A catalog record for this book is available from the Library of Congress

ISBN: 978-0-12-817945-1

For Information on all Elsevier publications
visit our website at https://www.elsevier.com/books-and-journals

Publisher: Candice Janco
Acquisition Editor: Louisa Munro
Editorial Project Manager: Redding Morse
Production Project Manager: Vijayaraj Purushothaman
Cover Designer: Matthew Limbert

Typeset by MPS Limited, Chennai, India

Contents

List of contributors

Juan José Alava Institute for the Oceans and Fisheries, University of British Columbia, Vancouver, BC, Canada; Fundación Ecuatoriana para el Estudio de Mamíferos Marinos (FEMM), Guayaquil, Ecuador

Edward H. Allison Nippon Foundation Nereus Program, School of Marine and Environmental Affairs, University of Washington, Seattle, WA, United States; CGIAR Research Program on FISH, WorldFish, Bayan Lepas, Malaysia

Rebecca G. Asch Department of Biology, East Carolina University, Greenville, NC, United States

Joey R. Bernhardt Institute for the Oceans and Fisheries, University of British Columbia, Vancouver, BC, Canada

Mike Bithell Department of Geography, University of Cambridge, Cambridge, United Kingdom

Robert Blasiak Stockholm Resilience Centre, Stockholm University, Stockholm, Sweden

Andre Boustany Principal Investigator, Fisheries, Monterey Bay Aquarium, CA, United Sates; Nicholas School of the Environment, Duke University, NC, United States

Richard Caddell School of Law and Politics, Cardiff University, Cardiff, United Kingdom

Brooke Campbell Australia National Centre for Ocean Resources and Security (ANCORS), University of Wollongong, Wollongong, NSW, Australia

Hing Man Chan Department of Biology, University of Ottawa, Ottawa, ON, Canada

Oai Li Chen Nippon Foundation Nereus Program, University of British Columbia, Vancouver, BC, Canada; Changing Ocean Research Unit, University of British Columbia, Vancouver, BC, Canada

William W.L. Cheung Institute for the Oceans and Fisheries, The University of British Columbia, Vancouver, BC, Canada; Changing Ocean Research Unit, The University of British Columbia, Vancouver, BC, Canada

Andrés M. Cisneros-Montemayor Nippon Foundation Nereus Program, Institute for the Oceans and Fisheries, The University of British Columbia, Vancouver, BC, Canada

Guillermo Ortuño Crespo Nicholas School of the Environment, Duke University, Beaufort, NC, United States

Larry B. Crowder Edward F. Ricketts Provostial Professor of Marine Ecology and Conservation at Hopkins Marine Station, Stanford University, CA, United states

Lisa Maria Dellmuth Department of Economic History and International Relations, Stockholm University, Stockholm, Sweden

B. Derrick *Sea Around Us*, Institute for the Oceans and Fisheries, University of British Columbia, Vancouver, BC, Canada

Hubert du Pontavice Ecology and Ecosystems Health, Agrocampus Ouest, Rennes, France and Nippon Foundation-Nereus Program, Institute for the Oceans and Fisheries, University of British Columbia, Vancouver, BC, Canada

Daniel C. Dunn Nicholas School of the Environment, Duke University, Beaufort, NC, United States; Centre for Biodiversity and Conservation Science, School of Earth and Environmental Sciences, University of Queensland, Brisbane, QLD, Australia

Tyler D. Eddy Institute for Marine & Coastal Sciences, University of South Carolina, Columbia, SC, United States; Fisheries & Marine Ecosystem Model Intercomparison Project (FishMIP)

Timothy H. Frawley Hopkins Marine Station, Stanford University, Pacific Grove, CA, United States

Thomas L. Frölicher Climate and Environmental Physics, Physics Institute, University of Bern, Bern, Switzerland; Oeschger Centre for Climate Change Research, University of Bern, Bern, Switzerland

Didier Gascuel Agrocampus Ouest, Ecology and Ecosystem Health Research Unit, Rennes, France

Kristen M. Green School of Earth, Energy, and Environmental Sciences, Stanford University, Stanford, CA, United States

Stephanie J. Green Hopkins Marine Station, Stanford University, Pacific Grove, CA, United States; Center for Ocean Solutions, Stanford University, Pacific Grove, CA, United States; Department of Biological Sciences, University of Alberta, Edmonton, AB, Canada

Solène A. Guggisberg Netherlands Institute for the Law of the Sea (NILOS), Utrecht University, Utrecht, The Netherlands; Nippon Foundation Nereus Program, University of British Columbia, Vancouver, BC, Canada

Patrick N. Halpin Nicholas School of the Environment, Duke University, Beaufort, NC, United States

Natasha Henschke Department of Earth, Ocean and Atmospheric Sciences, University of British Columbia, Vancouver, BC, Canada

L. Hood *Sea Around Us*—Indian Ocean, School of Biological Sciences, University of Western Australia, Crawley, WA, Australia

Tiff-Annie Kenny Department of Biology, University of Ottawa, Ottawa, ON, Canada

John N. Kittinger Center for Oceans, Conservation International, Honolulu, HI, United States; Center for Biodiversity Outcomes, Life Sciences Center, Julie Ann Wrigley Global Institute of Sustainability, Arizona State University, Tempe, AZ, United States; Conservation International, Betty and Gordon Moore Center for Science, Arlington, VA, United States

Vicky W.Y. Lam Nippon Foundation Nereus Program and Changing Ocean Research Unit, Institute for the Oceans and Fisheries, University of British Columbia, Vancouver, BC, Canada

Elizabeth J. Mansfield Hopkins Marine Station, Stanford University, Pacific Grove, CA, United States

Julia G. Mason Hopkins Marine Station, Stanford University, Pacific Grove, CA, United States

Chris McOwen UN Environment World Conservation Monitoring Centre (UNEP-WCMC), Cambridge, United Kingdom

Andrew Merrie Stockholm Resilience Centre, Stockholm University, Stockholm, Sweden

Erik J. Molenaar Netherlands Institute for the Law of the Sea, Utrecht University, Utrecht, The Netherlands; UiT The Arctic University of Norway, Tromsø, Norway

Josheena Naggea School of Earth, Energy, and Environmental Sciences, Stanford University, Stanford, CA, United States

Katrina Nakamura The Sustainability Incubator, Honolulu, HI, United States

William K. Oestreich Hopkins Marine Station, Stanford University, Pacific Grove, CA, United States

Henrik Österblom Stockholm Resilience Centre, Stockholm University, Stockholm, Sweden

Yoshitaka Ota Nippon Foundation Nereus Program, School of Marine and Environmental Affairs, University of Washington, Seattle, WA, United States

Muhammed A. Oyinlola Changing Ocean Research Unit, Institute for the Oceans and Fisheries, The University of British Columbia, Vancouver, BC, Canada

M.L.D. Palomares *Sea Around Us*, Institute for the Oceans and Fisheries, University of British Columbia, Vancouver, BC, Canada

D. Pauly *Sea Around Us*, Institute for the Oceans and Fisheries, University of British Columbia, Vancouver, BC, Canada

Matilda Tove Petersson Stockholm Resilience Centre, Stockholm University, Stockholm, Sweden

Colleen M. Petrik Program in Atmospheric and Oceanic Sciences, Princeton University, Princeton, NJ, United States; Department of Oceanography, Texas A&M University, College Station, TX, United States

Malin Pinsky Ecology, Evolution, and Natural Resources, Rutgers University, New Brunswick, NJ, United States

U. Rashid Sumaila Fisheries Economics Research Unit, Institute for the Oceans and Fisheries, The University of British Columbia, Vancouver, BC, Canada; UBC School of Public Policy and Global Affairs, The University of British Columbia, Vancouver, BC, Canada

Gabriel Reygondeau Changing Ocean Research Unit, Institute for the Oceans and Fisheries, University of British Columbia, Vancouver, BC, Canada

Sarah M. Roberts Duke University, Durham, NC, United States

Jorge L. Sarmiento Atmospheric and Oceanic Sciences, Princeton University, Princeton, NJ, United States

Rachel Seary Cambridge Coastal Research Unit, Department of Geography, University of Cambridge, Cambridge, United Kingdom; UN Environment World Conservation Monitoring Centre, Cambridge, United Kingdom

Rebecca Selden Ecology, Evolution, and Natural Resources, Rutgers University, New Brunswick, NJ, United States

Jennifer C. Selgrath Hopkins Marine Station, Stanford University, Pacific Grove, CA, United States

Katherine Seto Australia National Centre for Ocean Resources and Security (ANCORS), University of Wollongong, Wollongong, NSW, Australia

Gerald G. Singh Nippon Foundation Nereus Program, Institute for the Oceans and Fisheries, University of British Columbia, Vancouver, BC, Canada

Tom Spencer Cambridge Coastal Research Unit, Department of Geography, University of Cambridge, Cambridge, United Kingdom

Jessica Spijkers Stockholm Resilience Centre, Stockholm University, Stockholm, Sweden; ARC for Coral Reef Studies, James Cook University, Townsville, QLD, Australia

Charles A. Stock NOAA Geophysical Fluid Dynamics Laboratory, Princeton, NJ, United States

Elsie M. Sunderland Department of Environmental Health, Harvard T.H. Chan School of Public Health, Boston, MA, United States; Harvard John A. Paulson School of Engineering and Applied Sciences, Harvard University, Cambridge, MA, United States

Shannon S. Swanson School of Earth, Energy, and Environmental Sciences, Stanford University, Stanford, CA, United States

Wilf Swartz Nippon Foundation Nereus Program, Institute for Oceans and Fisheries, University of British Columbia, Vancouver, BC, Canada

Fernando González Taboada Atmospheric and Oceanic Sciences Program, Princeton University, Princeton, NJ, United States

Kisei R. Tanaka Postdoctoral Research Fellow, Atmospheric and Oceanic Sciences Program, Princeton University, Princeton, NJ, United States

Lydia C.L. Teh Nippon Foundation Nereus Program, Vancouver, BC, Canada; Institute for the Oceans and Fisheries, University of British Columbia, Vancouver, BC, Canada

Colin P. Thackray Harvard John A. Paulson School of Engineering and Applied Sciences, Harvard University, Cambridge, MA, United States

G. Tsui *Sea Around Us*, Institute for the Oceans and Fisheries, University of British Columbia, Vancouver, BC, Canada

Jose Urteaga School of Earth, Energy, and Environmental Sciences, Stanford University, Stanford, CA, United States

Marjo Vierros Nippon Foundation Nereus Program, Institute for the Oceans and Fisheries, AERL, University of British Columbia, Vancouver, BC, Canada

Colette C.C. Wabnitz Institute for the Oceans and Fisheries, University of British Columbia, Vancouver, BC, Canada

Timothy D. White Hopkins Marine Station, Stanford University, Pacific Grove, CA, United States

D. Zeller *Sea Around Us*—Indian Ocean, School of Biological Sciences, University of Western Australia, Crawley, WA, Australia

Preface

This volume is the culmination of eight years of research by the Nippon Foundation Nereus Program, a collaborative partnership of 18 institutes involving 23 principal investigators, 46 research fellows, and countless collaborators.

The Nippon Foundation Nereus Program began in 2011, in the immediate aftermath of the Great East Japan Earthquake. It was the most powerful earthquake ever recorded in Japan, and this tragic event cast its shadow at the onset of the program as we, the ocean science experts, struggled to understand the relevance of such an initiative when contrasted against the realities of life in coastal communities.

The Nereus Program was instituted to examine the concept of changing oceans—in terms of both environment and society—and to envision a future where sustainable oceans are not only an aspiration, but a possibility. For over eight years, we collaborated with researchers whose work is at the forefront of global environmental projections, from climate change impacts on ocean ecosystems to marine biodiversity and habitat loss, and what these projected futures mean for fisheries and coastal communities.

The first two sections of this volume focus on biophysical changes, using sophisticated simulation models that capture complex interactions in the oceans to address emerging issues such as the incidence of extreme events, pollution, and changing seasonality, and their effects on biodiversity. Building on these findings and insights, the next two sections examine subsequent impacts on fisheries, aquaculture, and other human activities that constitute the core relationship between coastal communities and peoples whose livelihoods are intertwined with the oceans. The final two sections address predominant challenges of ocean governance, from issues of transboundary cooperation to the compounding and sometimes conflicting demands that we place on our oceans. In each chapter, authors provide theories and evidence from their individual areas of expertise, highlight emerging issues and gaps in our understandings, and offer their perspectives as interdisciplinary ocean researchers. Each section also includes an introductory chapter that synthesizes findings and gives an overview of the section themes.

This volume is a comprehensive package of ocean science undertaken by the Nereus Program as the group explores what future oceans we can expect under current trajectories and identifies social issues that need urgent actions, such as human rights concerns linked to seafood production. Thus it is a unique display of the interdisciplinary culture of marine research. What you will be introduced to is not a kaleidoscope of the sciences of future oceans, but the voices of ocean researchers, passionate and sincere in their work and with genuine concerns over the future. I hope these chapters convince you that the future is not driven by only one thing, but will emerge through the accumulation of layers upon layers of environmental changes and by how we, people(s), respond to these changes and create new ideas and perspectives.

In the wake of that tragic day in Japan, or any place where the relationship between peoples and oceans has to be rebuilt, integrating science and policy is just as important as ensuring a humanitarian perspective. When coastal communities are faced with issues that are essential for their livelihoods and safety, solutions require a diversity of knowledge so that we can minimize any unintended consequences. Equally, we must understand how to represent the concerns of all people. Only by safeguarding equity for every member of the community, can we maintain our stewardship of the oceans.

The group of researchers who contributed to this volume has built a network that I believe can fulfill the role of experts as stewards of knowledge that can protect coastal livelihoods and fairness. This volume is an outcome of the continuous effort by a network of people who are willing to engage in interdisciplinary ways. Besides the book, this network is perhaps the most important contribution of the Nippon Foundation Nereus Program to future oceans and to the generations to come.

Dr. Yoshitaka Ota

Director, The Nippon Foundation Nereus Program,
Vancouver, BC, Canada

Acknowledgments

This work would not have been possible without the enormous effort and scholarship of the Nippon Foundation Nereus Program fellows, collaborators, and principal investigators, and their dedication to pursuing research that is cutting edge, transdisciplinary, and transformative. This book showcases the exemplary work of this highly diverse group of researchers, and we are eternally grateful for their commitment both to this scientific endeavor and to their fellow researchers. For many in the Nereus Program network, this book and their contributions reflect a unique time of growth and transition through professional careers, and we are grateful for their help in capturing that in a book that is both a display of current scientific understanding and a platform for future generations of researchers.

We thank the many reviewers who provided comments on chapters, and in particular Wilf Swartz for his help with the preparation of the book. We also thank the editorial team at Elsevier, particularly Louisa Hutchins, Redding Morse, Katerina Zaliva, and Vijayaraj Purushothaman, for their gracious support throughout the entire process.

This book is a product of the Nippon Foundation Nereus Program, a cross-disciplinary ocean research program bringing together the Nippon Foundation and 17 academic institutes across the world. We gratefully acknowledge these partner institutions and their additional support for Nereus Program fellows, principal investigators, and collaborators over the past decade. We would like to recognize that the Nereus Program reflects the vision of the Nippon Foundation's capacity-building efforts for ocean sustainability, led by its Chairman, Dr. Yohei Sasakawa, and his commitment to healthy oceans and communities. Furthermore, we acknowledge the continuous collaboration and advice of Mr. Mitsuyuki Unno throughout the duration of the program, with the strongest appreciation for his devotion to capacity building and innovation across research groups and fields.

It is our hope that the institutional, professional, and personal networks created throughout this process, and reflected in this book, will continue to grow to the benefit of coastal communities and future ocean sustainability.

SECTION 1

Predicting future oceans

CHAPTER 1

Predicting the future ocean: pathways to global ocean sustainability

William W.L. Cheung

Institute for the Oceans and Fisheries, The University of British Columbia, Vancouver, British Columbia, Canada

Chapter Outline

The Earth should be more accurately called the Ocean. Carl Sagan famously described our planet as the "pale blue dot" [1]. This view considers that the Earth is small relative to the size of the solar system while, more notably, the majority of our planet's surface is covered by the ocean. The ocean is key to the maintenance of the climatic condition of our planet, such as regulation of temperature and maintenance of water cycle, and makes it suitable for the vast diversity of life on Earth to survive, including humans [2,3]. The ocean and its biodiversity also provide many other services to people such as transportation, food, recreation, and culture. The importance of the ocean to us humans is largely undervalued historically [4], partly because of the remoteness of much of the ocean relative to what most people can see and experience every day. However, the tide is turning; in recent decades advancements in ocean natural and social sciences have helped us recognize the vital role of the ocean to the Earth system and humankind [3], along with a sense of the vast scale of our impacts on it.

Our awareness of the human impacts on the ocean and the knowledge about the ocean system is rapidly increasing (Fig. 1.1). Human activities have altered the biophysical properties of the ocean and the impacts on marine ecosystems are detectable by science and visible by the public at large [5] (Fig. 1.1). Such impacts have reduced the capacity of the ocean to support essential ecosystem services and the consequences have started to affect human wellbeing. Clear examples are overfishing and climate change driven by emission of massive amounts of greenhouse gas into the atmosphere that impacts marine ecosystems,

Predicting Future Oceans.
DOI: https://doi.org/10.1016/B978-0-12-817945-1.00001-0

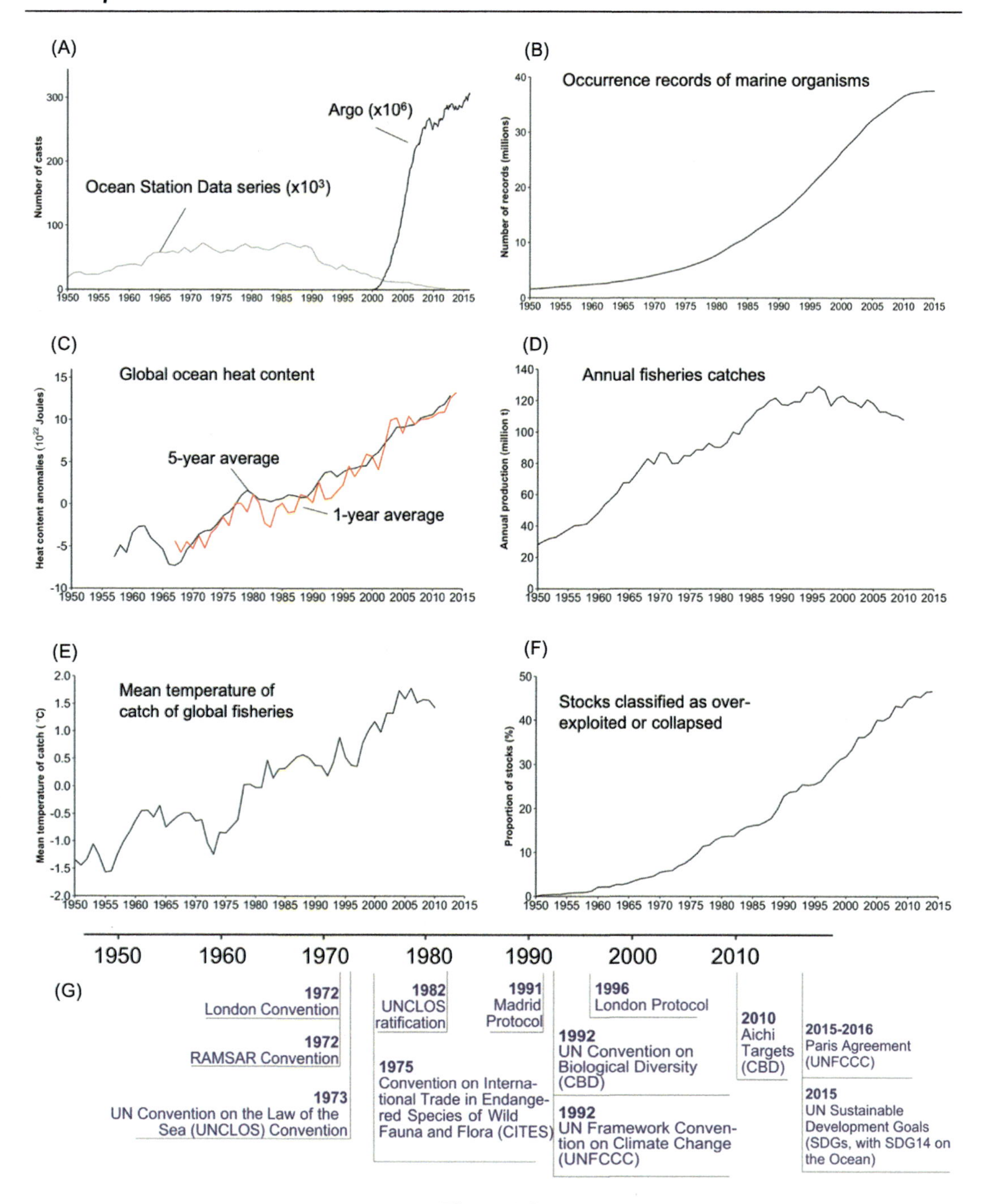

Figure 1.1

(Continued)

biodiversity, and coastal communities such as those in coral reefs [6–9]. Concurrently, over the last century, we have been rapidly generating and accumulating knowledge about the fundamental functioning of the ocean and its coupled natural and human systems. The advancement of technology and increasing international research collaborations have allowed us to go further, deeper, and spend more time observing the ocean. Moreover, new theories and models in oceanography, ecology, and the interrelationships between and within the ocean and human societies have helped us understand the role of the ocean to support human societies and the consequences of human activities on the ocean.

We are now at the crossroads of deciding what relationship between the ocean and people we would like to have; this will depend largely on the way we use, value, and govern the ocean in response to climate change and other anthropogenic stressors. Since the 1950s there have been major developments in the governance of the global ocean and regional seas (Fig. 1.1G). Examples of major ocean-related international agreements and governance approaches include the United Nations Convention on the Law of the Seas, Convention on Biological Diversity, the ecosystem approach to fisheries, the Paris Agreement (with specific mention of the ocean), and the United Nations Sustainable Development Goals (SDGs) (with a specific Goal for the ocean). International, regional, and local policy actions play an important role in determining the future of ocean(s). To make wise decisions, it is important to understand the consequences of actions and inactions, and identify available options for future ocean sustainability and their benefits, costs, and trade-offs.

◀ Time-series of indicators of ocean observation effort, consequences of human activities on marine environment, fish stocks, and fisheries, and key international marine policies: (A) number of ocean station data casts in the World Ocean Database ($\times 10^3$) and deployed from Argo[1] (including biogeochemical Argo) (10^6)—floats with automated instruments deployed to collect real-time ocean conditions data [10]; (B) number of records of occurrence of marine organisms in the Nererus-CORU marine biodiversity database (see Chapter 9: Current and future biogeography of exploited marine exploited groups under climate change); (C) global ocean heat content [11]; (D) global annual fisheries catch [12] (see Chapter 15: The *Sea Around Us* as provider of global fisheries catch and related marine biodiversity data to the Nereus Program and civil society); (E) mean temperature of global catches, computed from the mean of the temperature preference of species represented in the global fisheries weighted by their annual catch [13]; (F) number of fish stocks classified as overexploited or collapsed using the stock–status plot method (Sea Around Us: www.seaaroundus.org); and (G) year of signing or ratification of major international marine policies in relation to biodiversity, ecosystems, and ecosystem services. *Redrawn from J.-P. Gattuso, A.K. Magnan, L. Bopp, W.W.L. Cheung, C.M. Duarte, J. Hinkel, et al., Ocean solutions to address climate change and its effects on marine ecosystems, Front. Mar. Sci. 5 (2018) 337 [14].*

[1] Calculated based on an annual average casts per Argo float of 875,000 (https://oceanbites.org/oceantech-profiling-the-sub-surface-via-argo-floats/).

The use of scenarios and models is an important tool to integrate knowledge (including scientific, local, and traditional knowledge) and values to inform decisions and actions for the future ocean [15,16]. Scenarios are representations of possible futures for one or more components of a system under its drivers of changes, including alternative policy or management options. Models help describe the system qualitatively or quantitatively and can be used in association with scenarios to provide projections of plausible futures that are consistent with our current knowledge about the system. The use of scenarios and models requires a deep understanding of the functioning of coupled human–natural ocean systems, their past changes, current status, and future options—this encapsulates the essence of "Predicting the future ocean"[2]—an aspirational goal of the Nippon Foundation Nereus Program (hereafter called the Nereus Program). We set the time frame of the "prediction" [or more accurately projection[1]] to the mid-21st century (the 2050s). The contrast in outcomes of alternative decisions on actions and policies (e.g., climate mitigation) now may only start to become detectable robustly at that time frame, and yet a few decades into the future is a close enough time frame that people care about given that changes will be experienced by most of the current generation and their children.

This introduction chapter explains the key theoretical and analytical frameworks developed by the Nereus Program that contribute to the goal of "Predicting the future ocean" and thus provides the context and foundation for this interdisciplinary book. I also hope that the frameworks described in this chapter could inspire new ideas and be used or adapted by future studies. We chose to focus on four key components of "Predicting the future ocean": (1) characterizing the coupled human–natural marine system, (2) exploring the confidence and uncertainty in future ocean projections, (3) examining adaptation to the changing ocean, and (4) elucidating the linkages between the ocean and sustainable development. These four elements provide the framework to synthesize the understanding of the functioning of marine systems, using models and scenarios to generate projections of the future ocean to inform policies and decision-making, understanding the responses of human communities to the changing ocean, and viewing the ocean in the context of human society.

1.1 The coupled human–natural marine system

A first step that the Nereus Program took toward "predicting the future ocean" was the development of a framework for constructing scenarios and models for a coupled human–natural

[2] Scientific literature generally distinguishes the use of the term "projection" and "prediction." "Projection" tells us what could happen given a set of plausible, but not necessarily probable, circumstances. "Prediction" tells us what will happen, and thus implies more certainty than "projection." However, the Nippon Foundation Nereus Program chose "prediction" in the phasing of its goal because it is more understandable by the general public.

marine system [17]. Many previous efforts had been devoted to frameworks for models and scenarios for the biophysical components of marine systems. However, linkages of the biophysical system to parallel frameworks for the human dimension were less well-developed, except for the economic components of the human system. Given the strong and dynamic linkages and feedbacks between biophysical and human systems, it is important to consider the ocean as a coupled human–natural system. Frameworks that facilitate linkages and harmonization of scenarios and models across subsystems and scales are needed to fully understand the dynamics of marine systems and to generate knowledge to inform policy [18]. Such frameworks are relevant to the wide spectrum of knowledge generation approaches (e.g., from quantitative to qualitative).

Capitalizing on the wide range of expertise from natural to social sciences within the Nereus Program, an "end-to-end" framework for scenarios and models of the coupled human–natural marine system was developed (Fig. 1.2). The "natural system" component of the framework represents a typical conceptual model of marine ecosystems that includes the biogeochemical components of the environment, the marine food web, and human drivers that interact with these biophysical components, for example, greenhouse gas

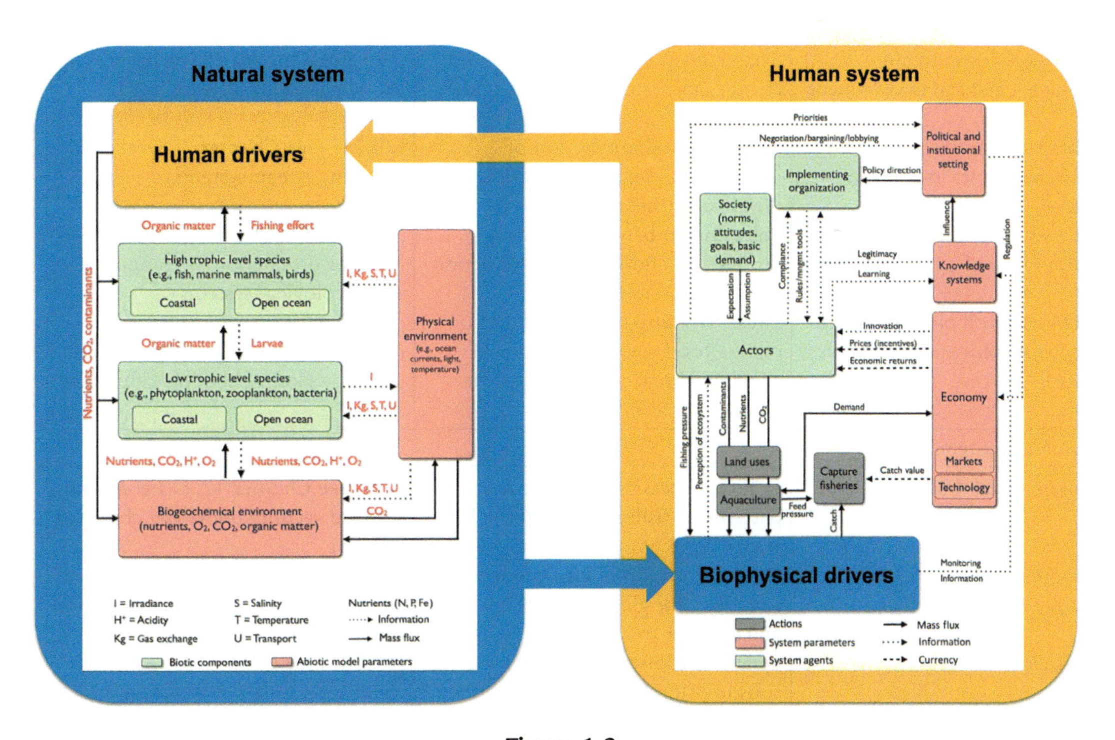

Figure 1.2

A schematic diagram depicting the coupled human–natural marine system developed at the beginning of the Nippon Foundation–Nereus Program. *Adapted from H. Österblom, A. Merrie, M. Metian, W.J. Boonstra, T. Blenckner, J.R. Watson, et al., Modeling social—ecological scenarios in marine systems, Bioscience 63 (9) (2013) 735–744.*

emissions, fishing, non-CO_2 pollution. Natural drivers are linked to the "human system" through human activities that directly interact with a natural system such as fishing and aquaculture. The human system includes a societal component with economic, knowledge, political, and institutional systems. A governance subsystem is specified to help represent policy decision-making and its implications for marine systems.

Much of the research in science and social sciences undertaken in the Nereus Program contributes to the improved understanding of the linkages and interrelationship between two or more components of this framework (Table 1.1). These research efforts contributed to answering overarching questions about the status, trends, and functioning of coupled human–natural marine systems, such as "how are ocean conditions changing?" "What factors determine the capacity of the ocean to produce fish?" "How are people dependent

Table 1.1: Examples of research undertaken by the Nereus Program that contribute to one or more components of the coupled human–natural system (Fig. 1.2).

Natural system		
System subcomponents	**Contribution**	**References**
Physical environment–biogeochemical environment	Characterization of projected changes in ocean conditions under scenarios of greenhouse gas emission	[19] (see Chapter 2: Synthesis: changing ocean systems)
Biogeochemical environment–lower trophic level–human drivers	Detection and attribution of different human and environmental drivers on fisheries catches	[20,21] (see Chapter 2: Synthesis: changing ocean systems)
Lower trophic level–upper trophic level	Relationship between primary production and fisheries production	[22] (see Chapter 3: Drivers of fisheries production in complex socioecological systems)
Physical and biogeochemical environment–upper trophic level	Effects of nutrient perturbation on marine food web; projections of climate change effects on marine biodiversity and fisheries	[23,24] (see Chapter 8: Marine biodiversity and ecosystem services: the large gloomy shadow of climate change)
Human system		
Biophysical drivers–fisheries	Environmental drivers of spatial patterns of fishing effort in the high seas	[25] (see Chapter 17: Fisheries and seafood security under the changing oceans)
Biophysical drivers–aquaculture	Characterization of environmental suitability for global marine aquaculture	[26] (see Chapter 17: Fisheries and seafood security under the changing oceans)
Fisheries–society	Quantification of global seafood consumption by coastal Indigenous communities; characterizing the interplay between fisheries and human conflict	[27,28] (see Chapter 25: Synthesis: changing social world of oceans; Chapter 33: Synthesis: the opportunities of changing ocean governance for sustainability)
Knowledge system–political/institutional settings	Using dynamic spatial approaches for marine conservation and fisheries management; the challenges to international marine law and policy under a changing ocean	[29,30] (see Chapter 40: Synthesis: oceans governance beyond boundaries: origins, trends, and current challenges)

on the ocean and fisheries?" "What principles and approaches are required to effectively govern the changing oceans?" The new knowledge generated from answering these questions helped improve the characterization of the coupled human–natural marine system and lay the foundation for credible projection of the future ocean.

1.2 Confidence and uncertainty in "predicting the future ocean"

Notwithstanding the usefulness of scenarios and models to inform and support decision-making for marine policies, projecting the complex coupled human–natural marine system decades into the future is associated with many uncertainties that must also be addressed and understood. The Nereus Program has extended a framework that is commonly used in assessing climate projection uncertainties [31,32] (Fig. 1.3, Chapter 7: Building confidence in projections of future ocean capacity). The framework categorizes uncertainties into three classes: scenario uncertainty, model uncertainty, and internal variability (Fig. 1.3). Scenario uncertainty consists of our uncertainties about the future trajectory of human and (or) biophysical drivers that are outside the bounds of the model being used to generate the projections, such as population growth, technology, and greenhouse gas emissions. Model uncertainty is associated with gaps in knowledge about the human–natural system being modeled, or the consequences of the choice of particular ways that the marine system is described in the model. Model uncertainty can be more specifically divided into parameter uncertainty and structural uncertainty, which involves choices about the way the

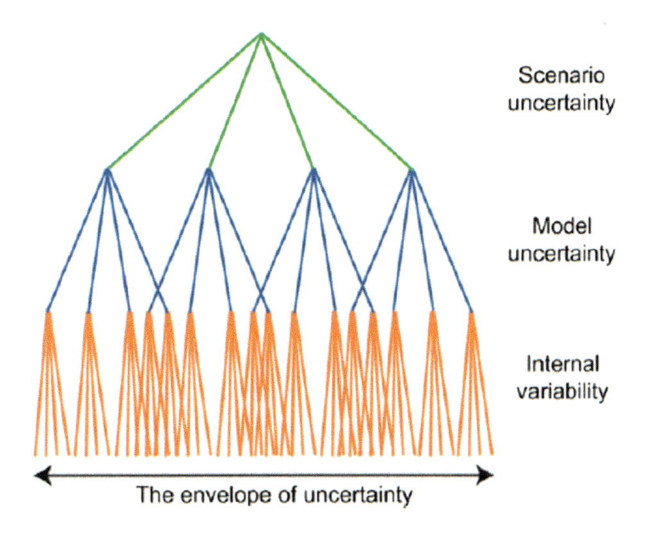

Figure 1.3
Framework that subdivides uncertainties about the future ocean into three classes based on their sources: scenario uncertainty, model uncertainty, and internal variability. *Redrawn from W.W.L. Cheung, T.L. Frölicher, R.G. Asch, M.C. Jones, M.L. Pinsky, G. Reygondeau, et al., Building confidence in projections of the responses of living marine resources to climate change, ICES J. Mar. Sci.73 (5) (2016) 1283–1296.*

human–natural system is being modeled (e.g., a size-structured fisheries model). Finally, internal variabilities are natural fluctuations generated from inherent processes of complex systems such as climate (e.g., El Niño Southern Oscillation), ecological, and social–economic systems. The three classes of uncertainties may be interrelated, and together form the envelope of projection uncertainties (Fig. 1.3).

Guided by the uncertainty framework, the scenarios and modeling work of the Nereus Program for marine systems, as described in various chapters in this book, includes three main components: (1) development of scenarios for future projections; (2) improvement of marine system understanding and systematic exploration of model uncertainties arising from the improvements in marine ecosystem understanding described in the previous section; and (3) exploration of the significance of system variability in addition to long-term mean changes.

The scenario works took a range of different forms, from qualitative [33] (see Chapter 48: Beyond prediction—radical ocean futures—a science fiction prototyping approach to imagining the future oceans) to quantitative [34], and for both the human and biophysical dimensions. Much of these explorations are based on archetypes of established scenarios used in global environmental assessments, such as the Shared Socioeconomic Pathways and the Representative Concentration Pathways [35]. The representation of the scenarios also took different forms, including "futuristic" visual descriptions of alternative futures and numerical projections of changes in biodiversity, fisheries and benefits to the society.

The Nereus Program applied existing approaches and frameworks to explore different aspects of model uncertainties, particularly through the use of multimodel approaches [36]. For example, we projected climate change effects on global marine ecosystems and fisheries using a wide variety of approaches such as species distribution modeling (including mechanistic, correlative and machine-learning) (Chapter 9: Current and future biogeography of exploited marine exploited groups under climate change), size-based modeling (Chapter 16: Life history of marine fishes and their implications for the future oceans), trophic spectrum-based modeling (Chapter 12: Changing biomass flows in marine ecosystems: from the past to the future), functional guild-based modeling [37], and empirical relationship-based modeling [22]. Uncertainties associated with the choice of parameters were then addressed during the development of each type of model. We also participated in community-based initiatives to compare global and regional scale marine ecosystem models, such as the Fisheries and Marine Ecosystem Impact Models Intercomparison Project (FISH-MIP) (Chapter 7: Building confidence in projections of future ocean capacity).

For the exploration of internal variability of marine ecosystems and fisheries, the Nereus Program focused on evaluating the implications of the variability of the climatic and oceanic system for marine ecosystems, biodiversity, and fisheries. Internal variability is characterized using projections from different ensemble members of one of the Earth system models (the NOAA–Geophysical Fluid Dynamics Laboratory Earth System Model 2M).

These ensemble projections were initialized using different sets of model parameters that generated different interannual variations of the climatic and oceanic system. Thus using these projections we were able to characterize, for example, the relative contribution of internal variability relative to other sources of uncertainties in projecting future changes in key ocean variables that are important to marine ecosystems [19]. The results of such analysis highlight the time frame for the "emergence" (i.e., long-term changes beyond historical variability) of changes in the key ecosystem stress. These changes, including warming, deoxygenation, acidification, and the occurrences of extreme events, such as heat waves, vary with these ocean variables and ocean basin [38]. These findings help inform the time frame and prioritize actions to address different types of impacts from the changing ocean.

1.3 Adaptation to the changing ocean

In the framework for the coupled human–natural marine system, we recognized the "adaptiveness" of the system, as it is dynamic and changing continuously in response to natural and human drivers [17]. One important aspect of the adaptive marine system that the Nereus Program focused on is adaptation to climate risks and impacts on marine biodiversity and fisheries at multiple levels (organization) and scales. We developed a framework to facilitate the organization of available studies and knowledge on climate adaptation in marine systems (Fig. 1.4) [39]. The framework recognizes that adaptation responses and actions are dependent on the contexts of specific levels and scales. For example, biological adaptation of fishes at an individual level to changing ocean conditions may ultimately relate to improving fitness (number of viable offspring) while adaptation actions at the individual human level may aim to maintain/improve livelihood or security (Fig. 1.4). The interplays of these different responses emerge into complex adaptation at the level of the coupled human–natural system [40].

Research of the Nereus Program contributed to the understanding of adaptive responses at different levels and scales of the coupled human–natural system. We synthesized knowledge about the adaptation of marine systems in published literature and found that there is a bias of knowledge toward natural systems and that information about responses that encapsulate both human and natural systems is limited [39]. Such knowledge gaps pose a barrier to future projections of the oceans that explicitly incorporate adaptation. This has thus informed the development of research projects that address these gaps using diverse approaches including laboratory studies to understand thermal adaptation of marine organisms (Chapter 8: Marine biodiversity and ecosystem services: the large gloomy shadow of climate change), participatory research to understand stakeholder responses to changes (Chapter 17: Fisheries and seafood security under the changing oceans), modeling of adaptive behavior of fish stocks and fisheries (Chapter 25: Synthesis: changing social world of oceans), and how fisheries and international policies are adapting to changing

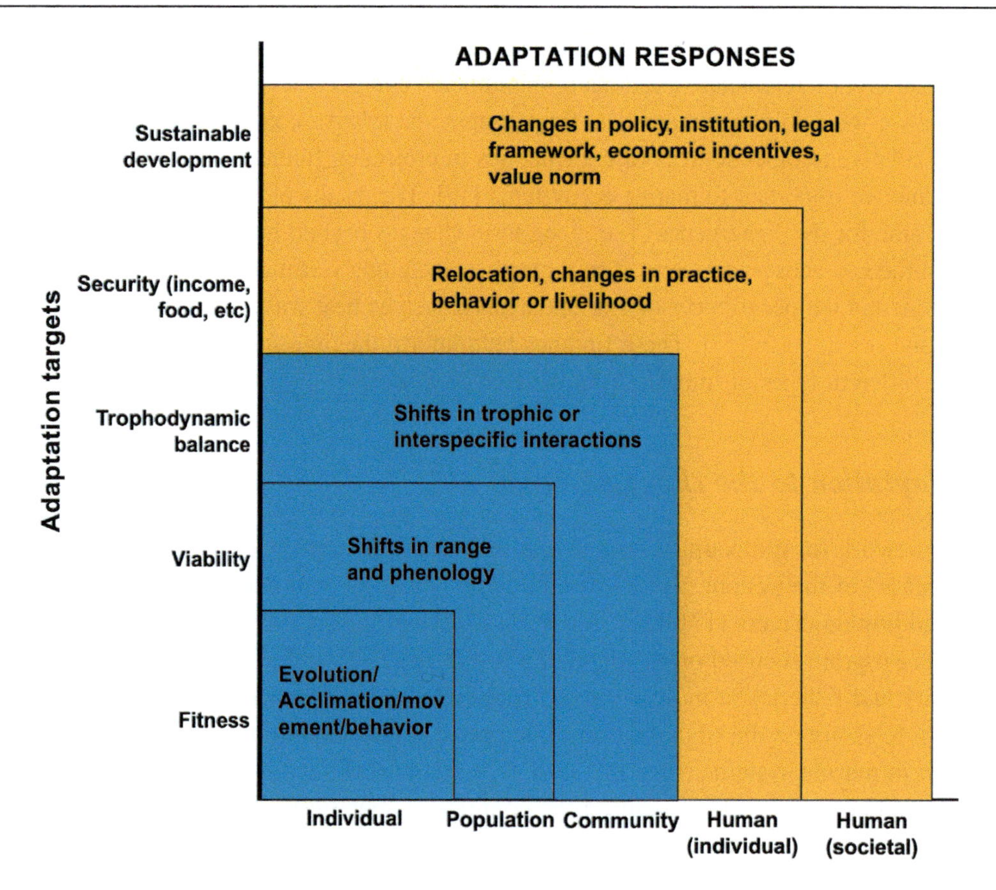

Figure 1.4

Adaptive responses to changing ocean (coupled human–natural system) at different levels of organization. *Blue*, natural system; *orange*, human system. *Adapted from D.D. Miller, Y. Ota, U.R. Sumaila, A.M. Cisneros-Montemayor, W.W.L. Cheung, Adaptation strategies to climate change in marine systems, Global Change Biol. 2017.*

oceans (Chapter 33: Synthesis: the opportunities of changing ocean governance for sustainability; Chapter 40: Synthesis: oceans governance beyond boundaries: origins, trends, and current challenges).

1.4 The linkages between the ocean, sustainable development, and policies

A main aspiration of Nereus' "predicting the future ocean" is to inform policies and pathways for the sustainable development of human society. Our research is rooted in an understanding of the human and natural components of the ocean systems as being closely linked. Thus changes in marine environments and our subsequent responses will have direct and indirect consequences on biodiversity, ecosystems, ecosystem services, and all dimensions of human wellbeing. The Nereus Program undertook a rapid appraisal of the potential linkages between ocean and sustainable development (as specified under the United Nations' SDGs) using expert knowledge and published literature to highlight these synergies and trade-offs within complex systems [41]. The resulting framework that links the ocean and SDGs provides a powerful way to identify opportunities and gaps in achieving sustainable development at both global and national scales (Chapter 32: Can aspirations lead us to the oceans we want?).

Overall the outputs of efforts to "predict the future ocean" are being used to inform international policies for ocean governance that ultimately contribute to securing sustainable development. Of particular focus in the Nereus Program is the governance of the high seas or areas beyond national jurisdiction (see Chapter 40: Synthesis: oceans governance beyond boundaries: origins, trends, and current challenges), which cover most of the global ocean. Although much of these areas are remote from us, they contribute substantially to the wellbeing of people (see Chapter 33: Synthesis: the opportunities of changing ocean governance for sustainability) and the Nereus Program has identified important policy gaps and ways in which these could be addressed (see Chapter 49: In conclusion: sustainable and equitable relationships between ocean and society).

The Nereus Program has aimed to spearhead innovative research, develop new knowledge, and offer new insights about the possible futures of the global ocean and its governance, as reflected partly in this book. In doing so, we also attempted to inform the development of future pathways for sustainable ocean development. The urgency of various current and emerging human pressures and threats on the ocean systems and the knowledge gaps that we identified highlight a continuing need for natural and social science research, but also existing opportunities and key strategies for adapting to ongoing change. In addition, one of the main goals of the Nereus Program is to develop human capacity for "predicting the future ocean;" the outcome of this is demonstrated in this book, in which most of the chapters are contributed by past and current postgraduate and postdoctoral fellows of the Nereus Program. In the final chapter (Chapter 49: In conclusion: sustainable and equitable relationships between ocean and society) we synthesize key findings from the interdisciplinary work of the Nereus Program and provide an outlook to secure the sustainability of our future ocean.

References

[1] C. Sagan, Pale Blue Dot: A Vision of the Human Future in Space, Random House Digital, Inc, 1997.
[2] C.H. Peterson, J. Lubchenco, et al., Marine Ecosystem Services, Island Press, Washington, DC, 1997.
[3] L. Inniss, A. Simcock, A.Y. Ajawin, A.C. Alcala, P. Bernal, H.P. Calumpong, et al., The First Global Integrated Marine Assessment: World Ocean Assessment I, United Nations, New York, 2016.

[4] E. Barbier, D. Moreno-Mateos, A. Rogers, J. Aronson, L. Pendleton, C. Van Dover, et al., Protect the deep sea, Nature. 505 (7484) (2014) 475–477.
[5] B.S. Halpern, M. Frazier, J. Potapenko, K.S. Casey, K. Koenig, C. Longo, et al., Spatial and temporal changes in cumulative human impacts on the world's ocean, Nat. Commun. 6 (2015) 7615.
[6] W.W.L. Cheung, D. Pauly, Impacts and effects of ocean warming on marine fishes, in: Explaining Ocean Warming: Causes, Scale, Effects and Consequences, 2016, pp. 239–253.
[7] M.A. MacNeil, N.A.J. Graham, J.E. Cinner, S.K. Wilson, I.D. Williams, J. Maina, et al., Recovery potential of the world's coral reef fishes, Nature 520 (7547) (2015) 341.
[8] T.P. Hughes, J.T. Kerry, A.H. Baird, S.R. Connolly, A. Dietzel, C.M. Eakin, et al., Global warming transforms coral reef assemblages, Nature 556 (7702) (2018) 492.
[9] T.J. Pitcher, W.W.L. Cheung, Fisheries: hope or despair? Mar. Pollut. Bull. 74 (2) (2013) 506–516.
[10] T.P. Boyer, J.I. Antonov, O.K. Baranova, C. Coleman, H.E. Garcia, A. Grodsky, et al., World Ocean Database, 2013, p. 2013.
[11] J. Blunden, D.S. Arndt, State of the climate in 2015, Bull. Am. Meteorol. Soc. 97 (8) (2016) Si–S275.
[12] D. Pauly, D. Zeller, Catch reconstructions reveal that global marine fisheries catches are higher than reported and declining, Nat. Commun. 7 (2016) 10244.
[13] W.W.L. Cheung, R. Watson, D. Pauly, Signature of ocean warming in global fisheries catch, Nature 497 (7449) (2013) 365.
[14] J.-P. Gattuso, A.K. Magnan, L. Bopp, W.W.L. Cheung, C.M. Duarte, J. Hinkel, et al., Ocean solutions to address climate change and its effects on marine ecosystems, Front. Mar. Sci. 5 (2018) 337.
[15] I.M.D. Rosa, H.M. Pereira, S. Ferrier, R. Alkemade, L.A. Acosta, H.R. Akcakaya, et al., Multiscale scenarios for nature futures, Nat. Ecol. Evol. 1 (10) (2017) 1416.
[16] IPBES, in: S. Ferrier, K.N. Ninan, P. Leadley, R. Alkemade, L.A. Acosta, H.R. Akçakaya, et al. (Eds.), Summary for Policymakers of the Methodological Assessment of Scenarios and Models of Biodiversity and Ecosystem Services of the Intergovernmental Science-Policy Platform on Biodiversity and Ecosystem Services, Secretariat of the Intergovernmental Science-Policy Platform on Biodiversity and Ecosystem Services, Bonn, Germany, 2016. 32 p.
[17] H. Österblom, A. Merrie, M. Metian, W.J. Boonstra, T. Blenckner, J.R. Watson, et al., Modeling social—ecological scenarios in marine systems, Bioscience 63 (9) (2013) 735–744.
[18] W.W.L. Cheung, C. Rondinini, R. Avtar, M. van den Belt, T. Hickler, J.P. Metzger, et al., Linking and harmonizing scenarios and models across scales and domains, The Methodological Assessment Report on Scenarios and Models of Biodiversity and Ecosystem Services, Secretariat of the Intergovernmental Science-Policy Platform on Biodiversity and Ecosystem Services, Born, Germany, 2016.
[19] T.L. Frölicher, K.B. Rodgers, C.A. Stock, W.W.L. Cheung, Sources of uncertainties in 21st century projections of potential ocean ecosystem stressors, Global Biogeochem. Cycles 30 (8) (2016) 1224–1243.
[20] C.J. McOwen, W.W.L. Cheung, R.R. Rykaczewski, R.A. Watson, L.J. Wood, Is fisheries production within Large Marine Ecosystems determined by bottom-up or top-down forcing? Fish Fish. 16 (4) (2015) 623–632.
[21] R.R. Rykaczewski, J.P. Dunne, W.J. Sydeman, M. García-Reyes, B.A. Black, S.J. Bograd, Poleward displacement of coastal upwelling-favorable winds in the ocean's eastern boundary currents through the 21st century, Geophys. Res. Lett. 42 (15) (2015) 6424–6431.
[22] C.A. Stock, J.G. John, R.R. Rykaczewski, R.G. Asch, W.W.L. Cheung, J.P. Dunne, et al., Reconciling fisheries catch and ocean productivity, Proc. Natl. Acad. Sci. U.S.A. 114 (8) (2017) E1441–E1449.
[23] K.A. Kearney, C. Stock, J.L. Sarmiento, Amplification and attenuation of increased primary production in a marine food web, Mar. Ecol. Prog. Ser. 491 (2013) 1–14.
[24] W.W.L. Cheung, G. Reygondeau, T.L. Frölicher, Large benefits to marine fisheries of meeting the 1.5°C global warming target, Science 354 (6319) (2016) 1591–1594.
[25] G.O. Crespo, D.C. Dunn, G. Reygondeau, K. Boerder, B. Worm, W. Cheung, et al., The environmental niche of the global high seas pelagic longline fleet, Sci. Adv. 4 (8) (2018) eaat3681.

[26] M.A. Oyinlola, G. Reygondeau, C.C.C. Wabnitz, M. Troell, W.W.L. Cheung, Global estimation of areas with suitable environmental conditions for mariculture species, PLoS One 13 (1) (2018) e0191086.
[27] A.M. Cisneros-Montemayor, D. Pauly, L.V. Weatherdon, Y. Ota, A global estimate of seafood consumption by Coastal Indigenous Peoples, PLoS One 11 (12) (2016) e0166681.
[28] J. Spijkers, T.H. Morrison, R. Blasiak, G.S. Cumming, M. Osborne, J. Watson, et al., Marine fisheries and future ocean conflict, Fish Fish. 19 (5) (2018) 98–806.
[29] D.C. Dunn, S.M. Maxwell, A.M. Boustany, P.N. Halpin, Dynamic ocean management increases the efficiency and efficacy of fisheries management, Proc. Natl. Acad. Sci. U.S.A. 113 (3) (2016) 668–673.
[30] M.L. Pinsky, G. Reygondeau, R. Caddell, J. Palacios-Abrantes, J. Spijkers, W.W.L. Cheung, Preparing ocean governance for species on the move, Science 360 (6394) (2018) 1189–1191.
[31] E. Hawkins, R. Sutton, The potential to narrow uncertainty in regional climate predictions, Bull. Am. Meteorol. Soc. [Internet] 90 (8) (2009) 1095–1107. Available from: https://doi.org/10.1175/2009BAMS2607.1.
[32] W.W.L. Cheung, T.L. Frölicher, R.G. Asch, M.C. Jones, M.L. Pinsky, G. Reygondeau, et al., Building confidence in projections of the responses of living marine resources to climate change, ICES J. Mar. Sci. 73 (5) (2016) 1283–1296.
[33] A. Merrie, P. Keys, M. Metian, H. Österblom, Radical ocean futures-scenario development using science fiction prototyping, Futures 95 (2018) 22–32.
[34] W.W.L. Cheung, M.C. Jones, V.W.Y. Lam, D.D. Miller, Y. Ota, L. Teh, et al., Transform high seas management to build climate resilience in marine seafood supply, Fish Fish. 18 (2) (2017) 254–263.
[35] D.P. van Vuuren, M.T.J. Kok, B. Girod, P.L. Lucas, B. de Vries, Scenarios in global environmental assessments: key characteristics and lessons for future use, Global Environ. Change [Internet] 22 (4) (2012) 884–895. Available from: http://www.sciencedirect.com/science/article/pii/S0959378012000635.
[36] W.W.L. Cheung, M.C. Jones, G. Reygondeau, C.A. Stock, V.W.Y. Lam, T.L. Frölicher, Structural uncertainty in projecting global fisheries catches under climate change, Ecol. Modell. 325 (2016) 57–66.
[37] V. Christensen, M. Coll, J. Buszowski, W.W.L. Cheung, T. Frölicher, J. Steenbeek, et al., The global oceans is an ecosystem: stimulating marine life and fisheries, Global Ecol. Biogeogr. 24 (5) (2015) 507–517.
[38] T.L. Frölicher, E.M. Fischer, N. Gruber, Marine heatwaves under global warming, Nature 560 (7718) (2018) 360.
[39] D.D. Miller, Y. Ota, U.R. Sumaila, A.M. Cisneros-Montemayor, W.W.L. Cheung, Adaptation strategies to climate change in marine systems, Global Change Biol. 24 (1) (2017) e1–e14.
[40] W.W.L. Cheung, The future of fishes and fisheries in the changing oceans, J. Fish. Biol. [Internet] 92 (3) (2018) 790–803. Available from: http://doi.wiley.com/10.1111/jfb.13558.
[41] G.G. Singh, A.M. Cisneros-Montemayor, W. Swartz, W. Cheung, J.A. Guy, T.-A. Kenny, et al., A rapid assessment of co-benefits and trade-offs among Sustainable Development Goals, Mar. Policy 93 (2017) 223–231.

SECTION 2

Changing Ocean Systems

CHAPTER 2

Changing ocean systems: A short synthesis

Charles A. Stock[1], William W.L. Cheung[2], Jorge L. Sarmiento[3] and Elsie M. Sunderland[4]

[1]NOAA Geophysical Fluid Dynamics Laboratory, Princeton, NJ, United States [2]Institute for the Oceans and Fisheries, The University of British Columbia, Vancouver, BC, Canada [3]Atmospheric and Oceanic Sciences, Princeton University, Princeton, NJ, United States [4]Department of Environmental Health, Harvard T.H. Chan School of Public Health, Boston, MA, United States

Chapter Outline

The ocean is often depicted as vast and unchanging. The first characterization is certainly true. Oceans cover over 70% of the Earth's surface with depths exceeding 10 km in some places. The second characterization is understandable to anyone who has stood on a beach and scanned the seascape as it stretches to the horizon, peered down on the ocean's expanse from an airplane, or listened to the unyielding rumble of waves breaking on the shore. The impression that the ocean is unchanging, however, is incorrect, or at least incomplete. While constant and unyielding in some respects, the ocean is in fact perpetually changing. This change can manifest dramatically in the passing of volatile storms or in rapid seasonal transitions. It can also be seen in more subtle ocean fluctuations and trends occurring over days to centuries that impact areas as small as local inlets and as large as ocean basins.

Fisheries and other marine resources exhibit profound responses to natural variations in climate and weather [1]. Such responses were evident even before the onset of industrial-scale fishing. Fish scales preserved in oxygen-deprived sediment layers off California, for example, reveal striking multidecadal shifts in dominance between Pacific sardine and northern anchovy that are correlated with climate-driven temperature fluctuations over the past 1600 years [2,3]. In the Bohuslän region of Southern Sweden, catch records and

Predicting Future Oceans.
DOI: https://doi.org/10.1016/B978-0-12-817945-1.00002-2

archeological evidence dating back 1000 years reveal periods of abundant herring during multidecadal cold periods and collapses of the local fishery during multidecadal warm events [4]. Similar changes can be seen in fluctuations in availability of species in the commercial seafood market today [5]. Striking examples arise during El Niño events, when anomalously weak easterly trade winds in the equatorial Pacific reduce upwelling and trigger an eastward propagation of warm, nutrient-poor western Pacific waters. This can lead to severe reductions in ocean productivity, reduced fisheries yields, and starvation of fish-reliant marine life in the central and eastern equatorial Pacific [6]. In 1972 a severe El Niño and subsequent overfishing contributed to the collapse of the world's largest fishery, the Peruvian anchoveta [7,8].

In today's ocean, the effect of ubiquitous "modes" of natural variability and fishing pressure has been compounded by other human-derived changes. The ocean's absorption of increasing atmospheric carbon dioxide (CO_2) associated with the burning of fossil fuels has acidified its waters [9]. Global warming arising from the accumulation of CO_2 and other greenhouse gases has warmed ocean waters and melted significant amounts of sea and land ice [10,11]. These ocean changes have been implicated in changes in seasonal ocean cycles, ocean productivity baselines, and ocean oxygen declines [12]. Effects associated with the accumulation of greenhouse gases are compounded by other anthropogenic stressors in the marine environment. Rapid development of coastal regions has led to enhanced nutrient inputs, eutrophication, and an increased frequency of harmful algal blooms in many coastal ecosystems [13,14]. Human activities release thousands of persistent organic pollutants and heavy metals to the atmosphere. Many of these chemicals biomagnify in food webs (i.e., are concentrated by a large degree with each increase in trophic level), and also pose risks to wildlife and seafood consumers [15,16].

Predicting the future ocean depends on understanding the response of ocean life to these myriad drivers and predicting future ocean changes. This chapter provides an overview of the ocean changes summarized above—ocean acidification, ocean warming and ice melt, changing ocean productivity baselines, ocean deoxygenation, changing coastlines, and ocean pollution. Contributions from the Nippon Foundation Nereus Program researchers and collaborators are highlighted. The chapters that follow in this Section provide deeper perspectives on a subset of these topics: ocean heatwaves and extremes (see Chapter 5, Extreme climatic events in the oceans), changes in oceans seasons (see Chapter 4, Changing seasonality of the sea), mercury pollution (see Chapter 6, Pathways of methylmercury accumulation), threatened vegetated coastal habitats (see Chapter 3, Drivers of ecosystem production in complex social-ecological systems), and efforts to project the impacts of changing oceans on fish populations (see Chapter 7, Building confidence in projections of future ocean capacity). Finally, this chapter concludes with a brief assessment of prospects for improved understanding of future ocean changes, and improved capacity to predict and adapt to them.

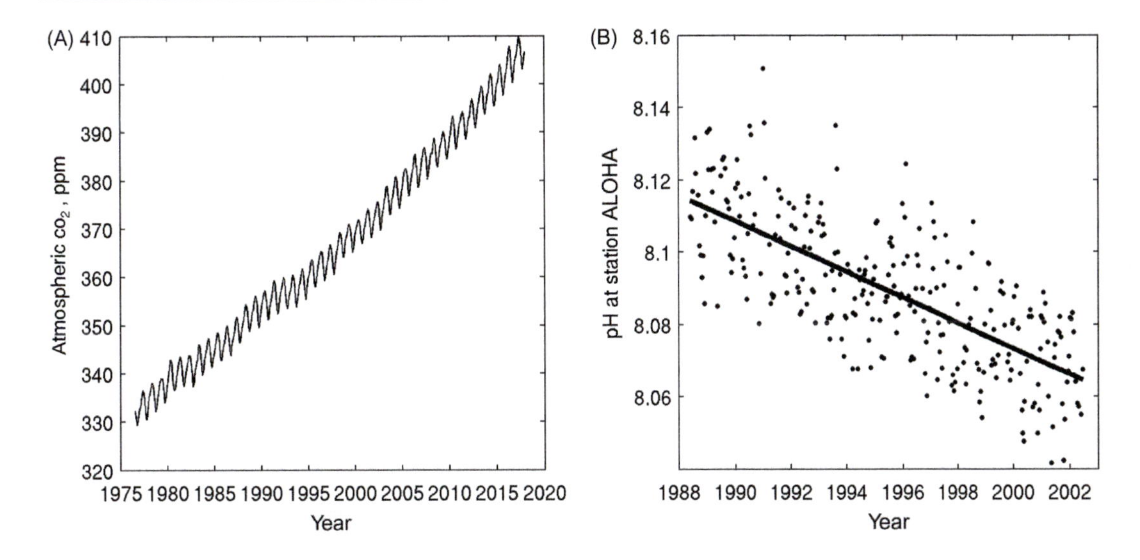

Figure 2.1
(A) The time-series of CO_2 observations observed at NOAA's Mauna Loa CO_2 observatory in Hawaii. Data are monthly averages of continuous surface flask measurements since 1976 described in Dlugokencky et al. [18]. (B) Ocean pH calculated from dissolved inorganic carbon, alkalinity, and other ocean measurements at station ALOHA [19]. *NOAA*, National Oceanic and Atmospheric Administration's.

2.1 Burning fossil fuels and ocean acidification

The extraction and subsequent burning of fossil fuels has released vast amounts of carbon dioxide into the atmosphere. This signal was first observed at the National Oceanic and Atmospheric Administration's Mauna Loa CO_2 observatory in Hawaii by David Keeling and colleagues [17] (Fig. 2.1A). Atmospheric CO_2 would be considerably larger were it not for the ocean. The ocean has absorbed approximately 30% of the combined CO_2 emissions from fossil fuels, cement production, and land use change since preindustrial times [20] and, given several thousand years, is capable of absorbing much more [21]. This capacity arises partly from the ocean's vast size, but also from the unique carbon chemistry of seawater. Most carbon dioxide absorbed from the atmosphere is rapidly partitioned into carbonate (CO_3^{2-}) and bicarbonate (HCO_3^-) ions that do not exchange with the atmosphere. Absorption of CO_2 by the ocean is assisted further by the "biological pump," whereby photosynthesis at the ocean's surface converts CO_2 to organic matter, some of which sinks before it can be respired back to CO_2. This creates a CO_2 deficit near the ocean surface relative to depth. If this biologically-driven surface CO_2 drawdown was removed, outgassing from the ocean would increase atmospheric CO_2 levels by $\sim$140 ppm [22].

The ocean has thus provided a great service to humankind by removing a large fraction of anthropogenic CO_2 from the atmosphere. This service, however, has come with a price. Formation of carbonic acid and its subsequent disassociation makes the ocean more acidic (i.e., it decreases the pH by releasing H^+ ions). The resulting "ocean acidification" has been observed at long-term ocean observatories such as the Bermuda Atlantic time-series (Fig. 2.1B) and surface ocean pH decreases in most areas are about 0.1 pH units. Future projections suggest that the ocean pH may decrease by up to 0.4 units under high CO_2 emissions scenarios [23], and work supported by the Nereus Program has shown that such changes rapidly surpass those originating from natural variability in most ocean regions [24].

Increasing CO_2 in the ocean may accelerate photosynthesis in some species of phytoplankton [25] but it is likely to negatively impact numerous other aspects of marine ecosystems [12,26–28]. These include hindering the biological formation and maintenance of coral reefs [29] and the shells of ecologically and commercially vital shellfish and crabs. Acidification also affects the survival of larval fish and the neurosensory responses of fishes [30,31]. This has the potential to affect the growth and population dynamics of exploited invertebrates and fish stocks [32–35]. Factors controlling ocean acidification and its effect on physiological and ecological processes remain an area of highly active investigation. Recent research has increasingly focused on the interactions between ocean acidification and other climatic (e.g., temperature and oxygen) and nonclimatic (e.g., chemical contaminants such as mercury) drivers [36,37].

2.2 Warming oceans, melting ice, and changing ocean circulation

The accumulation of CO_2 in the atmosphere detected at the Mauna Loa observatory (Fig. 2.1A) gains additional significance because CO_2 is a "greenhouse gas." That is, CO_2 absorbs outgoing infrared radiation emitted by the earth and reemits in all directions. This creates a net warming and, were it not for these greenhouse gases, the earth would be over 30°C cooler [38]. A corollary to this result is that an increase in atmospheric greenhouse gases could warm the earth. This possibility was first considered by the Swedish scientist Svante Arrhenius [39] and has been borne out by observations over the last century [10].

Water can absorb tremendous amounts of heat. It takes ~4000 Joules of energy to raise the temperature of one kilogram of water (about a liter) by 1°C, approximately 3500 times more than it takes to warm a similar volume of air. It is thus not surprising that over 90% of the excess heat in the earth system arising from the accumulation of greenhouse gases has been absorbed by the ocean [11,40]. This has resulted in a surface ocean warming rate of 0.11 (0.09–0.13)°C per decade over the past 40 years [11], with additional warming likely earlier in the century [41]. Arctic warming has led to the rapid reduction in summer sea ice [42], with nearly ice-free summers possible before the mid 21st century [43]. Projections suggest an additional surface warming of 2°C–3°C by the end of the 21st century under high emissions scenarios [44], pushing the ocean well outside the bounds of its natural variations in most regions [24].

Marine organisms have responded to the warming ocean, often shifting poleward and toward earlier seasonal timing of biological events [45]. Pinsky et al. [46], for example, used multidecadal surveys from across North America to demonstrate a robust tendency for the spatial distribution of fishes to follow preferred thermal windows as they shift over time. Warming has also been linked to increased dominance of warmwater affiliated species in global fish catch [47]. Such shifts are projected to continue [48] and have the potential to profoundly reshape regional fish assemblages. In some cases this has led to the shifts of fisheries across international boundaries, challenging management efforts [49]. The Nereus Program's research has scrutinized and refined these projections [50,51].

The warming ocean has also led to an increased prevalence of local temperature extremes, or "ocean heat waves" [52]. Record warm western Pacific temperatures in 2016, for example, led to coral bleaching of unprecedented extent on the Great Barrier Reef [53,54]. In the Northwest Pacific, a region of anomalous warmth ($>2.5°C$ above normal during its peak) developed in 2013, persisted for multiple years, and was linked to widespread ecosystem disruptions [55–57]. In New England, record temperatures in the Gulf of Maine led to highly anomalous early season lobster catch which collapsed prices, leading to an economic crisis for lobster fishermen [58]. Research arising from the Nereus Program has assessed the changing frequencies of these events under climate change [59,60] (see Chapter 5, Extreme climatic events in the oceans).

Lastly, the circulation of the atmosphere and ocean is intimately linked to the Earth's heat budget: atmospheric winds and ocean currents produce a net redistribution of heat from the tropics to the poles. Perturbing the heat budget through the introduction of greenhouse gases thus has far-reaching implications for ocean circulation [11]. Perhaps the most well-known example is the projected slowing of the North Atlantic overturning circulation [44]. Such a change would have global implications for ocean ecosystems [61] and localized effects, including making the Arctic more vulnerable to contaminants originating in temperate and subpolar regions [62]. Two additional changes of particular relevance for marine resources are the potential intensification of El Niño events [63] and the potential intensification of prolifically productive eastern boundary upwelling systems [64]. In the latter case, projected changes indicate the poleward displacement of atmospheric high-pressure systems responsible for upwelling winds under climate change [65].

2.3 Changing ocean productivity baselines

Photosynthesis, the creation of new organic material from inorganic precursors, underlies nearly all ocean life (including the marine resources from which society derives great benefit). The rate at which photosynthesis creates new organic material is thus referred to

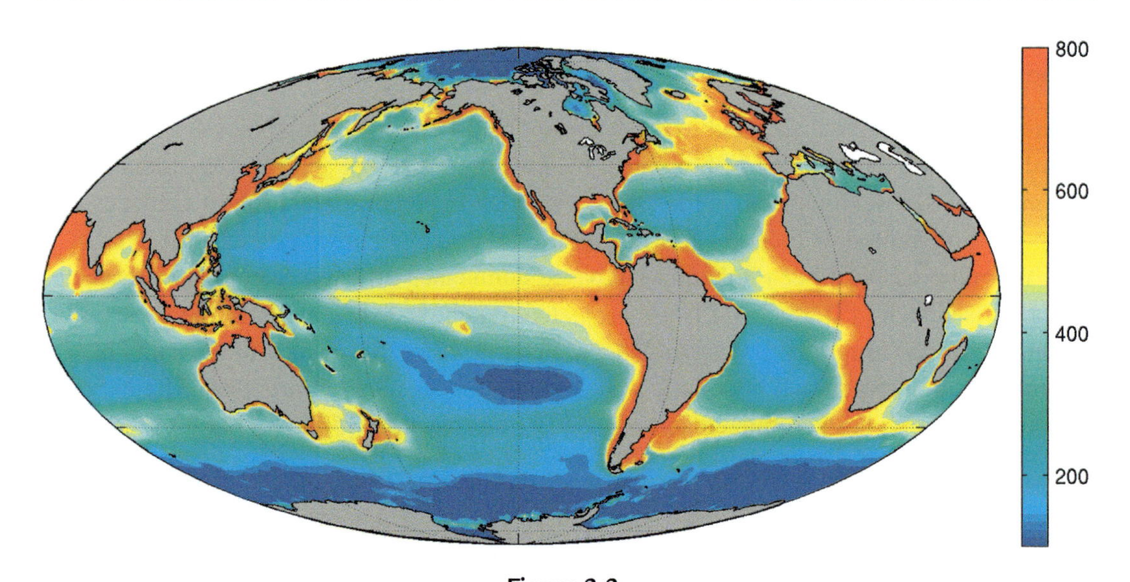

Figure 2.2
Global satellite-observed primary production (g C/m^2/year) from the VGPM [66]. The figure is the average of the VGPM and Eppley–VGPM models. *VGPM*, Vertically generalized production model.

as "primary production," reflecting its foundational role for marine ecosystems. The vast majority of primary production in the ocean is accomplished by microscopic phytoplankton in the well-lit layers of the surface ocean. In addition to phytoplankton, photosynthesis requires several key ingredients: CO_2, light, and nutrients. The first two of these ingredients are often in abundance in the surface ocean, but nutrients must generally be resupplied from deeper waters.

The primary barrier to nutrient resupply is stratification: the ocean's surface layers tend to be warmer, fresher, and, as a result, less dense than deeper waters. Physical processes must overcome stratification to bring nutrients to well-lit surface waters. Global satellite-based primary production estimates (Fig. 2.2) reveal the large-scale patterns in this confluence. At the equator and along the eastern boundaries of ocean basins wind-driven divergences in surface ocean currents create upwelling currents that elevate chlorophyll concentrations. At higher latitudes surface waters cooled by frigid, stormy winters mix deeply with underlying waters, replenishing nutrient supplies to support productive spring, summer, and fall seasons. The subtropical gyres, or "ocean deserts" (blue areas in Fig. 2.2), are deprived of nutrients due to year-round stratification and downwelling driven by wind-driven convergence in the surface ocean. Finally, elevated chlorophyll is evident in many of the coastal shelf seas, where energetic tides and currents flowing over complex bathymetry drive vigorous mixing of shallow and deep waters.

Global warming has strengthened ocean stratification [11] and this is projected to continue under climate change [67]. Stratification increases are augmented at high latitudes by the freshening of surface waters due to ice melt and an increasing surplus of precipitation over evaporation [68]. Ocean productivity trends resulting from increased stratification to date remain difficult to detect [69,70] but projections suggest that future modest to moderate declines (0%–20%) are likely for many low- and mid-latitude ocean areas [23]. Near the poles, light is more limiting than nutrients such that the increased stratification and melting ice expected under climate change may stimulate primary production.

Fisheries projections to date have suggested that ocean productivity declines expected under climate change will likely lead to declines in global potential fisheries' yields [71,72]. The energetic pathways connecting microscopic phytoplankton to fisheries, however, are complex [73,74]. Researchers and collaborators in the Nereus Program have contributed to understanding how food web factors may amplify changes in primary production to create larger changes in potential fish catch than one would infer from changes in productivity alone [72,75]. Nereus Program researchers have also contributed to the development and analysis of models capable of predicting the dynamic marine resource responses changing ocean productivity baselines [76–79]. Chapter 7, Building confidence in projections of future ocean capacity, of this volume describes global efforts to use multiple models to constrain future changes in potential fish yields [80]. Changes in fisheries' productivity may be compounded by potential shifts to less valuable fisheries species, including jellyfish, and Nereus Program researchers have developed models of how changes in ocean productivity along with other drivers may create such shifts [81]. Changes in the seasonal timing of productivity can also be critical for fish [82]. Research supported by the Nereus Program has elucidated how the different sensitivities of phytoplankton and fish to environmental drivers may drive mismatches between fish spawning and food resources that are detrimental to fisheries [83] (see Chapter 4, Changing seasonality of the sea).

2.4 Ocean deoxygenation

All primary production is eventually consumed and respired. For most organic matter this occurs within the well-lit surface layers of the ocean. A significant fraction, however, sinks below and is consumed at depth, decreasing oxygen in subsurface waters. Waters experiencing high organic matter fluxes and/or those that are persistently isolated from the surface ocean (where oxygen can be replenished) are thus naturally depleted in oxygen. The largest of these oceanic hypoxic zones are found along the equator, in highly stratified waters just below the wind-driven upwelling which elevates surface ocean productivity (Fig. 2.3).

Observations suggest that these ocean hypoxic zones are expanding [85,86], and projections under high CO_2 emissions scenarios suggest that such expansions are likely to continue [23]. These trends have been attributed to the combination of decreasing oxygen solubility

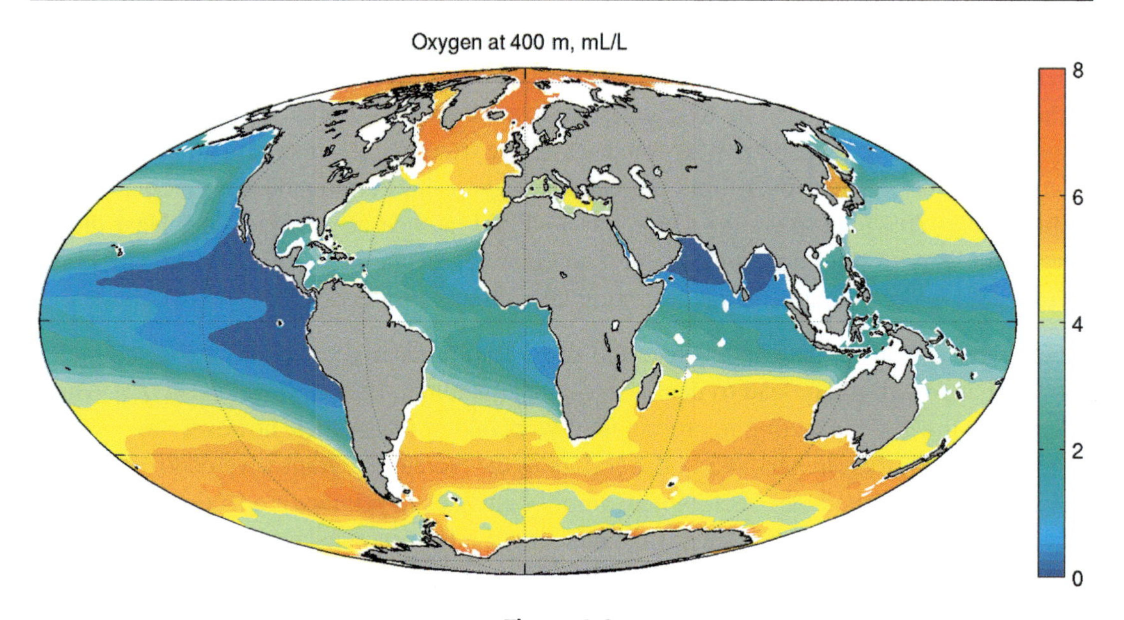

Figure 2.3
Ocean oxygen concentrations at 400 m; map adapted from the World Ocean Atlas [84]. Note the oxygen-depleted areas in equatorial regions.

in warming waters, and the tendency of stronger stratification to further isolate deep waters from the surface where oxygen can be resupplied. These large-scale patterns are compounded by enhanced hypoxia in coastal regions associated with enhanced nutrient inputs from fertilizers and land use changes [87]. These can lead to "dead zones" in some regions like the Gulf of Mexico and are associated with reduced species diversity, mortality, and disrupted fisheries due to insufficient oxygen for survival [88].

Aquatic "water-breathing" (respiring oxygen from water instead of air) organisms such as fish and invertebrates are impacted by declining ocean oxygen [12]. Through aerobic respiration oxygen plays a critical role in generating the energy for body functions such as movement, growth, and reproduction [89] and low oxygen can severely impair these processes [90]. Some organisms are able to tolerate lower oxygen levels than others, but they all have their limits, particularly if low oxygen conditions are sustained over a long period of time. Impacts include a reduction in growth, body size, and increase in mortality, and at larger scales, the reduction of suitable environments for the organisms to inhabit and for the interactions between species [91,92]. These effects lead to a reduction in the diversity and abundance of fish and invertebrates in hypoxic zones [93]. Ocean warming exacerbates the effect of deoxygenation, as invertebrates need more oxygen in warmer water. Deoxygenation effects may thus be particularly marked in the tropics, where oxygen demand is high and oxygen minimum zones are projected to expand in the 21st century [23,94].

2.5 Changing coastlines and ocean pollution

Ocean changes arising in the Anthropocene extend beyond the large-scale direct and indirect effects of increasing atmospheric CO_2 and other greenhouse gases emphasized in the preceding sections. The same carbon intensive fossil fuel sources responsible for releasing large quantities of CO_2 to the atmosphere coemit hundreds of other hazardous air pollutants, some of which are transported long-distances in the atmosphere and deposited in the global oceans [95]. The rapid growth of human populations, particularly in coastal regions, has generated an array of additional drivers of ocean change with both local and global imprints (coastal development, nutrients, pollution, resource exploitation). One large direct imprint on marine resources has been left by the evolution of fishing from a localized, artisanal imprint in freshwater and near-coastal regions to an industrial-scale effort spanning ocean basins [96]. Diverse aspects of the fishing industry and its relationship with marine resources are considered elsewhere in this volume (see Chapter 17, Fisheries and seafood security under changing oceans, and Section 4).

Coastal ecosystems have also been pressured by the buildup of coastal infrastructure [97,98], threatening saltmarsh and mangrove wetlands that provide critical nursery habitats for many coastal fisheries [99]. These risks are compounded by rising sea levels associated with a warming ocean and melting ice sheets [11]. Nereus Program researchers have been active in mapping the loss of mangrove habitats globally and quantifying the effect of these losses on local fisheries (see Chapter 3, Drivers of ecosystem production in complex social-ecological systems).

Human activities have also greatly enhanced nutrient inputs to the coastal ocean from rivers and atmosphere deposition [100,101]. The Haber–Bosch process, invented in 1908, has enabled humans to translate vast quantities of nitrogen from environmentally benign N_2 to ammonia (NH_3). This ammonia is then easily converted through oxidation into biologically available forms (e.g., nitrate NO_3^-), prompting the rapid expansion of synthetic nitrogen fertilizers in the mid-20th century. In the past four decades, there has been a four-fold increase in the use of nitrogen fertilizers and a three-fold increase in phosphate fertilizers [13]. These inputs are augmented, particularly in the tropics, by nitrogen mobilized through deforestation and land use change [102]. The delivery of this nitrogen and similarly rising phosphate inputs to the ocean are associated with prolific harmful algal blooms and acute coastal hypoxia, putting marine resources at risk [88].

Human activities have also perturbed the natural cycles of many heavy metals. The rise and fall of lead concentrations in the North Atlantic Ocean with the use and phaseout of leaded gasoline is well-documented [103]. This pollution entered the global thermohaline circulation and has been detected in the Southern Ocean [104]. Human activities have enriched atmospheric concentrations of mercury by three- to fivefold since the mid-1800s, with a resulting enrichment of concentrations in all ocean basins [105,106] (see Chapter 6, Pathways of methylmercury accumulation). Synthetic organic chemicals produced in large quantities for industry and other human applications have all been detected in remote ocean regions and wildlife [62,107–109]. Inputs of long-lived plastics have been accumulating in ocean waters since the 1960s and we are just beginning to fully understand their implications for marine life. An estimated 4.8–12.7 million MT of plastic waste entered the ocean in 2010, and this input rate could increase by an order of magnitude over the next decade without new management initiatives [110].

2.6 Prospects for understanding and predicting changing oceans

The ocean may appear immutable, but it is in a state of constant flux, interacting with a broad range of natural and unprecedented anthropogenic drivers. This chapter has attempted to provide a broad perspective on these changes and the response of marine organisms to them. As alluded to above, the chapters which follow in this Section, and throughout this book, provide more detailed perspectives on various critical areas of ocean change.

Much progress has been made in understanding the changing ocean and the response of marine life to it, but there is still much to learn. We observe only a fraction of the changing ocean. The response of organisms to ocean change, and to each other, is complex. Advances in ocean observations [111,112], ocean and earth system predictions [113,114], and dynamic management strategies [115,116] give cause for optimism that society may be able to track and adapt to changes as they occur. Making informed choices about the future ocean and formulating effective policies to reach desired goals, however, requires the capacity to integrate across ocean changes, organismal responses, management strategies, and societal outcomes. The development of such a comprehensive framework to outline future ocean choices, and the individuals with the capacity to harness and improve upon it, is a key outcome of the Nippon Foundation Nereus Program.

References

[1] P. Lehodey, J. Alheit, M. Barange, T. Baumgartner, G. Beaugrand, K. Drinkwater, et al., Climate variability, fish, and fisheries, J. Clim. 19 (20) (2006) 5009–5030.
[2] T. Baumgartner, A. Soutar, W. Riedel, Natural time scales of variability in coastal pelagic fish populations of the California Current over the past 1500 years: Response to global climate change and biological interaction, California Sea Grant Rep. for 1992–1995, California Sea Grant College, La Jolla, CA, 1996, pp. 31–37.
[3] A. Soutar, J.D. Isaacs, Abundance of pelagic fish during the 19th and 20th centuries as recorded in anaerobic sediment off the Californias, Fish. Bull. 72 (2) (1974) 257–273.
[4] J. Alheit, E. Hagen, Long-term climate forcing of European herring and sardine populations, Fish. Oceanogr. 6 (2) (1997) 130–139.
[5] E.M. Sunderland, M. Li, K. Bullard, Decadal changes in the edible supply of seafood and methylmercury exposure in the United States, Environ. Health Perspect. 126 (1) (2018) 017006.

[6] R.T. Barber, F.P. Chavez, Biological consequences of El Niño, Science 222 (4629) (1983) 1203–1210.
[7] F.P. Chavez, J. Ryan, S.E. Lluch-Cota, M. Ñiquen, From anchovies to sardines and back: multidecadal change in the Pacific Ocean, Science 299 (5604) (2003) 217–221.
[8] W. Clark, The lessons of the Peruvian anchoveta fishery, Calif. Coop. Oceanic Fish. Invest. Rep. 19 (1976) 57–63.
[9] S.C. Doney, V.J. Fabry, R.A. Feely, J.A. Kleypas, Ocean acidification: the other CO2 problem, Ann. Rev. Mar. Sci. 1 (2009) 169–192.
[10] T.F. Stocker, D. Qin, G.-K. Plattner, M. Tignor, S.K. Allen, J. Boschung, et al., IPCC, 2013: Climate Change 2013: The Physical Science Basis. Contribution of Working Group I to the Fifth Assessment Report of the Intergovernmental Panel on Climate Change, Cambridge University Press, 2013.
[11] M. Rhein, S.R. Rintoul, S. Aoki, E. Campos, D. Chambers, R.A. Feely, et al., Observations: Ocean, in: T. F. Stocker, D. Qin, G.-K. Plattner, M. Tignor, S.K. Allen, J. Boschung, A. Nauels, Y. Xia, V. Bex, P.M. Midgley (Eds.), Climate Change 2013: The Physical Science Basis. Contribution of Working Group I to the Fifth Assessment Report of the Intergovernmental Panel on Climate Change, Cambridge University Press, Cambridge; New York, NY, 2013, pp. 255–316.
[12] H.-O. Pörtner, D.M. Karl, P.W. Boyd, W.W.L. Cheung, S.E. Lluch-Cota, Y. Nojiri, et al., Ocean systems, in: C.B. Field, V.R. Barros, D.J. Dokken, K.J. Mach, M.D. Mastrandrea, T.E. Bilir, M. Chatterjee, K.L. Ebi, Y.O. Estrada, R.C. Genova, B. Girma, E.S. Kissel, A.N. Levy, S. MacCracken, P.R. Mastrandrea, L. L. White (Eds.), Climate Change 2014: Impacts, Adaptation, and Vulnerability. Part A: Global and Sectoral Aspects. Contribution of Working Group II to the Fifth Assessment Report of the Intergovernmental Panel on Climate Change, Cambridge University Press, Cambridge, United Kingdom and New York, NY, USA, 2014, pp. 411–484.
[13] P.M. Glibert, M.A. Burford, Globally changing nutrient loads and harmful algal blooms: recent advances, new paradigms, and continuing challenges, Oceanography 30 (1) (2017) 58–69.
[14] J. Heisler, P.M. Glibert, J.M. Burkholder, D.M. Anderson, W. Cochlan, W.C. Dennison, et al., Eutrophication and harmful algal blooms: a scientific consensus, Harmful Algae 8 (1) (2008) 3–13.
[15] M.L. Diamond, C.A. de Wit, S. Molander, M. Scheringer, T. Backhaus, R. Lohmann, et al., Exploring the planetary boundary for chemical pollution, Environ. Int. 78 (2015) 8–15.
[16] C.T. Driscoll, R.P. Mason, H.M. Chan, D.J. Jacob, N. Pirrone, Mercury as a global pollutant: sources, pathways, and effects, Environ. Sci. Technol. 47 (10) (2013) 4967–4983.
[17] C.D. Keeling, R.B. Bacastow, A.E. Bainbridge, C.A. Ekdahl Jr, P.R. Guenther, L.S. Waterman, et al., Atmospheric carbon dioxide variations at Mauna Loa observatory, Hawaii, Tellus 28 (6) (1976) 538–551.
[18] E.J. Dlugokencky, P.M. Lang, J.W. Mund, A.T. Crotwell, M.J. Crotwell, K.W. Thoning, Atmospheric Carbon Dioxide Dry Air Mole Fractions from the NOAA ESRL Carbon Cycle Cooperative Global Air Sampling Network, 1968-2017, Version: 2018-07-31, 2018.
[19] J.E. Dore, R. Lukas, D.W. Sadler, M.J. Church, D.M. Karl, Physical and biogeochemical modulation of ocean acidification in the central North Pacific, Proc. Natl. Acad. Sci. U.S.A. 106 (30) (2009) 12235–12240.
[20] C.L. Sabine, R.A. Feely, N. Gruber, R.M. Key, K. Lee, J.L. Bullister, et al., The oceanic sink for anthropogenic CO_2, Science 305 (5682) (2004) 367–371.
[21] D. Archer, H. Kheshgi, E. Maier-Reimer, Multiple timescales for neutralization of fossil fuel CO_2, Geophys. Res. Lett. 24 (4) (1997) 405–408.
[22] J.L. Sarmiento, N. Gruber, Ocean Biogeochemical Dynamics, Princeton University Press, 2013.
[23] L. Bopp, L. Resplandy, J.C. Orr, S.C. Doney, J.P. Dunne, M. Gehlen, et al., Multiple stressors of ocean ecosystems in the 21st century: projections with CMIP5 models, Biogeosciences 10 (2013) 6225–6245.
[24] T.L. Frölicher, K.B. Rodgers, C.A. Stock, W.W. Cheung, Sources of uncertainties in 21st century projections of potential ocean ecosystem stressors, Global Biogeochem. Cycles 30 (8) (2016) 1224–1243.
[25] U. Riebesell, K.G. Schulz, R. Bellerby, M. Botros, P. Fritsche, M. Meyerhöfer, et al., Enhanced biological carbon consumption in a high CO_2 ocean, Nature 450 (7169) (2007) 545.

[26] S.C. Doney, I. Lima, J.K. Moore, K. Lindsay, M.J. Behrenfeld, T.K. Westberry, et al., Skill metrics for confronting global upper ocean ecosystem-biogeochemistry models against field and remote sensing data, J. Mar. Syst. 76 (1−2) (2009) 95−112.
[27] V.J. Fabry, B.A. Seibel, R.A. Feely, J.C. Orr, Impacts of ocean acidification on marine fauna and ecosystem processes, ICES J. Mar. Sci. 65 (3) (2008) 414−432.
[28] K.J. Kroeker, R.L. Kordas, R. Crim, I.E. Hendriks, L. Ramajo, G.S. Singh, et al., Impacts of ocean acidification on marine organisms: quantifying sensitivities and interaction with warming, Global Change Biol. 19 (6) (2013) 1884−1896.
[29] B.D. Eyre, T. Cyronak, P. Drupp, E.H. De Carlo, J.P. Sachs, A.J. Andersson, Coral reefs will transition to net dissolving before end of century, Science 359 (6378) (2018) 908−911.
[30] A.C. Wittmann, H.-O. Pörtner, Sensitivities of extant animal taxa to ocean acidification, Nat. Clim. Change 3 (11) (2013) 995.
[31] I. Nagelkerken, P.L. Munday, Animal behaviour shapes the ecological effects of ocean acidification and warming: moving from individual to community-level responses, Global Change Biol. 22 (3) (2016) 974−989.
[32] W.W. Cheung, J. Dunne, J.L. Sarmiento, D. Pauly, Integrating ecophysiology and plankton dynamics into projected maximum fisheries catch potential under climate change in the Northeast Atlantic, ICES J. Mar. Sci. 68 (6) (2011) 1008−1018.
[33] S.R. Cooley, N. Lucey, H. Kite-Powell, S.C. Doney, Nutrition and income from molluscs today imply vulnerability to ocean acidification tomorrow, Fish Fish. 13 (2) (2012) 182−215.
[34] V.W. Lam, W.W. Cheung, U.R. Sumaila, Marine capture fisheries in the Arctic: winners or losers under climate change and ocean acidification? Fish Fish. 17 (2) (2016) 335−357.
[35] A.M. Queirós, J.A. Fernandes, S. Faulwetter, J. Nunes, S.P. Rastrick, N. Mieszkowska, et al., Scaling up experimental ocean acidification and warming research: from individuals to the ecosystem, Global Change Biol. 21 (1) (2015) 130−143.
[36] J.J. Alava, W.W. Cheung, P.S. Ross, U.R. Sumaila, Climate change−contaminant interactions in marine food webs: toward a conceptual framework, Global Change Biol. 23 (10) (2017) 3984−4001.
[37] K.J. Kroeker, R.L. Kordas, C.D. Harley, Embracing interactions in ocean acidification research: confronting multiple stressor scenarios and context dependence, Biol. Lett. 13 (3) (2017) 20160802.
[38] J. Marshall, R.A. Plumb, Atmosphere, Ocean and Climate Dynamics: An Introductory Text, Academic Press, 2016.
[39] S. Arrhenius, On the influence of carbonic acid in the air upon the temperature of the Earth, Publ. Astron. Soc. Pac. 9 (1897) 14.
[40] S. Levitus, J.I. Antonov, T.P. Boyer, O.K. Baranova, H.E. Garcia, R.A. Locarnini, et al., World ocean heat content and thermosteric sea level change (0−2000 m), 1955−2010, Geophys. Res. Lett. 39 (10) (2012) L10603.
[41] D. Roemmich, W.J. Gould, J. Gilson, 135 years of global ocean warming between the Challenger expedition and the Argo Programme, Nat. Clim. Change 2 (6) (2012) 425.
[42] J.C. Stroeve, M.C. Serreze, M.M. Holland, J.E. Kay, J. Malanik, A.P. Barrett, The Arctic's rapidly shrinking sea ice cover: a research synthesis, Clim. Change 110 (3−4) (2012) 1005−1027.
[43] M. Wang, J.E. Overland, A sea ice free summer Arctic within 30 years: an update from CMIP5 models, Geophys. Res. Lett. 39 (18) (2012) L18501.
[44] M. Collins, R. Knutti, J. Arblaster, J.-L. Dufresne, T. Fichefet, P. Friedlingstein, et al., Long-term Climate Change: Projections, Commitments and Irreversibility, in: T.F. Stocker, D. Qin, G.-K. Plattner, M. Tignor, S.K. Allen, J. Boschung, A. Nauels, Y. Xia, V. Bex, P.M. Midgley (Eds.), Climate Change 2013: The Physical Science Basis. Contribution of Working Group I to the Fifth Assessment Report of the Intergovernmental Panel on Climate Change, Cambridge University Press, Cambridge, United Kingdom and New York, NY, USA, 2013.
[45] E.S. Poloczanska, M.T. Burrows, C.J. Brown, J. García Molinos, B.S. Halpern, O. Hoegh-Guldberg, et al., Responses of marine organisms to climate change across oceans, Front. Mar. Sci. 3 (2016) 62.

[46] M.L. Pinsky, B. Worm, M.J. Fogarty, J.L. Sarmiento, S.A. Levin, Marine taxa track local climate velocities, Science 341 (6151) (2013) 1239–1242.
[47] W.W. Cheung, R. Watson, D. Pauly, Signature of ocean warming in global fisheries catch, Nature 497 (7449) (2013) 365.
[48] W.W. Cheung, V.W. Lam, J.L. Sarmiento, K. Kearney, R. Watson, D. Pauly, Projecting global marine biodiversity impacts under climate change scenarios, Fish Fish. 10 (3) (2009) 235–251.
[49] M.L. Pinsky, G. Reygondeau, R. Caddell, J. Palacios-Abrantes, J. Spijkers, W.W. Cheung, Preparing ocean governance for species on the move, Science 360 (6394) (2018) 1189–1191.
[50] W.W. Cheung, T.L. Frölicher, R.G. Asch, M.C. Jones, M.L. Pinsky, G. Reygondeau, et al., Building confidence in projections of the responses of living marine resources to climate change, ICES J. Mar. Sci. 73 (2016) 1283–1296.
[51] W.W. Cheung, M.C. Jones, G. Reygondeau, C.A. Stock, V.W. Lam, T.L. Frölicher, Structural uncertainty in projecting global fisheries catches under climate change, Ecol. Modell. 325 (2016) 57–66.
[52] A.J. Hobday, L.V. Alexander, S.E. Perkins, D.A. Smale, S.C. Straub, E.C. Oliver, et al., A hierarchical approach to defining marine heatwaves, Prog. Oceanogr. 141 (2016) 227–238.
[53] R.D. Stuart-Smith, C.J. Brown, D.M. Ceccarelli, G.J. Edgar, Ecosystem restructuring along the Great Barrier Reef following mass coral bleaching, Nature 560 (7716) (2018) 92.
[54] T.P. Hughes, J.T. Kerry, M. Álvarez-Noriega, J.G. Álvarez-Romero, K.D. Anderson, A.H. Baird, et al., Global warming and recurrent mass bleaching of corals, Nature 543 (7645) (2017) 373.
[55] N.A. Bond, M.F. Cronin, H. Freeland, N. Mantua, Causes and impacts of the 2014 warm anomaly in the NE Pacific, Geophys. Res. Lett. 42 (9) (2015) 3414–3420.
[56] E. Di Lorenzo, N. Mantua, Multi-year persistence of the 2014/15 North Pacific marine heatwave, Nat. Clim. Change 6 (11) (2016) 1042.
[57] M.G. Jacox, M.A. Alexander, N.J. Mantua, J.D. Scott, G. Hervieux, R.S. Webb, et al., Forcing of multiyear extreme ocean temperatures that impacted California current living marine resources in 2016, Bull. Am. Meteorol. Soc. 99 (1) (2018) S27–S33.
[58] K.E. Mills, A.J. Pershing, C.J. Brown, Y. Chen, F.-S. Chiang, D.S. Holland, et al., Fisheries management in a changing climate: lessons from the 2012 ocean heat wave in the Northwest Atlantic, Oceanography 26 (2) (2013) 191–195.
[59] T.L. Frölicher, E.M. Fischer, N. Gruber, Marine heatwaves under global warming, Nature 560 (7718) (2018) 360.
[60] T.L. Frölicher, C. Laufkötter, Emerging risks from marine heat waves, Nat. Commun. 9 (1) (2018) 650.
[61] A. Schmittner, Decline of the marine ecosystem caused by a reduction in the Atlantic overturning circulation, Nature 434 (7033) (2005) 628.
[62] X. Zhang, Y. Zhang, C. Dassuncao, R. Lohmann, E.M. Sunderland, North Atlantic Deep Water formation inhibits high Arctic contamination by continental perfluorooctane sulfonate discharges, Global Biogeochem. Cycles 31 (8) (2017) 1332–1343.
[63] W. Cai, G. Wang, B. Dewitte, L. Wu, A. Santoso, K. Takahashi, et al., Increased variability of eastern Pacific El Niño under greenhouse warming, Nature 564 (7735) (2018) 201.
[64] W. Sydeman, M. García-Reyes, D. Schoeman, R. Rykaczewski, S. Thompson, B. Black, et al., Climate change and wind intensification in coastal upwelling ecosystems, Science 345 (6192) (2014) 77–80.
[65] R.R. Rykaczewski, J.P. Dunne, W.J. Sydeman, M. García-Reyes, B.A. Black, S.J. Bograd, Poleward displacement of coastal upwelling-favorable winds in the ocean's eastern boundary currents through the 21st century, Geophys. Res. Lett. 42 (15) (2015) 6424–6431.
[66] M.J. Behrenfeld, P.G. Falkowski, Photosynthetic rates derived from satellite-based chlorophyll concentration, Limnol. Oceanogr. 42 (1) (1997) 1–20.
[67] A. Capotondi, M.A. Alexander, N.A. Bond, E.N. Curchitser, J.D. Scott, Enhanced upper ocean stratification with climate change in the CMIP3 models, J. Geophys. Res. Oceans 117 (C4) (2012) C04031.
[68] P.J. Durack, S.E. Wijffels, R.J. Matear, Ocean salinities reveal strong global water cycle intensification during 1950 to 2000, Science 336 (6080) (2012) 455–458.

[69] D.G. Boyce, M. Dowd, M.R. Lewis, B. Worm, Estimating global chlorophyll changes over the past century, Prog. Oceanogr. 122 (2014) 163–173.
[70] S.A. Henson, J.L. Sarmiento, J.P. Dunne, L. Bopp, I.D. Lima, S.C. Doney, et al., Detection of anthropogenic climate change in satellite records of ocean chlorophyll and productivity, Biogeosciences 7 (2010) 621–640.
[71] W.W. Cheung, G. Reygondeau, T.L. Frölicher, Large benefits to marine fisheries of meeting the 1.5°C global warming target, Science 354 (6319) (2016) 1591–1594.
[72] H.K. Lotze, D.P. Tittensor, A. Bryndum-Buchholz, T.D. Eddy, W.W. Cheung, E.D. Galbraith, et al., Global ensemble projections reveal trophic amplification of ocean biomass declines with climate change, Proceedings of the National Academy of Sciences 116 (26) (2019) 12907–12912.
[73] K.D. Friedland, C. Stock, K.F. Drinkwater, J.S. Link, R.T. Leaf, B.V. Shank, et al., Pathways between primary production and fisheries yields of large marine ecosystems, PLoS One 7 (1) (2012). Available from: https://doi.org/10.1371/journal.pone.0028945.
[74] J.H. Ryther, Photosynthesis and fish production in the sea, Science 166 (1969) 72–76.
[75] C.A. Stock, J.G. John, R.R. Rykaczewski, R.G. Asch, W.W. Cheung, J.P. Dunne, et al., Reconciling fisheries catch and ocean productivity, Proc. Natl. Acad. Sci. U.S.A. 114 (8) (2017) E1441–E1449.
[76] J.R. Watson, C.A. Stock, J.L. Sarmiento, Exploring the role of movement in determining the global distribution of marine biomass using a coupled hydrodynamic—size-based ecosystem model, Prog. Oceanogr. 138 (2015) 521–532.
[77] V. Christensen, M. Coll, J. Buszowski, W.W. Cheung, T. Frölicher, J. Steenbeek, et al., The global ocean is an ecosystem: simulating marine life and fisheries, Global Ecol. Biogeogr. 24 (5) (2015) 507–517.
[78] K.A. Kearney, C. Stock, K. Aydin, J.L. Sarmiento, Coupling planktonic ecosystem and fisheries food web models for a pelagic ecosystem: description and validation for the subarctic Pacific, Ecol. Modell. 237 (2012) 43–62. Available from: https://doi.org/10.1016/j.ecolmodel.2012.04.006.
[79] C.M. Petrik, C.A. Stock, K.H. Andersen, P.D. van Denderen, J.R. Watson, Bottom-up drivers of global patterns of demersal, forage, and pelagic fishes, Prog. Oceanogr. 176 (2019) 102–124. Available from: https://doi.org/10.1016/j.pocean.2019.102124.
[80] D.P. Tittensor, T.D. Eddy, H.K. Lotze, E.D. Galbraith, W. Cheung, M. Barange, et al., A protocol for the intercomparison of marine fishery and ecosystem models: Fish-MIPv1. 0, Geosci. Model Dev. 11 (4) (2018) 1421–1442.
[81] N. Henschke, C.A. Stock, J.L. Sarmiento, Modeling population dynamics of scyphozoan jellyfish (*Aurelia* spp.) in the Gulf of Mexico, Mar. Ecol. Prog. Ser. 591 (2018) 167–183.
[82] D. Cushing, Plankton production and year-class strength in fish populations: an update of the match/mismatch hypothesis, Advances in Marine Biology, Elsevier, 1990, pp. 249–293.
[83] R.G. Asch, C.A. Stock, J.L. Sarmiento, Climate change impacts on mismatches between phytoplankton blooms and fish spawning phenology, Global change biology (2019 May 31) 1–16. Available from: https://doi.org/10.1111/gcb.14650.
[84] H.E. Garcia, R.A. Locarnini, T.P. Boyer, J.I. Antonov, O.K. Baranova, M.M. Zweng, et al., World Ocean Atlas 2013, Volume 3: Dissolved Oxygen, Apparent Oxygen Utilization, and Oxygen Saturation, Rep. NOAA Atlas NESDIS, 2014, pp. 27.
[85] D. Breitburg, L.A. Levin, A. Oschlies, M. Grégoire, F.P. Chavez, D.J. Conley, et al., Declining oxygen in the global ocean and coastal waters, Science 359 (6371) (2018) eaam7240.
[86] S. Schmidtko, L. Stramma, M. Visbeck, Decline in global oceanic oxygen content during the past five decades, Nature 542 (7641) (2017) 335.
[87] R.J. Diaz, R. Rosenberg, Spreading dead zones and consequences for marine ecosystems, Science 321 (5891) (2008) 926–929.
[88] N.N. Rabalais, R.E. Turner, W.J. Wiseman Jr, Gulf of Mexico hypoxia, a.k.a. "The dead zone,", Annu. Rev. Ecol. Syst. 33 (1) (2002) 235–263.

[89] W.W.L. Cheung, D. Pauly, Impacts and effects of ocean warming on marine fishes, in: J.M. Baxter, D. Laffoley (Eds.), Explaining ocean warming: Causes, scale, effects and consequences, IUCN, Gland, Switzerland, 2016, pp. 239–254.
[90] B.A. Seibel, Critical oxygen levels and metabolic suppression in oceanic oxygen minimum zones, J. Exp. Biol. 214 (2) (2011) 326–336.
[91] C. Deutsch, A. Ferrel, B. Seibel, H.-O. Pörtner, R.B. Huey, Climate change tightens a metabolic constraint on marine habitats, Science 348 (6239) (2015) 1132–1135.
[92] K.E. Limburg, D. Breitburg, L.A. Levin, Ocean deoxygenation—a climate-related problem, Front. Ecol. Environ. 15 (9) (2017) 479.
[93] A.H. Altieri, K.B. Gedan, Climate change and dead zones, Global Change Biol. 21 (4) (2015) 1395–1406.
[94] L. Stramma, E.D. Prince, S. Schmidtko, J. Luo, J.P. Hoolihan, M. Visbeck, et al., Expansion of oxygen minimum zones may reduce available habitat for tropical pelagic fishes, Nat. Clim. Change 2 (1) (2012) 33.
[95] E.M. Sunderland, C.T. Driscoll Jr, J.K. Hammitt, P. Grandjean, J.S. Evans, J.D. Blum, et al., Benefits of Regulating Hazardous Air Pollutants from Coal and Oil-Fired Utilities in the United States, Environ. Sci. Technol. 50 (5) (2016) 2117–2120.
[96] C. Roberts, The Unnatural History of the Sea, Island Press, 2010.
[97] F. Bulleri, M.G. Chapman, The introduction of coastal infrastructure as a driver of change in marine environments, J. Appl. Ecol. 47 (1) (2010) 26–35.
[98] C.M. Crain, B.S. Halpern, M.W. Beck, C.V. Kappel, Understanding and managing human threats to the coastal marine environment, Ann. N.Y. Acad. Sci. 1162 (1) (2009) 39–62.
[99] M. Spalding, World Atlas of Mangroves, Routledge, 2010.
[100] J.W. Erisman, M.A. Sutton, J. Galloway, Z. Klimont, W. Winiwarter, How a century of ammonia synthesis changed the world, Nat. Geosci. 1 (10) (2008) 636.
[101] D. Fowler, M. Coyle, U. Skiba, M.A. Sutton, J.N. Cape, S. Reis, et al., The global nitrogen cycle in the twenty-first century, Philos. Trans. R. Soc. B: Biol. Sci. 368 (1621) (2013) 20130164.
[102] M. Lee, E. Shevliakova, C.A. Stock, S. Malyshev, P.C.D. Milly, Prominence of the tropics in the recent rise of global nitrogen pollution, Nat. Commun. 10 (2019) 1434.
[103] J. Wu, E.A. Boyle, Lead in the western North Atlantic Ocean: completed response to leaded gasoline phaseout, Geochim. Cosmochim. Acta 61 (15) (1997) 3279–3283.
[104] K. Ndungu, C.M. Zurbrick, S. Stammerjohn, S. Severmann, R.M. Sherrell, A.R. Flegal, Lead sources to the Amundsen Sea, West Antarctica, Environ. Sci. Technol. 50 (12) (2016) 6233–6239.
[105] E.M. Sunderland, R.P. Mason, Human impacts on open ocean mercury concentrations, Global Biogeochem. Cycles 21 (4) (2007) GB4022.
[106] H.M. Amos, D.J. Jacob, D.G. Streets, E.M. Sunderland, Legacy impacts of all-time anthropogenic emissions on the global mercury cycle, Global Biogeochem. Cycles 27 (2) (2013) 410–421.
[107] J. Dachs, R. Lohmann, W.A. Ockenden, L. Méjanelle, S.J. Eisenreich, K.C. Jones, Oceanic biogeochemical controls on global dynamics of persistent organic pollutants, Environ. Sci. Technol. 36 (20) (2002) 4229–4237.
[108] E. Jurado, F.M. Jaward, R. Lohmann, K.C. Jones, R. Simó, J. Dachs, Atmospheric dry deposition of persistent organic pollutants to the Atlantic and inferences for the global oceans, Environ. Sci. Technol. 38 (21) (2004) 5505–5513.
[109] C.A. McDonough, A.O. De Silva, C. Sun, A. Cabrerizo, D. Adelman, T. Soltwedel, et al., Dissolved organophosphate esters and polybrominated diphenyl ethers in remote marine environments: arctic surface water distributions and net transport through Fram Strait, Environ. Sci. Technol. 52 (11) (2018) 6208–6216.
[110] J.R. Jambeck, R. Geyer, C. Wilcox, T.R. Siegler, M. Perryman, A. Andrady, et al., Plastic waste inputs from land into the ocean, Science 347 (6223) (2015) 768–771.
[111] E. Boss, A. Waite, F. Muller-Karger, H. Yamazaki, R. Wanninkhof, J. Uitz, et al., Beyond chlorophyll fluorescence: the time is right to expand biological measurements in ocean observing programs, Limnol. Oceanogr. Bull. 27 (3) (2018) 89–90.

[112] A.M. Moore, M.J. Martin, S. Akella, H. Arango, M.A. Balmaseda, L. Bertino, et al., Synthesis of ocean observations using data assimilation for operational, real-time and reanalysis systems: a more complete picture of the state of the ocean, Front. Mar. Sci. 6 (2019) 90.

[113] G.B. Bonan, S.C. Doney, Climate, ecosystems, and planetary futures: the challenge to predict life in Earth system models, Science 359 (6375) (2018) eaam8328.

[114] Y. Kushnir, A.A. Scaife, R. Arritt, G. Balsamo, G. Boer, F. Doblas-Reyes, et al., Towards operational predictions of the near-term climate, Nat. Clim. Change 9 (2019) 94–101.

[115] D. Tommasi, C.A. Stock, A.J. Hobday, R. Methot, I.C. Kaplan, J.P. Eveson, et al., Managing living marine resources in a dynamic environment: the role of seasonal to decadal climate forecasts, Prog. Oceanogr. 152 (2017) 15–49.

[116] A.J. Hobday, C.M. Spillman, J. Paige Eveson, J.R. Hartog, Seasonal forecasting for decision support in marine fisheries and aquaculture, Fish. Oceanogr. 25 (2016) 45–56.

CHAPTER 3

Drivers of fisheries production in complex socioecological systems

Chris McOwen[1], Tom Spencer[2] and Mike Bithell[3]

[1]UN Environment World Conservation Monitoring Centre (UNEP-WCMC), Cambridge, United Kingdom [2]Cambridge Coastal Research Unit, Department of Geography, University of Cambridge, Cambridge, Cambridgeshire, United Kingdom [3]Department of Geography, University of Cambridge, Cambridge, Cambridgeshire, United Kingdom

Chapter Outline

Identifying the factors that drive the production of fish biomass and the quantity that can be extracted for human use is essential if we are to sustainably manage fisheries. The development of effective policies to achieve this aim depends on being able to understand enough of system dynamics to have confidence that the desired outcomes will be approached, against a backdrop of high uncertainties. Due to the complexity of the systems involved, this requires the formulation, calibration, and testing of what might be seen as "sufficiently good" models.

For decades researchers have sought to unpick marine ecosystems in order to ensure models are specified accurately and their subsequent projections of fisheries production and catch are reliable. Broadly speaking, the majority of attention to date has focused on two areas: (1) better understanding the dynamics of the underlying system in order to simulate "bottom-up forcing" and (2) determining the role of predators (including fishing) in order to simulate "top-down forcing." Historically, each component has been studied in isolation, for example, exploring the relationship between primary productivity and fisheries catch, or the role fishing has in shaping marine ecosystems and influencing their capacity to produce biomass. However, increasingly it has become apparent that the interdependency of these two forcings means that they cannot be treated in isolation [1]. Indeed, an extensive literature has demonstrated that a combination of bottom-up and top-down drivers are often at play. Rather than one dominating over another, it is now apparent that their relative importance varies with context. These contexts include ecosystem type (e.g., whether a region is tropical or temperate), geographical location, and the species targeted. In addition, these controls are not fixed in time. One driver may alter a

Predicting Future Oceans.
DOI: https://doi.org/10.1016/B978-0-12-817945-1.00003-4

system (either temporally or permanently) in such a way that its sensitivity to other drivers increases or decreases. For example, the targeting of larger bodied fish can modify the size structure and functioning of fish assemblages and alter their productivity and sensitivity to bottom-up forcing. In addition, changes in environmental conditions, for example, as a result of climate change, may have similar impacts—decreasing the body size of fish species and altering their sensitivity to environmental change and fishing pressure.

Whilst considerable progress has been made in this area, there remain a number of gaps in our understanding due to the narrow focus of our research activities. For example, there are a number of externalities to the system which are poorly understood, and often absent or inadequately represented, in current models. This is particularly true in coastal ecosystems, which are strongly influenced by local topography, terrestrial systems, and the presence (or absence) of benthic habitats. Of these, vegetated coastal habitats, including seagrass beds, seaweed meadows (including kelp forests), mangroves, and salt marshes, are perhaps the biggest omission. Coastal vegetated systems play a critical role in supporting marine biodiversity and are important drivers of fisheries production in many regions of the world, benefitting a range of actors from subsistence foragers to commercial offshore fisheries. Yet the dynamics of such systems, and the important role they play in modifying coastal ecosystems and supporting fisheries, are generally excluded from large-scale models [2]. Such omissions not only limit our capacity to understand and model present-day conditions but also mean that a large amount of uncertainty is not accounted for when making projections of fisheries production and catch into the future. There are two principal omissions in this regard. First, the interlinkages between vegetated systems and fisheries productivity are rarely taken into account; they include the outflow of nutrients to coastal waters, and the role they play as nursery grounds for juvenile fish and as foraging grounds for predators [3,4]. Second, the consideration that coastal habitats are not static in time and space and are themselves complex systems, driven by a range of factors. For example, there is widespread uncertainty as to how mangroves will respond to sea level rise during the 21st century [5]. They are complex systems which are influenced by multiple drivers such as the supply of sediment, changing rates of saltwater intrusion, shifting ocean and air temperature limits (and changes in ENSO and storm tracks), interspecific changes in growth rates and productivity, and the availability of space for mangroves and salt marshes to migrate inland which is strongly influenced by human activities in the coastal zone [6,7]. Cumulatively, uncertainty stemming from both of these omissions will have significant implications for how accurate our projections are for the totality of fisheries in coastal, vegetative regions, and limit our capacity to develop effective fisheries policies and formulate mitigation/adaptation plans. Whilst this will likely have minimal impact on the projections made for the most economical or widespread fish stocks (e.g., tuna), over 100 million people around the world, principally in developing countries, live within 10 km of large mangrove forests, benefiting from their fisheries.

Furthermore, when formulating policy for management purposes or to meet sustainability targets, the dynamics of oceanographic and ecological systems must be seen alongside the dynamics of human societies [8–10]. Economic processes, governance regimes, and social and cultural interactions may contribute to the building of social cohesion and ecological health (i.e., the building of resilience). Conversely, they may create lock-ins to unsustainable extraction of resources and low biodiversity (i.e., the erosion of resilience). Furthermore, it is possible to imagine a situation in which resilience building and erosion are occurring simultaneously but at varying rates across different space- and timescales. Over the last decade, modeling of society has progressed to include not just economics but behavioral and social effects, based on the idea of "bottom-up" representations that include individual humans as the unit of computation. These are known as agent-based models (ABM) [11,12]. However, while the potential of ABM is clear, until relatively recently their applications have been limited by computational power and lack of technical development. However, it is now possible to run ABM models at a global scale with the inclusion of both vegetation and animals, both on land and in the ocean [13]. In addition, a new generation of "earth system models" will need to include full representation of people as drivers of environmental change (e.g., [14]). Such models are ideally placed to incorporate human dynamics and can therefore allow coupled socioecological models to be developed and interrogated. This will provide a unique means to explore the dynamics of fisheries biomass beyond the net primary production from phytoplankton, and allow trophic structure (food chains) to emerge as a property of the dynamics, including social, technological, political, and cultural change. This means that the impact of fisheries on a dynamic ecosystem structure can be studied at the global scale, allowing the heterogeneity of human activity and circumstances to be coupled to the consequences of policy and environmental change in unprecedented detail.

Looking forward to 2050 and beyond, this implies a coordinated international effort that goes beyond the climate focus of the Intergovernmental Panel on Climate Change and the ecosystem focus of the Intergovernmental science-policy Platform on Biodiversity and Ecosystem Services. The UN Sustainable Development Goals imply a need to understand the functioning of societies (not just economies) across the range of scale from small communities to international organizations, and their accompanying interlinkages. This is a program for the long term: over the last 50 years the first simple atmospheric models have evolved, through the work of many, with increasing sophistication and steadily improving skill, into highly complex but very effective computational systems. While the resulting modern earth system models still have many uncertainties, our current global social-ecological systems models have arguably yet to reach the level of the early atmospheric models of the 1950s. Yet the prospects for rapid progress are good: understanding of modeling techniques, data availability, and computing power are orders of magnitude better than in the mid-20th century.

Combining new model paradigms with machine learning techniques for analysis of model output should provide us with the tools to make truly meaningful progress in the understanding of the drivers of fisheries production.

References

[1] C.J. McOwen, W.W. Cheung, R.R. Rykaczewski, R.A. Watson, L.J. Watson, Is fisheries production within Large Marine Ecosystems determined by bottom-up or top-down forcing? Fish and Fisheries 16 (2) (2015) 623–632. Available from: https://doi:10.1111/faf.12082.

[2] M. Sheaves, R. Baker, I. Nagelkerken, R. Connolly, True value of estuarine and coastal nurseries for fish: incorporating complexity and dynamics, Estuaries Coasts 38 (2) (2015) 401–414. Available from: https://doi.org/10.1007/s12237-014-9846-x.

[3] J. Hutchison, M. Spalding, P. zu Ermgassen, The role of mangroves in fisheries enhancement, The Nature Conservancy and Wetlands International (2014). 54 pp.

[4] P. Saenger, D. Gartside, S. Funge-Smith, RAP Publication 2013/09 A Review of Mangrove and Seagrass Ecosystems and their Linkage to Fisheries and Fisheries Management, FAO, Bangkok, 2013. p. 75.

[5] J.C. Ellison, Vulnerability assessment of mangroves to climate change and sea-level rise impacts, Wetlands Ecol. Manage. 23 (2015) 115–137. Available from: https://doi.org/10.1007/s11273-014-9397-8.

[6] C.E. Lovelock, D.R. Cahoon, D.A. Friess, G.R. Guntenspergen, K.W. Krauss, R. Reef, et al., The vulnerability of Indo-Pacific mangrove forests to sea-level rise, Nature 526 (2015) 559–563. Available from: https://doi.org/10.1038/nature15538.

[7] M. Schuerch, T. Spencer, S. Temmerman, M.L. Kirwan, C. Wolff, D. Lincke, et al., Future response of global coastal wetlands to sea-level rise, Nature 561 (2018) 231–234. Available from: https://doi.org/10.1038/s41586-018-0476-5.

[8] M.R. Evans, M. Bithell, S.J. Cornell, S.R.X. Dall, S. Diaz, S. Emmott, et al., Predictive systems ecology, Proc. R. Soc. Lond. 280B (2013). Available from: https://doi.org/10.1098/rspb.2013.1452.

[9] B. Neumann, A.T. Vafeidis, J. Zimmermann, R.J. Nicholls, Future coastal population growth and exposure to sea-level rise and coastal flooding – a global assessment, PLoS One 10 (2015) e0118571. Available from: https://doi.org/10.1371/journal.pone.0118571.

[10] D. Purves, J.P.W. Scharlemann, M. Harfoot, T. Newbold, D.P. Tittensor, J. Hutton, et al., Time to model all life on Earth, Nature 493 (2013) 295–297. Available from: https://doi.org/10.1038/493295a.

[11] M. Bithell, J. Brasington, K. Richards, Discrete-element, individual-based and agent-based models: tools for interdisciplinary enquiry in geography? Geoforum 39 (2008) 625–642. Available from: https://doi.org/10.1016/j.geoforum.2006.10.014.

[12] M.M. Waldrop, Free agents, Science 360 (2018) 144–147. Available from: https://doi.org/10.1126/science.360.6385.144.

[13] M.B. Harfoot, T. Newbold, D.P. Tittensor, S. Emmott, J. Hutton, V. Lyutsarev, et al., Emergent global patterns of ecosystem structure and function from a mechanistic general ecosystem model, PLoS Biol. 12 (4) (2014). Available from: https://doi.org/10.1371/journal.pbio.1001841.

[14] J.D. Farmer, C. Hepburn, P. Mealy, A. Teytelboym, A third wave in the economics of climate change, Environ. Resour. Econ. 62 (2) (2015) 329–357. Available from: https://doi.org/10.1007/s10640-015-9965-2.

CHAPTER 4

Changing seasonality of the sea: past, present, and future

Rebecca G. Asch

Department of Biology, East Carolina University, Greenville, NC, United States

Chapter Outline

Phenology refers to the study of recurring, biological events and the effects of climate and weather on those events. Focusing primarily on the seasonal timescale, examples of phenological events that occur in marine environments include plankton blooms, seasonal spawning aggregations, and recurrent migrations. Phenology is important for structuring ecological interactions because predators, prey, and interspecific competitors need to be in the same place at the same time for those interactions to occur. Phenology can also influence many ecosystem services that people rely upon since seasonal processes influence agriculture and aquaculture productivity. A primary concern is that climate change may alter the cues signaling when seasonal behaviors should be undertaken. Since different organisms may rely on distinct cues, such environmental changes could unravel important ecological interactions as seasonal events that previously occurred in synchrony now become mistimed. Compared to terrestrial ecosystems where networks dedicated to phenological research date back to 1750 [1], less research has investigated marine phenology since it can be challenging to monitor marine organisms in distant ocean regions and at great depths on a daily to weekly basis. This chapter principally focuses on present-day phenology of plankton and marine fishes and their projected future changes. This is because both sets of organisms are well studied and interactions between these groups are posited to influence recruitment of young-of-year fishes to fisheries.

Predicting Future Oceans.
DOI: https://doi.org/10.1016/B978-0-12-817945-1.00004-6

4.1 Past: a brief history of phenology research in marine ecosystems[1]

Victor Henson, the scientist who coined the term "plankton," conducted the first study of seasonality focusing on pelagic ecosystems [3]. Between 1883 and 1886 Henson and his colleagues organized 34 monthly cruises off Kiel, Germany to study the seasonal cycle of phytoplankton. A spring peak in diatom abundance and an autumn maximum in dinoflagellate concentration were noted. However, Henson believed that these seasonal maxima reflected sampling error, because he had assumed that the seasonality in the ocean would be analogous to that on land [3]. Franz Schütt, an oceanographer who collected many of these plankton samples for Henson, did recognize the seasonal pattern associated with spring and fall plankton blooms and wrote about them in the book *Analytische Plankton-Studien* [3,4]. Schütt [4] described the seasonal succession of different phytoplankton species, stating, "... with just as absolute certainty as the cherries bloom before the sunflowers, so *Skeletonemas* arrive at their yearly peak earlier than the *Ceratiums*."

Another early effort to understand seasonal patterns of plankton abundance was undertaken by Sir William Herdman and his colleagues, who collected macroplankton samples 6 days a week between 1907 and 1921 in the Irish Sea [5]. Not only did this effort document the seasonal succession of 38 plankton species, it also provided the first account of interannual variability in marine phenology. For example, Johnstone et al. [5] noted peaks in the diatom *Chaetoceros* occurred in March during some years and in May during other years. These authors suspected that this variability reflected hydrographic conditions, but they did not have the necessary oceanographic measurements to test this hypothesis.

In 1953 Harald Sverdrup developed the critical depth hypothesis to explain the timing of the spring phytoplankton bloom observed each year in temperate ocean waters [6]. Critical depth is defined as the location where depth-integrated primary production is equal to depth-integrated community respiration. Respiration rates were assumed by Sverdrup to be constant throughout the water column, whereas photosynthesis decreases exponentially with depth due to light limitation. As a result, critical depth is a function of solar irradiance and the transparency of the water column. Sverdrup [6] argued that a phytoplankton bloom cannot occur when the mixed layer depth (MLD) exceeds the critical depth since depth-integrated respiration would be greater than primary production. Such conditions exist in winter when MLD reaches its seasonal maximum. As winter ends, the MLD shoals due to increased stratification reflecting surface warming of the water column and/or decreased salinity due to melting of snow and sea ice. At the same time, critical depth deepens as solar irradiance increases in the spring. Based on this hypothesis, bloom initiation timing is predicted to vary as a function of changes in MLD and seasonal irradiance patterns.

[1] Parts of this section have been modified from Ref. [2].

The satellite era has vastly expanded our knowledge of phytoplankton phenology due to the availability of daily, global observations. While gaps due to cloud cover can interfere with detection of phenological patterns from satellites [7,8], satellite data have allowed for the description of large-scale biogeographical patterns. Temperate latitudes are characterized by spring blooms (Fig. 4.1), with some ecosystems displaying a subsequent fall bloom. Reflecting increased light limitation on photosynthesis, bloom timing is delayed at higher latitudes, resulting in summer blooms. The oligotrophic subtropical gyres exhibit increases in chlorophyll during fall and winter since storms during these seasons mix the water column sufficiently to bring nutrients into the euphotic zone. The tropics are typically subject to a dampened seasonal cycle of primary production. These biogeographic patterns have been documented in numerous studies (including Refs. [9–13]).

Observations of blooms beginning prior to the shoaling of MLD have prompted biological oceanographers to develop new hypotheses beyond the classical critical depth hypothesis. Proposed mechanisms explaining variations in bloom timing include reduced encounter rates between phytoplankton and microzooplankton grazers due to MLD deepening in winter [14], reductions in light limitation on photosynthesis as water column mixing slows coincident with changes in air–sea heat flux [15], increases in the doubling rate of phytoplankton as temperature warms seasonally [16], and eddy-driven slumping of density gradients allowing for earlier onset of stratification and reduced mixing of phytoplankton below the euphotic depth [17]. Disagreements about what mechanisms are responsible for the spring bloom may stem in part from use of different phenological methods and metrics. Examples of different categories of methods for analyzing phenological changes among phytoplankton include rate of change, threshold, and cumulative sum methods [7], whereas phenological metrics that can be calculated with many of these methods include the start of

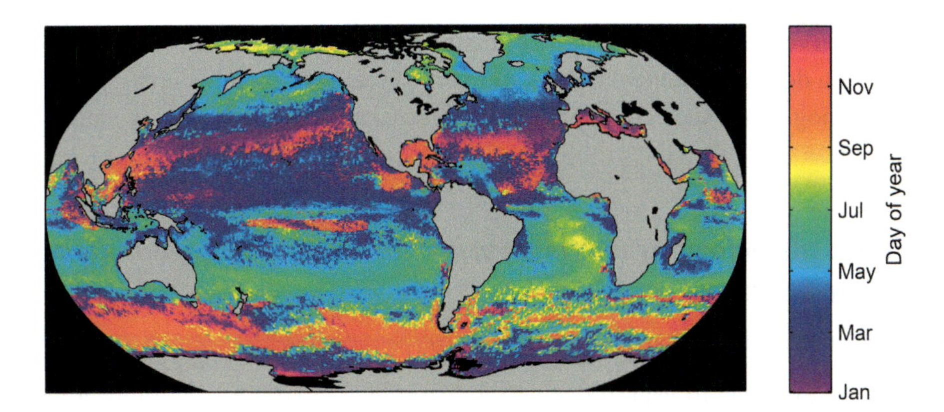

Figure 4.1

Global map of bloom initiation dates based on data from the SeaWiFS satellite averaged over the years 1998–2007. Only the first bloom occurring during the year is shown. The threshold algorithm described in Asch [2] was used to process the data shown here.

the seasonal increase in phytoplankton abundance, the bloom peak, bloom mid-point, bloom end date, and bloom duration [18]. Metrics and methods that produce earlier or later bloom start dates may affect whether these dates coincide with changes in oceanographic forcing [2,7]. Moreover, Chiswell et al. [19] argued that all of these hypotheses about bloom phenology are valid in different regions of the ocean or at different times of the year.

In fisheries science, interest in seasonal cycles stemmed from the development of the match–mismatch hypothesis by David Cushing in 1974, as well as its antecedents based on the work of Johann Hjort [20,21]. The match–mismatch hypothesis seeks to explain how order-of-magnitude variations in fish recruitment are connected to fluctuations in ocean climate [22]. Cushing observed that spawning among several fishes typically peaked coincident with the spring phytoplankton bloom, but the bloom exhibited substantial interannual variability in its seasonal occurrence. This would occasionally lead to seasonal mismatches between plankton production and fish egg and larvae production, which could result in poor feeding conditions for larvae and increased fish mortality [22]. Reduced survival of larvae can lead to lower recruitment once that year class becomes susceptible to capture by fisheries. At the time of the match–mismatch hypothesis's development, information on plankton production was rarely available on the spatial scale of the full range inhabited by fish stocks, making it challenging to definitively test the hypothesis. Nevertheless, Cushing [23] identified several case studies from fisheries around the world that were provisionally supportive of his hypothesis. Again the advent of remotely sensed ocean color was transformative because it allowed for synoptic observations of the spring bloom across the spatial scale over which fisheries operate, allowing the match–mismatch hypothesis to be more thoroughly assessed. Support for the match–mismatch is now available from a variety of commercially important fishes, including cod [24], haddock [25], herring [26], and salmon [27,28]. However, it is widely recognized that, while mismatches often result in poor recruitment, matches between phytoplankton blooms and larval production do not guarantee high recruitment since recruitment is controlled by a number of oceanographic and ecological factors acting on several life history stages of fishes [29].

4.2 Present: phenology as a "fingerprint" of climate change impacts on marine ecosystems

Across ecosystems, climate change has reinvigorated interest in phenology since warming temperatures are expected to lead to the earlier occurrence of spring conditions, a longer duration of summer, a delayed onset of fall, and shortened winter conditions. An analysis of seasonal temperatures observed with 38 years of satellite data confirms that changes such as these can already be observed and attributed to climate change [30]. Along with shifts in species distribution, changes in organismal phenology are considered to be one of the two major fingerprints of the ecological impacts of climate change that can be detected with time series from ecological research programs [31,32]. In terrestrial ecosystems, phenological changes have been characterized as "the most widely reported and probably the most easily detectable" ecological impact of climate change [33], but, as discussed earlier, this phenomenon has been understudied in marine systems [34]. This is demonstrated by the fact that $\leq 4\%$ of the species examined in two early global meta-analyses of climate change impacts occupied marine habitats [31,32].

Since the time when these early global meta-analyses were published, the marine research community has been galvanized to investigate changing phenology. Studies have now been published examining global trends in the phenology of marine phytoplankton, zooplankton, and seabirds. In particular, global changes in phytoplankton phenology have been well studied. The first paper to explore these global trends was Kahru et al. [35] who found that earlier melting of sea ice in the Arctic is tied to early peaks in chlorophyll concentration. Racault et al. [11] detected a trend toward a decreasing duration of phytoplankton blooms during the satellite era. This contrasts with the results of Friedland et al. [13] who identified a trend toward longer bloom duration at low-to-mid latitudes, as well as a generalized pattern of earlier bloom onset. Differences between the results of these studies may be attributable to the examination of different time periods and use of different bloom detection algorithms. Across several regions of the Atlantic and Pacific, there has been a tropicalization of phytoplankton phenology since the early 1980s accompanied by a poleward movement of biogeographical zones described based on phenology [36]. Since all of these trends rely on ocean color time series that are <20 years long, it is uncertain the extent to which these changes can truly be attributed to climate change versus interannual and decadal climate variability. Indeed, Henson et al. [37] estimated that >30 years of data are needed in most ocean biomes to definitely attribute changes in phytoplankton phenology to climate change.

An intercomparison of zooplankton phenology across 16 regions revealed several patterns relevant to how this trophic level is likely to respond to climate change [38]. First, most zooplankton taxa exhibited a high degree of interannual variability in phenology (i.e., 1–3 months). Much of this variation could be explained by interannual changes in temperatures to which zooplankton were exposed. However, variations in zooplankton phenology exceeded the rate expected if these changes were solely due to a thermal acceleration of physiological rates. This suggests that zooplankton may use temperature as a sensory cue for triggering developmental or behavioral activities. Interestingly, in many regions, a poor correlation between phytoplankton and zooplankton phenology was noted [38], suggesting that different oceanographic or physiological processes control the phenology of organisms in each of these trophic levels. If different drivers influence their phenology, it is likely that these taxa will display different responses to climate change, creating more frequent seasonal mismatches between trophic levels.

To the best of my knowledge, a comprehensive study of observed, global changes in fish phenology has not been undertaken. However, studies of fish reproductive and migration phenology from several regions suggest that fishes are capable of undergoing phenological changes that are of a similar magnitude to those exhibited by zooplankton (Fig. 4.2) [39–45]. Understanding the mechanisms that control fish phenology is essential for making reliable future projections of phenological change. The most common determinants of variations in fish phenology are temperature and photoperiod [46]. Photoperiod often triggers initial onset of reproductive development. Since most fishes are poikilotherms, temperature then acts to accelerate the rate of physiological processes associated with reproduction [47], often leading to earlier reproductive phenology during warm years. As a result the cumulative degree days of temperature experienced by a fish is a good predictor of reproductive timing across diverse species [43,48–50]. In contrast, when photoperiod is the predominant control on fish phenology, such as occurs at high latitudes where light can limit fish foraging success [51], less interannual variability in phenology is anticipated both now and continuing into the future. Beyond the widespread influences of temperature and photoperiod on fish phenology, other factors that can affect the phenology of individual fish species include:

- Fish size and age distribution: larger and older fishes reproduce earlier in the year among Atlantic cod [52,53], Atlantic mackerel [54], Pacific herring [48], capelin [55],

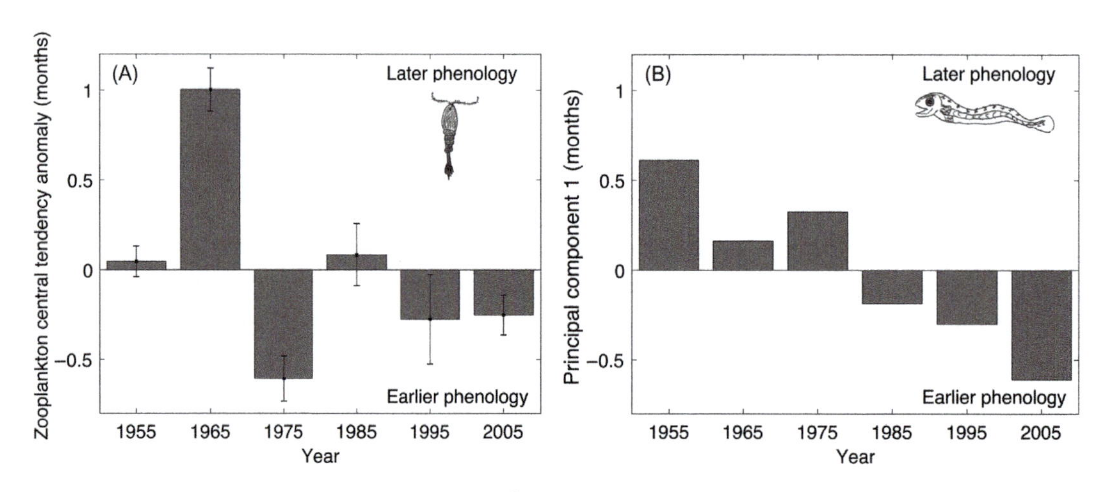

Figure 4.2

Decadal variability in the phenology of zooplankton and larval fishes in the southern California Current ecosystem are of a comparable magnitude, although decadal trends in phenology differ between these trophic levels. (A) Central tendency anomalies (± standard error) of monthly mesozooplankton displacement volume. (B) Eigenvectors from a principal components analysis performed on the phenology of 43 species of larval fishes. Details of both analyses are described by Asch [44].

and other species. This can reflect the fact that larger individuals are capable of quicker swimming speeds, allowing them to arrive earlier at spawning grounds. However, migration patterns also change with age and fish size, which can result in later arrival times at spawning grounds for large individuals among some species [56]. Effects of fish age and size on phenology are the more pronounced when the population is made up of only a small number of age classes [55].

- Genetic structure: genetics have a strong influence on salmon migration timing [57]. In salmon runs composed of multiple stocks spawning in different river tributaries, interannual variations in migration phenology often reflect changes in the abundance of one stock versus another.
- Migratory behavior: some fish species will use different overwintering habitats depending on whether it is a cold or warm year. This can affect migration phenology, as well as the timing of when fish are ready to reproduce after the end of overwintering [40,58,59].
- Hydrography: fish spawning phenology has been linked to tides, river flow, upwelling, and other oceanic features that influence transport of planktonic fish eggs and larvae [44,46,57,60,61]. Among anadromous fishes, river flow also impacts phenology by either facilitating downstream migration or impeding upstream migration.
- Prey availability: most fishes initiate reproductive development weeks to months in advance of spawning, making it difficult to precisely align spawning time with the future availability of food for their offspring in a variable environment. However, some species act as "income breeders" whose reproductive output directly reflects food availability for adult fishes [62]. For example, batch-spawning northern anchovy can reproduce as frequently as once a week if sufficient food is available [63].
- Social cues: Pankhurst and Porter [46] hypothesize that social cues may be particularly influential on fish phenology in tropical environments.

Among organisms in upper trophic levels the phenology of seabirds has been studied most extensively. In a global meta-analysis Keogan et al. [64] found that on average the breeding phenology of seabirds has not exhibited any substantial trends, nor is it closely correlated with oceanic temperature. This pattern does vary across some species and regions. For example, seabirds that do not undergo distant migrations and those that live in upwelling regions are both characterized by greater phenological variability [64]. Since these results indicate that seabirds exhibit less phenological plasticity than fishes or zooplankton, this suggests that mismatches between trophic levels could increase in the future since these organisms will respond differently to climate change.

Averaged across all trophic levels the phenology of marine organisms is trending earlier at a rate of 4.4 days/decade [42]. This rate is faster than that detected in global meta-analyses that used comparable methods but focused predominantly on terrestrial species [31,65]. The faster rate of phenological change among marine ecosystems has been hypothesized to be related to the fact that seasonal temperature gradients in marine systems are smaller than those in terrestrial systems [66]. Due to these differences, marine organisms need to undergo larger changes in phenology to ensure that thermally sensitive activities occur at the same temperature as they did in the past. Other studies have arrived at conflicting conclusions regarding whether the phenology of marine organisms is truly changing quicker than that of terrestrial and aquatic species. For example, Cohen et al. [67] found marine organisms have a faster mean rate of phenological change than terrestrial organisms, but these differences were not statistically significant. However, since marine species made up 3.8% of the taxa examined in this study, their results could have been possibly affected by the unbalanced sample sizes.

In addition, there are questions as to whether cross-trophic level patterns in phenological change are similar in marine and terrestrial environments. Thackeray et al. [68,69] found that the fastest rates of phenological change occur among primary producers, with slower rates observed at progressively higher trophic levels. While these authors argue that this pattern holds across ecosystems, their data suggest that this might not be the case in aquatic and marine systems. For example, Thackeray et al. [68] found that phytoplankton, which is the principal primary producer in marine ecosystems, had the slowest rate of phenological change among any organismal group studied. Also results in Thackeray et al. [68] indicated that rates of phenological change were on average similar across marine plants, invertebrates, and vertebrates.

4.3 What may the future hold?

Compared to changes in the distribution of marine species which have been extensively studied through the Nippon Foundation Nereus Program and other initiatives [70,71], fewer projections have been developed to understand how phenology may change in the future under climate change. To date the most extensive study examining future changes in phytoplankton phenology is Henson et al. [37], which investigated how the amplitude of the seasonal cycle of primary production and the month of maximum primary production changed under the Representative Concentration Pathway 8.5 (RCP8.5) scenario across six earth system models (ESMs). Note that RCP8.5 is a high greenhouse gas emissions scenario; changes in phytoplankton phenology have not yet been studied under lower emissions scenarios that would be congruent with commitments made under the Paris Agreement. Henson et al. [37] projected that, under climate change, the seasonal peak in primary production would become earlier by 0.5–1 months by the end of the 21st century across much of the globe. However, the oligotrophic subtropical gyres are projected to experience delays in phenology. These contrasting patterns relate to the fact that increases in stratification under climate change will have different effects on ocean biomes. Heightened stratification would lead to earlier easing of light limitation on primary production at high latitudes. Since nutrients limit primary production in the subtropical gyres, increased stratification would require greater winter mixing to replenish nutrients to the eutrophic zone, leading to delays in the seasonal cycle of primary production.

A potential limit of this initial study is that it examined data with a monthly resolution, which may be insufficient for examining phenological changes. This issue was addressed by a subsequent study by Henson et al. [72], which projected advances in bloom timing occurring in the Arctic, Southern Ocean, and some equatorial regions. Delays in bloom phenology were projected in some subpolar areas and were attributed to an increased prevalence of fall blooms relative to spring blooms.

Among higher trophic levels, climate change projections of fish reproductive phenology have been made for a limited number of species across a few regions. Neuheimer and MacKenzie [43] used cumulative degree days to model the spawning time of 21 stocks of Atlantic cod. Examining scenarios in which bottom temperatures warmed by either 0.5°C or 2.0°C, spawning time was projected to advance by anywhere between 1 and 129 days depending on the stock and climate scenario. Since cod stocks occurring at higher latitudes were projected to undergo larger phenological changes, latitudinal gradients in the cod spawning season may be reduced in the future. Asch [44] produced projections of future changes in the phenology for 43 species of larval fishes occurring in the southern California Current system. Over 60 years 39% of these species have exhibited advances in their phenology related to seasonal changes in sea surface temperature (SST) and zooplankton volume, whereas 18% of species have displayed delays in phenology connected to SST and upwelling. Using an ensemble of 30 ESMs, Asch [44] projected that species in the former group would continue to advance their spawning phenology over the 21st century. In contrast, for the species whose phenology is influenced by upwelling, different climate models projected different outcomes. This likely reflects the fact many ESMs do not have adequate spatial resolution to resolve coastal upwelling dynamics [73]. Neither of these studies on future fish phenology directly address the question of whether trophic mismatches are likely to become more pronounced in the future. However, this is a subject to be addressed by a recently published Nereus Program study [74]. Here, a model simulating climate change impacts on both fish reproductive phenology and phytoplankton bloom timing found that the latter process was more sensitive to changing temperatures, leading to an increased frequency of mismatches between trophic levels under the RCP8.5 climate change scenario. However, poleward migration by fishes in response to climate change minimized these mismatches. This suggests that fishes that are not currently shifting their distribution poleward in response to climate change may be particularly vulnerable to experiencing phenological mismatches with lower trophic levels [74].

In conclusion, a few additional questions need to be accounted for when considering how future changes in phenology will impact marine ecosystems and fisheries. Shifts in phenology can constitute a form of behavioral plasticity that can allow fishes to adapt to changing conditions and persist at a high level of abundance. However, in cases where the cues that evoke phenological responses are no longer aligned with the conditions that promote growth and survival of fishes and their offspring, seasonal mismatches in phenology can occur. When such mismatches affect the survival of larval fishes, at times density dependence during the juvenile stage can compensate for high mortality during earlier life history stages [75,76]. At other times no compensation occurs and declines in recruitment are likely. Also it is possible that phenotypic plasticity in fish phenology can delay the onset of selection pressure that leads to genetic adaptation [77]. In such cases, if the rates of environmental change are fast enough and genetic adaptation does not occur rapidly, populations may not be able to persist [77,78]. As a result, understanding the trade-offs between rates of environmental change, phenotypic plasticity, and the amount of genetic variance in fish populations may be essential for predicting how changing marine seasonality will impact fish population dynamics [77].

References

[1] J. Terhivuo, E. Kubin, J. Karhu, Phenological observation since the days of Linné in Finland, Ital. J. Agrometeorol. 1 (2009) 45–49.
[2] R.G. Asch, Interannual-to-Decadal Changes in Phytoplankton Phenology, Fish Spawning Habitat, and Larval Fish Phenology (Ph.D. dissertation), Scripps Institution of Oceanography, University of California San Diego, San Diego, CA, 2013.
[3] E.L. Mills, Biological Oceanography. An Early History, 1870-1960, Cornell University Press, Ithaca, NY, 1989.
[4] F. Schütt, Analytische plankton-studien. Methoden und anfangs-resultate der quantitativ-analytischen planktonforschung, Lipsius and Tischer, Kiel, 1892.
[5] J. Johnstone, A. Scott, H.C. Chadwick, The Marine Plankton, The University Press of Liverpool, London, 1924.
[6] H.U. Sverdrup, On conditions for the vernal blooming of phytoplankton, J. Cons. Int. Explor. Mer. 18 (1953) 287–295.
[7] S.R. Brody, M.S. Lozier, J.P. Dunne, A comparison of methods to determine phytoplankton bloom initiation, J. Geophys. Res. Oceans 118 (2013) 2345–2357.
[8] H. Cole, Henson, S.A. Martin, A. Yool, Mind the gap: the impact of missing data on the calculation of phytoplankton phenology metrics, J. Geophys. Res. 117 (2012) C08030.
[9] D.A. Siegel, S.C. Doney, J.A. Yoder, The North Atlantic spring phytoplankton bloom and Sverdrup's critical depth hypothesis, Science 296 (2002) 730–733.
[10] S.A. Henson, J.P. Dunne, J.L. Sarmiento, Decadal variability in North Atlantic phytoplankton bloom, J. Geophys. Res. 114 (2009) C04013.
[11] M.F. Racault, C. Le Quéré, E. Buitenhuis, S. Sathyendranath, T. Platt, Phytoplankton phenology in the global ocean, Ecol. Indic. 14 (2012) 152–163.
[12] M.R.P. Sapiano, C.W. Brown, S. Schollaert Uz, M. Vargas, Establishing a global climatology of marine phytoplankton phenological characteristics, J. Geophys. Res. 117 (2012) C08026. 2012.
[13] K.D. Friedland, C.B. Mouw, R.G. Asch, A.S.A. Ferreira, S. Henson, K.J.W. Hyde, et al., Phenology and time series trends of the dominant seasonal phytoplankton bloom across global scales, Glob. Ecol. Biogeogr. 27 (2018) 551–569.

[14] M.J. Behrenfeld, Abandoning Sverdrup's critical depth hypothesis on phytoplankton blooms, Ecology 91 (4) (2010) 977–989.
[15] J.R. Taylor, R. Ferrari, Shutdown of turbulent convection as a new criterion for the onset of spring phytoplankton blooms, Limnol. Oceanogr. 55 (6) (2011) 2293–2307.
[16] K.R. Hunter-Cevera, M.G. Neubert, R.J. Olson, A.R. Solow, A. Shalapyonok, H.M. Sosik, Physiological and ecological drivers of early spring blooms on a coastal phytoplankter, Science 354 (2016) 326–329.
[17] A. Mahadevan, E. D'Asaro, C. Lee, M.J. Perry, Eddy-driven stratification initiates North Atlantic spring phytoplankton blooms, Science 337 (2012) 54–58.
[18] R. Ji, M. Edwards, D.L. Mackas, J.A. Runge, A.C. Thomas, Marine plankton phenology and life history in a changing climate: current research and future directions, J. Plankton. Res. 32 (10) (2010) 1355–1368.
[19] S.M. Chiswell, P.H.R. Calil, P.W. Boyd, Spring blooms and annual cycles of phytoplankton: a unified perspective, J. Plankton. Res. 37 (3) (2015) 500–508.
[20] J. Hjort, Fluctuations in the great fisheries of Northern Europe, Rapp. Int. Cons. Explor. Mer. 20 (1914) 1–228.
[21] J. Hjort, Fluctuations in the year classes of important food fishes, J. Cons. Int. Explor. Mer. 1 (1926) 5–38.
[22] D.H. Cushing, The natural regulation of fish populations, in: F.R.H. Jones (Ed.), Sea Fisheries Research, John Wiley & Son, New York, 1974, pp. 399–412.
[23] D.H. Cushing, Plankton production and year-class strength in fish populations: an update of the match/mismatch hypothesis, Adv. Mar. Biol. 26 (1990) 249–293.
[24] G. Beaugrand, K.M. Brander, J.A. Lindley, S. Souissi, P.C. Reid, Plankton effect on cod recruitment in the North Sea, Nature 426 (2003) 661–664.
[25] T. Platt, C. Fuentes-Yaco, K.T. Frank, Spring algal bloom and larval fish survival, Nature 423 (2003) 398–399.
[26] J.F. Schweigert, M. Thompson, C. Fort, D.E. Hay, T.W. Therriault, L.N. Brown, Factors linking pacific herring (*Clupea pallasi*) productivity and the spring plankton bloom in the strait of Georgia, British Columbia, Canada, Prog. Oceanogr. 115 (2013) 103–110.
[27] C.M. Chittenden, J.A. Jensen, D. Ewart, S. Anderson, S. Balfry, E. Downey, et al., Recent salmon declines: result of lost feeding opportunities due to bad timing? PLoS One 5 (8) (2010) e12423.
[28] M.J. Malick, S.P. Cox, F.J. Mueter, R.M. Peterman, Linking phytoplankton phenology to salmon productivity along a north-south gradient in the Northeast Pacific Ocean, Can. J. Fish. Aquat. Sci. 72 (2015) 697–708.
[29] E.D. Houde, Emerging from Hjort's Shadow, J. Northwest Atl. Fish. Sci. 41 (2008) 53–70.
[30] B.D. Santer, S. Po-Chedley, M.D. Zelinka, I. Cvijanovic, C. Bonfils, P.J. Durack, et al., Human influence on the seasonal cycle of tropospheric temperature, Science 361 (2018) eaas8806.
[31] C. Parmesan, G. Yohe, A globally coherent fingerprint of climate change impacts across natural systems, Nature 421 (2003) 37–42.
[32] T.L. Root, J.T. Price, K.R. Hall, S.H. Schneider, C. Rosenweig, J.A. Pounds, Fingerprints of global warming on wild animals and plants, Nature 421 (2) (2003) 57–60.
[33] A.J. Miller-Rushing, R.B. Primack, Global warming and flowering times in Thoreau's Concord: a community perspective, Ecology 89 (2) (2008) 332–341.
[34] A.J. Richardson, E.S. Poloczanska, Under-resourced, under threat, Science 320 (2008) 1294–1295.
[35] M. Kahru, V. Brotas, M. Manzano-Sarabia, B.G. Mitchell, Are phytoplankton blooms occurring earlier in the Arctic? Glob. Change Biol. 17 (2011) 1733–1739.
[36] F. D'Ortenzio, D. Antoine, E. Martinez, M. Ribera d'Alcalà, Phenological changes of oceanic phytoplankton in the 1980s and 2000s as revealed by remotely sensed ocean-color observations, Global Biogeochem. Cycles 26 (2012) GB4003.
[37] S. Henson, H. Cole, C. Beaulieu, A. Yool, The impact of global warming on seasonality of ocean primary production, Biogeosciences 10 (2013) 4357–4369.

[38] D.L. Mackas, W. Greve, M. Edwards, S. Chiba, K. Tadokoro, D. Eloire, et al., Changing zooplankton seasonality in a changing ocean: comparing time series of zooplankton phenology, Prog. Oceanogr. 97-100 (2012) 31–62.

[39] M. Edwards, A.J. Richardson, Impact of climate change on marine pelagic phenology and trophic mismatch, Nature 430 (2004) 881–884.

[40] M.J. Genner, N.C. Halliday, S.D. Simpson, A.J. Southward, S.J. Hawkins, D.W. Sims, Temperature-driven phenological changes within a marine larval fish assemblage, J. Plankton. Res. 32 (5) (2010) 699–708.

[41] W. Greve, S. Prinage, H. Zidowitz, J. Nast, F. Reiners, On the phenology of North Sea ichthyoplankton, ICES J. Mar. Sci. 2005 (62) (2005) 1216–1223.

[42] E.S. Poloczanska, C.J. Brown, W.J. Sydeman, W. Kiessling, D.S. Schoeman, P.J. Moore, et al., Global imprint of climate change on marine life, Nat. Clim. Change 3 (2013) 919–925.

[43] A.B. Neuheimer, B.R. MacKenzie, Explaining life history variation in a changing climate across a species' range, Ecology 95 (12) (2014) 3364–3375.

[44] R.G. Asch, Climate change and decadal shifts in the phenology of larval fishes in the California Current Ecosystem, Proc. Natl. Acad. Sci. U.S.A. 112 (30) (2015) E4065–E4074.

[45] G. Reygondeau, J.C. Molinero, S. Coombs, B.R. MacKenzie, D. Bonnet, Progressive changes in the Western English Channel foster a reorganization in the plankton food web, Prog. Oceanogr. 137 (2015) 524–532.

[46] N.W. Pankhurst, M.J.R. Porter, Cold and dark or warm and light: variations on the theme of environmental control of reproduction, Fish. Physiol. Biochem. 28 (2003) 385–389.

[47] N.W. Pankhurst, P.L. Munday, Effects of climate change on fish reproduction and early life history stages, Mar. Freshw. Res. 62 (2011) 1015–1026.

[48] D.M. Ware, R.W. Tanasichuk, Biological basis of maturation and spawning waves in Pacific herring (*Clupea harengus pallasi*), Can. J. Fish. Aquat. Sci. 46 (1989) 1776–1784.

[49] U. Lange, W. Greve, Does temperature influence the spawning time, recruitment and distribution of flatfish via its influence on the rate of gonadal maturation? Dtsch Hydrogr Z 49 (2) (1997) 251–263.

[50] C. Gillet, P. Quétin, Effect of temperature change on the reproductive cycle of roach in Lake Geneva from 1983 to 2001, J. Fish. Biol. 69 (2006) 518–534.

[51] O. Varpe, O. Fiksen, Seasonal plankton-fish interactions: light regime, prey phenology, and herring foraging, Ecology 91 (2) (2010) 311–318.

[52] K. Wieland, A. Jarre-Teichmann, K. Horbowa, Changes in the timing of spawning of Baltic cod: possible causes and implications for recruitment, ICES J. Mar. Sci. 57 (2000) 452–464.

[53] R.S. Millner, G.M. Pilling, S.R. McCully, H. Hoie, Changes in the timing of otolith zone formation in North Sea cod from otolith records: an early indicator of climate-induced temperature stress? Mar. Biol. 158 (2011) 21–30.

[54] T. Jansen, H. Gislason, Temperature affects the timing of spawning and migration of North Sea mackerel, Cont. Shelf Res. 31 (2011) 64–72.

[55] J. Carscadden, B.S. Nakashima, K.T. Frank, Effects of fish length and temperature on the timing of peak spawning in capelin (*Mallotus villosus*), Can. J. Fish. Aquat. Sci. 54 (1997) 781–787.

[56] J.L. Callihan, J.E. Harris, J.E. Hightower, Coastal migration and homing of Roanoke River striped bass, Mar. Coast. Fish.: Dyn. Manage. Ecosyst. Sci. 7 (1) (2015) 301–315.

[57] J.J. Anderson, W.N. Beer, Oceanic, riverine, and genetic influences on spring Chinook salmon migration timing, Ecol. Appl. 19 (8) (2009) 1989–2003.

[58] J.A. Hutchings, R.A. Myers, Timing of cod reproduction: interannual variability and the influence of temperature, Mar. Ecol. Prog. Ser. 108 (1994) 21–31.

[59] D.W. Sims, V.J. Wearmouth, M.J. Genner, A.J. Southward, S.J. Hawkins, Low-temperature-driven early spawning migration of a temperate marine fish, J. Anim. Ecol. 73 (2004) 333–341.

[60] R.H. Parrish, C.S. Nelson, A. Bakun, Transport mechanisms and reproductive success of fishes in the California Current, Biol. Oceanogr. 1 (2) (1981) 175–203.

[61] A.M. Kaltenberg, R.L. Emmett, K.J. Benoit-Bird, Timing of forage fish seasonal appearance in the Columbia River plume and link to ocean conditions, Mar. Ecol. Prog. Ser. 419 (2010) 171–184.

[62] O. Varpe, C. Jorgensen, G.A. Tarling, O. Fiksen, The adaptive value of energy storage and capital breeding in seasonal environments, Oikos 118 (2009) 363–370.

[63] PFMC (Pacific Fishery Management Council), Coastal Pelagic Species Fishery Management Plan, Pacific Fishery Management Council, Portland, OR, 1998.

[64] K. Keogan, F. Daunt, S. Wanless, R.A. Phillips, C.A. Walling, P. Agnew, et al., Global phenological insensitivity to shifting ocean temperatures among seabirds, Nat. Clim. Change 8 (2018) 313–318.

[65] C. Parmesan, Influences of species, latitudes and methodologies on estimates of phenological response to global warming, Glob. Change Biol. 13 (2007) 1860–1872.

[66] M.T. Burrows, D.S. Schoeman, L.B. Buckley, P. Moore, E.S. Poloczanska, K.M. Brander, et al., The pace of shifting climate in marine and terrestrial ecosystems, Science 334 (2011) 652–655.

[67] J.M. Cohen, M.J. Lajeunesse, J.R. Rohr, A global synthesis of animal phenological responses to climate change, Nat. Clim. Change 8 (2018) 224–228.

[68] S.J. Thackeray, T.H. Sparks, M. Frederiksen, S. Burthes, P.J. Bacon, J.R. Bell, et al., Trophic level asynchrony in rates of phenological change for marine, freshwater and terrestrial environments, Glob. Change Biol. 2010 (16) (2010) 3304–3313.

[69] S.J. Thackeray, P.A. Henrys, D. Hemming, J.R. Bell, M.S. Botham, S. Burthe, et al., Phenological sensitivity to climate across taxa and trophic levels, Nature 535 (2016) 241–245.

[70] J.G. Molinos, B.S. Halpern, D.S. Schoeman, C.J. Brown, W. Kiessling, P.J. Moore, et al., Climate velocity and the future global redistribution of marine biodiversity, Nat. Clim. Change 6 (2015) 83–88.

[71] W.W.L. Cheung, G. Reygondeau, T.L. Frölicher, Large benefits to marine fisheries of meeting the 1.5°C global warming target, Science 354 (2016) 1591–1594.

[72] S.A. Henson, H.S. Cole, J. Hopkins, A.P. Martin, A. Yool, Detection climate change-driven trends in phytoplankton phenology, Glob. Change Biol. 24 (2018) e101–e111.

[73] C.A. Stock, M.A. Alexander, N.A. Bond, K.M. Brander, W.W.L. Cheung, E.N. Curchitser, On the use of IPCC-class models to assess the impact of climate on Living Marine Resources, Prog. Oceanog. 88 (2011) 1–27.

[74] R.G. Asch, C.A. Stock, J.L. Sarmiento, Climate change impacts on mismatches between phytoplankton blooms and fish spawning phenology, Glob. Change Biol. (2019). Available from: https://doi.org/10.1111/gcb.14650.

[75] H.W. Van der Veer, R. Berghahn, J.M. Miller, A.D. Rijnsdorp, Recruitment in flatfish, with special emphasis on North Atlantic species: progress made by the Flatfish Symposia, ICES J. Mar. Sci. 57 (2000) 202–215.

[76] T.E. Reed, V. Grøtan, S. Jenouvrier, B.E. Saether, M.E. Visser, Population growth in a wild bird is buffered against phenological mismatch, Science 340 (2013) 488–491.

[77] J.J. Anderson, E. Gurarie, C. Bracis, B.J. Burke, K.L. Laidre, Modeling climate change impacts on phenology and population dynamics of migratory marine species, Ecol. Modell. 264 (2013) 83–97.

[78] C.A. Botero, F.J. Weissing, J. Wright, D.R. Rubenstein, Evolutionary tipping points in the capacity to adapt to environmental change, Proc. Natl. Acad. Sci. U.S.A. 112 (2015) 184–189.

CHAPTER 5

Extreme climatic events in the ocean

Thomas L. Frölicher[1,2]

[1]Climate and Environmental Physics, Physics Institute, University of Bern, Bern, Switzerland
[2]Oeschger Centre for Climate Change Research, University of Bern, Bern, Switzerland

Chapter Outline

In large parts of the Northern Hemisphere, the summer of 2018 was exceptionally hot and dry. Europe, North America, and parts of East Asia were swept almost simultaneously by an exceptional heat wave that lasted several weeks, set record high temperatures, caused drought and wildfires, and affected the health of many people [1]. In contrast to earlier heat waves, such as the European heat wave in 2003, this large-scale heat wave did not catch the climate community by surprise as it has been known for more than a decade that such extreme events become more likely under global warming [2]. But it was not until recently that a similar dynamic has emerged and has been documented in the ocean with far-reaching consequences for marine ecosystems [3]. In fact some of the recently observed marine heatwaves (MHWs) revealed the high vulnerability of marine ecosystems and fisheries to such extreme temperature events in the ocean. MHWs are periods of extremely high temperatures that can last for days to months, can extend up to thousands of kilometers, and can penetrate multiple hundreds of meters into the deep ocean [4,5].

MHWs have been observed in all ocean basins over the past two decades Fig. 5.1. One of the first documented MHWs was the Mediterranean Sea 2003 MHW with sea surface temperatures up to 3°C above average [6]. Another well-documented heatwave was the Western Australian 2011 MHW. It was characterized by record high sea surface

Predicting Future Oceans.
DOI: https://doi.org/10.1016/B978-0-12-817945-1.00005-8

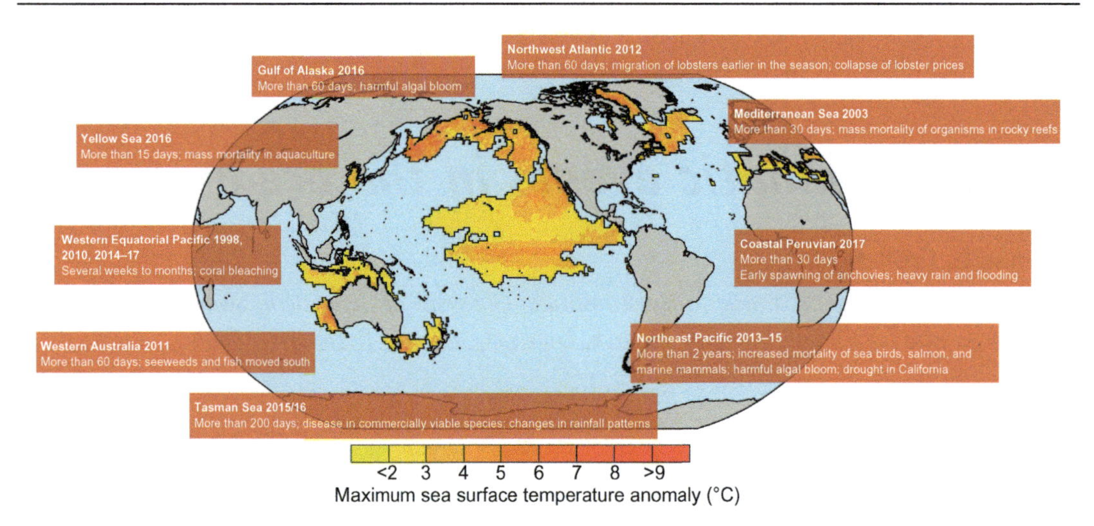

Figure 5.1
Spatial extension and maximal intensity of documented marine heatwaves over the last two decades. Yellow to red colors show the observed maximum temperature anomalies during the marine heatwaves. The orange boxes indicate the location and year of the marine heatwave occurrence and highlight the duration and some examples of observed impacts on natural and human systems. *Modified from T.L. Frölicher, C. Laufkötter, Emerging risks from marine heat waves, Nat. Commun. 9 (2018) 650. doi:10.1038/s41467-018-03163-6.*

temperatures of up to 5°C above average that persisted for more than 10 weeks in early 2011 [7]. The perhaps most famous heatwave was "The Blob" in the Northeast Pacific from 2013 to 2016 [8]. The heatwave had transiently a diameter of up to 1600 km with water temperatures of up to 6°C above average off Southern California [9]. MHWs have also been observed in the Gulf of Alaska, the Northwest Atlantic, the Tasman Sea, off the coast of Peru, and in the Yellow Sea. The western tropical Pacific including the Coral Sea has experienced multiple heatwaves (1998, 2010, 2014–17).

5.1 What drives marine heatwaves?

Over land, atmospheric blocking conditions often generate prolonged periods of very high temperatures that are often amplified by soil-moisture deficits [2]. In the ocean there are many processes that may trigger heat waves. These processes are less well known and quantified than in the terrestrial realm. The most important global driver of MHWs are El Niño events [10]. During El Niño years the sea surface temperatures, in particular of the central and eastern equatorial Pacific, are anomalously warm. El Niño is the result of a strong coupling between the atmosphere and the ocean, and the weaker than normal trade winds reduce the upwelling of cold subsurface waters in the eastern equatorial Pacific. Some MHWs are triggered by atmospheric-driven perturbations, such as stable weather conditions or changes in wind patterns that can be amplified through positive feedbacks with the warm ocean water. For example, a persistent atmospheric high pressure system possibly amplified by feedback processes with the anomalous warm ocean surface water in the North Pacific may have caused "The Blob" in 2013–16 [11]. This persistent high pressure system blocked the prevailing mid-latitude westerlies and led to lower than normal rates of heat loss from the ocean to the atmosphere. The Western Australia 2011 MHW was caused by a shift in wind patterns over the Indo-Pacific Ocean that strengthened and shifted the warm Leeuwin current southward resulting in warmer than normal waters off the coast of Australia [7]. Heat waves over land and/or ocean turbulence can also induce extreme anomalies in ocean temperatures.

5.2 The warming oceans

The global ocean plays a central role in regulating climate and in mitigating climate change, because of its immense volume and the large heat capacity of seawater. In fact, the largest amount of the extra heat that has been accumulated in the Earth system due to the increase in greenhouse gas concentrations has been taken up by the ocean. Between 1970 and 2010 the ocean stored approximately 93% or 274 ZJ (1 ZJ = 10^{21} J) of the extra heat [12]. Only 7% of the excess energy is distributed within the atmosphere and the land, and has caused ice melt. As a direct result of the excess heat uptake, the ocean is warming at the surface and throughout the deeper layers. The near surface layers of the global ocean have warmed at a rate of about 0.1°C per decade since the mid-20th century [13], albeit with pronounced regional and seasonal variability and with greater warming in the world's coastal regions [14]. But also deeper layers have warmed over the last few decades and the abyssal ocean (below 4000 m) continues to warm in the Southern Hemisphere [12,15].

5.3 Increase in marine heatwaves

Superimposed onto the long-term ocean warming trend are short-term extreme hot temperature events, so-called MHWs, during which ocean temperatures are anomalously high [3,4]. Analysis of daily satellite-based measurements of sea surface temperature covering the period 1982–2016 reveal that the number of MHW days exceeding the 99th percentile, calculated over the 1982–2016 period, has doubled globally since 1982 [10,16]. In other words, MHWs that occurred twice a year in 1982 are now (i.e., year 2016) occurring four times a year. MHWs are not only getting more frequent, they are also increasing in extent, duration, and intensity. As a result of the record high sea surface temperatures in 2015 and 2016, one-quarter of the worlds' oceans experienced either the longest or most intense events since 1982 in 2015 and 2016 [4]. On a regional scale MHWs have become more common in 38% of the world's coastal oceans over the last few decades [14].

What has driven this large increase in MHWs over the last few decades? To test whether the observed multidecadal increase in the number of MHW days over the satellite data-taking period is different from what would be expected from natural variability, such as El Niño Southern Oscillation or the Meridional Overturning Circulation, Frölicher et al. [16] compared Earth system model simulations that are forced with anthropogenic climate change with simulations not including anthropogenic climate change. They found that the observed trend toward more frequent MHW days is much larger than what can be expected from natural variability alone. In fact 87% of the MHWs occurring today are attributable to human-caused global warming [16], and some recent MHWs such as the Alaskan Sea 2016 MHW [17] and the extensive warming over the Great Barrier Reef in 2016 [18] have nearly been fully attributed to anthropogenic forcing. In other words, such events are very rarely found or are absent in preindustrial climate model simulations.

5.4 Future changes

Given current trends in greenhouse gas emissions and the major challenge posed by a transformation to a fossil-fuel-free society, it is very likely that global warming will continue to increase over the next few decades [19]. It is therefore expected that MHW days will also continue to increase with unabated global warming.

Earth system model simulations suggest that if global atmospheric surface temperature were to rise by 1.5°C relative to preindustrial levels by the end of the 21st century, as has been pledged in the Paris accord, the average number of MHW days would be 16 times higher than in preindustrial times when using the 99th preindustrial percentile threshold definition (blue dots for "$>99\%$" in Fig. 5.2A) [16]. If temperature were to rise by 2°C, the number of MHW days would be 23 times larger (yellow dots for "$>99\%$" in Fig. 5.2A), and under a 3.5°C rise they would be 41 times larger (red dots for "$>99\%$" in Fig. 5.2A). In other words, MHWs occurred every 100th day at preindustrial times, under a 1.5°C rise every sixth day, under a 2°C rise every fourth day, and under a 3.5°C rise every second day. In general, the probability ratio (i.e., the relative increase in the number of MHW days) increases the most for very rare extremes (Fig. 5.2A). For example, the probability ratio is 23 for moderate MHWs (defined as the 99th preindustrial percentile) and 890 for the very rare MHWs (99.99th preindustrial percentile) under 2°C global warming. But not only the number of MHW days is increasing. The heatwaves are also becoming longer lasting and spatially more extensive. Under 3.5°C global warming, the duration of a MHW, defined as the 99th preindustrial percentile threshold, would increase to 112 days (red dots for "$>99\%$" in Fig. 5.2B) and the spatial extent would increase to 94.5×10^5 km^2—equivalent to the total area of China. As a comparison, at preindustrial times, a MHW lasted on average 11 days (black dots for "$>99\%$" in Fig. 5.2B) and had a spatial extent of 4.2×10^5 km^2, the area of Switzerland.

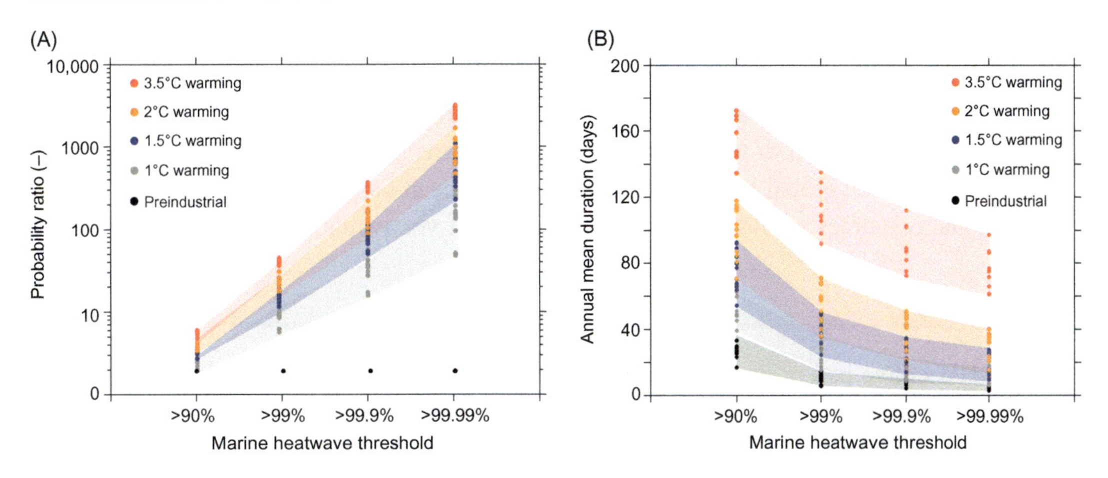

Figure 5.2

Simulated changes in the (A) probability ratio (i.e., relative increase in marine heatwave days; logarithmic scale) and the (B) duration of marine heatwaves for different global warming levels and different marine heatwave thresholds. Results are from 12 CMIP5 Earth system model simulations over the historical period and following the RCP8.5 scenario. The shaded areas indicate the maximum range and the points the individual model results should be points instead of point. *Modified from T.L. Frölicher, E.M. Fischer, N. Gruber, Marine heatwaves under global warming, Nature 560 (2018) 360–364. doi:10.1038/s41586-018-0383-9.*

All ocean regions will experience an increase in MHW days. The regional differences are caused by the heterogenous warming of the ocean surface. The largest changes are projected to occur in the Arctic Ocean and the western tropical Pacific. In the Arctic Ocean, the number of MHW days will increase 50-fold, in the western tropical Pacific even 70-fold. In the Arctic Ocean, the retreat of sea ice leads to an overproportional increase in temperature and therefore MHW days. In the western tropical Pacific, the seasonal and annual changes in sea surface temperature are small, and a relatively small increase in sea surface temperature can lead to a large increase in MHW days. In the Southern Ocean, however, only a small increase in MHW days is projected. There, the water upwells from very deep and cold layers and therefore the water at the surface does not warm up rapidly.

Interestingly, the number of MHWs is increasing more rapidly than the number of heat waves over land, even though the land generally warms more than the ocean. The reason is that the temperature variability is much smaller in the water than in the atmosphere. The probability of MHWs, therefore, increases disproportionally to comparatively small temperature increases [3].

A comparison between satellite measurements and model projections of MHWs suggests that the models adequately represent the trend in the number of MHW days over the last 35 years [16]. However, models have difficulties in simulating the duration and the spatial extent of the heatwaves, possibly caused by the relatively coarse resolution of the ocean (and atmosphere) models. The horizontal resolution of the ocean models is typically about 100 km and therefore too coarse to resolve mesoscale processes that may be critical to improve the representation of the duration and spatial extent of MHWs. To run long simulations with such global high-resolution, coupled ocean–atmosphere models enormous computational resources would be needed which are not yet available.

5.5 Impacts

Recent MHWs have had profound impacts on marine organisms (Fig. 5.1). Warm-water corals, for example, are an ecosystem that reacts very sensitively to elevated ocean temperatures. The prolonged MHW from 2014 to 2017 in the tropics and subtropics caused mass bleaching of corals, the third global-scale event in the past two decades. The heat stress during this event caused bleaching at 75% of global reefs and mortality at 30% [20].

Apart from the strong impact on corals, MHWs also strongly impacted other ecosystems. Reported biological impacts range from geographical species shifts and widespread changes in species composition to harmful algal blooms, mass stranding of mammals and mass mortalities of particular species. The Western Australia 2011 MHW, for example, led to a collapse of the temperate kelp forest off Australia and also to a shift in community composition with an increase in herbivorous tropical fishes that prevent the reestablishment of the kelp forest [21]. In several cases MHWs also strongly affected the fishery industry and tourism. The Northeast Pacific 2013–15 MHW, for example, caused a coast-wide harmful algal bloom [22] and the closing of beaches and commercially important fisheries and aquaculture industry [23].

5.6 Outlook

Whereas the long-term changes in ocean temperatures and the associated rise in sea level have been subject to intensive research for decades, with the exception of tropical coral reef systems, little focus has been given to MHWs and their impact on natural and human systems. The abovementioned examples demonstrate that a range of organisms and ecosystems can be impacted by MHWs with cascading risks for other natural and human systems. As MHW days are predicted to increase with continued global warming, it is therefore likely that this will result in profound impacts on natural and human systems. Especially organisms that are sessile and which cannot adapt to higher temperature will face very high risk of impacts.

Observations and model simulations also demonstrate that other drivers such as deoxygenation and acidification are putting additional stress on marine organisms and ecosystems [24]. Of particular concern are "Compound Events" [25], which correspond to extreme events with multiple concurrent and consecutive drivers (e.g., MHWs co-occurring with very low oxygen and pH levels) resulting in extreme consequences for marine ecosystems. Although there are a few studies on individual compound events in the ocean, the underlying drivers and the degree to which they can be represented in current climate models are currently unknown, making it difficult to design appropriate adaptation strategies. In order to better understand the impact of such compound events on individual organisms and entire ecosystems, continued interdisciplinary collaborations are needed.

Acknowledgment

The author acknowledges support from the Swiss National Science Foundation under grant PP00P2_170687.

References

[1] WMO, July sees extreme weather with high impacts. <http://t1p.de/5yhy>, 2018 (accessed 01.08.18).

[2] S.I. Seneviratne, N. Nicholls, D. Easterling, C.M. Goodess, S. Kanae, J. Kossin, et al., Changes in climate extremes and their impacts on the natural physical environment, Manage. Risks Extrem. Events Disasters Adv. Clim. Chang. Adapt. (2012) 109–230. Available from: https://doi.org/10.1017/CBO9781139177245.006.

[3] T.L. Frölicher, C. Laufkötter, Emerging risks from marine heat waves, Nat. Commun. 9 (2018) 650. Available from: https://doi.org/10.1038/s41467-018-03163-6.

[4] A.J. Hobday, L.V. Alexander, S.E. Perkins, D.A. Smale, S.C. Straub, E.C.J. Oliver, et al., A hierarchical approach to defining marine heatwaves, Prog. Oceanogr. 141 (2016) 227–238. Available from: https://doi.org/10.1016/j.pocean.2015.12.014.

[5] H. Scannell, A. Pershing, A.M. Alexander, A.C. Thomas, K.E. Mills, Frequency of marine heatwaves in the North Atlantic and North Pacific since 1950, Geophys. Res. Lett. 43 (2016) 2069–2076. Available from: https://doi.org/10.1002/2015GL067308.Received.

[6] A. Olita, R. Sorgente, A. Ribotti, S. Natale, S. Gaberšek, Effects of the 2003 European heatwave on the Central Mediterranean Sea surface layer: a numerical simulation, Ocean Sci. 3 (2007) 273–289. Available from: https://doi.org/10.5194/osd-3-85-2006.

[7] A.F. Pearce, M. Feng, The rise and fall of the "marine heat wave" off Western Australia during the summer of 2010/2011, J. Mar. Syst. 111–112 (2013) 139–156. Available from: https://doi.org/10.1016/j.jmarsys.2012.10.009.

[8] N.A. Bond, M.F. Cronin, H. Freeland, N. Mantua, Causes and impacts of the 2014 warm anomaly in the NE Pacific, Geophys. Res. Lett. 42 (2015) 3414–3420. Available from: https://doi.org/10.1002/2015GL063306.Received.

[9] C.L. Gentemann, M.R. Fewings, M. García-Reyes, Satellite sea-surface temperatures along the west coast of the United States during the 2014-2016 northeast Pacific marine heat wave, Geophys. Res. Lett. 44 (2017) 312–319. Available from: https://doi.org/10.1002/2016GL071039.

[10] E.C.J. Oliver, M.G. Donat, M.T. Burrows, P.J. Moore, D.A. Smale, L.V. Alexander, et al., Longer and more frequent marine heatwaves over the past century, Nat. Commun. 9 (2018) 1324. Available from: https://doi.org/10.1038/s41467-018-03732-9.

[11] E. Di Lorenzo, N. Mantua, Multi-year persistence of the 2014/15 North Pacific marine heatwave, Nat. Clim. Change 6 (2016) 1–7. Available from: https://doi.org/10.1038/nclimate3082.

[12] L. Cheng, K.E. Trenberth, J. Fasullo, T. Boyer, J. Abraham, J. Zhu, Improved estimates of ocean heat content from 1960 to 2015, Sci. Adv. 3 (2017) 1–11. Available from: https://doi.org/10.1126/sciadv.1601545.

[13] M. Rhein, S.R. Rintoul, S. Aoki, E. Campos, D. Chambers, R.A. Feely, et al., Observations: Ocean (2013). Available from: https://doi.org/10.1017/CBO9781107415324.010.

[14] F.P. Lima, D.S. Wethey, Three decades of high-resolution coastal sea surface temperatures reveal more than warming, Nat. Commun. 3 (2012) 1–13. Available from: https://doi.org/10.1038/ncomms1713.

[15] S.G. Purkey, G.C. Johnson, Warming of global abyssal and deep southern ocean waters between the 1990s and 2000s: contributions to global heat and sea level rise budgets, J. Clim. 23 (2010) 6336–6351. Available from: https://doi.org/10.1175/2010JCLI3682.1.

[16] T.L. Frölicher, E.M. Fischer, N. Gruber, Marine heatwaves under global warming, Nature 560 (2018) 360–364. Available from: https://doi.org/10.1038/s41586-018-0383-9.

[17] J.E. Walsh, R. Thomas, L. Bhatt, P.A. Bienik, B. Brettschneider, M. Brubaker, et al., The high latitude marine heat wave of 2016 and its impacts on Alaska, Bull. Am. Meteorol. Soc. 98 (2018) 39–43. Available from: https://doi.org/10.1175/BAMS-D-17-0105.1.

[18] M. Newman, A.T. Wittenberg, L. Cheng, G.P. Compo, C.A. Smith, The extreme 2015/16 El Nino, in the context of historical climate variability and change, Bull. Am. Meteorol. Soc. 98 (2018) 16–20.

[19] C. Le Quéré, R.M. Andrew, P. Friedlingstein, S. Sitch, J. Pongratz, A.C. Manning, et al., Global Carbon Budget 2017, Earth Syst. Sci. Data 10 (2018) 405–448. Available from: https://doi.org/10.5194/essd-10-405-2018.

[20] T.P. Hughes, K.D. Anderson, S.R. Connolly, S.F. Heron, J.T. Kerry, J.M. Lough, et al., Spatial and temporal patterns of mass bleaching of corals in the Anthropocene, Science 359 (2018) 80–83. Available from: http://science.sciencemag.org/content/359/6371/80.abstract.

[21] T. Wernberg, S. Bennett, R.C. Babcock, T. De Bettignies, K. Cure, M. Depczynski, et al., Climate-driven regime shift of a temperate marine ecosystem, Science 353 (2016) 169–172. Available from: https://doi.org/10.1126/science.aad8745.

[22] R.M. McCabe, B.M. Hickey, R.M. Kudela, K.A. Lefebvre, N.G. Adams, B.D. Bill, et al., An unprecedented coastwide toxic algal bloom linked to anomalous ocean conditions, Geophys. Res. Lett. 43 (2016) 10,366–10,376. Available from: https://doi.org/10.1002/2016GL070023.

[23] L. Cavole, A. Demko, R. Diner, A. Giddings, I. Koester, C. Pagniello, et al., Biological impacts of the 2013–2015 warm-water anomaly in the Northeast Pacific: winners, losers, and the future, Oceanography 29 (2016) 273–285. Available from: https://doi.org/10.5670/oceanog.2016.32.

[24] H.O. Pörtner, D.M. Karl, P.W. Boyd, W.W.L. Cheung, S.E. Lluch-Cota, Y. Nojiri, et al., Ocean systems, in: Clim. Chang. 2014 Impacts, Adapt. Vulnerability. Part A Glob. Sect. Asp. Contrib. Work. Gr. II to Fifth Assess. Rep. Intergov. Panel Clim. Chang. (2014) 411–484.

[25] J. Zscheischler, S. Westra, B.J.J.M. van den Hurk, S.I. Seneviratne, P.J. Ward, A. Pitman, et al., Future climate risk from compound events, Nat. Clim. Change 8 (2018) 469–477. Available from: https://doi.org/10.1038/s41558-018-0156-3.

CHAPTER 6

Seafood methylmercury in a changing ocean

Colin P. Thackray and Elsie M. Sunderland

Harvard John A. Paulson School of Engineering and Applied Sciences, Harvard University, Cambridge, MA, United States

Chapter Outline

6.1 Introduction

Methylmercury (MeHg) is a neurotoxic and bioaccumulative contaminant that is found at varying concentrations in all seafood, due to processes discussed in this chapter. Sensitivity to MeHg exposure is greatest for children and fetuses during neurodevelopmental periods, resulting in neurocognitive deficits. The concentration of MeHg in human blood is strongly correlated with seafood consumption, with both the type and amount of seafood consumed being important factors. This has led many regional, national, and international agencies such as the World Health Organization to recommend limited consumption of high-mercury seafood such as swordfish, shark, and some varieties of tuna (e.g., yellowfin and albacore tuna) by pregnant women and children [1]. Some people, such as Inuit and Canadian First Nations People in the Arctic, and coastal aboriginal people generally, experience much higher exposure than average due to their reliance on the consumption of marine mammals and seafood without access to alternatives [2].

Mercury is emitted to the environment by mining, fossil fuel combustion, volcanoes, and gradual releases from the Earth's crust [3]. The dominant modern sources of mercury are the burning of coal and artisanal and small-scale gold mining in developing countries [4]. Mercury is found as an impurity in coal and is released to the atmosphere upon combustion. Historically,

Predicting Future Oceans.
DOI: https://doi.org/10.1016/B978-0-12-817945-1.00006-X

the dominant anthropogenic source of mercury to the environment was gold mining [3]. Liquid mercury (e.g., mined and refined from cinnabar) was used to create an amalgam with the gold that could be easily separated and then burned, releasing the mercury to the atmosphere and leaving pure gold for the miner. During the 19th century, this took place during the Gold Rush in North America, and currently still takes place in artisanal gold mines in South America and Africa. Mercury is also released naturally from the Earth's crust, for example, during volcanic eruptions. Despite the natural source and the comparatively short history of anthropogenic releases (antiquity to present), the human contribution to environmental mercury levels is major. Mercury is subject to long-range atmospheric transport and the vast majority of mercury in the atmosphere and ocean ecosystems is anthropogenic [4].

6.2 The steps that lead to methylmercury accumulation in fish

Most mercury in the atmosphere and ocean was released from the large repository in the Earth's crust by human activities (e.g., mining and fossil fuel combustion). The emitted mercury cycles through the environment among the atmosphere, terrestrial, and estuarine ecosystems, and oceans. In the oceans, inorganic mercury can be methylated by microbes and converted into an organic form of mercury (MeHg) that is bioaccumulative and makes its way into food webs. A more detailed description of these processes follows.

- *Mercury, from emission to the ocean*

  Most of the mercury that is found in ocean ecosystems can be traced to atmospheric emissions. Gaseous mercury found in the atmosphere by burning mercury-containing substances, such as coal or the amalgam used in gold mining, stays in the atmosphere for months to years on average and is therefore found all across the globe and in the air over oceans far from the emissions sources. Mercury goes from the atmosphere into the oceans by gaseous deposition, particle settling, and via rain. Once in the ocean, soluble forms of mercury are transported throughout the ocean by currents, eddies, and mixing processes, and particle-bound mercury sinks into the deep oceans and into sediments. This distribution of mercury takes decades to centuries depending on the ocean basin, making ocean mercury a globally distributed and long-lasting problem. If anthropogenic emissions of mercury were to stop today, it would be many years before concentrations in the subsurface oceans fully reflected this change. The mercury being emitted today will be present in the environment that we interact with for at least decades to come.
- *Mercury methylation*

  Inorganic mercury deposited into the ocean from the atmosphere does not itself bioaccumulate. Inorganic mercury is converted to bioaccumulative MeHg by the process of methylation. Methylation of mercury is a microbial process that occurs under specific biogeochemical conditions that can be found in sediments, wetlands, and in marine waters. In marine waters the methylation coincides with the microbial activity that breaks down the organic matter provided by the sinking of dead plankton from near the ocean surface [5].

- *MeHg uptake by plankton*

  Marine MeHg enters the food web at its base, through uptake by plankton. Available data suggest MeHg uptake occurs mostly through passive diffusion and therefore is mainly affected by the cell surface area and MeHg forming compounds in seawater with other ligands. This results in the smallest phytoplankton having approximately 100 times greater MeHg concentration than the largest phytoplankton, meaning that plankton community size distribution plays an important role in food web bioaccumulation [6]. Concentrations in phytoplankton are 1000–100,000 times higher than those in the surrounding seawater, decreasing with dissolved organic carbon concentration and under eutrophic conditions [6].

- *Bioenergetics and bioaccumulation in heterotrophs*

  While fish and other predators also passively uptake MeHg from the surrounding water, this makes up $<5\%$ of their MeHg because of the large MeHg concentration difference between the surrounding water and the prey that they eat [7]. As predators consume their prey, they also consume the MeHg present in the prey. This dietary source of MeHg is therefore proportional to the amount of prey consumed, as well as the MeHg concentration of the prey. While MeHg is eliminated from the bodies of fish over time, this elimination is slower than the uptake, leading to accumulation of MeHg [7]. This makes the MeHg concentration in a fish of a given species generally proportional to its age, as a fish that has lived longer has had more time for MeHg accumulation. Since MeHg accumulates in many kinds of tissue, particularly muscle and protein-rich tissues (not just fat like some contaminants such as persistent organic pollutants), trophic position of a species is generally a very good predictor of MeHg concentration because of the biomagnification at each step along the food chain [8]. To summarize simply, a fish's MeHg concentration is primarily driven by two factors: how much it has to eat, and the MeHg concentration of its food.

6.3 Fish methylmercury in a changing ocean

The MeHg concentration in seafood is connected to many aspects of the ocean environment which are likely to change in the future. The abundance of mercury in seawater itself will evolve because of emissions that have already taken place cycling in the environment, as well as the trajectory of future emissions, and is likely to increase for the next many years at least, due to increased emissions from Asia and small-scale gold mining in the Southern Hemisphere [9]. The methylation that makes this mercury available for bioaccumulation depends on biogeochemical factors undergoing changes that may alter the rate of methylation. Warmer temperatures and more productive ecosystems could also increase the rate of methylmercury formation from inorganic mercury. While the oceans get warmer, fish living in the same location will accumulate more mercury than they did in the past due to the temperature increase, while migration and differential migration could alter the food web structure, leading to increases and decreases in fish MeHg concentration depending on the location. Some fish may delay the onset of these changes or force different changes through alteration of their foraging depth or changing their diet choices. A summary of these changes is illustrated in Fig. 6.1, and the rest of this chapter discusses each of these changes in detail.

- *Changes to seawater mercury and methylation*

  Because of the timescales involved in the geochemical cycling of mercury and our history of emissions, the seawater concentration of mercury will increase over the next 50 years unless the most aggressive actions to reduce mercury emissions take place, including global adherence to the Minamata Convention on Mercury. Mercury methylation is thought to be associated with the remineralization of carbon during the

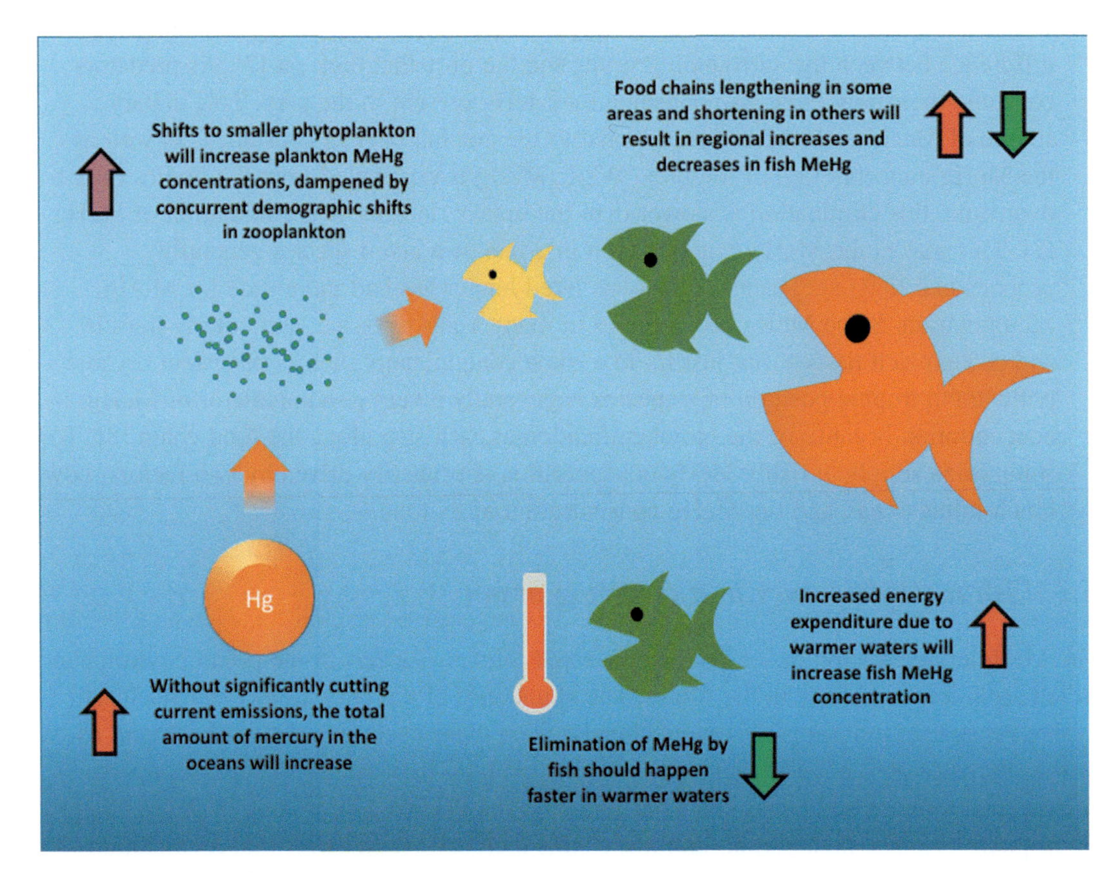

Figure 6.1
Cartoon summary of causes of future changes to fish MeHg. Upward arrows represent factors that will increase fish MeHg and downward arrows represent factors that will decrease fish MeHg.

sinking of organic matter in the ocean [5]. The oceans are expected to change in many ways that could affect this process, including becoming increasingly stratified, inhibiting calcifying plankton sinking because of ocean acidification, and shifting plankton community composition toward smaller plankton that do not sink as readily [10]. Not enough is currently known about this complicated system and its relation to mercury methylation to project changes in mercury methylation at this time. In the absence of complicating ecosystem changes the increasing seawater mercury will lead to a proportional increase in seafood MeHg, so the changes discussed below will take place relative to this backdrop of increasing MeHg.

- *Changes to plankton community distribution and structure*

  Due to the expansion of low-productivity regions and increased stratification in polar regions, the global population of phytoplankton may shift toward smaller species [11]. Smaller phytoplankton species have up to 100 times greater MeHg concentration than larger phytoplankton under the same conditions (due largely to their greater surface area to volume ratio) [6], meaning that such a demographic shift in the plankton community would have a strong impact on MeHg concentrations, though they are likely to be highly spatially heterogeneous. At the same time, it is possible that increased temperature and CO_2 could increase the growth rate within species of phytoplankton, leading to a partial offset of the MeHg increase. The effects of the changing temperature, CO_2 and nutrient concentrations, and pH of the oceans on phytoplankton species and community structures is highly uncertain and a function of complex interactions that are presently not well enough understood. While changes in phytoplankton MeHg concentrations could be large, they will be dampened by the associated demographic shifts in zooplankton. This reduced effect at the top of the plankton food chain will mean that the largest changes in seafood MeHg will occur because of changes affecting the fish themselves.

- *Bioaccumulation and biomagnification*

  The structure of the food web is very important for fish MeHg, especially for those predators at higher trophic levels which have the highest MeHg concentrations. The MeHg concentration of a fish grows exponentially with its trophic level; this is the definition of biomagnification. Each full trophic level increase results on average in an eightfold increase in MeHg concentration [8]. This means that longer food chains will strongly increase the MeHg concentration of the top predators, and shorter food chains will strongly decrease it. The importance of the food chain starts at the microscale, where changes in the zooplankton food web due to changing contents of riverine discharges have been shown to strongly affect zooplankton MeHg bioaccumulation [12].

  Food chains shorten in more productive regions. The predicted pattern of changes in ocean productivity are nonuniform, with some regions increasing and others decreasing. Many areas projected to see increases in food chain length, and therefore increases in MeHg in top predators, due to productivity changes are also projected to have lower fish populations in the future. Conversely, in open-ocean regions that are projected to have more fish, food chains should shorten due to increased productivity. If we assume for simplicity that fisheries catch is proportional to the amount of available fish, this will lead to a decrease in MeHg concentration in seafood in the absence of other changes, as fishing moves on average to higher productivity, shorter food chain areas.

The strong effect of food web structure changes on fish MeHg concentrations is illustrated in the change of diet between the 1970s and 2000s in Gulf of Maine (GoM) cod and dogfish. Because of large changes in prey stocks the diet of GoM cod shifted from clupeids to macroinvertebrates, which would result in 10%–20% increases in MeHg concentration. On the other hand, GoM dogfish over the same time period shifted from eating cephalopods to eating clupeids, with associated 25%–40% decreases in MeHg concentration [7]. The strong differences between specific species in the same environment highlight the complexity of the interactions involved, and the difficulty in predicting changes in seafood MeHg overall [7].

- *Temperature and fish diet*

  Temperature changes in the seawater that fish live in directly impact their MeHg concentration due to the relationship between their consumption of prey and the ambient temperature. At higher temperatures, fish must eat more to maintain a given growth rate than they would at lower temperatures. Eating more means consuming more MeHg, and since they are consuming more MeHg but achieving the same body mass, their MeHg concentration will be higher than it would be at lower temperatures [7]. Put another way, at the same consumption rate but at higher temperatures, fish body mass will shrink but MeHg consumption will remain the same, leading to higher MeHg concentrations (bioamplification) [13]. Ocean temperatures have been increasing and will continue to increase in the future. This means that the effect of future temperatures on seafood methylmercury will be to increase it. This also means that reductions in seafood MeHg because of future efforts to curb global Hg emissions will be offset by the ocean temperature changes that will almost certainly occur due to carbon emissions that have already taken place.

- *Migration and energy expenditure*

  An important element of the bioaccumulation of MeHg in tuna and other large predators is their energy expenditure as fast, migratory fish. The added energy expenditure of movement, and its resulting increased consumption at a constant body mass, explain tunas' increased MeHg concentration compared to the expected value based purely on their trophic position [7]. On a basic level, more migration will lead to higher MeHg concentrations and less migration will lead to lower MeHg. However, the future of the migration habits of even a specific species of tuna (e.g., Atlantic bluefin tuna) is uncertain, and tuna migration itself will be linked to many of the factors that themselves affect MeHg. Tuna migration is linked to seawater temperature, the availability of specific prey species, and other factors that affect larval survivability.

These intertwined factors could lead to changes in tuna migration that are not easily predictable. While we can predict the direction of the effects on MeHg of individual factors contributing to migration, since they are the same factors discussed above, the overall impact of migration changes on tuna MeHg will have to be the focus of future study.

- *Temperature and MeHg elimination*

  The elimination of MeHg by fish is influenced by the temperature of their surroundings, with warmer waters leading to faster elimination. While this relationship is significant over large temperature differences (a doubling of elimination rate for approximately 7ºC increase [14]), near-term likely ocean temperature increases would cause approximately a 30% increase in MeHg elimination rate. This effect could provide a partial global offset for the potential sources of MeHg increases in the future, but would be very unlikely to be sufficiently large to cancel them out entirely.

6.4 Conclusion

Overall changes in seafood MeHg in the future will be the combination of many competing effects. Without severe reductions in global mercury emissions, the rising concentration of seawater mercury will act to increase seafood MeHg concentrations. On top of this, rising ocean temperatures will increase MeHg bioaccumulation and further the increase of MeHg in fish. Complicating this overall increasing trend will be species-specific changes in habitat, prey availability, and trophic position, which could add more to increasing MeHg concentrations or partially offset it. Since both carbon emissions and mercury emissions in the present day will have effects on the ocean for the coming decades to centuries, the problem of seafood MeHg will be a persistent one for the foreseeable future.

References

[1] JECFA, Methylmercury, Safety Evaluation of Certain Food Additives and Contaminants. Report of the 61st Joint FAO/WHO Expert Committee on Food Additives, WHO Technical Report Series 922, World Health Organization, International Programme on Chemical Safety, Geneva, 2004, pp. 132–139.

[2] A.M. Cisneros-Montemayor, D. Pauly, L.V. Weatherdon, Y. Ota, A global estimate of seafood consumption by coastal Indigenous peoples, PLoS One 11 (12) (2016) e0166681.

[3] D.G. Streets, H.M. Horowitz, D.J. Jacob, Z. Lu, L. Levin, A.F.H. ter Schure, et al., Total mercury released to the environment by human activities, Environ. Sci. Technol. 51 (2017) 5969–5977.

[4] H.M. Horowitz, D.J. Jacob, H.M. Amos, D.G. Streets, E.M. Sunderland, Historical mercury releases from commercial products: global environmental implications, Environ. Sci. Technol. 48 (2014) 10242–10250.

[5] E.M. Sunderland, D.P. Krabbenhoft, J.W. Moreau, S.A. Strode, W.M. Landing, Mercury sources, distribution, and bioavailability in the North Pacific Ocean: Insights from data and models, Global Biochem. Cycl. 23 (2009) GB2010.

[6] A.T. Schartup, A. Qureshi, C. Dassuncao, C.P. Thackray, G. Harding, E.M. Sunderland, A model for methylmercury uptake and trophic transfer by marine plankton, Environ. Sci. Technol. 52 (2018) 654–662.

[7] A.T. Schartup, C.P. Thackray, A. Qureshi, C. Dassuncao, K. Gillespie, A. Hanke, E.M. Sunderland, Climate Change and Overfishing Increase Neurotoxicant in Marine Predators, Nature, 2019 (in press).
[8] R.A. Lavoie, T.D. Jardine, M.M. Chumchal, K.A. Kidd, L.M. Campbell, Biomagnification of mercury in aquatic food webs: a worldwide meta-analysis, Environ. Sci. Technol. 47 (2013) 13385–13394.
[9] D.G. Streets, H.M. Horowitz, Z. Lu, L. Levin, C.P. Thackray, E.M. Sunderland, Global and regional trends in mercury emissions and concentrations, 2010–2015, Atmos. Environ. 201 (2019) 417–427.
[10] K.L. Smith Jr., H.A. Ruhl, B.J. Bett, D.S.M. Billett, R.S. Lampitt, R.S. Kaufmann, Climate, carbon cycling, and deep-ocean ecosystems, Proc. Natl. Acad. Sci. U.S.A. 107 (2009) 19211–19218.
[11] C.A. Stock, J.G. John, R. Rykaczewski, R.G. Asch, W.W.L. Cheung, J.P. Dunne, et al., Reconciling fisheries catch and ocean productivity, Proc. Natl. Acad. Sci. U.S.A. 114 (2017) E1441–E1449.
[12] S. Jonsson, A. Andersson, M.B. Nilsson, U. Skyllberg, E. Lundberg, J.K. Schaefer, et al., Terrestrial discharges mediate trophic shifts and enhance methylmercury accumulation in estuarine biota, Sci. Adv. 3 (1) (2017) e1601239.
[13] J.J. Alava, A.M. Cisneros-Montemayor, U.R. Sumaila, W.W. Cheung, Projected amplification of food web bioaccumulation of MeHg and PCBs under climate change in the Northeastern Pacific, Sci. Rep. 8 (1) (2018) 13460.
[14] M. Trudel, J.B. Rasmussen, Modeling the elimination of mercury by fish, Environ. Sci. Technol. 31 (1997) 1716–1722.

CHAPTER 7

Building confidence in projections of future ocean capacity

Tyler D. Eddy[1,2]

[1]*Institute for Marine & Coastal Sciences, University of South Carolina, Columbia, SC, United States*
[2]*Fisheries & Marine Ecosystem Model Intercomparison Project (FishMIP)*

Chapter Outline

7.1 Ocean capacity and ecosystem services

The oceans provide us with ecosystem services such as food provision from fisheries and aquaculture, carbon sequestration, flood control and waste detoxification for people living in coastal communities, and biodiversity provision [1]. These services play a direct role in the composition of the Earth's atmosphere that regulates our weather and climate [2]. The capacity of the oceans to provide these ecosystem services can change over time due to human activities such as fishing, emission of greenhouse gases, pollution, and coastal development [3]. We can measure components of nature to quantify the capacity of the ocean to provide ecosystem services and how capacity changes with time. One component of ocean capacity is the extent and quality of habitats such as coral reefs, mangroves, sea grasses, and kelp forests, which are directly related to the abundance and biomass of species and biodiversity associated with each habitat [4]. Habitats, biomass, and biodiversity respond to human activities such as fishing, habitat destruction, pollution, and sedimentation as well as to environmental conditions that are affected by climate change such as temperature, nutrient flux, pH, and oxygen levels of the oceans. By tracking these

Predicting Future Oceans.
DOI: https://doi.org/10.1016/B978-0-12-817945-1.00007-1

key aspects of ocean capacity through time, we can understand how ecosystem services respond to human activities and climate change.

7.2 Climate change impacts on ocean capacity

Humans have been impacting the oceans for millennia through fishing, which has reduced the abundance and biomass of many different species worldwide, in turn affecting the structure and function of marine food webs and ecosystems [5,6]. These activities have altered the capacity of the oceans to provide ecosystem services such as food provision. More recently, since the mid-1800s and the industrial revolution, the emission of greenhouse gases has played a transformative role in the Earth's biogeochemical cycle and climate. These effects are already observed in coastal communities, as fishers have noticed traditionally fished species are absent and new species arriving to their waters or that they have to travel farther to fish species than they used to [7,8]. Reducing greenhouse gas emissions has been addressed by policies, technologies, local and regional initiatives, and international agreements such as the Paris Agreement. However, it remains unclear at what rate greenhouse gases will be emitted in the future, and for that reason scientists have been working with a range of greenhouse gas scenarios to make projections about what we might expect for the future capacity of the oceans to provide ecosystem services. Projections about how the oceans, marine biodiversity, marine aquaculture, and fisheries will respond to climate change are used by policy makers, businesses, natural resource producers and users, and insurance companies to plan for future environmental and ecological conditions.

The Intergovernmental Panel on Climate Change (IPCC) 5th Assessment Report summarizes the state of the art scientific research addressing how climate change is projected to impact the oceans [3]. This report presented an approach to project how catch potential—the amount of fish in the ocean available to be caught—is expected to change in the future. This analysis relied on a modeling approach developed by Cheung et al. [9], referred to as the Dynamic Bioclimate Envelope Model (DBEM). As with every model, this model is based on assumptions and simplifications of how the world works, and is subject to various uncertainties, biases, limitations, and sensitivities [10]. As a result, projections about the future ocean capacity made by this model may differ from projections made by other models. DBEM is based on a species distribution approach, whereby if environmental conditions for a particular species are favorable, the model will predict that it will occur there. This model does not take into account species interactions, but some models such as EcoOcean [11], a global extension of the popular and foundational Ecopath with Ecosim regional model [12], specify species interactions via predator–prey relationships. At the other end of the spectrum of model architecture are size-based models which do not resolve species or functional groups, but instead focus on size classes of marine life [13,14].

Generally, DBEM and size-based models project greater overall declines in global fish biomass under climate change when compared to EcoOcean [15].

Acknowledging the need to improve our understanding of uncertainties in projections of climate change impacts on the world's oceans, a research group composed of fisheries and ecosystem modelers at both regional and global scales as well as members from the climate and Earth system modeling community was created to address this question. The Fisheries and Marine Ecosystem Model Intercomparison Project (FishMIP) was formed in 2013 to bring together a wide array of modeling approaches and compare them in a meaningful way in order to understand how models project climate change impacts and quantify associated uncertainties, biases, limitations, and sensitivities for model projections of future ocean capacity. The founding coordinators were Heike Lotze, Derek Tittensor, Eric Galbraith, William Cheung, and myself. FishMIP maintains a membership that is open to all who wish to participate, and at present includes more than 50 members.

7.3 The model intercomparison project experience: model ensembles

The Model Intercomparison Project (MIP) approach of using an ensemble of models to run standardized simulations has been applied to many different sectors and fields. An ensemble of models is a group of models that are all capable of making projections about the same thing—such as future land surface temperature or sea surface temperature—but may be built differently. The rationale for using an ensemble of models forced with the same data is because, as above, each model has its own structure, assumptions, uncertainties, biases, limitations, and sensitivities. Most models give different answers when asked the same question, and learning why is important to improve and refine our understanding of how we model systems. Using an ensemble of models accounts for a greater envelope of variability or level of agreement in model projections. Under the IPCC framework to evaluate evidence, the two quantitative criteria that are applied are: the amount of evidence and evidence agreement [3]. To allow for meaningful comparisons among models, it is necessary to standardize how simulations are run as much as possible, thereby elucidating differences that are due to the models themselves. Once a general understanding of ensemble variability under different scenarios has been achieved, whether all models predict the same direction of change (increase or decrease), and how different the magnitude of change varies, a more detailed and controlled experimental design of simulations and analyses of factors contributing to the envelope of variability can be undertaken. For these reasons, the MIP approach has been taken up by many modeling communities worldwide.

The Ocean MIP "aims to provide a framework for evaluating, understanding, and improving the ocean, sea ice, tracer, and biogeochemical components of global climate and Earth system models" contributing to the Coupled MIP (CMIP) which also represents terrestrial processes. The fifth iteration of CMIP (CMIP5) was used widely in the IPCC 5th

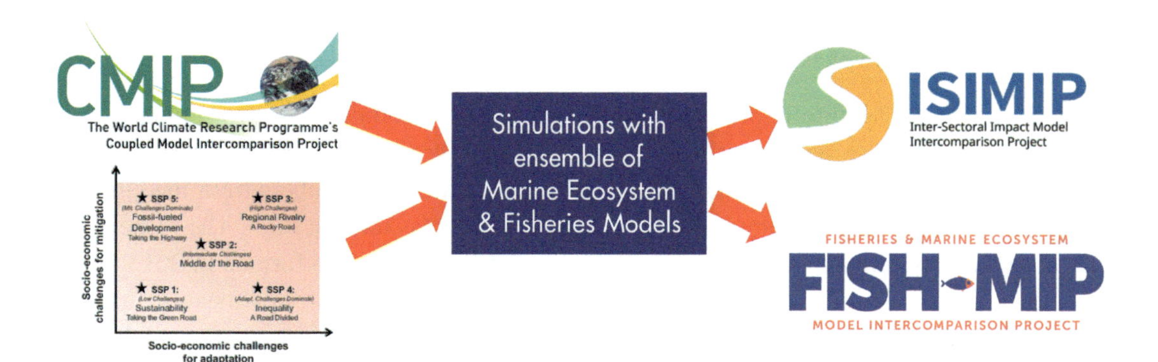

Figure 7.1
Methodological approach of the FishMIP to make projections about future ocean capacity. *FishMIP*, Fisheries and Marine Ecosystems Model Intercomparison Project.

Assessment Report to detail future projections about land and sea temperatures, primary productivity, precipitation, winds and storms, ocean circulation, salinity, acidity, and oxygen concentration under different carbon emissions scenarios [3]. The Inter-Sectoral Impact MIP (ISIMIP) was formed to interface with CMIP in a downstream manner using CMIP outputs as ISIMIP inputs. Outputs from the CMIP ensemble describing projections for physical and environmental variables under climate change scenarios are used as inputs for impact MIPs from different sectors such as agriculture, biomes, coastal infrastructure, energy, forests, water, human health, lakes, permafrost, terrestrial biodiversity, and fisheries and marine ecosystems (Fig. 7.1; [16]). The aim of ISIMIP is not only to standardize climate change simulations among models within sectors but also to standardize simulations among sectors to allow for a broader comparison of climate change impacts [17,18]. In this standardization of simulations lies the challenge of using a MIP approach; by including a wide variety of model structures, and their required inputs, and outputs, compromises have to be made by some models in order to have a lowest common denominator that is inclusive of all models.

7.4 Fisheries and Marine Ecosystem Model Intercomparison Project: projecting future ocean capacity

FishMIP was conceived in response to a question posed by researchers working in the agricultural climate change community about whether the oceans could potentially make up for food losses that were projected to happen on land under climate change. After four years of FishMIP, we were able to answer this question by comparing FishMIP fish projections to crop projections from the Agricultural MIP (AgMIP) under climate change. Unfortunately, countries in the tropics that are projected to experience the biggest losses in agricultural production on land are also projected to have the largest losses in fisheries catch production in the ocean [13,14].

The process of developing a simulation protocol for FishMIP that could be applied by a diverse set of fisheries and marine ecosystem models took many years, workshops, emails, and conference calls and was largely a consensus process among modelers who were interested in the project (https://www.isimip.org/gettingstarted/marine-ecosystems-fisheries/). Models ranged greatly in their data requirements (Table 7.1) and resolution of marine organisms, from size-based modeling approaches such asthe BiOeconomic mArine Trophic Size-spectrum (BOATS) model that resolves all fish in the oceans into three size categories [19] to the DBEM which resolves more than 1000 different species [9]. As a result of this heterogeneity, FishMIP outputs from all models have been disaggregated into three size categories [16,20–22]. An additional level of complexity of FishMIP was that models of both global scale and regional scale were included. While it would have been more straightforward to only include models of global scale, we wanted to be able to address how projections by global models based on first principles and driven by bottom-up processes for individual regions compared to corresponding regional models that were parameterized with local biomass survey and fisheries data and driven by top-down processes such as

Table 7.1: Description of common model inputs and outputs employed by Fisheries and Marine Ecosystem Model Intercomparison Project (FishMIP) model ensemble, as well as the standardized model outputs provided by all models participating in FishMIP.

Common model inputs	Units
Ocean current speed	m/s
Sea temperature or potential temperature	K
Dissolved O_2 concentration	mol/m^3
Primary organic carbon productivity	$mol/m^3/s$
Zooplankton carbon concentration	mol/m^3
pH	Unitless
Salinity	psu
Common model outputs	
Fish species and functional group carbon biomass density	$g/m^3/month$
Fisheries metrics	Various
Relative species and functional group abundances	Unitless
Trophic level	Unitless
Production of carbon	$g/m^3/month$
Production and biomass ratio	Unitless
Mortality rate	$month^{-1}$
FishMIP standardized outputs	
Total system carbon biomass	g/m^2
Total consumer carbon biomass density	g/m^2
Carbon biomass density of consumers > 10 cm	g/m^2
Carbon biomass density of consumers > 30 cm	g/m^2
Total catch (all commercial functional groups or size classes)	g wet biomass/m^2
Total landings (all commercial functional groups or size classes)	g wet biomass/m^2

fishing. Additionally, we included different socioeconomic scenarios, often referred to as shared socioeconomic pathways (SSPs) of future fishing pressure, not to try to predict the future, but to provide an exploratory approach to a range of possible ocean futures (Fig. 7.1).

During each iteration of the FishMIP process, we learned more and more about other models in the project, the model(s) that we work with and develop directly, and why they give different answers. Initial results indicate that global mean fisheries productivity is projected to decrease under climate change, with greater declines under higher carbon emissions scenarios [15,22]. There is however, a lot of spatial variation in projected changes, and there is less agreement among regional and global models for specific regions compared to agreement among global models. We are now in a position to run controlled experiments of changes in the two primary climate drivers—sea surface temperature and primary productivity—to pinpoint the mechanisms in each model that lead to variability in projections. For example, in the AgMIP, the main factor leading to differences in projections of crop production under climate change was if models included CO_2 fertilization or not (increased crop production due to increased CO_2 in the atmosphere under climate change; [23]). These comparison exercises lead to a better understanding of how different representations of models respond to climate change drivers, to further refine projections of future ocean capacity.

7.5 Socioeconomic drivers of future ocean capacity

It has been shown in a number ofinstances that while climate change may reduce future fisheries productivity in some regions, the most important factors to consider when projecting future fish biomass are socioeconomic in nature. For example, a recent study that used satellite tracks from vessels and machine learning algorithms to differentiate fishing behavior from transit behavior quantified the global footprint of fishing effort–and found that the largest reduction in fishing effort annually occurred during the holidays of Chinese New Year [24]. A fishery that is aiming to fish at maximum sustainable yield typically reduces the biomass of the stock by about half. While some species may be more susceptible to climate change impacts than others, typically the projected changes in biomass due to climate are much less severe when compared to potential changes due to fishing activities–even for sustainable fisheries reference points. Therefore any projections of climate change impacts on fisheries need to take into account how fishing effort will change in the future.

While there have been some efforts to qualitatively map out future exploratory ocean scenarios related to the SSPs [25,26], at present there is a lack of the types of socioeconomic scenarios that can be run with FishMIP models. For these reasons, FishMIP used two different fishing scenarios in combination with different climate emissions scenarios: hold fishing constant at 2005 levels to be consistent with ISIMIP protocols (and due to data availability) and a no-fishing scenario. While highly unrealistic, the no-fishing scenario is a control run that allows for an analysis of how much climate change affects future ocean capacity in the absence of fishing. In future iterations of FishMIP we aim to be able to incorporate more detailed socioeconomic scenarios. The goal of these exploratory scenarios is not to predict the future, but to try to encompass the extremes in terms of the range of how fishing effort might change to understand the range of future ocean capacity under different socioeconomic and climate scenarios.

7.6 Summary

Overall, an ensemble or MIP approach to project ocean capacity has its strengths in being able to partition uncertainty according to choice of the Earth system model, fisheries/ ecosystem model, climate scenario, and socioeconomic scenario. We can also find out how dependent our projections are on different components of model projections, such as choice of Earth system model, ecosystem model, climate scenario, and fishing scenario. By comparing projections of models of varying philosophy and structure, we can learn more about the models themselves, where their sensitivities lie, what mechanisms and processes lead to variation in projections, how to improve them, and where to focus efforts to build confidence in ocean capacity projections moving forward. This process has been developed with the goals of not only improving modeling of ocean ecosystems and fisheries, but to provide information for developing management and adaptation strategies and policies to climate change.

References

[1] B. Worm, E.B. Barbier, N. Beaumont, J.E. Duffy, C. Folke, B.S. Halpern, et al., Impacts of biodiversity loss on ocean ecosystem services, Science 314 (2006) 787–790.

[2] IPBES, The methodological assessment report on scenarios and models of biodiversity and ecosystem services, in: S. Ferrier, K.N. Ninan, P. Leadley, R. Alkemade, L.A. Acosta, H.R. Akçakaya, et al. (Eds.), Secretariat of the Intergovernmental Science-Policy Platform on Biodiversity and Ecosystem Services, IPBES, Bonn, Germany, 2016. 348 pp.

[3] IPCC, in: O. Edenhofer, R. Pichs-Madruga, Y. Sokona, E. Farahani, S. Kadner, K. Seyboth, et al. (Eds.), Climate Change 2014: Mitigation of Climate Change. Contribution of Working Group III to the Fifth Assessment Report of the Intergovernmental Panel on Climate Change, Cambridge University Press, Cambridge, United Kingdom and New York, 2014.

[4] S. Drakare, J.J. Lennon, H. Hillebrand, The imprint of the geographical, evolutionary and ecological context on species–area relationships, Ecol. Lett. 9 (2006) 215–227.

[5] J.B.C. Jackson, M.X. Kirby, W.H. Berger, K.A. Bjorndal, L.W. Botsford, B.J. Bourque, et al., Historical overfishing and the recent collapse of coastal ecosystems, Science 293 (2001) 629–638. Available from: https://doi.org/10.1126/science.1059199.

[6] H.K. Lotze, H.S. Lenihan, B.J. Bourque, R.H. Bradbury, R.G. Cooke, M.C. Kay, et al., Depletion, degradation, and recovery potential of estuaries and coastal seas, Science 312 (2006) 1806–1809.

[7] A.J. Pershing, M.A. Alexander, C.M. Hernandez, L.A. Kerr, A. Le Bris, K.E. Mills, et al., Slow adaptation in the face of rapid warming leads to collapse of the Gulf of Maine cod fishery, Science 350 (2015) 809–812.

[8] M.L. Pinsky, G. Reygondeau, R. Caddell, J. Palacios-Abrantes, J. Spijkers, W.W.L. Cheung, Preparing ocean governance for species on the move, Science 360 (2018) 1189–1191.
[9] W.W.L. Cheung, et al., Large-scale redistribution of maximum fisheries catch potential in the global ocean under climate change, Glob. Change Biol. 16 (2010) 24–35.
[10] M.R. Payne, M. Barange, H.P. Batchelder, W.W.L. Cheung, X. Cormon, T.D. Eddy, et al., Uncertainties in projecting climate change impacts in marine ecosystems, ICES J. Mar. Sci. 73 (2016) 1272–1282.
[11] V. Christensen, M. Coll, J. Buszowski, W.W.L. Cheung, T. Frölicher, J. Steenbeek, et al., The global ocean is an ecosystem: simulating marine life and fisheries, Global Ecol. Biogeogr. 24 (2015) 507–517.
[12] V. Christensen, C.J. Walters, Ecopath with Ecosim: methods, capabilities and limitations, Ecol. Modell. 172 (2004) 109–139.
[13] J.L. Blanchard, R.A. Watson, E.A. Fulton, R.A. Cottrell, K.L. Nash, A. Bryndum-Buchholz, et al., Linked sustainability challenges and trade-offs among fisheries, aquaculture and agriculture, Nat. Ecol. Evol. 1 (2017) 1240–1249.
[14] J.L. Blanchard, R.F. Heneghan, J.D. Everett, R. Trebilco, A.J. Richardson, From bacteria to whales: using functional size spectra to model marine ecosystems, Trends Ecol. Evol. 32 (2017) 174–186.
[15] H.K. Lotze, D.P. Tittensor, A. Bryndum-Buchholz, T.D. Eddy, W.W.L. Cheung, E.D. Galbraith, et al., Ensemble Projections Reveal Consistent Declines of Global Fish Biomass Under Climate Change. bioRxiv preprint: 467175, 2018.
[16] D.P. Tittensor, T.D. Eddy, H.K. Lotze, E.D. Galbraith, W.W.L. Cheung, M. Barange, et al., A protocol for the intercomparison of marine fishery and ecosystem models: FishMIPv1.0, Geosci. Model. Dev. 11 (2018) 1421–1442.
[17] K. Frieler, R. Betts, E. Burke, P. Ciais, S. Denvil, D. Deryng, et al., Assessing the impacts of 1.5°C global warming—simulation protocol of the Inter-Sectoral Impact Model Intercomparison Project (ISIMIP2b). 2017, Geosci. Model Dev. 10 (2017) 4321–4345.
[18] J. Schewe, S.N. Gosling, C. Reyer, F. Zhao, P. Ciais, J. Elliott, et al., State-of-the-art global models underestimate impact from climate extremes, Nature Communications 10 (2019) 1005.
[19] E.D. Galbraith, D.A. Carozza, D. Bianchi, A coupled human-Earth model perspective on long-term trends in the global marine fishery, Nature Communications 8 (2017) 14884.
[20] Eddy T.D., Bulman C.M., Cheung W.W.L., Coll M., Fulton E.A., Galbraith E.D., et al., 2018. ISIMIP2a Simulation Data from Fisheries & Marine Ecosystems (Fish-MIP; regional) Sector. GFZ Data Services. https://doi.org/10.5880/PIK.2018.004
[21] Tittensor D.P., Lotze H.K., Eddy T.D., Galbraith E.D., Cheung W.W.L., Bryndum-Buchholz A., et al., 2018. ISIMIP2a Simulation Data from Fisheries & Marine Ecosystems (Fish-MIP; global) Sector. GFZ Data Services. https://doi.org/10.5880/PIK.2018.005
[22] Lotze H.K., Tittensor D.P., Bryndum-Buchholz A., Eddy T.D., Cheung W.W.L., Galbraith E.D., et al. In press. Global ensemble projections reveal trophic amplification of ocean biomass declines with climate change. Proceedings of the National Academy of Sciences of the United States of America.
[23] C. Rosenzweig, J. Elliott, D. Deryng, A.C. Ruane, C. Müller, A. Arneth, et al., Assessing agricultural risks of climate change in the 21st century in a global gridded crop model intercomparison, Proc. Natl. Acad. Sci. U.S.A. 111 (2013) 3268–3273.
[24] D.A. Kroodsma, J. Mayorga, T. Hochberg, N.A. Miller, K. Boerder, F. Ferretti, et al., Tracking the Global Footprint of Fisheries. 2018. Tracking the global footprint of fisheries, Science 359 (2018) 904–908.
[25] O. Maury, L. Campling, H. Arrizabalaga, O. Aumont, L. Bopp, G. Merino, et al., From shared socio-economic pathways (SSPs) to oceanic system pathways (OSPs): building policy-relevant scenarios for global oceanic ecosystems and fisheries, Global Environ. Change 45 (2017) 203–216.
[26] CERES, Exploratory socio-political scenarios for the fishery and aquaculture sectors in Europe, in: J.K. Pinnegar, G.H. Engelhard (Eds.), Deliverable D1.1—Glossy 'Report Card' Aimed at Stakeholders, Centre for Environment, Fisheries & Aquaculture Science (Cefas), Lowestoft, 2016. 8 pp.

SECTION 3

Changing Marine Ecosystems and Biodiversity

CHAPTER 8

Marine biodiversity and ecosystem services: the large gloomy shadow of climate change

Didier Gascuel[1] and William W.L. Cheung[2]

[1]Agrocampus Ouest, Ecology and Ecosystem Health Research Unit, Rennes, France [2]Changing Ocean Research Unit, The University of British Columbia, Vancouver, BC, Canada

Chapter Outline

8.1 Introduction

For decades, if not centuries, biodiversity in the global ocean has been under increasing pressure from human activities. Over the last century fisheries have been the main driver of global change in the sea, leading not only to the depletion of exploited stocks [1–3], but also to the degradation of the integrity of habitats in the seafloor [4,5], the truncation of age structure of exploited populations, and the modifications of species assemblages [6]. The impacts of fishing have affected the functioning of marine food webs [7,8] such as ecosystem productivity and the resilience of ecosystems [9,10]. In the last few decades the effects of climate change on the structure and the functioning of marine ecosystems have already been detected [11]. The poleward migration of fish stocks has been documented worldwide [12,13] and is expected to continue in the 21st century [14]. Moreover, ocean warming affects ocean primary production which consequently alters potential fisheries catches [15,16]. Ocean acidification and deoxygenation are expected to affect the productivity, abundance, and distribution of marine species, including those that are exploited by fisheries [17]. Altogether fishing and climate change are compromising ecosystem services provided by the ocean.

Predicting Future Oceans.
DOI: https://doi.org/10.1016/B978-0-12-817945-1.00008-3

The chapters in this section review various aspects of the impact of climate change on marine biodiversity and ecosystems. This synthesis chapter for this section aims to summarize the main findings from the various contributions on the effects of global change on ecosystem functions and services, and put it in the context of the broader understanding of the changing ocean biodiversity and ecosystems under climate change.

8.2 Changing marine ecosystem and biodiversity: seven highlights

The biogeography of marine biodiversity is strongly shaped by the biology of marine organisms, their relationship with the environmental conditions, and their changes. In this section Reygondeau (Chapter 9: Current and future biogeography of exploited marine exploited groups under climate change) analyzes the distribution of marine biodiversity and how it will be impacted by future changes in the environment. Using niche models he notably projects the distribution by 2100 of a large number of marine species, including mammals and some invertebrates, thus highlighting the expected changes in global marine biogeography. The results highlight large areas of the ocean where most marine species would likely be threatened by climate change. Du Pontavice (Chapter 12: Changing biomass flows in marine ecosystems) discusses the effects of ocean warming on the functioning of marine food webs. Using trophic models he shows that warmer oceans are expected to result in less efficient and faster trophic transfers, leading to a global drop in biomass and production, up to −20% by 2100 in the worse climate change scenario. Bernhardt (Chapter 10: Linking individual performance to population persistence) underlines the importance of understanding the thermal biology of organisms in explaining and projecting changes at the biodiversity and ecosystem levels. Particularly, she shows that the use of metabolic theory, informed by experimentation and model simulations, may help to identify the potential of natural adaptation of organisms in the context of future oceans.

Two chapters discuss the importance of observations, their availability, and accuracy in understanding and projecting the changing marine ecosystems and biodiversity, and improving fisheries management in response to these changes. Pauly et al. (Chapter 11: The *Sea Around Us* as provider of global fisheries catch and related marine biodiversity data) underline the importance of improving global statistics on fisheries catches. They present the reconstruction of global fisheries catches conducted by the *Sea Around Us* project. The reconstructed catch data help improve our understanding of the changes in global catches and their relationship with fishing and environmental drivers since the 1950s. Such new knowledge helps paint a more accurate picture of human impacts on marine ecosystems, and offers a realistic baseline for management and assessing future changes in ecosystem services. Gonzalez (Chapter 15: Understanding variability in marine fisheries) discusses patterns of population variability in marine fisheries and the role played by environmental fluctuations. He argues that new data provided by satellite remote sensing or by automatic platforms of floats and gliders, as well as high-quality forecast products derived from Earth system modeling, provide the opportunity to develop more realistic probabilistic models and novel approaches to implement dynamics fisheries management strategies.

Henschke (Chapter 14: Jellyfishes in a changing ocean) discusses how the study of jellyfish can be used as a lens to understand the effects of global change on marine ecosystems. Jellyfish populations are commonly expected to benefit from global change as a result of overfishing of jellyfish predators and competitors (planktivorous fish), expansion of invasive jellyfish through translocation by the rapid growth of shipping activities, and expansion of eutrophication events or hypoxic water masses under climate change—jellyfish are generally more resilient to these adverse conditions than many other species. However, the potential expansion of jellyfish in the future may be more uncertain than we envisioned. Henschke shows, using a size-structured population model, that the increase in medusa biomass would be small by 2100, and mainly driven by the increase in zooplankton biomass expected in the shallow waters of northern regions. She highlights the role of jellyfish in marine ecosystems and the need to manage their fisheries in a flexible and localized way under global change.

Roberts (Chapter 13: The role of cyclical climate oscillations in species distribution shifts) discusses current research to understand the influence of benthic substrate and cyclical ocean oscillations, as environmental constraints of the poleward shift of commercial marine species. Using some species of the South Atlantic Bight as case studies, and based on trawl data, she especially shows that links to benthic habitats may lower the expected shifts in fish distribution, while the North Atlantic Oscillation would determine the relationship between species presence and bottom temperature.

8.3 Global impact of climate change on marine ecosystems and the biodiversity

The chapters in this section and numerous recently published papers have provided strong evidence that climate change affects marine biodiversity. Overall, ocean warming, acidification, deoxygenation, and associated oceanographic changes such as increased stratification of large areas of the ocean, are threatening almost all marine life and biodiversity [18,19]. Marine ecosystem functions such as primary productivity will also be impacted with regional differences that are driven partly by region-specific changes in magnitude and ratio of nutrient supply [20,21]. In general, the global organic matter production is expected to increase in high latitude and to decrease in low latitude, with implications for all the pelagic and seafloor ecosystems [22]. The emergence of novel ocean conditions is already driving changes in ecophysiology, biogeography, and ecology of all organisms, from plankton to mammals [13,23], while scope for adaptation appears limited for many organisms [24]. Thus the observed and projected changes result in a population decline in their lower latitudes ranges [25], an expansion in their poleward boundaries [12], and an earlier timing of biological events [26]. Such changes in species biogeography imply an overall shift in community structure and a decrease in global animal biomass [27]. High trophic levels are expected to be particularly affected [13].

The interactions between climate change and other human impacts from land and ocean may exacerbate the stresses on coastal marine ecosystems [18]. Habitats erosion, sea level rise, and more severe storms are leading to losses of vegetation, especially in sandy beaches and salt marshes [28,29] while herbivory intensified by warming should add a cumulative stress on coastal vegetation and reduce their productivity [30,31]. Extreme events such as heat waves and storms are also exacerbating the rate of ecosystem changes, particularly in kelp forests and seagrass meadows [32]. All in all, compounding effects of warming, deoxygenation, acidification, and changes in nutrient supplies will exacerbate the decrease in species richness and spatial heterogeneity in coastal ecosystems [14].

Some marine ecosystems such as coral reefs and deep-sea habitats are particularly vulnerable to climate change impacts. Coral reef ecosystems appeared especially vulnerable to climate change and the already observed massive reef destruction is expected to amplify in the coming decades [33]. Ocean acidification and the related aragonite undersaturation will especially affect deep-water coral reefs, through dissolution and intensified bioerosion of the nonliving matrix [18,34]. Additionally, benthic communities in deep-sea habitats will experience structural and functional changes that affect the carbon cycle [35]. Thus, much of the abyssal seafloor is expected to experience declines in food supply that will diminish benthic biomass, change community structure, and change rates of carbon burial [36,37].

8.4 Impacts on marine ecosystem services

Changes in the structure and functioning of marine ecosystems will have large impacts on all the services they provide. Seafood provision from fisheries will likely be among the most affected services [19,38]. Fisheries' catches and their composition are already affected by the effects of warming, deoxygenation, and changes in primary production on growth, reproduction, and survival of fish stock [12,17,18], while the reduction in the availability and quality of habitats of fish and invertebrates populations is expected to amplify the decrease in the diversity and productivity of many marine populations that support fisheries [15]. The level of impacts will vary regionally, with tropical coastal ecosystems being the most at risk relatively to those in mid-latitude regions [14,39]. In the 21st century, potential global fisheries catches are projected by multiple models to decrease [40,41]. Although the changes in realized catch will strongly depend on fishing intensity [42], this might severely reduce the revenue from fisheries sectors, thus impacting the livelihood of the dependent communities and food security of vulnerable people [43]. In addition, seafood provision from aquaculture is and will be increasingly impacted by climate change [19,44]. Shellfish aquaculture is especially sensitive to ocean acidification [45], while farmed species will be exposed to increased risk of disease and harmful algal blooms [46].

Other ecosystem services are being threatened by climate change. In particular, the reduction in nutrient cycling in the deep seafloor ecosystems induced by warming is compromising supporting services essential for all the marine life [28]. On the other hand, the ocean ecosystems' role in climate regulation may be impaired by the reduction in carbon stocking and sequestration [47], for instance, in salt marshes as a result of sea level rise, and more generally as a consequence of the loss and degradation of biodiversity and ecosystems' functions [48]. Protection services provided by many coastal ecosystems, such as mangroves, salt marshes, or coral reefs, are important too, and losses will increase the exposure of coastal communities to storms, erosion, and saltwater intrusions [49]. Finally, culture and recreational services that are important for human well-being are also threatened in many ecosystems. Overall, climate and biodiversity changes are expected to result in losses of opportunities for using ocean ecosystems for education and the relationship with some Indigenous knowledge and culture.

In conclusion, based on the contributions from this section, the future of marine biodiversity and ecosystems, under "business-as-usual" can be envisioned within four dimensions of ecological deterioration. We expect marine ecosystems to be (1) less ecologically productive as a result of changing ocean primary production, the degradation and loss of habitats for marine species, the effects of changing ocean biogeochemistry on organisms, and the decrease in food web efficiency; (2) less stable as a result of an environment becoming more variable, and the weakening of biological regulations and feedback mechanisms [50]; (3) less reversible because of the higher risk of large shifts in ecosystem structure that often result in an alternative stable state, while the erosion of genetic diversity may accelerate; and (4) less resilient, because of the reduction in functional marine biodiversity that is an important factor supporting ecological resilience. This section sets the stage for the subsequent sections to discuss how human societies can respond in view of the worrying future of the ocean system.

References

[1] D. Pauly, R. Watson, J. Alder, Global trends in world fisheries: impacts on marine ecosystems and food security, Phil. Trans. R. Soc. B 360 (2005) 5–12.

[2] L. Tremblay-Boyer, D. Gascuel, R. Watson, V. Christensen, D. Pauly, Modelling the effects of fishing on the biomass of the world's oceans from 1950 to 2006, Mar. Ecol. Prog. Ser. 442 (2011) 169–185.

[3] B. Worm, R. Hilborn, J.K. Baum, T.A. Branch, J.S. Collie, C. Costello, et al., Rebuilding global fisheries, Science 325 (5940) (2009) 578–585.

[4] B.S. Halpern, S. Walbridge, K.A. Selkoe, C.V. Kappel, F. Micheli, C. D'Agrosa, et al., A global map of human impact on marine ecosystems, Science 319 (5865) (2008) 948–952.

[5] J.G. Hiddink, et al., Global analysis of depletion and recovery of seabed biota after bottom trawling disturbance, Proc. Natl. Acad. Sci. U.S.A. 114 (2017) 8301–8306.

[6] S. Jennings, S.P.R. Greenstreet, J.D. Reynolds, Structural change in an exploited fish community: a consequence of differential fishing effects on species with contrasting life histories, J. Anim. Ecol. 68 (1999) 617–627.
[7] P.V. Anh, G. Everaert, P. Goethals, C.T. Vinh, F. De Laender, Production and food web efficiency decrease as fishing activity increases in a coastal ecosystem, Estuar. Coast. Shelf Sci. 165 (2015) 226–236.
[8] A. Maureaud, D. Gascuel, M. Colléter, M.L. Palomares, H. Du Pontavice, D. Pauly, et al., Global change in the trophic functioning of marine food webs, PLoS One 12 (2017) e0182826.
[9] C. Folke, S. Carpenter, B. Walker, M. Scheffer, T. Elmqvist, L. Gunderson, et al., Regime shifts, resilience, and biodiversity in ecosystem management, Annu. Rev. Ecol. Evol. Syst. 15 (35) (2004) 557–581.
[10] J.B. Jackson, et al., Historical overfishing and the recent collapse of coastal ecosystems, Science 293 (2001) 629–638.
[11] H.O. Pörtner, D. Karl, P.W. Boyd, W.W.L. Cheung, S.E. Lluch-Cota, Y. Nojiri, et al., IPCC Fifth Assessment Report Working Group II: Ocean Systems, 2014, 138 pp.
[12] W.W.L. Cheung, R. Watson, D. Pauly, Signature of ocean warming in global fisheries catch, Nature 497 (7449) (2013) 365.
[13] E.S. Poloczanska, M.T. Burrows, C.J. Brown, J. García Molinos, B.S. Halpern, O. Hoegh-Guldberg, et al., Responses of marine organisms to climate change across oceans, Front. Mar. Sci. 3 (2016) 62.
[14] M.C. Jones, W.W.L. Cheung, Multi-model ensemble projections of climate change effects on global marine biodiversity, ICES J. Mar. Sci. 72 (3) (2015) 741–752.
[15] J.K. Moore, W. Fu, F. Primeau, G.L. Britten, K. Lindsay, M. Long, et al., Sustained climate warming drives declining marine biological productivity, Science 359 (6380) (2018) 1139–1143.
[16] C.A. Stock, J.G. John, R.R. Rykaczewski, R.G. Asch, W.W.L. Cheung, J.P. Dunne, et al., Reconciling fisheries catch and ocean productivity, Proc. Natl. Acad. Sci. U.S.A. 114 (8) (2017) E1441–E1449.
[17] E. Olsen, et al., Ocean futures under ocean acidification, marine protection and changing fishing pressures explored using a worldwide suite of ecosystem models, Front. Mar. Sci. 5 (2018) 64.
[18] J.-P. Gattuso, et al., Contrasting futures for ocean and society from different anthropogenic CO_2 emissions scenarios, Science 349 (2015) aac4722.
[19] M. Barange, et al. Impacts of climate change on fisheries and aquaculture: synthesis of current knowledge, adaptation and mitigation options, in: FAO Fisheries and Aquaculture Technical Paper No. 629, 2018.
[20] G.I. Hagstrom, S.A. Levin, A. Martiny, Balance between resource supply and demand determines nutrient limitation of primary productivity in the ocean, preprint available on: bioRxiv 064543. Available from: https://doi.org/10.1101/064543.
[21] C.M. Moore, et al., Processes and patterns of oceanic nutrient limitation, Nat. Geosci. 6 (9) (2013) 701–710. ngeo1765.
[22] L. Kwiatkowski, et al., Emergent constraints on projections of declining primary production in the tropical oceans, Nat. Clim. Change 7 (5) (2017) 355–358.
[23] D. Breitburg, et al., Declining oxygen in the global ocean and coastal waters, Science 359 (2018) 6371.
[24] P. Gienapp, J. Merilä, Evolutionary responses to climate change, in: D.A. Dellasala, M.I. Goldstein (Eds.), Encyclopedia of the Anthropocene, Elsevier, Oxford, 2018, pp. 51–59.
[25] H.K. Lotze, et al., Ensemble projections of global ocean animal biomass with climate change, Preprint available on: bioRxiv 467175 (2018). Available from: https://doi.org/10.1101/467175.
[26] M. Edwards, A.J. Richardson, Impact of climate change on marine pelagic phenology and trophic mismatch, Nature 430 (7002) (2004) 881–884. nature02808-884.
[27] C.A. Stock, J.P. Dunne, J.G. John, Drivers of trophic amplification of ocean productivity trends in a changing climate, Biogeosciences 11 (24) (2014) 7125.
[28] B. Duarte, et al., Modelling sea level rise (SLR) impacts on salt marsh detrital outwelling C and N exports from an estuarine coastal lagoon to the ocean (Ria de Aveiro, Portugal), Ecol. Modell. 289 (2014) 36–44.

[29] D. Hubbard, J. Dugan, N. Schooler, S. Viola, Local extirpations and regional declines of endemic upper beach invertebrates in southern California, Estuar. Coast. Shelf Sci. 150 (2014) 67–75.
[30] G.A. Hyndes, et al., Accelerating tropicalization and the transformation of temperate seagrass meadows, BioScience 66 (11) (2016) 938–948.
[31] E.B. Watson, et al., Wetland loss patterns and inundation-productivity relationships prognosticate widespread salt marsh loss for southern New England, Estuaries Coasts 40 (3) (2017) 662–681.
[32] D. Reed, et al., Extreme warming challenges sentinel status of kelp forests as indicators of climate change, Nat. Commun. 7 (2016) 1–7.
[33] N.A. Graham, S. Jennings, M.A. MacNeil, D. Mouillot, S.K. Wilson, Predicting climate-driven regime shifts versus rebound potential in coral reefs, Nature 518 (2015) 94–97.
[34] A. Ordoñez, C. Doropoulos, G. Diaz-Pulido, Effects of ocean acidification on population dynamics and community structure of crustose coralline algae, Biol. Bull. 226 (2014) 255–268.
[35] K.L. Smith, C.L. Huffard, A.D. Sherman, H.A. Ruhl, Decadal Change in Sediment Community Oxygen Consumption in the Abyssal Northeast Pacific, Aquat. Geochem. 22 (5) (2016) 401–417.
[36] A.K. Sweetman, et al., Major impacts of climate change on deep-sea benthic ecosystems, Elementa Sci. Anthropocene 5 (2017) 4.
[37] L. Bopp, et al., Multiple stressors of ocean ecosystems in the 21st century: projections with CMIP5 models, Biogeosciences 10 (10) (2013) 6225–6245.
[38] U.R. Sumaila, W.W.L. Cheung, V.W.Y. Lam, D. Pauly, S. Herrick, Climate change impacts on the biophysics and economics of world fisheries, Nat. Clim. Change 1 (2011) 449–456.
[39] M. Barange, et al., Impacts of climate change on marine ecosystem production in societies dependent on fisheries, Nat. Clim. Change 4 (2014) 211.
[40] W.W.L. Cheung, G. Reygondeau, T.L. Frolicher, Large benefits to marine fisheries of meeting the 1.5°C global warming target, Science 354 (6319) (2016) 1591–1594.
[41] J.L. Blanchard, R.A. Watson, E.A. Fulton, et al., Linked sustainability challenges and trade-offs among fisheries, aquaculture and agriculture, Nature ecology & evolution 1 (9) (2017) 1240–1249.
[42] W.W.L. Cheung, M.C. Jones, G. Reygondeau, T.L. Frölicher, Opportunities for climate-risk reduction through effective fisheries management, Global Change Biol. 2108 (2018) 1–15.
[43] C.D. Golden, et al., Nutrition: Fall in fish catch threatens human health, Nature 534 (7607) (2016) 317–320.
[44] M. Ruckelshaus, et al., Securing ocean benefits for society in the face of climate change, Mar. Policy 40 (2013) 154–159.
[45] N. Narita, K. Rehdanz, R.S.J. Tol, Economic costs of ocean acidification: a look into the impacts on global shellfish production, Clim. Change 113 (2012) 1049–1063.
[46] C.A. Burge, et al., Climate change influences on marine infectious diseases: implications for management and society, Annu. Rev. Mar. Sci. 6 (2014) 249–277.
[47] U. Riebesell, A. Körtzinger, A. Oschlies, Sensitivities of marine carbon fluxes to ocean change, Proc. Natl. Acad. Sci. U.S.A. 106 (2009) 20602–20609.
[48] C.M. Duarte, J.J. Middelburg, N. Caraco, Major role of marine vegetation on the oceanic carbon cycle, Biogeosciences 2 (1) (2005) 1–8.
[49] M.D. Spalding, et al., The role of ecosystems in coastal protection: adapting to climate change and coastal hazards, Ocean Coast. Manage. 90 (2014) 50–57.
[50] D.G. Boyce, K.T. Frank, B. Worm, W.C. Leggett, Spatial patterns and predictors of trophic control in marine ecosystems, Ecol. Lett. 18 (2015) 1001–1011.

CHAPTER 9

Current and future biogeography of exploited marine exploited groups under climate change

Gabriel Reygondeau

Changing Ocean Research Unit, Institute for the Oceans and Fisheries, University of British Columbia, Vancouver, BC, Canada

Chapter Outline

9.1 Introduction

Quantifying species spatial distribution and biodiversity patterns represent one of the pillars of ecology [1]. The ocean and human society are in a closed interaction loop, with the ocean providing benefits such as food provision and humans influencing the natural state of the ocean either by direct pressures such as fisheries or indirect pressures such as modification of climate. Ecosystem services provided by the ocean are structured at local or regional scales by the biodiversity pool of species occurring, which define the trophodynamics of the ecosystem. Therefore the study of species distributions and their interactions allows for the characterization of ecosystem functioning and the quantification of the potential benefits of the ocean to human society.

Predicting Future Oceans.
DOI: https://doi.org/10.1016/B978-0-12-817945-1.00009-5

The discipline of biogeography is defined as:

> The study of the spatial distributions of organisms, both past and present, understanding all patterns of geographic variation in nature—from genes to entire communities and ecosystems—elements of biological diversity that vary across geographic gradients including those of area, isolation, latitude, depth, and elevation
>
> ***From Lomolino et al. [2].***

This discipline developed in terrestrial ecology to study species distribution and macroecological patterns of fauna. Marine biogeography has for a long time not been studied due to the large extent and volume of the water mass of the ocean along with the relative technological limitations to sampling the marine realm. In recent years the development of international collaboration, online databases, and statistical tools has boosted the discipline, allowing the community to determine the factors setting the extent of species geographic ranges and identifying marine macroecological patterns. Moreover, statistical tools such as species distribution models (SDMs) help to define current and future patterns of biodiversity. In the context of global climate change, driven largely by greenhouse gas emissions from anthropogenic activities, that has modified the physical and biological compartment of the oceans, it is now crucial to quantify current and future patterns to adapt proper measures for biodiversity conservation and the management of stocks of exploited species.

9.2 Biotic partition of the ocean

9.2.1 Historical background on marine biotic biogeography

Historical approaches in biogeography based on spatial distributions of species for a long time had inherent difficulties in the ocean compared with the terrestrial realm [3]. In comparison, in 1870 terrestrial ecologists had already accurately located "Wallace's line" in the Indo-Pacific archipelago, separating Australian from Asian fauna [1]. The high cost of sampling at sea, the high level of migration of pelagic species, and the near impossibility of sampling exhaustively the three-dimensional distributions associated with the high dynamic properties of the ocean are among the most serious difficulties for marine biogeographers. But that is not to say it has not been attempted. Mary Somerville, a science writer and one of the first female members of the Royal Astronomical Society, proposed the first division of the global oceans in 1872. Based on the limited physical and chemical knowledge available to her at the time, Somerville's idea was to separate cold from warm water (latitudinal divisions) by creating "Homozoic zones" [4].

It is only in the mid-20th century that studies in marine biogeography emerged in the literature based on field observations and growing knowledge in marine biology and oceanography. In those early years, the emphasis was on surface temperature as the primary determinant of pelagic species distributions, then progressively the attention was directed to global circulation patterns [5]. Several regional divisions were proposed based on different taxonomic groups: copepod distribution [6], clupeid fish [7], phytoplankton [8], euphosiids [9], chaetognaths [10], and aggregated taxonomic groups [11,12]. Whatever the basis of classification, all authors agreed that the distributions of species represented a random pattern at the ocean surface rather than a simple gradient from poles to equator, but fell into a small number of discrete classes that are related to distance from coasts and to latitude. After almost 150 years of investigation of the pelagic fauna, Backus summarized the accumulated research and formally developed a partitioning system based on the linkage between physical geography and biogeography [13]. Backus argued for the retention of a nine-zone system: Arctic and Antarctic zones, sub-Arctic and sub-Antarctic zones, North and South Temperate (or transitional) zones, North and South subtropical zones, and a tropical zone. He suggested that the temperate zones are bounded equatorward by the subtropical convergence of each hemisphere and that the oceanic polar front is the poleward limit of the sub-Antarctic zone and sub-Arctic zone.

9.2.2 Current ocean partition based on species distribution and diversity

Until the emergence of global marine biogeographic databases, the spatial coverage of marine species observations was insufficient to develop an objective biotic division of the ocean. Consequently, biogeographers have for a long time attempted to examine how well the existing abiotic ocean partitions describe the distribution of marine ecosystems. The approach of Longhurst [1], inspired by the Yentsh and Garside reflection [14], a bottom-up structuring of the ecosystem ([15,16]; Fig. 9.1), appeared to be a perfect candidate for such a test. Assuming that the average and natural seasonal fluctuations of environmental conditions within an identified biogeochemical province (BGCP) structured the diversity and the composition of the phytoplankton pool [17], the physiology, reproduction, and feeding behavior and therefore the spatial distribution of subsequent trophic levels should be defined by trophic linkages with the primary producer pool and environmental forcings (Fig. 9.1). Consequently, by studying the spatial distribution of several marine species at different trophic levels, boundaries between adjacent BGCPs should be detected as a shift in the community (Fig. 9.2).

At a regional or basin scale, several studies have investigated the ecological relevance of BGCP by examining the spatial distribution of marine species. These studies have focused on several taxonomic groups, such as bacteria species [18], plankton species [19–21], and marine exploited species [16]. Overall, the authors concluded that there is a relatively good spatial match between the BGCP and species abundance, associations, and biodiversity. The studies concluded that BGCPs represent specific environmental conditions that directly affect the abundance of species at lower trophic levels due to their low physiological

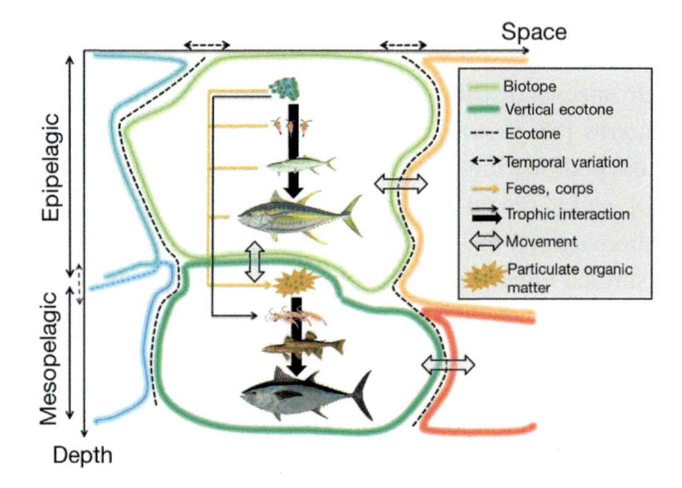

Figure 9.1
Diagram of conceptualized ecosystem within an identified ecological unit.

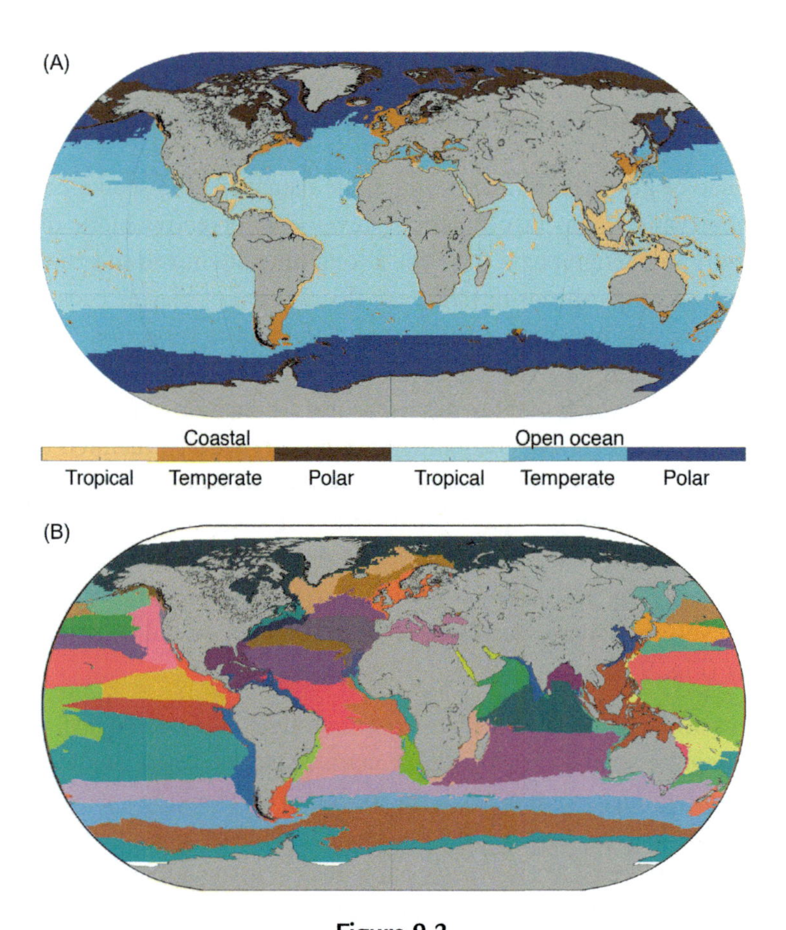

Figure 9.2
Map of (A) biomes and (B) BGCPs proposed by Reygondeau et al. [23]. *BGCPs*, Biogeochemical provinces.

tolerance to abiotic variations [22] and at higher trophic levels due to the trophodynamic dependency for the energy budget.

While some attempts have been made to identify pelagic ecological units using species distribution, most of the studies have focused on a specific group of species or at community levels. The first attempt to provide a biotic division of the global ocean using up to date global records with a large biodiversity spectrum was undertaken by Costello et al. [24]. The authors gathered spatial records of over 65,000 species and computed an endemic rate following the assumption that spatial change in the endemic pool of species will delineate a change in environmental forcing and hence ecological unit (Fig. 9.1). The authors describe more than 30 distinct ecological units named "oceanic realms" that summarize changes in environmental conditions as well as potential regional speciation patterns.

9.3 Distribution of exploited marine biodiversity

9.3.1 Historical and current limitations in studying marine diversity patterns

For a long period the spatial coverage of these studies was limited to local or regional scales, preventing researchers from evaluating hypotheses of marine macroecological patterns stemming from terrestrial ecology. The root of marine macroecology (based on observations) can be traced back to studies from the Scripps Institute of Oceanography during the 1980s. Scripps' researchers relied on increasing international collaboration and global data sharing processes to collect a database of samples with sufficient spatial coverage to support a biogeographic study of the Pacific Ocean.

In recent years the development of free online databases of marine species observations (and hence international collaboration), such as the Ocean Biogeographic Information System (OBIS) or the Global Biodiversity Information Facility (GBIF), have allowed biogeographers to study marine macroecological patterns at a global scale. These databases aggregate observations from a wide variety of sources (ranging from museum collections to data from independent oceanographic cruises), and on a variety of species (ranging from bacteria to top predators). However, several caveats in the use of these databases have been highlighted by Webb et al. [25] and Chaudhary et al. [26]:

1. Most of the records are sampled in the first 100 m of the water column [25] and over the continental shelf (<100 nautical miles with a depth <200 m). This pattern can be mainly explained by the physical and fiscal difficulties associated with sampling the open ocean and deep sea.
2. The high difference in sampling between the northern and southern hemisphere is associated with strong differences in research capacity.

3. The high variation in the representation of marine species. Low trophic levels species such as plankton, bacteria, or mollusks that have a higher theoretical diversity [27] are less represented in species number and full distribution coverage than large, high trophic levels species such as fish or mammals.

9.3.2 Global patterns in marine diversity

As previously shown, the large extent and volume of the water mass of the ocean along with the relative technological limitations to sampling the marine realm do not allow one to define the full spatial distribution of quasi all marine species and biases the study of species/diversity spatial patterns. Due to these limitations, marine macroecological studies have either focused on patterns of well-studied species or restricted geographical locations that were well sampled (North Sea and other coast of the Atlantic continental shelf).

To tackle this issue, one of the solutions employs the application of the SDM or environmental niche model (ENM). The methodology rests on the concept of ecological niche defined by Hutchinson [28]. The ecological niche theory states that a species can occur in a multidimensional environmental interval that has been defined by evolution shaping their physiological and competition characteristics to compete for resources and survive in their endemic environment. These numerical approaches can be broadly summarized as various exploratory or deterministic statistical methodologies that detangle the relationship between species observation records and the environment in order to quantify the multidimensional environmental interval where the species could occur. These approaches are highly adapted for marine species, which are composed of mostly ectotherms, and thus their spatial distribution is assumed to be driven by environmental conditions.

In order to identify the distribution of most of the exploited marine groups, we first gathered the occurrence records of 7597 species of fish, 107 species of mammals, 271 species of cephalopods, 1079 species of echinoderms, 52 species of lobster, 649 species of crabs, and 357 species of shrimp. Records of species occurrence were collated from the following publicly accessible databases: the OBIS (www.iobis.org), the Intergovernmental Oceanographic Commission (ioc-unesco.org), the GBIF (www.gbif.org), Fishbase (www.fishbase.org), and the International Union for the Conservation of Nature (http://www.iucnredlist.org/technical-documents/spatial-data).

Following the methodologies of Asch et al. [29], we used a multimodel approach to best approximate the environmental niche of each species using four ENMs: Bioclim and Boosted Regression Trees models from the Biomod2 R package [30], Maxent [31], and the Non-Parametric Probabilistic Ecological Niche model [32]. The spatial distribution of each species from the four ENMs, extrapolated from the environmental conditions modeled using the Geophysical Fluid Dynamics Laboratory's Earth System Model, were averaged over

1970–2000 and 2080–2100. Species richness (i.e., alpha diversity) was computed by numerically identifying a probability threshold of confirmed occurrence of a species based on a receiver operating characteristic curve using species occurrence and species modeled distribution for the period 1970–2000. If the habitat suitability index was greater than or equal to the threshold, then the species was considered present. Species richness was subsequently computed by summing the number of species present in each grid cell over each time period (Fig. 9.3).

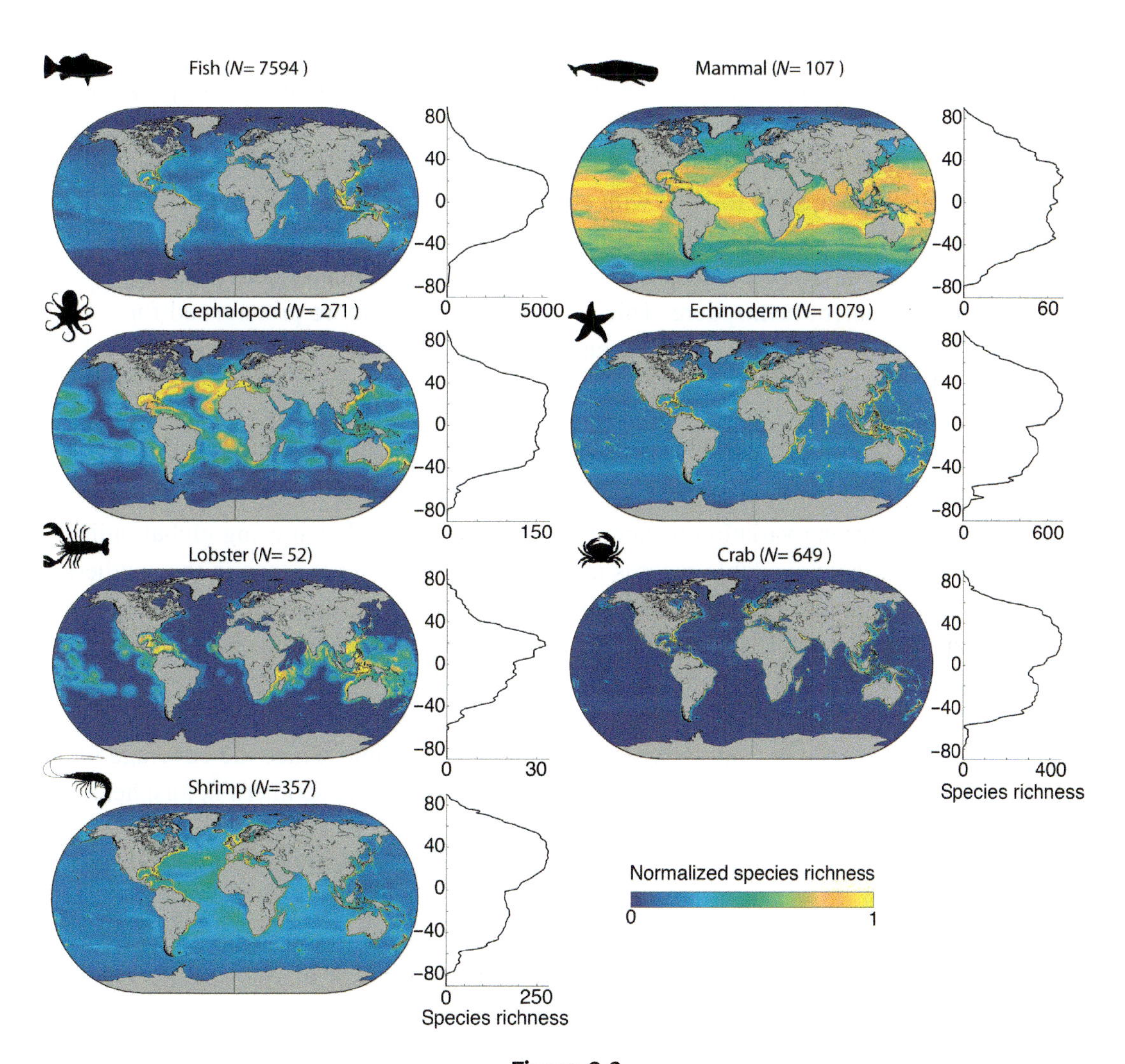

Figure 9.3

Map of the modeled species richness for fish, mammals, cephalopods, echinoderms, lobster, crab, and shrimp for the period 1970–2000. Number of species modeled for each group is presented as *N*. Map of species richness is normalized between 0 and 1 by dividing the maximal values of the raw diversity to the whole map. For each map of species richness, the latitudinal diversity gradient (LDG) is computed and presented on the right side of the map.

Several marine macroecological studies have attempted to identify global spatial patterns for known taxa such as phytoplankton [33,34], zooplankton [20], exploited marine species [35], emblematic species, or all known marine species [25,26]. These studies revealed several common marine diversity patterns among taxa groups including LDGs and the effects of distance to the coast. The LDG, with species richness peaking around the tropics and tailing off toward the poles, appears as the emergent pattern. This pattern has been observed, with only minor variation, across various types of species from primary producers to highly migratory species such whales or tunas [36]. However, contrary to terrestrial findings with respect to LDG, marine LDG displays a bimodal distribution with peaks of richness located around 20–40 degrees of latitude in both hemispheres [26], more marked for coastal than open ocean species (Fig. 9.3). Indeed LDGs vary among species taxa based on their physiological capacity to cope with environmental variations related to their movement abilities and the habitat that they occupy (i.e., pelagic, benthic, or demersal). Marine LDGs also exhibit a second specificity, which is an asymmetry between the northern and the southern hemisphere. This asymmetry appears more pronounced for invertebrate benthic or demersal groups, such as lobster, cephalopods, crabs, and echinoderms (Fig. 9.3).

To explain this asymmetrical bimodality in the ocean realm, several hypotheses have been put forward:

1. *Sampling bias.* Strong variation in sampling effort has been shown using global online databases, with more available records in the northern hemisphere than in the southern hemisphere. However, this hypothesis has been rejected as several authors [26,37,38] have statistically normalized the sampling effort for several types of organisms and revealed that the bimodality persisted.
2. *Continental shelf coverage.* It is well known that higher species richness is found over the continental shelf where more types of habitat are available. Since the coverage of continental shelf is less important in the equatorial regions and southern hemisphere, the asymmetrical bimodality could have emerged due to decreased availability of marine habitat resulting in a peak of coastal associated species diversity in the northern tropical region.
3. *Influence of temperature on marine organism and diversity.* Temperature plays a pivotal role in shaping observed spatial distribution patterns of marine species [39,40]. Most marine species are ectothermic and exhibit distinctive thermal strategies, resulting from their physiological and morphological adaptations that determine in part, their spatial ranges [41]. Previous studies have demonstrated significant positive correlations between seawater temperature and patterns of species richness. Consequently, the marine LDGs could emerge from a nonuniform latitudinal gradient of the mean and seasonal fluctuations in temperature between the northern and southern hemispheres.

9.4 Effect of climate change on marine species biogeography

9.4.1 Climate-driven change in the distribution of marine species

Since the industrial revolution, the human population has exponentially emitted massive volumes of greenhouse gases by burning fossil energies. This increase of greenhouse gases in the atmosphere has altered the thermodynamic properties of the troposphere, the radiative equilibrium of the atmosphere, and the exchange of energy with the ocean [42]. Recent studies have shown that the increase of the heat stock in the atmosphere in the past decades has been mainly absorbed by the ocean [43], resulting in global warming [44] that has affected the physical and biological compartments of the ocean [45]. The modifications of the environmental conditions characterized by a speed and amplitude never before observed in the biosphere are expected to increase in the near future, affecting the global biogeographical partition of the ocean and the associated biodiversity [46].

Marine organisms are highly dependent on environmental conditions, especially the ectothermic species representing 90% of the marine biodiversity. Indeed environmental parameters strongly influence the physiology and the biological interaction of marine organisms by regulating their metabolic activity (growth, photosynthesis, excretion, cellular cycle, hormone production) and, by extension, key biological parameters (nutrition, reproduction, and development time). Therefore any variation in the range of environmental parameters expressed by the ecosystem can strongly affect the life cycle dynamics of species, the abundance of a whole population, and the biodiversity of the area [47]. Consequently, predictions based on a global increase of temperature from 0.6°C to 4°C have shown a potential alteration of the natural range of many species. Indeed marine organisms have no other choice but a spatial [48,49] and/or vertical [50] migration in their distribution to follow their optimal environmental conditions and maintain their populations. Spatial shifts in several trophic levels have been observed in a range of marine species—from mesozooplankton [48] to fish [35]—with subpolar species retreating poleward and more temperate species replacing them. This modification in species distribution will of course vary among and within taxa depending on the types of organisms and it will depend on their physiological ability to adapt to the novel environmental conditions occurring in their natural range (Fig. 9.4).

The shift in marine species' distributions has direct consequences on marine diversity gradients and hot spots of diversity (Fig. 9.4, i.e., the change in species richness in 2090–99 compared to 1970–2000; Fig. 9.3). First, results show an important local species extirpation in tropical areas where the amplitude of temperatures has reached values over the natural range of most species. Consequently, diversity in marine species groups decreased from 15% to more than 70% in these regions even for the climate mitigation scenarios (RCP 2.6). Marine species will tend to perform a latitudinal migration with

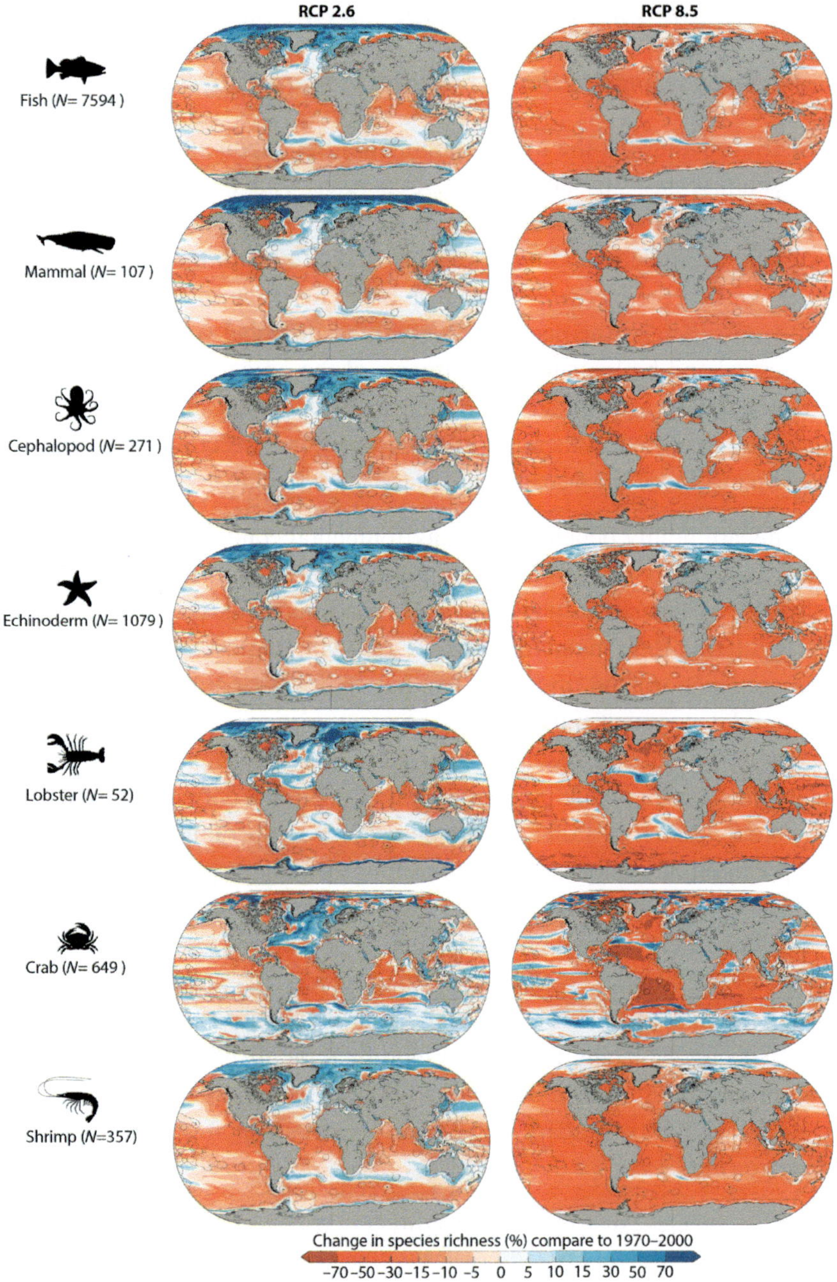

Figure 9.4

Map of change in species richness (%) at the end of the century (2090–99) compared to a reference period 1970–2000 (Fig. 9.3) for the scenario RCP 2.6 and 8.5.

warming water, resulting in an increase of nonendemic species in temperate and polar areas [38], compensating for potential local loss of species. However, in a "business as usual" scenario (RCP 8.5), the high degree of environmental change occurring at the end of the century will allow for the existence of only a few refuge areas where an observed increase of species richness can be found (Fig. 9.4). Within the RCP 8.5 scenario, the majority of ocean regions are likely to suffer from a decrease in species richness. Consequently, species composition of several marine ecosystems will be modified in the future with a simplification of the tropical food webs where some of the endemic species will perform a poleward shift and only euryecious species (i.e., species with a wide niche) may belong in the area. In parallel, temperate and polar domains will experience an important species reorganization due to the arrivals of alien species, creating a modification in the biodiversity pool and trophodynamics.

9.4.2 Implication for marine ecosystem services

Global environmental change is now known to alter the distributions of most of the species in the global ocean. The species most affected by environmental changes are endemic species with a narrow spatial distribution that occur in tropical regions. These species are adapted to the low seasonal fluctuations typically found in tropical or polar climates and often to specific habitats, such as corals or seagrass beds. Thus their tolerance range of environmental conditions will be narrow due to hyperspecialization. In contrast, highly migratory species, such as tunas and billfishes, found in tropical, temperate, and polar ecosystems, are typically characterized by a wide spatial distribution tolerating a large range of environmental conditions. The more widely distributed a species is, the more likely that the effects of climate change will have a minimal influence on its distribution due to a more evolved physiology that could cope with environmental variation such as pseudoendothermic regulation.

While most of the marine species' distributions seem to be affected to various degrees by climate change (Fig. 9.4), the implementation of concrete conservation measures of marine biodiversity is now crucial. One of the main tools to conserve biodiversity efficiently is marine protected areas (MPAs). Despite a clear international effort in recent years to implement large MPAs over the global ocean, serious concerns have been raised in the ability to conserve endemic biodiversity within the context of climate change [51]. Indeed while MPAs are often located in fixed positions, endemic fauna targeted for conservation might shift their distribution in the near future (Fig. 9.4) to track their optimal environmental conditions. Three solutions are therefore available to reach the goals set by the Convention of Biological Diversity and Millennium Ecosystem Assessment for good management of future marine diversity: (1) use current MPAs to protect invasive species spreading under climate change [52]; (2) develop a network of novel MPAs following shifts in distribution of endemic marine species; and/or (3) implement novel MPAs in areas known as climate refuge zones. Climate refuge zones can be defined as areas where local extinction is moderate and local invasion of climatic invasive species is increasing. These areas will most likely either conserve a similar number of species or have an increasing biodiversity pool (Fig. 9.4). While the last solution appears as optimal for conservation, the majority of these areas are located in the high seas where jurisdiction and legislation are contested or not applied. Therefore the implementation of clear international agreements on the management of international waters is now needed [53].

One of the greatest contributions of the ocean to human society rests on the provision of food (i.e., fisheries). As one can expect, shifts in distribution of the majority of exploited species in the ocean will reorganize the distribution of the potential fisheries catch [54]. As an important extirpation of marine species is expected in the tropical regions, a massive decrease in their abundance and catch is also expected. In contrast, high latitudinal regions that will host species with altered distributions are expected to have an increase in potential catch. Decreases in species richness and catch potential can affect incomes, livelihoods, food security, and political stability. Pelagic fishery operations, including catches, fish processors, and licensing fees for foreign fleets, contribute considerably to the gross domestic product and government revenues of many tropical countries. A decrease in catch potential would reduce revenues generated from these fisheries and associated businesses [55]. Small-scale coastal fisheries also contribute substantially to subsistence and livelihoods in these regions; such contributions may have been largely underestimated in official statistics. The projected local extinction of several species may substantially reduce seafood availability to coastal communities in the regions who are generally highly nutritionally vulnerable [56]. Moreover, ecotourism is an important industry in several tropical regions [57]. Decreases in diversity and the degradation of habitats from projected warming and ocean acidification are likely to reduce the attractiveness of the tropical Pacific to international tourists. In addition, the stocks of exploited species in the tropical regions straddle the Exclusive Economic Zones (EEZs) of multiple nations and territories. The projected shift in distribution and abundance of exploited species could destabilize current international transboundary fisheries agreements and management, resulting in international disputes and conflict in fisheries resource sharing. Therefore an international update over the agreement on the management of transboundary exploited species is now needed to deal with the near future reorganization of regional marine biodiversity pools.

The results presented here, in combination with recent studies, indicate that the biodiversity changes that would occur in the global ocean in a "business as usual" RCP 8.5 scenario would lead to a drastic reorganization of global marine biogeography and associated trophic networks. These changes would include the emergence of wide regions where most marine species would likely be threatened. If the global climate is not kept below 2°C warming, this process will start in the very near future. In such a scenario, most marine areas will suffer from a decrease of diversity of most of the major exploited taxonomic groups (Fig. 9.4), ranging from 5% to more than 70% of local loss, with few areas having an increased potential diversity pool. These numbers would be drastically reduced at the end of the 21st century in the RCP 2.6 scenario. Mitigating CO_2 emissions at a level sufficient to reach the Paris Agreement targets would therefore substantially reduce the risk of species shifts, and the dramatic consequences that such large-scale ecological changes would entail for tropical marine biodiversity, associated fisheries, and the human communities that they support [58].

References

[1] A. Longhurst, Ecological Geography of the Sea, second ed., Academic Press, London, 2007, p. 390.

[2] M.V. Lomolino, B.R. Riddle, J.H. Brown, third ed., Biogeography, vol. 1, Sinauer Associates, Inc., Sunderland, MA, 2005, p. 845.

[3] L.F. Beaufort, Zoogeography of the Land and Inland Waters, Sidgwick & Jackson, Ltd., London, 1951, p. 208.

[4] M. Sommerville, Physical Geography, Blanchard, Boston, MA, 1872.

[5] K.V. Beklemishev, Ekologiya i biogeografiya pelagiali (Ecology and Biogeography of the Pelagial), Nauka, Moscow, 1969.

[6] A. Steuer, Zur planmässigen Erforschung der geographischen Verbreitung des Haliplanktons, besonders. der Copepoden, Zoogeographica 1 (3) (1933) 269–302.

[7] J.H. Rosa, T. Laevastu, Comparison of Biological and Ecological Characteristics of Sardine and Related Species—A Preliminary Study, 1960.

[8] R. Margalef, Correlations entre certains caractères synthétiques des populations de phytoplancton, Hydrobiologia. 18 (1) (1961) 155–164.

[9] E. Brinton, The Distribution of Pacific Euphausiids, vol. 8, 1962, University of California Press Berkeley and Los Angeles.

[10] A. Alvariño, Chaetognaths, Allen and Unwin, 1965.

[11] J.A. McGowan, Oceanic Biogeography of the Pacific, Zn BM Funnell W R Riedel reds, Micropaleontol Ocean, Cambridge, 1971, pp. 3–74.

[12] J.L. Reid, E. Brinton, A. Fleminger, E.L. Venrick, J.A. McGowan, Ocean circulation and marine life, Adv. Oceanogr. (1978) 65–130.

[13] R.H. Backus, Biogeographic boundaries in the open ocean. Pelagic biogeography, UNESCO Tech. Pap. Mar. Sci. 49 (1986) 9–13.

[14] C.S. Yentsch, J.C. Garside, Patterns of phytoplankton abundance and biogeography, Pelagic Biogeogr. 278 (1986) 278–284.

[15] P.M. Cury, Y.-J. Shin, B. Planque, J.M. Durant, J.-M. Fromentin, S. Kramer-Schadt, et al., Ecosystem oceanography for global change in fisheries, Trends Ecol. Evol. [Internet] 23 (6) (2008) 338–346. Jun [cited 2014 Oct 25] Available from: http://www.ncbi.nlm.nih.gov/pubmed/18436333.

[16] G. Reygondeau, O. Maury, G. Beaugrand, J.M. Fromentin, A. Fonteneau, P. Cury, Biogeography of tuna and billfish communities, J. Biogeogr. 39 (1) (2012) 114–129.

[17] H.U. Sverdrup, On conditions for the vernal blooming of phytoplankton, ICES J. Mar. Sci. 18 (1953) 287–295.

[18] W.K.W. Li, E.J.H. Head, W. Glen Harrison, Macroecological limits of heterotrophic bacterial abundance in the ocean, Deep Sea Res., I Oceanogr. Res. Pap. [Internet] 51 (11) (2004) 1529–1540. Nov [cited2014 Nov 17]; Available from: http://linkinghub.elsevier.com/retrieve/pii/S0967063704001530.

[19] M.J. Gibbons, Pelagic biogeography of the South Atlantic Ocean, Mar. Biol. [Internet] 129 (4) (1997) 757–768. Available from: http://link.springer.com/10.1007/s002270050218.

[20] Beaugrand G., Ibañez F., Lindley J.A., Reid P.C. Diversity of calanoid copepods in the North Atlantic and adjacent seas: species associations and biogeography. Mar. Ecol. Prog. Ser. 2002;232:179–195.
[21] R.S. Woodd-Walker, P. Ward, A. Clarke, Large-scale patterns in diversity and community structure of surface water copepods from the Atlantic Ocean, Mar. Ecol. Prog. Ser. 236 (2002) 189–203.
[22] A.J. Richardson, D.S. Schoeman, Climate impact on plankton ecosystems in the northeast Atlantic, Science 305 (2004) 1609–1612.
[23] G. Reygondeau, A. Longhurst, E. Martinez, G. Beaugrand, D. Antoine, O. Maury, Dynamic biogeochemical provinces in the global ocean, Global Biogeochem. Cycl. 27 (4) (2013) 1046–1058.
[24] M.J. Costello, P. Tsai, P.S. Wong, A.K.L. Cheung, Z. Basher, C. Chaudhary, Marine biogeographic realms and species endemicity, Nat. Commun. 8 (1) (2017) 1057.
[25] T.J. Webb, E. Vanden Berghe, R. O'Dor, Biodiversity's big wet secret: the global distribution of marine biological records reveals chronic under-exploration of the deep pelagic ocean, PLoS One 5 (8) (2010) e10223.
[26] C. Chaudhary, H. Saeedi, M.J. Costello, Bimodality of latitudinal gradients in marine species richness, Trends Ecol. Evol. 31 (9) (2016) 670–676.
[27] C. Mora, D.P. Tittensor, S. Adl, A.G.B. Simpson, B. Worm, How many species are there on earth and in the ocean? PLoS Biol. (2011) e1001127.
[28] G.E. Hutchinson, Concluding remarks, Cold Spring Harb. Symp. Quant. Biol. 22 (1957) 415–427.
[29] R.G. Asch, W.W.L. Cheung, G. Reygondeau, Future marine ecosystem drivers, biodiversity, and fisheries maximum catch potential in Pacific Island countries and territories under climate change, Mar. Policy 88 (2018) 285–294.
[30] W. Thuiller, B. Lafourcade, R. Engler, M.B. Araújo, BIOMOD—a platform for ensemble forecasting of species distributions, Ecography (Cop). 32 (3) (2009) 369–373.
[31] S.J. Phillips, R.P. Anderson, R.E. Schapire, Maximum entropy modeling of species geographic distributions, Ecol. Modell. 190 (3) (2006) 231–259.
[32] G. Beaugrand, S. Lenoir, F. Ibañez, C. Manté, A new model to assess the probability of occurrence of a species, based on presence-only data, MEPS 424 (2011) 175–190.
[33] H. Demarcq, G. Reygondeau, S. Alvain, V. Vantrepotte, Monitoring marine phytoplankton seasonality from space, Remote Sens. Environ. 117 (2012) 211–222.
[34] M.J. Follows, S. Dutkiewicz, S. Grant, S.W. Chisholm, Emergent biogeography of microbial communities in a model ocean, Science [Internet] 315 (5820) (2007) 1843–1846. Mar 30 [cited 2014 Jul 10] Available from: http://www.ncbi.nlm.nih.gov/pubmed/17395828.
[35] W.W.L. Cheung, V.W.Y. Lam, J.L. Sarmiento, K. Kearney, R. Watson, D. Pauly, Projecting global marine biodiversity impacts under climate change scenarios, Fish Fish. 10 (3) (2009) 235–251.
[36] G. Reygondeau, O. Maury, G. Beaugrand, J.M. Fromentin, A. Fonteneau, P. Cury, Biogeography of tuna and billfish communities, J. Biogeogr. 39 (1) (2012) 114–129.
[37] I. Rombouts, G. Beaugrand, F. Ibanez, S. Gasparini, S. Chiba, L. Legendre, Global latitudinal variations in marine copepod diversity and environmental factors, Proc. Biol. Sci. [Internet] 276 (1670) (2009) 3053–3062. Sep 7 [cited 2014 Nov 17] Available from: http://www.pubmedcentral.nih.gov/articlerender.fcgi?artid = 2817135&tool = pmcentrez&rendertype = abstract.
[38] S. Rutherford, S. D'Hondt, W. Prell, Environmental controls on the geographic distribution of zooplankton diversity, Nature 400 (6746) (1999) 749–753.
[39] D.P. Tittensor, C. Mora, W. Jetz, H.K. Lotze, D. Ricard, E.V. Berghe, et al., Global patterns and predictors of marine biodiversity across taxa, Nature [Internet] 466 (7310) (2010) 1098–1101. Nature Publishing Group, Aug 26 [cited 2014 Jul 10] Available from: http://www.ncbi.nlm.nih.gov/pubmed/20668450.
[40] G. Beaugrand, M. Edwards, L. Legendre, Marine biodiversity, ecosystem functioning, and carbon cycles, Proc. Natl. Acad. Sci. U.S.A. [Internet] 107 (22) (2010) 10120–10124. Jun 1 [cited 2014 Nov 5] Available from: http://www.pubmedcentral.nih.gov/articlerender.fcgi?artid = 2890445&tool = pmcentrez&rendertype = abstract.

[41] J.M. Sunday, A.E. Bates, N.K. Dulvy, Global analysis of thermal tolerance and latitude in ectotherms, Proc. R. Soc. B: Biol. Sci. 278 (2011) 1823–1830.

[42] T.R. Karl, K.E. Trenberth, Modern globalclimate change, Science 302 (5651) (2003) 1719.

[43] T.P. Barnett, D.W. Pierce, K.M. AchutaRao, P.J. Gleckler, B.D. Santer, J.M. Gregory, et al., Penetration of human-induced warming into the world's oceans, Science 309 (2005) 284–287.

[44] S. Levitus, J. Antonov, T. Boyer, Warming of the world ocean, 1955-2003, Geophys. Res. Lett. 32 (2005) L02604.

[45] C. Parmesan, G. Yohe, A globally coherent fingerprint of climate change impacts across natural systems, Nature [Internet] 421 (6918) (2003) 37–42. Available from: http://www.ncbi.nlm.nih.gov/pubmed/12511946.

[46] K. Brander, Impacts of climate change on marine ecosystems and fisheries, J. Mar. Biol. Assoc. India 51 (2009) 1–13.

[47] G. Beaugrand, P.C. Reid, Long-term changes in phytoplankton, zooplankton and salmon linked to climate change, Glob. Change Biol. 9 (2003) 801–817.

[48] G. Beaugrand, P.C. Reid, F. Ibañez, J.A. Lindley, M. Edwards, Reorganisation of North Atlantic marine copepod biodiversity and climate, Science 296 (2002) 1692–1694.

[49] A.L. Perry, P.J. Low, J.R. Ellis, J.D. Reynolds, Climate change and distribution shifts in marine fishes, Science [Internet] 308 (5730) (2005) 1912–1915. Jun 24 [cited 2014 Jul 11] Available from: http://www.ncbi.nlm.nih.gov/pubmed/15890845.

[50] N.K. Dulvy, S.I. Rogers, S. Jennings, V. Stelzenmüller, S.R. Dye, H.R. Skjoldal, Climate change and deepening of the North Sea fish assemblage: a biotic indicator of warming seas, J. Appl. Ecol. 45 (4) (2008) 1029–1039.

[51] T.E. Davies, S.M. Maxwell, K. Kaschner, C. Garilao, N.C. Ban, Large marine protected areas represent biodiversity now and under climate change, Sci. Rep. 7 (2017) 9569.

[52] B. Gallardo, D.C. Aldridge, P. González-Moreno, J. Pergl, M. Pizarro, P. Pyšek, et al., Protected areas offer refuge from invasive species spreading underclimate change, Glob. Change Biol. 23 (12) (2017) 5331–5343.

[53] G. Reygondeau, D. Dunn, Encyclopedia of Ocean Sciences, second ed., Elsevier Ltd., 2018.

[54] W.W.L. Cheung, V.W.Y. Lam, J.L. Sarmiento, K. Kearney, R. Watson, D. Zeller, et al., Large-scale redistribution of maximum fisheries catch potential in the global ocean under climate change, Glob. Change Biol. [Internet] 16 (1) (2010) 24–35. Jan [cited 2014 Jul 9] Available from: http://doi.wiley.com/10.1111/j.1365-2486.2009.01995.x.

[55] V.W.Y. Lam, W.W.L. Cheung, G. Reygondeau, U. Rashid Sumaila, Projected change in global fisheries revenues under climate change, Sci. Rep. 6 (2016) 32607.

[56] C.D. Golden, E.H. Allison, W.W.L. Cheung, M.M. Dey, B.S. Halpern, D.J. McCauley, et al., Nutrition: fall in fish catch threatens human health, Nature 534 (7607) (2016) 317–320.

[57] C.C.C. Wabnitz, A.M. Cisneros-Montemayor, Q. Hanich, Y. Ota, Ecotourism, climate change and reef fish consumption in Palau: benefits, trade-offs and adaptation strategies, Mar. Policy 88 (2017) 323–332.

[58] W.W.L. Cheung, G. Reygondeau, T.L. Frölicher, Large benefits to marine fisheries of meeting the 1.5°C global warming target, Science 354 (6319) (2016) 1591–1594.

CHAPTER 10

Linking individual performance to population persistence in a changing world

Joey R. Bernhardt

Institute for the Oceans and Fisheries, University of British Columbia, Vancouver, BC, Canada

Chapter Outline

10.1 Scaling metabolism from cells to ecosystems

A central goal of ecology is to understand what drives the abundance, distribution, and diversity of life on Earth. For centuries, biologists have been addressing these questions from a variety of perspectives, and yet we still lack a coherent and mechanistic explanation of what drives patterns of abundance and distribution. One process that is shared by all of life on Earth is metabolism. Every living thing must uptake, modify, and allocate resources toward growth, maintenance, and reproduction. The rate at which these processes occur is the metabolic rate. The metabolic theory of ecology (MTE) posits that temperature and body size impose universal constraints on the flux, storage, and turnover of energy, matter, and information [1,2]. At its core, the premise is that principles of chemistry, physics, and biology operating on metabolic processes at lower levels of biological organization (i.e., molecules and cells) constrain organismal performance and give rise to predictable patterns at higher levels (i.e., populations, communities, and ecosystems). In this way, the focus on

Predicting Future Oceans.
DOI: https://doi.org/10.1016/B978-0-12-817945-1.00010-1

the metabolic processes shared by all living things has created new possibilities for a unified and predictive science of ecology.

10.2 Temperature constrains the pace of life

The development and testing of MTE has revealed remarkable generality in the way that organisms respond to temperature [3–5]. Syntheses across broad ranges of taxonomic diversity show that temperature has highly conserved and predictable effects on organismal metabolic rates [4–7]. Patterns of size and temperature scaling (where "scaling" refers to a relationship between two quantities, where a relative change in one quantity results in a proportional relative change in the other, independent of the initial size of those quantities) are some of the most broadly observed patterns in nature [8–11]. In organisms ranging from bacteria to salmon, warming causes metabolic rates to speed up. As metabolic rates accelerate with warming, so do the rates of other important physiological processes, including growth rates and development rates [12]. In this way, the temperature dependence of biochemical systems within cells scales up to organisms, communities, and ecosystems. The relationship between temperature and metabolism has been applied on land and in the ocean to predict patterns of global primary productivity [13], offspring size in the oceans [14], and range shifts [15].

10.3 Linking individual performance to the dynamics of populations

A critical challenge for MTE has been to link individual metabolic rates to patterns of population abundance and distribution. One approach is to trace demography back to energy budgets of individuals [16]. For a population to increase in abundance, the individuals in the population must capture more energy than they use for maintenance, and convert the excess into new biomass (growth) and individuals (reproduction). MTE postulates that the temperature dependence of widely shared metabolic rates (photosynthesis and respiration) drives temperature dependence of demographic rates (birth, death), leading to predictable effects of temperature on population growth and abundance (Fig. 10.1). While MTE predicts a relationship between individual metabolism and the dynamics of populations, this relationship is complicated by evidence that any signal of a general metabolic temperature dependence is potentially overwhelmed by the complexity and contingency in how temperature affects physiological traits, demographic processes and their interactions [17–19].

Cross-species syntheses suggest that temperature-dependent metabolism scales up to affect the temperature dependence of population dynamics [16]. Experimental evidence shows that the temperature dependence of population carrying capacity reflects the temperature dependence of metabolism and body size [20]. Theoretical models have shown that the

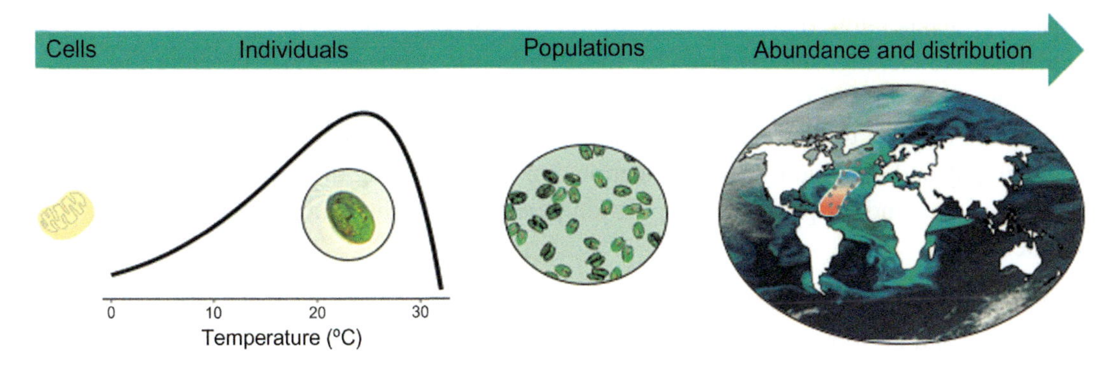

Figure 10.1
The MTE posits that temperature constrains rates of metabolic processes within cells, and these constraints emerge at higher levels of biological organization, such as individuals, populations, and species interactions, ultimately influencing patterns of abundance and distribution. *MTE*, Metabolic theory of ecology.

effects of temperature on the outcomes of species interactions such as predator–prey interactions are not directly proportional to the effects of temperature on metabolism because species interactions are governed not only by feedbacks within individuals but also among them. However, empirical evidence suggests that while the effects of temperature on species interactions may not be directly proportional to the effects on organismal performance, they are still predictable. Experiments have shown that MTE predicts the temperature dependence of within-host parasite dynamics [21] and consumer-resource dynamics [22].

10.4 Integrating evolutionary theory with metabolic scaling theory

While MTE has revealed remarkable predictability in the way that biological systems respond to temperature, it takes temperature dependence as a given, and does not include any explanation for how thermal traits evolve. In the context of a changing climate, understanding the potential for adaptive evolution in thermal traits is a major open question. While efforts are underway to predict how species will respond to changing climate conditions, these efforts tend to overlook the capacity for evolutionary adaptation, which can greatly alter patterns of vulnerability across a species' range. Until recently, evolution was seen as a slow, varying constraint on species traits, rather than a dynamic feedback mediating shifts in abundance and distribution. But now, evolutionary processes are seen as an important component of ecological dynamics [23].

Evolutionary theory predicts that the evolution of thermal performance should be constrained by generalist–specialist trade-offs, and trade-offs between performance at low versus high temperature (hot–cold trade-offs) [24–28], but careful empirical tests of these predictions are lacking, and evidence for these trade-offs is mixed [29]. Therefore the fundamental question of what facilitates or constrains the evolution of temperature dependence remains unanswered. Answering this question is a critical challenge in biogeography and evolution, with implications for understanding the origin and distribution of life on Earth, and for predicting population persistence in the context of a changing climate.

Thermal performance curves (Fig. 10.1), which describe the relationship between temperature and performance are a widely-used tool in physiological ecology because they describe the fundamental thermal niche of a species that determines the geographical ranges of species and population persistence in the face of a changing climate [30–33]. The evolutionary trajectories of thermal performance curves may be constrained by low genetic variance and strong genetic correlations [34,35] but few studies have quantified how the genetic architecture of thermal performance curves varies within and among populations [36,37]. As a result, making predictions of how genetic covariances will shape evolutionary responses of thermal reaction norms in the face of a changing climate remains a major challenge. More generally, the factors that influence the capacity for evolutionary adaptation in temperature dependence include generation time, the amount of standing genetic variation in a population, and potential trade-offs with selection on other traits.

Recent empirical evidence from laboratory experimental evolution studies using phytoplankton shows that shifts in the thermal performance curve are possible within ~350 generations [38,39]. These findings contrast with findings from comparative studies of thermal niches among species in their native and invaded ranges, which show little evidence of thermal niche evolution [40,41]. One possible explanation for these opposing findings could be that most laboratory experimental evolution studies are done with plentiful resource supply, whereas resources are often limiting in nature, and trade-offs may exist between thermal performance and the ability to acquire limiting resources. Indeed, theory predicts that the height and position of the thermal performance curve should shift down and to lower temperatures under conditions of resource limitation [42]. Since performance at very high temperatures comes at a physiological cost [43], it is possible that the ability to evolve a higher thermal tolerance is limited by resource supply, but this hypothesis has not yet been tested empirically.

10.5 Outlook

In an era of rapid climate change, ecologists are now urgently challenged to improve our ability to predict changes in the functioning of the ecosystems on which human well-being depends. This is nothing less than a challenge to our most basic understanding of what drives the abundance and distribution of life on Earth. Despite major advances in ecology based on flows of energy and materials across scales, leading theories in ecology, including MTE, have yet to formally integrate existing knowledge of the evolution of biological responses to the environment. MTE can be a powerful framework to project future ecological states from a theoretical basis with a broad empirical domain. Future work should incrementally increase the complexity of the systems in which these predictions are tested to continue to test the theory and to refine and expand its predictions. Finally, no understanding of the temperature and mass dependence of processes across levels of biological organization will be complete without knowledge of the factors that limit or facilitate the evolution of temperature dependence. Answering this question is critical to achieving a coherent picture of what generates and maintains current and future patterns of abundance, distribution, and diversity.

References

[1] J.H. Brown, J.F. Gillooly, A.P. Allen, V.M. Savage, G.B. West, Toward a metabolic theory of ecology, Ecology 85 (2004) 1771–1789. Available from: https://doi.org/10.1890/03-9000.
[2] J.H. Brown, R.M. Sibly, A. Kodric-Brown, Introduction: Metabolism as the Basis for a Theoretical Unification of Ecology. Metabolic Ecology, John Wiley & Sons, Ltd, Chichester, UK, 2012, pp. 1–6. Available from: https://doi.org/10.1002/9781119968535.ch.
[3] B.J. Enquist, J.H. Brown, G.B. West, Allometric scaling of plant energetics and population density, Nature 395 (1998) 163.
[4] J.F. Gillooly, J.H. Brown, G.B. West, V.M. Savage, E.L. Charnov, Effects of size and temperature on metabolic rate, Science 293 (2001) 2248–2251.
[5] M.E. Dillon, G. Wang, R.B. Huey, Global metabolic impacts of recent climate warming, Nature 467 (2010) 704–706. Available from: https://doi.org/10.1038/nature09407.
[6] S.L. Chown, S. Nicolson, Insect physiological ecology: mechanisms and patterns, Biology (2004) 254. Available from: https://doi.org/10.3987/Contents-12-85-7.
[7] Á. López-Urrutia, E. San Martin, R.P. Harris, X. Irigoien, Scaling the metabolic balance of the oceans, Proc. Natl. Acad. Sci. U.S.A. 103 (2006) 8739–8744.
[8] J.S. Huxley, Constant differential growth-ratios and their significance, Nature 114 (1924) 895–896.
[9] M. Kleiber, Body size and metabolism, Hilgardia 6 (1932) 11.
[10] K. Schmidt-Nielsen, Scaling: Why Is Animal Size So Important? Cambridge University Press, 1984.
[11] G.B. West, B.J. Enquist, J.H. Brown, A general quantitative theory of forest structure and dynamics, Proc. Natl. Acad. Sci. U.S.A. 106 (2009) 7040–7045.
[12] M.I. O'Connor, J.F. Bruno, S.D. Gaines, B.S. Halpern, S.E. Lester, B.P. Kinlan, et al., Temperature control of larval dispersal and the implications for marine ecology, evolution, and conservation, Proc. Natl. Acad. Sci. U.S.A. 104 (2007) 1266–1271. Available from: https://doi.org/10.1073/pnas.0603422104.
[13] S.T. Michaletz, D. Cheng, A.J. Kerkhoff, B.J. Enquist, Convergence of terrestrial plant production across global climate gradients, Nature 512 (2014) 39–43. Available from: https://doi.org/10.1038/nature13470.
[14] A.K. Pettersen, C.R. White, R.J. Bryson-Richardson, D.J. Marshall, Linking life-history theory and metabolic theory explains the offspring size-temperature relationship, Ecol. Lett. 22 (2019) 518–526. Available from: https://doi.org/10.1111/ele.13213.
[15] W.W.L. Cheung, V.W.Y. Lam, J.L. Sarmiento, K. Kearney, R.E.G. Watson, D. Zeller, et al., Large-scale redistribution of maximum fisheries catch potential in the global ocean under climate change, Glob. Change Biol. 16 (2010) 24–35.
[16] V.M. Savage, J.F. Gillooly, J.H. Brown, G.B. West, E.L. Charnov, Effects of body size and temperature on population growth, Am. Nat. 163 (2004) 429–441.

[17] D. Tilman, Niche tradeoffs, neutrality, and community structure: a stochastic theory of resource competition, invasion, and community assembly, Proc. Natl. Acad. Sci. U.S.A. 101 (2004) 10854–10861. Available from: https://doi.org/10.1073/pnas.0403458101.

[18] A.I. Dell, S. Pawar, V.M. Savage, Temperature dependence of trophic interactions are driven by asymmetry of species responses and foraging strategy, J. Anim. Ecol. 83 (2014) 70–84. Available from: https://doi.org/10.1111/1365-2656.12081.

[19] M.I. O'Connor, J.R. Bernhardt, The metabolic theory of ecology and the cost of parasitism, PLoS Biol. 16 (2018) e2005628. Available from: https://doi.org/10.1371/journal.pbio.2005628.

[20] J.R. Bernhardt, J.M. Sunday, M.I. O'Connor, Metabolic theory and the temperature-size rule explain the temperature dependence of population carrying capacity, Am. Nat. 192 (2018) 687–697. Available from: https://doi.org/10.1086/700114.

[21] D. Kirk, N. Jones, S. Peacock, J. Phillips, P.K. Molnár, M. Krkošek, et al., Empirical evidence that metabolic theory describes the temperature dependency of within-host parasite dynamics, PLoS Biol. 16 (2018) e2004608. Available from: https://doi.org/10.1371/journal.pbio.2004608.

[22] M.I. O'Connor, B. Gilbert, C.J. Brown, Theoretical predictions for how temperature affects the dynamics of interacting herbivores and plants, Am. Nat. 178 (2011) 626–638. Available from: https://doi.org/10.1086/662171.

[23] L. Govaert, E.A. Fronhofer, S. Lion, C. Eizaguirre, D. Bonte, M. Egas, et al., Eco-evolutionary feedbacks—theoretical models and perspectives, Funct. Ecol. 29 (2018) 107. Available from: https://doi.org/10.1111/1365-2435.13241.

[24] R. Levins, Evolution in Changing Environments: Some Theoretical Explorations, Princeton University Press, 1968.

[25] R.B. Huey, J.G. Kingsolver, Evolution of resistance to high temperature in ectotherms, Am. Nat. 142 (1993) S21–S46. Available from: https://doi.org/10.1086/285521.

[26] R. Gomulkiewicz, M. Kirkpatrick, Quantitative genetics and the evolution of reaction norms, Evolution 46 (1992) 390–411. Available from: https://doi.org/10.1111/j.1558-5646.1992.tb02047.x.

[27] G.W. Gilchrist, Specialists and generalists in changing environments. 1. Fitness landscapes of thermal sensitivity, Am. Nat. 146 (1995) 252–270. Available from: https://doi.org/10.1086/285797.

[28] R. Lande, Evolution of phenotypic plasticity and environmental tolerance of a labile quantitative character in a fluctuating environment, J. Evol. Biol. 27 (2014) 866–875. Available from: https://doi.org/10.1111/jeb.12360.

[29] D. Berger, E. Postma, W.U. Blanckenhorn, R.J. Walters, Quantitative genetic divergence and standing genetic (co)variance in thermal reaction norms along latitude, Evolution 67 (2013) 2385–2399. Available from: https://doi.org/10.1111/evo.12138.

[30] C.A. Deutsch, J.J. Tewksbury, R.B. Huey, K.S. Sheldon, C.K. Ghalambor, D.C. Haak, et al., Impacts of climate warming on terrestrial ectotherms across latitude, Proc. Natl. Acad. Sci. U.S.A. 105 (2008) 6668–6672. Available from: https://doi.org/10.1073/pnas.0709472105.

[31] J.M. Sunday, A.E. Bates, N.K. Dulvy, Thermal tolerance and the global redistribution of animals, Nat. Clim. Change 2 (2012) 686–690. Available from: https://doi.org/10.1038/nclimate1539.

[32] R. Izem, J.G. Kingsolver, Variation in continuous reaction norms: quantifying directions of biological interest, Am. Nat. 166 (2005) 277–289. Available from: https://doi.org/10.1086/431314.

[33] J.R. Bernhardt, J.M. Sunday, P.L. Thompson, M.I. O'Connor, Nonlinear averaging of thermal experience predicts population growth rates in a thermally variable environment, Proc. Biol. Sci. 285 (2018). Available from: https://doi.org/10.1098/rspb.2018.1076. 20181076.

[34] M. Blows, B. Walsh, Spherical cows grazing in flatland: constraints to selection and adaptation, in: J. van der Werf, H.-U. Graser, R. Frankham, C. Gondro (Eds.), Adaptation and Fitness in Animal Populations: Evolutionary and Breeding Perspectives on Genetic Resource Management, Springer, Dordrecht, The Netherlands, 2009, pp. 83–101. Available from: https://doi.org/10.1007/978-1-4020-9005-9_6.

[35] M.W. Blows, A.A. Hoffmann, A reassessment of genetic limits to evolutionary change, Ecology 86 (2005) 1371–1384.

[36] J.L. Knies, R. Izem, K.L. Supler, J.G. Kingsolver, C.L. Burch, The genetic basis of thermal reaction norm evolution in lab and natural phage populations, PLoS Biol. 4 (2006) e201. Available from: https://doi.org/10.1371/journal.pbio.0040201.

[37] K. Yamahira, M. Kawajiri, K. Takeshi, T. Irie, Inter- and intrapopulation variation in thermal reaction norms for growth rate: evolution of latitudinal compensation in ectotherms with a genetic constraint, Evolution 61 (2007) 1577–1589. Available from: https://doi.org/10.1111/j.1558-5646.2007.00130.x.

[38] D. Padfield, G. Yvon-Durocher, A. Buckling, S. Jennings, G. Yvon-Durocher, Rapid evolution of metabolic traits explains thermal adaptation in phytoplankton, Ecol. Lett. 19 (2016) 133–142. Available from: https://doi.org/10.1111/ele.12545.

[39] D.R. O'Donnell, C.R. Hamman, E.C. Johnson, C.T. Kremer, C.A. Klausmeier, E. Litchman, Rapid thermal adaptation in a marine diatom reveals constraints and trade-offs, Global Change Biol. 24 (2018) 4554–4565. Available from: https://doi.org/10.1111/gcb.14360.

[40] B. Petitpierre, C. Kueffer, O. Broennimann, C. Randin, C. Daehler, A. Guisan, Climatic niche shifts are rare among terrestrial plant invaders, Science 335 (2012) 1344–1348. Available from: https://doi.org/10.1126/science.1215933.

[41] M.B. Araújo, F. Ferri-Yáñez, F. Bozinovic, P.A. Marquet, F. Valladares, S.L. Chown, Heat freezes niche evolution, Ecol. Lett. 16 (2013) 1206–1219. Available from: https://doi.org/10.1111/ele.12155.

[42] M.K. Thomas, M. Aranguren-Gassis, C.T. Kremer, M.R. Gould, K. Anderson, C.A. Klausmeier, et al., Temperature–nutrient interactions exacerbate sensitivity to warming in phytoplankton, Global Change Biol. 23 (2017) 3269–3280. Available from: https://doi.org/10.1111/gcb.13641.

[43] M.J. Angilletta, R.S. Wilson, C.A. Navas, R.S. James, Tradeoffs and the evolution of thermal reaction norms, Trends Ecol. Evol. 18 (2003) 234–240. Available from: https://doi.org/10.1016/S0169-5347(03)00087-9.

CHAPTER 11

The Sea Around Us *as provider of global fisheries catch and related marine biodiversity data to the Nereus Program and civil society*

D. Pauly[1], M.L.D. Palomares[1], B. Derrick[1], G. Tsui[1], L. Hood[2] and D. Zeller[2]

[1]Sea Around Us, *Institute for the Oceans and Fisheries, University of British Columbia, Vancouver, BC, Canada* [2]Sea Around Us—*Indian Ocean, School of Biological Sciences, University of Western Australia, Crawley, WA, Australia*

Chapter Outline

11.1 Introduction

The catch of fisheries is their most important characteristic. People fish to generate a catch, whether they are on the deck of a megatrawler in the frigid waters of the North Pacific, in a canoe along an African coast, or even walking with their children on a reef flat collecting invertebrates for their next meal on an island in the Indian Ocean. Thus, reliable information on current and past catches are the foundation for understanding fisheries [1,2] and are crucial baseline data for any attempt to project or predict future catches (see, for example, [3]).

Nowadays, fisheries are globally integrated, not because fish move, as asserted by many, but rather because fishing fleets switch between fishing grounds and ocean basins. Thus, globally mobile fishing fleets integrate fisheries much more than tuna or other "highly migratory fish" ever could [4].

Predicting Future Oceans.
DOI: https://doi.org/10.1016/B978-0-12-817945-1.00015-0

Local, regional, and national fisheries studies can generally be conducted using national or subnational data sets, often even with data that the investigator(s) may have contributed to. Thus the "local" or situational knowledge of the investigators will ensure a high likelihood of awareness about possible issues or challenges with the data sets being used.

However, such local "context" or awareness is lost in the fisheries catch data submitted annually to the Food and Agriculture Organization of the United Nations (FAO) by its member countries, and which the FAO, after some harmonization, disseminates as the world's capture statistics [5]. These fisheries statistics, even though they have been and continue to be used largely unchallenged by many (e.g., Ref. [6]), suffer from numerous biases, of which the following may be the most important:

1. Several countries do not submit figures derived from the catch realized by their fisheries, but of the quantities they plan or anticipate to catch (see Ref. [7] for Myanmar; or Ref. [8] for China).
2. The catch of artisanal (i.e., small-scale commercial) fisheries is often underrepresented by the reporting agencies in both developed and developing countries [9].
3. The catches of noncommercial subsistence and recreational fisheries are largely unreported, even though they can be considerable in various countries (see Ref. [10] for subsistence fisheries and Ref. [11] for recreational fisheries).
4. The discarding of fish, a common practice in certain industrial fisheries, especially trawling, although well covered in FAO publications (see, for example, Ref. [12]), is explicitly excluded from consideration in FAO fisheries statistics, which therefore are comprised of landings, and are not *catch* statistics [13].
5. No attempt is made to account for illegally caught fish, which obviously are not officially reported, unless laundered, for example, via transshipment at sea.

While item (1) often leads to catch *over*estimation, items (2–5) will lead to catches being *under*estimated.

11.2 Catch reconstructions

Over the last 15 years, the *Sea Around Us* has collaborated with hundreds of colleagues throughout the world to complete "catch reconstructions" (sensu [14,15]) in all maritime countries of the world. These catch reconstructions are based on the notion that the deficiencies in (1–5) can be overcome, or at least mitigated by the systematic acquisition and analysis of secondary data [15]. Such data come from various sources, ranging from local studies of fishing villages by anthropologists (e.g., Refs. [16,17]), or localized case studies of fishing subsectors (e.g., Refs. [18,19]), or seafood purchasing receipts by restaurants and hotels [20], or national Household Income and Expenditure Surveys [21], to international databases on general food consumption [22].

The philosophy behind catch data reconstructions using secondary data rests on two conceptual pillars:

1. Fisheries never operate in a social vacuum; because they are a social activity, they throw a "shadow" on the society and economy in which they are embedded [1]. Thus it is almost always possible to infer a catch from some indirect measures of fishing activity, such as fuel use, employment, and direct sales to restaurants and hotels (e.g., Ref. [20]).
2. If a fishery operates somewhere, it generates a non-zero catch. Thus if in the absence of detailed data on this catch, a government official decides not to enter an approximate figure for the catch (which may or may not be correct), the catch of that fishery will be precisely 0 ± 0 in the official national data reported to FAO. While this is a very precise estimate, it is guaranteed to be wrong.

Thus, catch reconstructions involve replacing precise but erroneous estimates of zero catch by imprecise but roughly accurate estimates of the catch of the hitherto undocumented fisheries. The results we derived from the about 200 reconstruction studies that were performed for all maritime countries and their territories were presented in Pauly and Zeller [9]. The title of that contribution, "Catch reconstructions reveal that global marine fisheries catches are higher than reported and declining" summarizes the situation that we find ourselves in: we catch far more than we thought, and we are seeing declining trends in catches, mainly due to overfishing and overfished stocks (Fig. 11.1A).

One of the major points made in Pauly and Zeller [9] is that the strong decline observed in the reconstructed total catches since the mid-1990s is somewhat masked in the data reported by FAO due to what is now called "presentist bias" [23]. This bias is the inadvertent by-product of efforts by countries to regularly improve their national data collection and reporting systems, which is a commendable endeavor. Unfortunately, such improvement efforts often overlook the need to also comprehensively correct historical data back to 1950 for any changes in new data being incorporated into data reporting systems. Hence the focus on the *present* at the expense of the *past*, which is the essence of the presentist bias [23]. Thankfully, there are signs that the FAO has recognized the importance of this bias (p. 8 in Ref. [7]) and the utility of catch reconstructions and other retroactive data corrections (p. 93, Box 5 in Ref. [7]). The catch data assembled through our massive catch reconstruction effort are now publicly available through our website (www.seaaroundus.org) and cloud data servers that are optimized for the delivery of large data sets.

The data sets are very large because they present annual marine catch data from 1950 to 2014 (with updates underway) for the 273 individual exclusive economic zone (EEZ) entities of all maritime countries and territories of the world as well as a global reconstruction and harmonization of the industrial tuna and large pelagic fisheries

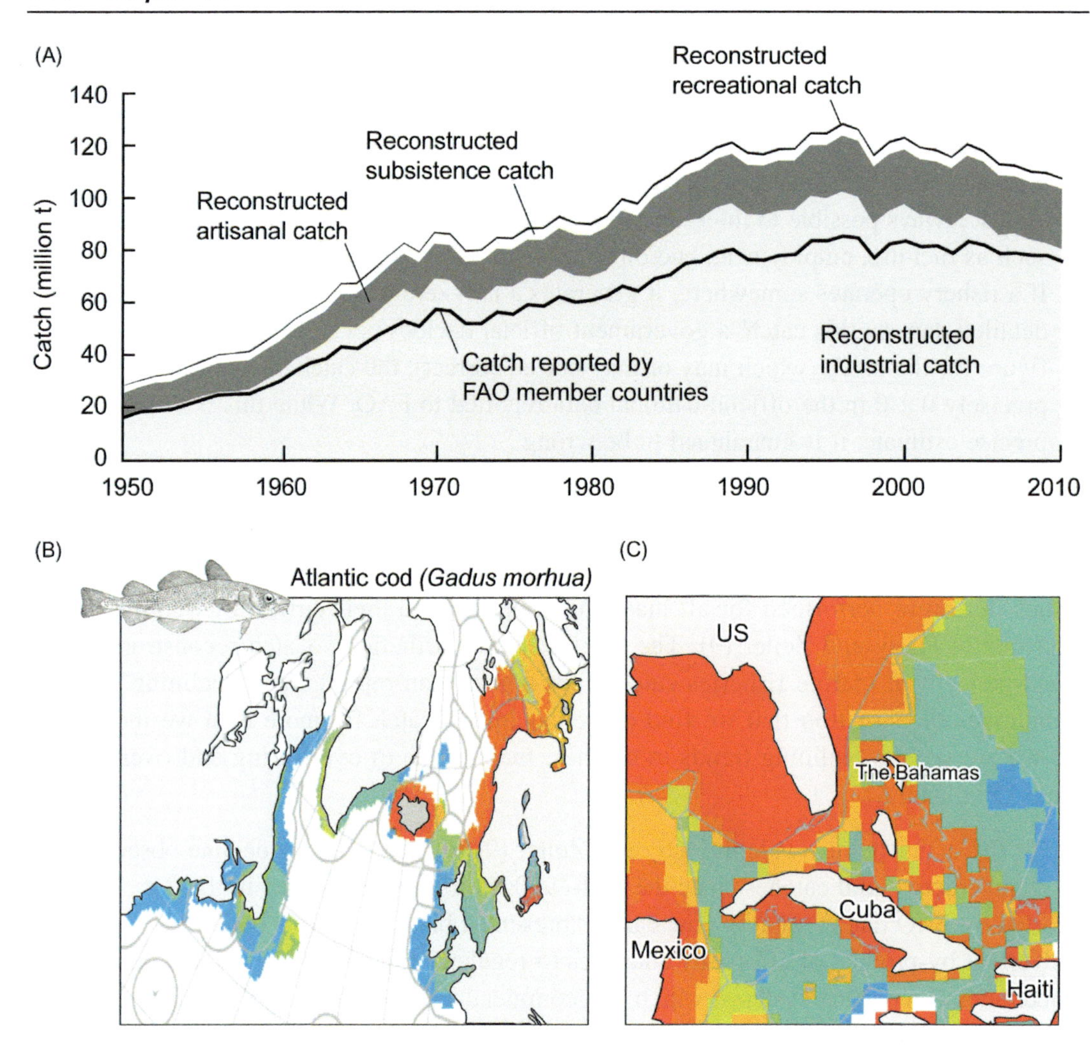

Figure 11.1

Major products of the *Sea Around Us*, as can be downloaded from its website (www.seaaroundus.org). The graph in (A) shows global reconstructed and officially reported marine catch by sectors since 1950; similar time series are available by country exclusive economic zones, by large marine ecosystems, and other geographies; also the catches are available by taxon, gear, and other criteria, in tonnes or in terms of their ex-vessel value. The map in (B) documents the distribution ranges of Atlantic cod (*Gadus morhua*) as used, along with over 2000 such (downloadable) maps to allocate catches to special cells; the map in (C) gives an example of the resulting catch maps, which also can be downloaded.

conducted in High Seas waters under the auspices of several Regional Fisheries Management Organizations [24]. Indeed these data sets are so large because they also present these data by fishing sector (industrial, artisanal, subsistence, and recreational), by taxon (over 2700 species, genera, families, or higher groups), by fishing country, by reporting status (reported or unreported catches), by type of catch (landed or discarded), by major fishing gear (trawls, purse seines, longlines, etc.), and by the end use of the catch (direct human consumption, fish meal, etc.).

In addition to being more comprehensive than officially reported data, the reconstructed catch data of the *Sea Around Us* are allocated to over 150,000 half-degree latitude/longitude marine spatial cells in a manner that is both ecologically viable and politically feasible. This is achieved by analytically intersecting the catch data with biological probability distributions of occurrence for each of the over 2700 taxa in the catch data sets ([25]; Fig. 11.1B), and permitting access to countries' EEZ waters only to those fishing countries that are known to access these waters, either via fishing access agreements or observed access [15]. This analytical intersection between catch data, biological distributions, and fishing access information is obtained through allocation algorithms that are being constantly improved [26]. Thus the availability of the *Sea Around Us* catch data at biologically and politically refined spatial resolutions that are much smaller than those of the global FAO landings data that are reported by 19 very large FAO statistical areas allows us to draw meaningful maps of spatial fishing impacts (Fig. 11.1C).

The catch-by-cell approach implemented by the *Sea Around Us* implies that it is very straightforward to present, evaluate, and analyze data for virtually any geography. Thus, we present catches not only by EEZs and High Seas areas, but also by the 66 Large marine ecosystems defined by the U.S. National Oceanic and Atmospheric Administration (NOAA) [27], by the 19 large and ecologically uninformative FAO Statistical Areas, by the management areas of the 17 Regional Fisheries Management Organizations [28], and, most recently, by the 232 Marine Ecoregions of the World [29].

The latter also provides a geographic framework for stock assessments, that is, estimates of the biomass trends in exploited fish and invertebrates populations, as can be obtained from catch and ancillary data using the CMSY method [30,31]. Hopefully our global application of this stock assessment method, which relies primarily on catch time series, will finally lay to rest the misguided insistence by some (e.g., Ref. [32], but see Ref. [2]) that catch data cannot tell us anything useful about the abundance or biomass of stocks and hence the status of fisheries.

The comprehensive and parameter rich cell-based catch data of the *Sea Around Us* are also the main currency for exchange and collaboration with other research groups and institutions, notably with the Nippon Foundation Nereus Program (see examples in Table 11.1). Based on the fruitful cooperation with the Nereus Program and similar collaborations with other groups (e.g., Convention on Biological Diversity [47,48]; NOAA [46]; EU Parliament [49,50]; Minderoo Foundation [51]), we anticipate that the *Sea Around Us* databases and web portal will become a major data and information resource for fisheries scientists and conservation groups around the world, and especially in underresourced developing countries.

Table 11.1: Examples of Nereus contributions that used *Sea Around Us* data and expertise.

Topic/study area	Data type (core *Sea Around Us* reference)	Nereus product
Climate change effects on fisheries in 147 countries	Landing data by fishing sectors (i.e., Ref. [9])	Blasiak et al. (2017) [52]
Small pelagic fisheries in major upwelling systems	Global reconstructed catch (i.e., Ref. [9])	Checkley et al. (2017) [53]
Reliability of modeling	Catch of sablefish in the Gulf of Alaska LME (i.e., Ref. [33])	Cheung et al. (2016) [54]
Canadian marine fisheries	Canadian catch in weight and value (i.e., Refs. [34–36])	Cisneros-Montemayor et al. (2017) [55]
Use of unreported catch in Panama	Reconstructed catch in Panama (i.e., Ref. [37])	Cisneros-Montemayor et al. (2018) [56]
Biodiversity in 66 LMEs	Species distribution maps (i.e., Ref. [25])	Cosme et al. (2017) [57]
Future of Bangladesh fisheries	Bangladesh catch, for example, hilsa and Bombay duck (i.e., Ref. [38])	Fernandes et al. (2016) [58]
Global High Sea	High Sea catches (i.e., Ref. [24])	Merrie et al. (2014) [59]
Future of the South China Sea	Fisheries catch by sector and related information (i.e., Ref. [9])	Sumaila and Cheung (2018) [60]
Expansion of global Soviet fisheries	Reconstructed catches of ex-Soviet Republics (i.e., Ref. [39–44])	Österblom and Folke (2014) [61]
Potential for global mariculture expansion	Database on time series of mariculture data (i.e., Ref. [45])	Oyinlola et al. (2018) [62]
Catch versus ocean productivity	Catch by 66 LMEs (i.e., Ref. [46])	Stock et al. (2017) [63]

LME, Large marine ecosystem.

More specifically, the availability of reliable global fisheries catch and related data starting as early as 1950, before the post-WWII worldwide expansion of industrial fishing, should provide a reliable baseline for projections that may frame a future ocean.

Acknowledgments

The authors thank the *Sea Around Us* and *Sea Around Us*—Indian Ocean. All *Sea Around Us* activities are currently supported by a number of philanthropic foundations, notably the Oak Foundation, the Marisla Foundation, the MAVA Foundation, the David and Lucile Packard Foundation, and the Paul M. Angell Family Foundation.

References

[1] D. Pauly, Rationale for reconstructing catch time series, EC Fish. Cooper Bull. 11 (2) (1998) 4–7.
[2] D. Pauly, Does catch reflect abundance? Yes, it is a crucial signal, Nature 494 (2013) 303–306.
[3] W.W.L. Cheung, V.W.Y. Lam, J.L. Sarmiento, K. Kearney, R. Watson, D. Zeller, et al., Large-scale redistribution of maximum fisheries catch potential in the global ocean under climate change, Global Change Biol. 16 (2010) 24–35.
[4] D.K. Tickler, J.J. Meeuwig, K. Bryant, F. David, J.A.H. Forrest, E. Gordon, et al., Modern slavery and the race to fish, Nat. Commun. 9 (2018) 4643.

[5] L. Garibaldi, The FAO global capture production database: a six-decade effort to catch the trend, Mar. Pol. 36 (2012) 760–768.
[6] C. Costello, D. Ovando, R. Hilborn, S.D. Gaines, O. Deschenes, S.E. Lester, Status and solutions for the world's unassessed fisheries, Science 338 (6106) (2012) 517–520.
[7] FAO, *The State of World Fisheries and Aquaculture 2018 (SOFIA): Meeting the Sustainable Development Goals*, Food and Agriculture Organization, Rome, 2018. xiii + 210 p.
[8] R. Watson, D. Pauly, Systematic distortions in world fisheries catch trends, Nature 414 (2001) 534–536.
[9] D. Pauly, D. Zeller, Catch reconstructions reveal that global marine fisheries catches are higher than reported and declining, Nat. Commun. (2016). Available from: https://doi.org/10.1038/ncomms10244. 9 p.
[10] D. Zeller, S. Harper, K. Zylich, D. Pauly, Synthesis of under-reported small-scale fisheries catch in Pacific-island waters, Coral Reefs 34 (1) (2015) 25–39.
[11] K. Kleisner, C. Brennan, A. Garland, S. Lingard, S. Tracey, P. Sahlqvist, et al., Australia: Reconstructing Estimates of Total Fisheries Removal, 1950-2010. Fisheries Centre Working Paper #2015-02, University of British Columbia, Vancouver, 2015. 26 p.
[12] K. Kelleher, Discards in the world's marine fisheries. An update, in: FAO Fisheries Technical Paper (470), 2005, 131 pp.
[13] D. Zeller, T. Cashion, M.L.D. Palomares, D. Pauly, Global marine fisheries discards: a synthesis of reconstructed data, Fish Fish. 19 (1) (2018) 30–39.
[14] D. Zeller, S. Booth, G. Davis, D. Pauly, Re-estimation of small-scale fishery catches for U.S. flag-associated island areas in the western Pacific: the last 50 years, Fish. Bull. 105 (2) (2007) 266–277.
[15] D. Zeller, M.L.D. Palomares, A. Tavakolie, M. Ang, D. Belhabib, W.W.L. Cheung, et al., Still catching attention: *Sea Around Us* reconstructed catch data, their spatial expression and public accessibility, Mar. Policy 70 (2016) 145–152.
[16] Y. Ota, Custom and Fishing: Cultural Meaning and Social Relations of Pacific Fishing, Republic of Palau, Micronesia (Ph.D.), University College of London, London, UK, 2006. 253 pp.
[17] Y. Ota, An anthropologist in Palau, Sea Around Us Newsl., May/June 35 (2006) 1–3.
[18] R.C. Wass, The shoreline fishery of American Samoa: past and present, in: J.L. Munro (Ed.), Marine and Coastal Processes in the Pacific: Ecological Aspects of Coastal Zone Management, Papers presented at a UNESCO seminar held at Motupore Island Research Centre, University of Papua New Guinea, United Nations Educational Scientific and Cultural Organization, Jakarta Pusat, 1980, pp. 51–83.
[19] N.J.F. Rawlinson, D.A. Milton, S.J.M. Blaber, A. Sesewa, S.P. Sharma, A Survey of the Subsistence and Artisanal Fisheries in Rural Areas of Viti Levu, Fiji. ACIAR Monograph #35, Australian Centre for International Agricultural Research, Canberra, 1996.
[20] N.S. Smith, D. Zeller, Unreported catch and tourist demand on local fisheries of small island states: the case of The Bahamas 1950-2010, Fish. Bull. 114 (1) (2016) 117–131.
[21] Anon, Republic of Palau Household Income and Expenditure Survey, Office of Planning & Statistics, Bureau of Budget & Planning, Ministry of Finance, Ngerulmud, Republic of Palau, 2014, 98 pp.
[22] S. Khatibzadeh, M. Saheb-Kashaf, M. Micha, S. Fahimi, P. Shi, I. Elmadfa, et al., A global database of food and nutrient consumption, Bull. World Health Organ. 94 (12) (2016) 931–934.
[23] D. Zeller, D. Pauly, The 'presentist bias' in time-series data: implications for fisheries science and policy, Mar. Policy 90 (2018) 14–19.
[24] F. Le Manach, R. Chavance, A.M. Cisneros-Montemayor, A. Lindop, A. Padilla, L. Schiller, et al., Global catches of large pelagic fishes, with emphasis on the High Seas, in: D. Pauly, D. Zeller (Eds.), Global Atlas of Marine Fisheries: A Critical Appraisal of Catches and Ecosystem Impacts, Island Press, Washington, DC, 2016, pp. 34–45.
[25] M.L.D. Palomares, W.W.L. Cheung, V.W.Y. Lam, D. Pauly, The distribution of exploited marine biodiversity, in: D. Pauly, D. Zeller (Eds.), Global Atlas of Marine Fisheries: A Critical Appraisal of Catches and Ecosystem Impacts, Island Press, Washington, DC, 2016, pp. 46–58.
[26] V.W.Y. Lam, A. Tavakolie, M.L.D. Palomares, D. Pauly, D. Zeller, The *Sea Around Us* catch database and its spatial expression, in: D. Pauly, D. Zeller (Eds.), Global Atlas of Marine Fisheries: A Critical Appraisal of Catches and Ecosystem Impacts., Island Press, Washington, DC, 2016, pp. 59–67.

[27] K. Sherman, A.M. Duda, Large marine ecosystems: an emerging paradigm for fishery sustainability, Fisheries 24 (12) (1999) 15–26.

[28] S. Cullis-Suzuki, D. Pauly, Global evaluation of high seas fisheries management, in: D. Pauly, D. Zeller (Eds.), Global Atlas of Marine Fisheries: A Critical Appraisal of Catches and Ecosystem Impacts., Island Press, Washington, DC, 2016, pp. 79–85.

[29] M.D. Spalding, H.E. Fox, G.R. Allen, N. Davidson, Z.A. Ferdaña, M. Finlayson, et al., Marine ecoregions of the world: a bioregionalization of coastal and shelf areas, Bioscience 57 (2007) 573–583. Available from: https://doi.org/10.1641/b570707.

[30] R. Froese, N. Demirel, G. Coro, K.M. Kleisner, H. Winker, Estimating fisheries reference points from catch and resilience, Fish Fish. 18 (3) (2017) 506–526.

[31] S. Martell, R. Froese, A simple method for estimating MSY from catch and resilience, Fish Fish. 14 (4) (2013) 504–514.

[32] R. Hilborn, T.A. Branch, Does catch reflect abundance? No, it is misleading, Nature 494 (2013) 303–306.

[33] B. Doherty, D. Gibson, Y. Zhai, A. McCrea-Strub, K. Zylich, D. Zeller, et al., Reconstruction of marine fisheries catches for subarctic Alaska, 1950-2010, in: Fisheries Centre Working Paper #2015-82, University of British Columbia, Vancouver, 2015. 31 p.

[34] D. Zeller, S. Booth, E. Pakhomov, W. Swartz, D. Pauly, Arctic fisheries catches in Russia, USA and Canada: baselines for neglected ecosystems, Polar Biol. 34 (7) (2011) 955–973.

[35] E. Divovich, D. Belhabib, D. Zeller, D. Pauly, Eastern Canada, "a fishery with no clean hands": Marine fisheries catch reconstruction from 1950 to 2010, in: Fisheries Centre Working Paper #2015-56, University of British Columbia, Vancouver, 2015. 37 pp.

[36] C.H. Ainsworth, British Columbia marine fisheries catch reconstruction: 1873 to 2011, BC Stud. 188 (2016) 81–90.

[37] S. Harper, H.M. Guzmán, K. Zylich, D. Zeller, Reconstructing Panama's total fisheries catches from 1950 to 2010: highlighting data deficiencies and management needs, Mar. Fish. Rev. 76 (1–2) (2014) 51–65.

[38] H. Ullah, D. Gibson, D. Knip, K. Zylich, D. Zeller, Reconstruction of total marine fisheries catches for Bangladesh: 1950-2010, in: Fisheries Centre Working Paper #2014-15, Fisheries Centre, University of British Columbia, Vancouver, 2014. 10 pp.

[39] D. Zeller, P. Rossing, S. Harper, L. Persson, S. Booth, D. Pauly, The Baltic Sea: estimates of total fisheries removals 1950-2007, Fish. Res. 108 (2011) 356–363.

[40] E. Divovich, B. Jovanović, K. Zylich, S. Harper, D. Zeller, D. Pauly, Caviar and politics: a reconstruction of Russia's marine fisheries in the Black Sea and Sea of Azov from 1950 to 2010, in: Fisheries Centre Working Paper #2015-84, University of British Columbia, Vancouver, 2015. 24 pp.

[41] B. Jovanović, E. Divovich, S. Harper, D. Zeller, D. Pauly, Preliminary estimate of total Russian fisheries catches in the Barents Sea region (ICES subarea I) between 1950 and 2010, in: Fisheries Centre Working Paper #2015-59, University of British Columbia, Vancouver, 2015. 16 pp.

[42] A. Ulman, E. Divovich, The marine fishery catch of Georgia (including Abkhazia), 1950-2010, in: Fisheries Centre Working Paper #2015-88, University of British Columbia, Vancouver, 2015. 25 pp.

[43] A. Ulman, V. Shlyakhov, S. Jatsenko, D. Pauly, A reconstruction of the Ukraine's marine fisheries catches, 1950-2010, J. Black Sea Mediterr. Environ. 21 (2) (2015) 103–124.

[44] A. Sobolevskaya, E. Divovich, The Wall Street of fisheries: the Russian Far East, a catch reconstruction from 1950 to 2010, in: Fisheries Centre Working Paper #2015-45, University of British Columbia, Vancouver, 2015. 64 pp.

[45] B. Campbell, D. Pauly, Mariculture: a global analysis of production trends since 1950, Mar. Policy 39 (2013) 94–100.

[46] D. Pauly, J. Alder, S. Booth, W.W.L. Cheung, V. Christensen, C. Close, et al., Fisheries in large marine ecosystems: descriptions and diagnoses, in: K. Sherman, G. Hempel (Eds.), The UNEP Large Marine Ecosystem Report: A Perspective on Changing Conditions in LMEs of the World's Regional Seas, Nairobi, Kenya, UNEP Regional Seas Reports and Studies No. 182, 2008, pp. 23–40.

[47] D. Pauly, R. Watson, Background and interpretation of the 'Marine Trophic Index' as a measure of biodiversity, Philos. Trans. R. Soc.: Biol. Sci. 360 (2005) 415–423.

[48] K. Kleisner, H. Mansour, D. Pauly, Region-based MTI: resolving geographic expansion in the Marine Trophic Index, Mar. Ecol. Prog. Ser. 512 (2014) 185–199.

[49] D. Pauly, D. Belhabib, W.W.L. Cheung, A. Cisneros-Montemayor, S. Harper, V. Lam, et al., Catches [of the Chinese distant- water fleet], in: R. Blomeyer, I. Goulding, D. Pauly, A. Sanz, K. Stobberup (Eds.), The role of China in World Fisheries, European Parliament, Directorate General for Internal Policies, Policy Department B: Structural and Cohesion Policies—Fisheries, Brussels, 2012, pp. 21–29 & 81–85.

[50] D. Pauly, D. Belhabib, R. Blomeyer, et al., China's distant-water fisheries in the 21st century, Fish Fish. 15 (3) (2014) 474–488.

[51] D.K. Tickler, J.J. Meeuwig, D. Pauly, M.L.D. Palomares, D. Zeller, Far from home: distance patterns of global fishing fleets, Sci. Adv. 4 (2018) eaar3279.

[52] R. Blasiak, J. Spijkers, K. Tokunaga, J. Pittman, N. Yagi, H. Österblom, Climate Change and Marine Fisheries: Least Developed Countries Top Global Index of Vulnerability, in: B.R MacKenzie (Eds.), PloS One 12 (6) (2017) e0179632. Available from: https://doi.org/10.1371/journal.pone.0179632.

[53] D.M. Checkley, R.G. Asch, R.R. Rykaczewski, Climate, Anchovy, and Sardine, Annual Review of Marine Science 9 (1) (2017) 469–493. Available from: https://doi.org/10.1146/annurev-marine-122414-033819.

[54] W.W.L. Cheung, T.L. Frölicher, R.G. Asch, M.C. Jones, M.L. Pinsky, G. Reygondeau, et al., Building Confidence in Projections of the Responses of Living Marine Resources to Climate Change, ICES J. Mar. Sci 73 (5) (2016) 1283–1296. Available from: https://doi.org/10.1093/icesjms/fsv250.

[55] A.M. Cisneros-Montemayor, W.W.L. Cheung, K. Bodtker, L. Teh, N. Steiner, M. Bailey, et al., Towards an Integrated Database on Canadian Ocean Resources: Benefits, Current States, and Research Gaps, Can. J. Fish. Aquat. Sci. 74 (1) (2017) 65–74. Available from: https://doi.org/10.1139/cjfas-2015-0573.

[56] A.M. Cisneros-Montemayor, S. Harper, T.C. Tai, The Market and Shadow Value of Informal Fish Catch: A Framework and Application to Panama, Nat. Resour. Forum 42 (2) (2018) 83–92. Available from: https://doi.org/10.1111/1477-8947.12143.

[57] N. Cosme, M.C. Jones, W.W.L. Cheung, H.F. Larsen, Spatial Differentiation of Marine Eutrophication Damage Indicators Based on Species Density, Ecol. Indic. 73 (2017) 676–685. Available from: https://doi.org/10.1016/j.ecolind.2016.10.026.

[58] J.A. Fernandes, S. Kay, M.A. R. Hossain, M. Ahmed, W.W. L. Cheung, A.N. Lazar, et al., Projecting Marine Fish Production and Catch Potential in Bangladesh in the 21st Century under Long-Term Environmental Change and Management Scenarios, ICES J. Mar. Sci 73 (5) (2015) 1357–1369. Available from: https://doi.org/10.1093/icesjms/fsv217.

[59] A. Merrie, D.C. Dunn, M. Metian, A.M. Boustany, Y. Takei, A. Oude Elferink, et al., An Ocean of Surprises – Trends in Human Use, Unexpected Dynamics and Governance Challenges in Areas beyond National Jurisdiction, Global Environ. Chang 27 (2014) 19–31. Available from: https://doi.org/10.1016/j.gloenvcha.2014.04.012.

[60] U.R. Sumaila, W.W.L. Cheung, Boom or Bust. The Future of Fish in the South China Sea, OceanAsia Report (2015). Available from: https://drive.google.com/file/d/0B_oUJE4kCTZrbVI4N2tTVjlpYTA/view.

[61] H. Österblom, C. Folke, Globalization, Marine Regime Shifts and the Soviet Union, Philosophical Transactions of the Royal Society B: Biological Sciences 370 (1659) (2014). Available from: https://doi.org/10.1098/rstb.2013.0278. 20130278–20130278.

[62] M.A. Oyinlola, G. Reygondeau, C.C.C. Wabnitz, M. Troell, W.W.L. Cheung, Global Estimation of Areas with Suitable Environmental Conditions for Mariculture Species, PloS One 13 (1) (2018) e0191086. Available from: https://doi.org/10.1371/journal.pone.0191086.

[63] C.A. Stock, J.G. John, R.R. Rykaczewski, R.G. Asch, W.W.L. Cheung, J.P. Dunne, et al., Reconciling Fisheries Catch and Ocean Productivity, Proceedings of the National Academy of Sciences 114 (8) (2017) E1441–E1449. Available from: https://doi.org/10.1073/pnas.1610238114.

CHAPTER 12

Changing biomass flows in marine ecosystems: from the past to the future

Hubert du Pontavice

Ecology and Ecosystems Health, Agrocampus Ouest, Rennes, France and Nippon Foundation-Nereus Program, Institute for the Oceans and Fisheries, University of British Columbia, Vancouver, BC, Canada

Chapter Outline

12.1 Introduction

Throughout the world's oceans anthropogenic stressors, such as fishing [1], pollution [2], degradation of habitats, and climate change [3,4], have led to drastic changes not only in the abundance of exploited species, but also in the species assemblages, the structure and functioning of food webs, and ultimately in the productivity, stability, and resilience of marine ecosystems. Over the last century, fishing was the main driver of global changes in the ocean. However, many scientific works have showed that climate change has already had very significant effects which should play a key role in the future, with a displacement of species toward the poles and a huge impact on the structure and functioning of ecosystems.

Ecological units such as marine ecosystems are being increasingly studied, notably to better understand their food webs dynamics and energy transfer. However, the effects of increasing anthropogenic impacts remain poorly understood in comparison to terrestrial ecosystems. Further efforts are crucially needed to understand various aspects of marine ecosystem functioning. A key aspect of the marine ecosystem functioning is the trophic

Predicting Future Oceans.
DOI: https://doi.org/10.1016/B978-0-12-817945-1.00012-5

dynamics (i.e., the food web dynamics) and the driving question is: who eats whom? Besides, the challenge is also to identify the factors and especially the anthropogenic stressors affecting the food webs.

To facilitate analyses of food webs, ecosystem modeling constitutes a simplifying tool to detangle trophic dynamics of marine ecosystems and to evaluate the impact of anthropogenic activities. Nowadays, ecosystem models are being increasingly used to develop global-scale approaches, and estimate the marine biomass, its distribution, and catch potential for fisheries [5–10]. This also enables reaching a better understanding of the fishing impacts [10–12] and the potential impacts of climate change [13,14] following various theoretical scenarios.

The majority of ecosystem models aim to simplify complex biotic interactions by summarizing marine food webs as a network of functional compartments. Each compartment represents a species or a group of species that is interconnected by trophic links (i.e., who eats whom). A complementary approach to represent the food web is to view it as a flow of energy from the base (phytoplankton) to the top (top predators like tunas or certain shark species; Fig. 12.1). This food web representation provides a simplistic overview of the trophic linkages within the ecosystem based on the biomass flow [15]. EcoTroph (ET) represents an approach articulated around the trophic level (TL) concept [15,16] which summarizes the position of an organism in the food web. The trophic functioning of aquatic ecosystems is here modeled as a continuous flow of biomass surging up the food web, from lower to higher TLs, through predation and ontogenic (life history stages) processes.

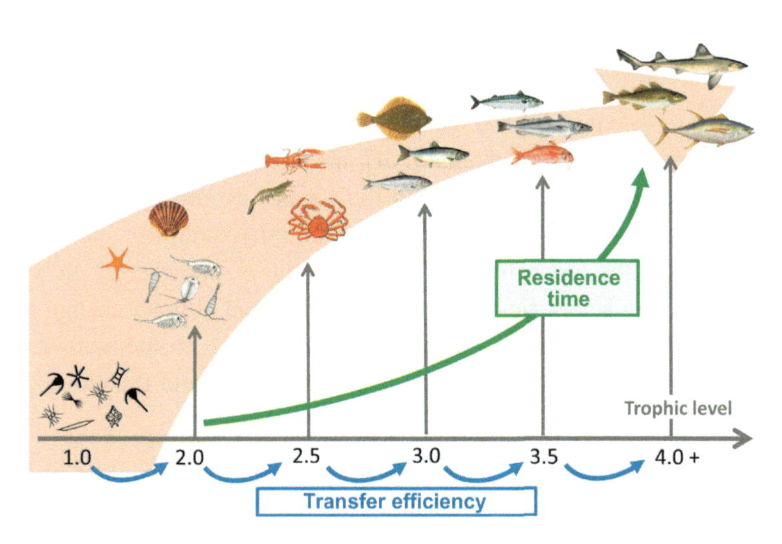

Figure 12.1
Synthetic representation of a biomass flow in EcoTroph model with the two trophic transfer indicators.

Such an approach, wherein "species" as such disappear and are instead combined into classes based only on their TL (i.e., their position in the food web), may be viewed as the final stage in the use of the TL metric for ecosystem modeling. It provides a simplified representation of food web functioning to evaluate impacts of fishing and climate change. ET has been used either in theoretical contexts based on virtual ecosystems, in specific case studies to assess the fishing impacts at the ecosystem scale in various ecosystems [17,18], at the global scale to estimate the ocean living biomass and fishing impacts [19], or to validate estimates of the mesopelagic fish biomass in the open ocean [9].

According to theoretical ecology, disturbances may especially lead to faster and less efficient biomass flows in the food web, and thus to more unstable and less productive ecosystems. The expected global environmental changes are therefore a major issue for the future of fisheries and beyond for services provided by marine ecosystems.

12.2 Development of indicators to analyze biomass flows and trends since 1950

Maureaud et al. [20] proposed two indicators to quantify and explore the changes in the biomass flow in marine food webs, the *transfer efficiency* (TE) and the *residence time* (RT) of the biomass. The TE is the fraction of biomass or energy transferred through predation from the base of the food web to the top, and the RT of biomass is the time a unit of biomass spends at a given position in the food web before moving up the food web through predation and/or ontogeny. In other words, these indicators measure how much of the biomass flow is transferred from one TL (one position in the food web) to the next (TE), and how fast the biomass transfer occurs (RT); these parameters depend on the ecosystem structure and functioning and the species making up the food web. For instance, Maureaud et al. [20] noticed that the relative abundance of small pelagic species in the Humboldt Current (along the western coast of South America) led to variations in TE and RT of biomass in the entire ecosystem. In the early 1990s, when the Peruvian anchoveta (*Engraulis ringens*) was abundant in this ecosystem, the TE increased and the RT of biomass decreased since this species is twice as efficient in converting food into energy to grow and reproduce than the South American pilchard (*Sardinops sagax*) that temporarily replaced it due to environmental drivers.

These indicators (TE, RT) are measured using the biomass trophic spectrum, that is, the biomass distribution along the biomass flow to measure the quantity of biomass at each level TL. We estimated the two indicators using marine fisheries catches data reconstructed by the Sea Around Us project (http://www.seaaroundus.org), assuming they can be used as proxy of the true features of the food web. Then the biological characteristics (growth rate, length, etc.) of every species making up the marine communities were compiled (FishBase www.fishbase.org and SeaLifeBase https://www.sealifebase.ca) and used to build the indicators.

The first step of the analysis was to examine the past trends of these indicators since 1950. These indicators were measured for 56 large marine ecosystems (LMEs) [20] and for all $1° \times 1°$ (latitude $\times$ longitude) coastal cells (5526 cells) around the world. Based on the measurement of these indicators at these two geographical scales, we showed that biomass flow has become both more efficient (higher TE) and faster (lower RT) since 1950, implying a global average shift of species assemblages towards smaller, faster growing species with shorter life spans. The results of Maureaud et al. [20] in LMEs showed that the decrease in residence of the biomass in the food web is at least partly due to the global increase in fishing pressure. In several ecosystems, a "fishing down the food web" syndrome [21] can be the consequence of a high fishing pressure, with sequential overexploitation of higher TLs (starting with top predators). Results also suggest that the increase in efficiency of biomass transfer was mainly driven by increases in fishing pressure in almost all LMEs, leading to adaptive responses throughout the food web to this ecosystem perturbation.

12.3 Temperature effects on biomass flows in marine ecosystems

Preliminary results and Maureaud et al. [20] suggest that the functioning of the biomass flow varies along a latitudinal gradient, with efficient and slow biomass flows in polar ecosystems and less efficient and faster biomass flows in tropical ecosystems. We confirmed this finding in the $1° \times 1°$ coastal cells by showing that TE and RT vary significantly among ecosystem types: around 10% of the biomass is transferred from one TL to the next in polar ecosystems, while a lower fraction of the biomass (around 4%) is transferred in tropical ecosystems. The rest of the biomass which is not transferred to the upper TL is lost in the process due mainly to natural mortality and metabolism of species transforming the ingested biomass (respiration and excretion). At the same time the RT of the biomass in the ecosystem is higher in polar ecosystems where a unit of biomass spends 4 years on average in polar ecosystems, compared to only 2 years in tropical ecosystems.

The latitudinal gradient in the two biomass flow indicators has suggested a direct or indirect link between sea water temperature and food web functioning. Besides, many authors, including the work of Cheung et al. [22], showed that ocean warming will affect species geographic distributions throughout the oceans, and subsequently species compositions in marine communities. These changes will likely lead to species replacement, introduction, or extinction and consequently modify the entire food web functioning with new species compositions and new interactions. In order to understand how ocean warming will affect food web functioning in marine ecosystems, we have studied the present and past biomass flows along food webs to determine the temperature effect on marine ecosystems.

Our models revealed that sea temperature is negatively correlated to TE and RT of biomass in marine ecosystems. That means that biomass transfers are slower (because the units of biomass spend less time in the food web) and more efficient in cold coastal waters, and conversely the biomass is transferred faster and less efficiently in warm waters. The thermal sensitivity of TE and RT imply that species characteristics vary along a sea temperature gradient. In other words, efficient and slow-growing species in cold ecosystems are replaced by less efficient and fast-growing species in warm ecosystems.

Considering the relationship between the two trophic transfer indicators (TE and RT) and the seawater temperature, we projected TE and RT by 2100 using the predicted changes in sea temperature. Several institutes around the world have developed complex models to predict changes in the atmosphere and oceans by 2100, taking into account various carbon emission scenarios (also called representative concentration pathways (RCP)) from the more optimistic (RCP 2.6—the scenario to keep global mean atmosphere temperature below 2°C) to the more pessimistic (RCP 8.5—the business as usual scenario where greenhouse gases continue to rise throughout the 21st century). In our analysis, we used the predicted changes in seawater temperature produced by these two contrasted climate change scenarios (RCP 2.6 and RCP 8.5) to encompass the consequences of ocean warming on the biomass transfers in marine ecosystems. Our results show that the rise in sea temperature should lead to a decrease in TE and in RT and consequently an increase in speed of biomass flow (Fig. 12.2). Furthermore, the magnitude of the ocean warming effects on biomass transfers through the marine food web will depend on our ability to mitigate climate change.

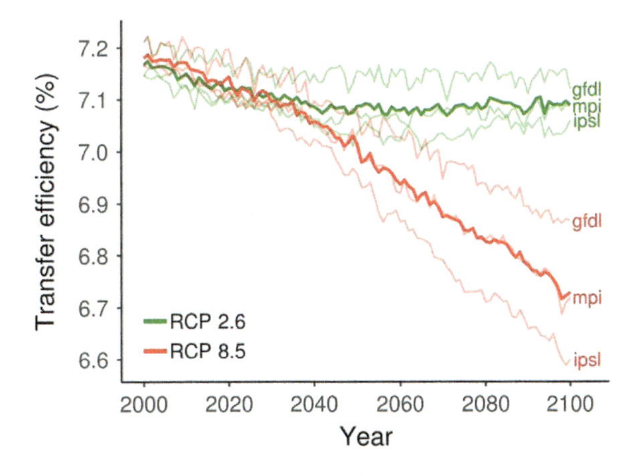

Figure 12.2
Projection of transfer efficiency at the global scale between 2000 and 2100 for two climate change scenarios, RCP 2.6 (optimistic scenario) and RCP 8.5 (business as usual) in green and red, respectively. The dark lines are the mean values of the trophic transfer parameters and the light lines are the values of each Earth system model GFDL, geophysical fluid dynamics laboratory; MPI, max plank institute; IPSL, institut pierre simon laplace.

12.4 A biomass flow model to project the effect of climate change on production and biomass in the world ocean

The next steps of this research will be to use the ET model to evaluate the effects of climate change on biomass and fish production in the world ocean. According to ET theory, knowing the biomass at the base of the food web (primary production) and how the biomass is transferred from the base to the top predators of this web, we are able to infer the amount of biomass at each TL. The purpose of the study is to analyze how these changes in primary production (i.e., phytoplankton) and in sea temperature will affect the biomass and the production in the marine ecosystem. First, the changes in primary production will directly affect the amount of biomass fuelling the food web. Second, changes in sea temperature will alter the biomass transfer, affecting the TE and the RT of the biomass (as explained previously).

This research will enter into the broader framework of FISH-MIP (Fisheries and Marine Ecosystem Model Intercomparison Project) which is a network of marine ecosystem modelers. The goal of this project is to bring together various ecosystem models based on various assumptions, drivers, and processes "to better understand and forecast the long-term impacts of climate change on fisheries and marine ecosystems" (https://www.isimip.org/gettingstarted/marine-ecosystems-fisheries/). The ET model adds a new perspective by summarizing food web functioning into a biomass flow. Moreover, the ET structure and computation make it easy and time-efficient to test hypotheses on processes and drivers, with predictions of spatial and temporal changes fully consistent with other FISH-MIP models.

12.5 Conclusion

The functioning of marine food webs has shifted during the last half century with biomass transfers becoming faster and slightly more efficient, especially due to increased fishing pressure. Even if some the climate change effects on marine communities have already been shown at the global scale, our study did not clearly exhibit the climate change impact on biomass flows over the past 60 years. One reason may be that climate change effects on trophic functioning are harder to detect because of synergistic interactions with fisheries. Several studies show that changes in community structure and ecosystem productivity in the recent decades have been driven by both climate and fisheries [23,24]. Another hypothesis is that fishing has been the predominant factor, affecting trophic functioning at the global scale over the past period, thereby masking the increasing the climate change effects on marine ecosystems. However, we identified a clear temperature effect on the functioning of the marine ecosystem, suggesting that warmer oceans are likely to result in faster and less efficient biomass flows; these effects may be considerable in polar and tropical ecosystems.

Using this new knowledge on the functioning of the food web and the ET model, we have projected a global drop in biomass and productivity (secondary productivity) in the ocean by more than 20% if carbon emissions continue to rise throughout the 21st century and 5% if carbon emissions start to decrease in 2020. Moreover, preliminary results suggest that the effect of climate change might vary a lot throughout the world, with productivity gains in some regions and losses in others. These projections must be interpreted with caution because they come from models incorporating only part of the ecological processes in marine ecosystems and are based on strong assumptions. However, they allow us to evaluate the magnitude of the coming changes and understand the underlying processes responsible for these changes. Management and governance need to be adaptive and global to face future challenges, such as shifts in fish stocks or decreases in fishing catch potential. Recognizing climate change effects must be at the heart of efforts to manage the marine ecosystem, taking into consideration the impacts of climate change on the ocean and its potential synergistic interactions with other anthropogenic pressures such as fishing and pollution.

References

[1] D. Pauly, R. Watson, J. Alder, Global trends in world fisheries: impacts on marine ecosystems and food security, Philos. Trans. R. Soc. B: Biol. Sci. 360 (1453) (2005) 5–12.

[2] J.R. Jambeck, R. Geyer, C. Wilcox, T.R. Siegler, M. Perryman, A. Andrady, et al., Plastic waste inputs from land into the ocean, Science 347 (6223) (2015) 768–771.

[3] B.S. Halpern, M. Frazier, J. Potapenko, K.S. Casey, K. Koenig, C. Longo, et al., Spatial and temporal changes in cumulative human impacts on the world's ocean, Nat. Commun. [Internet] 6 (1) (2015) [cité 9 juill 2018] Disponible sur: http://www.nature.com/articles/ncomms8615.

[4] B.S. Halpern, S. Walbridge, K.A. Selkoe, C.V. Kappel, F. Micheli, C. D'Agrosa, et al., A Global map of human impact on marine ecosystems, Science 319 (5865) (2008) 948–952.

[5] V. Christensen, M. Coll, J. Buszowski, W.W.L. Cheung, T. Frölicher, J. Steenbeek, et al., The global ocean is an ecosystem: simulating marine life and fisheries: modelling life and fisheries in the global ocean, Global Ecol. Biogeogr. 24 (5) (2015) 507–517.

[6] V. Christensen, D. Pauly, Trophic models of aquatic ecosystems. International Center for Living Aquatic Resources Management, International Council for the Exploration of the Sea, DANIDA, editors. 1993, 390 pp. (ICLARM Conference Proceedings).

[7] M.J. Fogarty, A.A. Rosenberg, A.B. Cooper, M. Dickey-Collas, E.A. Fulton, N.L. Gutiérrez, et al., Fishery production potential of large marine ecosystems: a prototype analysis, Environ. Dev. 17 (2016) 211–219.

[8] M.B.J. Harfoot, T. Newbold, D.P. Tittensor, S. Emmott, J. Hutton, V. Lyutsarev, et al., Emergent global patterns of ecosystem structure and function from a mechanistic general ecosystem model, PLoS Biol. 12 (4) (2014) e1001841.

[9] X. Irigoien, T.A. Klevjer, A. Røstad, U. Martinez, G. Boyra, J.L. Acuña, et al., Large mesopelagic fishes biomass and trophic efficiency in the open ocean, Nat. Commun. [Internet] (2014) 5 [cité 5 juill 2017] Disponible sur: http://www.nature.com/doifinder/10.1038/ncomms4271.

[10] S. Jennings, K. Collingridge, Predicting consumer biomass, size-structure, production, catch potential, responses to fishing and associated uncertainties in the world's marine ecosystems, PLoS One 10 (7) (2015) e0133794.

[11] V. Christensen, M. Coll, C. Piroddi, J. Steenbeek, J. Buszowski, D. Pauly, A century of fish biomass decline in the ocean, Mar. Ecol. Prog. Ser. 512 (2014) 155–166.

[12] K. Enberg, C. Jørgensen, E.S. Dunlop, Ø. Varpe, D.S. Boukal, L. Baulier, et al., Fishing-induced evolution of growth: concepts, mechanisms and the empirical evidence: fishing-induced evolution of growth, Mar. Ecol. 33 (1) (2012) 1–25.

[13] W.W.L. Cheung, G. Reygondeau, T.L. Frölicher, Large benefits to marine fisheries of meeting the 1.5°C global warming target, Science 354 (6319) (2016) 1591–1594.

[14] M. Barange, G. Merino, J.L. Blanchard, J. Scholtens, J. Harle, E.H. Allison, et al., Impacts of climate change on marine ecosystem production in societies dependent on fisheries, Nat. Clim. Change 4 (3) (2014) 211–216.

[15] D. Gascuel, D. Pauly, EcoTroph: modelling marine ecosystem functioning and impact of fishing, Ecol. Modell. 220 (21) (2009) 2885–2898.

[16] D. Gascuel, S. Guenette, D. Pauly, The trophic-level-based ecosystem modelling approach: theoretical overview and practical uses, ICES J. Mar. Sci. 68 (7) (2011) 1403–1416.

[17] F. Moullec, D. Gascuel, K. Bentorcha, S. Guénette, M. Robert, Trophic models: what do we learn about Celtic Sea and Bay of Biscay ecosystems? J. Mar. Syst. 172 (2017) 104–117.

[18] M. Colléter, D. Gascuel, J.-M. Ecoutin, L. Tito de Morais, Modelling trophic flows in ecosystems to assess the efficiency of marine protected area (MPA), a case study on the coast of Sénégal, Ecol. Modell. 232 (2012) 1–13.

[19] L. Tremblay-Boyer, D. Gascuel, R. Watson, V. Christensen, D. Pauly, Modelling the effects of fishing on the biomass of the world's oceans from 1950 to 2006, Mar. Ecol. Prog. Ser. 442 (2011) 169–185.

[20] A. Maureaud, D. Gascuel, M. Colléter, M.L. Palomares, H. Du Pontavice, D. Pauly, et al., Global change in the trophic functioning of marine food webs, PLoS One 12 (8) (2017) e0182826.

[21] B. Bhathal, D. Pauly, 'Fishing down marine food webs' and spatial expansion of coastal fisheries in India, 1950–2000, Fish. Res. 91 (1) (2008) 26–34.

[22] W.W.L. Cheung, V.W.Y. Lam, J.L. Sarmiento, K. Kearney, R. Watson, D. Pauly, Projecting global marine biodiversity impacts under climate change scenarios, Fish Fish. 10 (3) (2009) 235–251.

[23] R.I. Perry, P. Cury, K. Brander, S. Jennings, C. Möllmann, B. Planque, Sensitivity of marine systems to climate and fishing: concepts, issues and management responses, J. Mar. Syst. 79 (3-4) (2010) 427–435.

[24] R.R. Kirby, G. Beaugrand, J.A. Lindley, Synergistic effects of climate and fishing in a marine ecosystem, Ecosystems 12 (4) (2009) 548–561.

CHAPTER 13

The role of cyclical climate oscillations in species distribution shifts under climate change

Sarah M. Roberts
Duke University, Durham, NC, United States

Chapter Outline

Over the last several decades, anthropogenic greenhouse gas emissions have driven substantial increases in global and regional ocean temperatures and salinity contrasts [1]. As ocean temperatures warm, researchers have determined that marine species' geographic ranges have shifted, and will likely continue to shift as climate change persists [2–5]. While global efforts to quantify the impacts of climate change on species' ranges continue, more nuanced and specific questions emerge. For example, researchers are beginning to identify the effects of climate change on food web structure [6] and species assemblages [7]. Additionally, recent research has identified more regional patterns that are inconsistent with a general poleward shift in marine species' ranges [5]. As the field of climate science expands, questions remain about the importance of other variables, such as cyclical climate oscillations on species shifts in a changing climate. This chapter discusses current research to understand the influence of cyclical climate oscillations as well as anthropogenic warming on fish communities along the coast of the southeast United States, which is commonly referred to as the South Atlantic Bight (SAB) (Fig. 13.1).

Previous studies have shown that, as a consequence of increasing temperatures, species have generally shifted poleward and deeper in search for cooler waters [3], while more recent research proposes that biological differences among species lead to heterogeneous range shifts [8]. Along the east coast of the United States, researchers have indicated that species will shift distributions under ocean warming scenarios depending on their preferred thermal habitat [9]. Species with smaller thermal habitats are likely to shift more under ocean warming, as their range of preferred temperatures limits where those species are more likely to survive.

Predicting Future Oceans.
DOI: https://doi.org/10.1016/B978-0-12-817945-1.00011-3

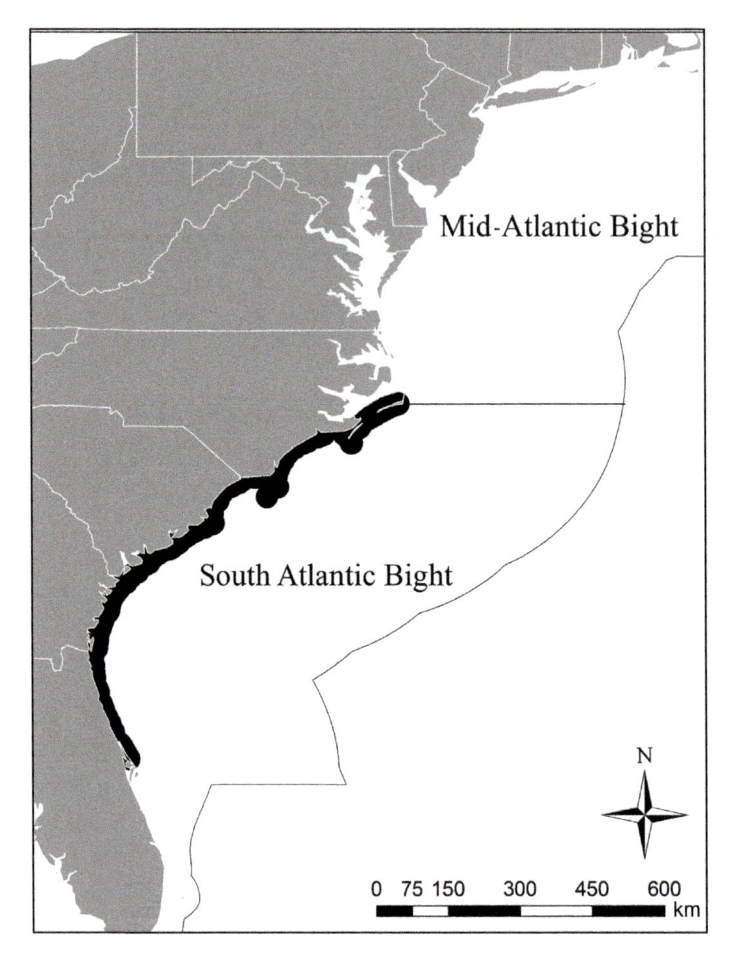

Figure 13.1
Extent of the original trawl surveys from 1990–2014 (black).

In the SAB Morley et al. have identified that winter temperatures are particularly useful for determining interannual variations in species distributions [10]. More comprehensive research by Morley et al. showed that predicted temperature changes in the SAB will cause species to shift towards the Mid-Atlantic in the future [9]. In general, the South Atlantic has experienced relatively week ocean warming compared to the northeast United States, but the area is projected to experience increased ocean warming under plausible emissions scenarios such as Representative Concentration Pathways (RCPs) 8.5 and 2.6 [9].

In conjunction with anthropogenic warming, natural climate variations affecting ocean conditions, such as the Atlantic Multidecadal Oscillation (AMO) and the North Atlantic Oscillation (NAO), have been shown to influence marine species ranges and ecosystem-wide shifts [3,11]. The NAO is a pressure and circulation pattern, which is a measurement of the sea surface pressure differences between the subtropical or Azores High and subpolar or Icelandic Low [12]. The subtropical high is a semipermanent system of high atmospheric pressure over the North Atlantic that influences weather and climatic patterns over North Africa, southern Europe, and the southeastern United States [13]. The subpolar low is a semipermanent system of low pressure between Greenland and Iceland [14]. When the NAO is in its positive phase an intense Icelandic Low, combined with a strong Azores High, leads to increased pressure differences and stronger westerly winds [15]. A positive NAO phase is associated with warmer temperatures in the eastern United States, with the opposite pattern occurring in negative phases [12].

The NAO's influence on the eastern United States oceanography and regional climate has also impacted fish species abundances [15]. For example, temperature fluctuations associated with the NAO have been shown to influence the recruitment of several fish stocks [16], and the growth and survival of fish larvae, as well as altering the size and composition of zooplankton in the Atlantic Basin [17] through temperature fluctuations and changes in ocean currents [18]. Research suggests that the NAO influences wind stress, which can alter gyre circulations [19] and perhaps vary nutrient availability for commercially important fish species. In Narragansett Bay, Rhode Island, a positive NAO phase has been associated with decreases in chlorophyll concentrations with a negative phase associated with increases in chlorophyll concentrations [20]. These fluctuations in chlorophyll could have implications on the ecological community as a whole, through shifts in phytoplankton abundance that influence higher trophic levels. Finally, the NAO has been shown to influence coastal community compositions through changes in prey abundance as a result of changing ocean circulations [21]. In the SAB, the NAO influences temperatures, precipitation, the transport of the North Atlantic current, and the strength of the gyre circulation [21]. Considering the studied influences of climate variations on marine ecology and the impacts of the NAO on the oceanography of the SAB, researchers must understand the combined influence of the NAO as well as anthropogenic warming on species distributions throughout this region.

Roberts et al. [22] recently found that in the SAB, historical commercially harvested groundfish species' presence responses to changing ocean bottom temperatures (BTs) and salinities differ depending on the NAO or winter NAO (WNAO) phase. For certain species, such as Atlantic croaker (*Micropogonias undulatus*), butterfish (*Peprilus triacanthus*), and spot (*Leiostomus xanthurus*), the phase of the NAO (positive or negative) changes the relationship (positive or negative) between the species' historical presence and BT and salinity. For example, butterfish presence during a negative WNAO phase (gray) is negatively associated with BT, and during a positive WNAO phase (black) there is a positive relationship between BT and butterfish presence. This pattern is reversed for salinity, with a negative WNAO phase (gray) associated with a positive relationship between butterfish presence and salinity (Fig. 13.2). For Atlantic croaker, butterfish, and

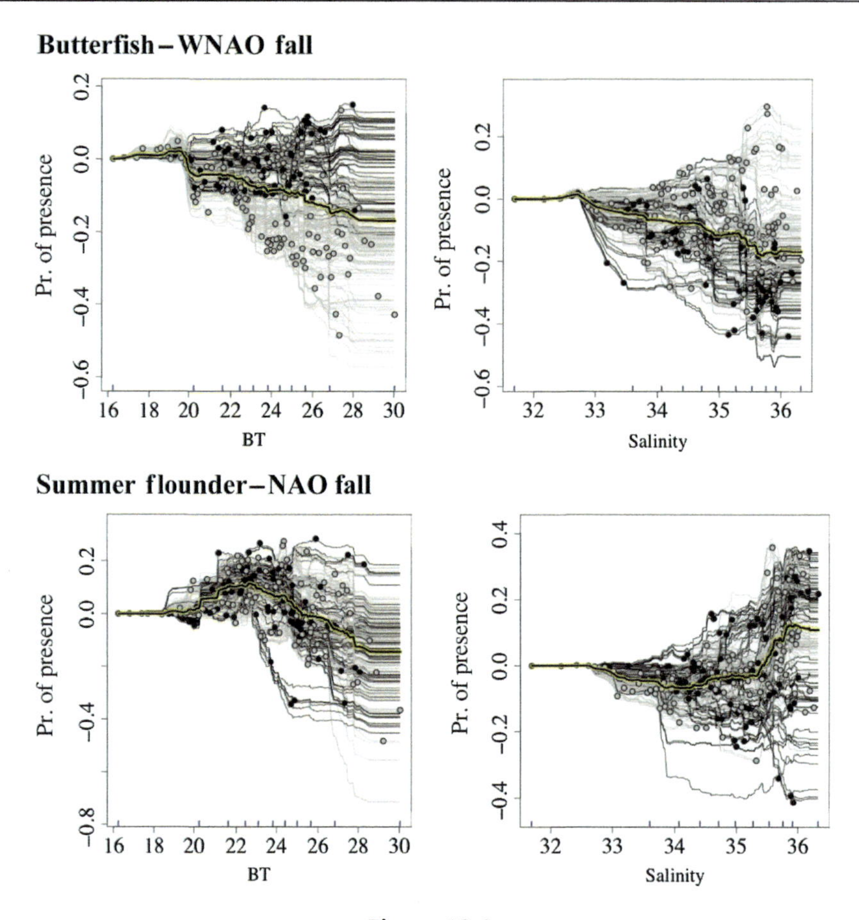

Figure 13.2
Centered individual conditional expectation partial dependence plots depicting salinity and bottom temperature influence on butterfish and summer flounder presence colored by winter North Atlantic Oscillation or average annual North Atlantic Oscillation sign (gray = negative, black = positive) for the fall season. Points represent observations.

spot, the cyclical oscillation has the highest or second-highest relative importance for predicting the presence of that species (as determined by a random forest statistical model). Thus, these species' responses to climate change (increasing ocean temperatures and salinity) may differ according to the NAO phase. In contrast, the presence of summer flounder in the fall is most influenced by BT and salinity. As a result the centered partial dependence plots show little difference in trend depending on WNAO phase (Fig. 13.2), and summer flounder's response to climate change will depend less on the phase of this cyclical oscillation.

These results suggest that the direction that butterfish, for example, will shift under warming will depend on the NAO phase. Importantly, the NAO is not currently predictable beyond 1−2 years, meaning it is not accounted for under climate change models [23]. Thus, the predictability of certain species' shifts may depend on their relationship with cyclical climate oscillations, and species distributions that are more associated with climate oscillations will prove more difficult to predict under a changing climate. Additionally, these results suggest that other processes associated with changes in the NAO but not reflected in the local BTs and salinities appear to have important impacts on species habitat distributions. Such properties might include phytoplankton or zooplankton abundances, currents, wind patterns, or the frequency or intensity of storms wind stress [21]. Two plausible hypotheses are outlined below. First, the influence of the NAO on wind and water circulation patterns may affect larval advection or predator abundance which may influence the relationship between BT and species distribution in different phases of the NAO. In other work, researchers have found that other oscillations, such as the AMO, have affected Atlantic menhaden larval transport and larval ingress in the Northwest Atlantic, but conclude that the underlying mechanistic linkages between the AMO and larval recruitment dynamics remain obscure and may be a combination of interacting processes [24]. Perhaps the NAO's influence on water circulation and oceanographic processes is manipulating the larval transport of pelagic species. The second hypothesis relates to the NAO's influence on habitat quality. Nutrient supply and primary production differ between one phase of the NAO to the next [25], which may affect the statistical relationship between BT and butterfish presence by influencing prey abundance. For example, the productivity of oceanic habitats may be influenced by NAO phases as well as temperature which could have cascading influences on larval and juvenile survival. These results could also suggest that the time and spatial scale of environmental variability associated with the NAO may influence butterfish's relationship with local BTs and salinities.

Understanding the historic shifts in ranges of commercial fish species, and how these relate to changing environmental conditions and cyclical oscillations could have significant management implications. Management can prioritize fisheries vulnerability to climate change by identifying which species are more influenced by cyclical oscillations compared to anthropogenic warming. Species that respond to local BTs and salinities may shift linearly as a result of ocean warming and salinity changes. It will be harder to predict future distributions under climate change for species that are more influenced by the NAO, as the NAO is currently not predictable. To complicate things further, the association of changes in distribution with large-scale climate variables such as the NAO may suggest that distribution shifts will persist, as positive NAO phases may be more frequent with tropospheric warming [26].

These results, in conjunction with work examining predicted shifts in species distributions under anthropogenic warming, requires further research. Understanding the potential for species to shift distributions and expand and contract ranges could have serious management implications. As species shift out of traditional areas, fishers may be faced with lost access to stocks, and managing bodies may have to reexamine existing, static management strategies. In the case of the Pacific Salmon fishery, a lack of understanding of climate-related influences on stock abundance and migrations led to a breakdown of a cooperative harvesting agreement between the United States and Canada [27]. In Alaska, the North Pacific Fishery Management Council has taken a risk-averse approach to managing their fisheries under a changing climate by closing the Arctic to commercial fishing and developing adaptive management strategies for areas where research has identified the linkage between climate change and shifting fisheries distributions [28]. In the Mid-Atlantic, plans for an ecosystem-based management approach have documented the necessity of including climate change impacts on species and ecosystem distributions, and these efforts are in their preliminary stages [29]. Considering the precautionary lessons in the Pacific Salmon fishery and the recent call for adaptive fisheries management under climate change, the above results present a relevant and necessary study on the relative role of the NAO on species presences along the Mid-Atlantic Bight and SAB. Although modeling future projections encourages uncertainty, steps must be taken to hypothesize future distribution shifts before they happen, giving management adequate time to respond.

References

[1] S. Levitus, J.I. Antonov, T.P. Boyer, O.K. Baranova, H.E. Garcia, R.A. Locarnini, et al., World ocean heat content and thermosteric sea level change (0–2000 m), 1955–2010, Geophys. Res. Lett. 39 (10) (2012) L10603. n/a-n/a.

[2] W.W.L. Cheung, V.W.Y. Lam, J.L. Sarmiento, K. Kearney, R. Watson, D. Pauly, Projecting global marine biodiversity impacts under climate change scenarios, Fish Fish. 10 (3) (2009) 235–251.

[3] J.A. Nye, J.S. Link, J.A. Hare, W.J. Overholtz, Changing spatial distribution of fish stocks in relation to climate and population size on the Northeast United States continental shelf, Mar. Ecol. Prog. Ser. 393 (2009) 111–129.

[4] A.L. Perry, P.J. Low, J.R. Ellis, J.D. Reynolds, Climate change and distribution shifts in marine fishes, Science 308 (2005) 1912.

[5] M.L. Pinsky, B. Worm, M.J. Fogarty, J.L. Sarmiento, S.A. Levin, Marine taxa track local climate velocities, Science 341 (6151) (2013) 1239–1242.

[6] J.L. Blanchard, A rewired food web, Nature 527 (2015) 173.

[7] A.E. Magurran, M. Dornelas, F. Moyes, N.J. Gotelli, B. McGill, Rapid biotic homogenization of marine fish assemblages, Nat. Commun. 6 (2015) 8405.

[8] A.L. Angert, L.G. Crozier, L.J. Rissler, S.E. Gilman, J.J. Tewksbury, A.J. Chunco, Do species' traits predict recent shifts at expanding range edges? Ecol. Lett. 14 (7) (2011) 677–689.

[9] J.W. Morley, R.L. Selden, R.J. Latour, T.L. Frölicher, R.J. Seagraves, M.L. Pinsky, Projecting shifts in thermal habitat for 686 species on the North American continental shelf, PLoS One 13 (5) (2018) e0196127.

[10] J.W. Morley, R.D. Batt, M.L. Pinsky, Marine assemblages respond rapidly to winter climate variability, Global Change Biol. 23 (7) (2017) 2590–2601.

[11] L.C. Stige, G. Ottersen, K. Brander, K.-S. Chan, N.C. Stenseth, Cod and climate: effect of the North Atlantic Oscillation on recruitment in the North Atlantic, Mar. Ecol. Prog. Ser. 325 (2006) 227–241.

[12] J.W. Hurrell, Decadal trends in the North Atlantic oscillation: regional temperatures and precipitation, Science 269 (5224) (1995) 676–679.

[13] W. Li, L. Li, R. Fu, Y. Deng, H. Wang, Changes to the North Atlantic subtropical high and its role in the intensification of summer rainfall variability in the Southeastern United States, J. Clim. 24 (5) (2010) 1499–1506.

[14] M.K. Flatau, L. Talley, P.P. Niiler, The North Atlantic oscillation, surface current velocities, and SST changes in the subpolar North Atlantic, J. Clim. 16 (14) (2003) 2355–2369.

[15] G. Ottersen, B. Planque, A. Belgrano, E. Post, P.C. Reid, N.C. Stenseth, Ecological effects of the North Atlantic oscillation, Oecologia 128 (1) (2001) 1–14.

[16] J. Santiago, The North Atlantic oscillation and recruitment of temperate tunas, ICCAT Col. Vol. Sci. Pap. 48 (3) (1998) 240–249.

[17] K.M. Brander, Cod recruitment is strongly affected by climate when stock biomass is low, ICES J. Mar. Sci. 62 (3) (2005) 339–343.

[18] R.A. Myers, K. Drinkwater, The influence of Gulf Stream warm core rings on recruitment of fish in the northwest Atlantic, J. Mar. Res. 47 (3) (1989) 635–656.

[19] S. Häkkinen, P.B. Rhines, Decline of subpolar North Atlantic circulation during the 1990s, Science 304 (5670) (2004) 555–559.

[20] Y. Li, T.J. Smayda, A chlorophyll time series for Narragansett Bay: assessment of the potential effect of tidal phase on measurement, Estuaries 24 (3) (2001) 328–336.

[21] G. Ottersen, N.C. Stenseth, Marine Ecosystems and Climate Variation: The North Atlantic: A Comparative Perspective, Oxford University Press, Incorporated, New York, 2005.

[22] S. Roberts, A. Boustany, P. Halpin, R. Rykaczewski, Cyclical climate oscillation alters species statistical relationships with local habitat, Mar. Ecol. Prog. Ser. 614 (2019) 159–171. Available from: https://doi.org/10.3354/meps12890.

[23] J.W. Hurrell, Y. Kushnir, G. Ottersen, M. Visbeck, An overview of the North Atlantic oscillation, The North Atlantic Oscillation: Climatic Significance and Environmental Impact, American Geophysical Union, 2013, pp. 1–35.

[24] A. Buchheister, T.J. Miller, E.D. Houde, D.H. Secor, R.J. Latour, Spatial and temporal dynamics of Atlantic menhaden (*Brevoortia tyrannus*) recruitment in the Northwest Atlantic Ocean, ICES J. Mar. Sci. 73 (4) (2016) 1147–1159.

[25] A. Oschlies, NAO-induced long-term changes in nutrient supply to the surface waters of the North Atlantic, Geophys. Res. Lett. 28 (9) (2001) 1751–1754.

[26] D. Rind, J. Perlwitz, P. Lonergan, AO/NAO response to climate change: 1. Respective influences of stratospheric and tropospheric climate changes, J. Geophys. Res. Atmos. 110 (D12) (2005) D12107–D12115.

[27] K.A. Miller, G.R. Munro, Climate and cooperation: a new perspective on the management of shared fish stocks, Mar. Resour. Econ. 19 (3) (2004) 367–393.

[28] D.L. Stram, D.C.K. Evans, Fishery management responses to climate change in the North Pacific, ICES J. Mar. Sci. 66 (7) (2009) 1633–1639.

[29] MAFMC, Ecosystem approach to fisheries management guidance document, Mid-Atlantic Fisheries Management 800 North State Street, Suite 201, Dover, DE 19901 <https://static1.squarespace.com/static/511cdc7fe4b00307a2628ac6/t/589a2b61d2b8575c64fe05ff/1486498674225/EAFM_Guidance + Doc_2017-02-07.pdf>, 2016.

CHAPTER 14

Jellyfishes in a changing ocean

Natasha Henschke

Department of Earth, Ocean and Atmospheric Sciences, University of British Columbia, Vancouver, BC, Canada

Chapter Outline

14.1 Review of current knowledge

Jellyfish (cnidarians and ctenophores; Fig. 14.1) form dense blooms that are a natural feature of healthy pelagic ecosystems [1]. They are a food source for over 150 species of fish, seabirds, and other marine animals, with 11 fish species and the endangered leatherback turtle, *Dermochelys coriacea*, exclusively feeding on jellyfish [2,3]. Yet, due to the rapid digestion of jellyfish tissue, levels of predation on jellyfish are likely to be underestimated [4]. Jellyfish also provide a predation refuge for some juvenile fish species, significantly increasing their survival [5]. Humans have been consuming jellyfish for more than 1700 years [6], with jellyfish fisheries occurring in at least 18 countries and in excess of 1 million tonnes/year caught globally [7]. Apart from human consumption, jellyfish and their by-products are used in a variety of industries including agriculture, cosmetics, and pharmaceuticals [8].

Despite the widespread benefits of jellyfish, the negative impacts of jellyfish blooms have led to a human perception of jellyfish as a nuisance species. Deleterious consequences of jellyfish blooms include losses in tourist revenue at beaches [9], reduction in commercial fish abundance through competition and/or predation, and losses in fisheries revenue due to burst fishing nets and contaminated catches [10]. In extreme cases, blooms of the giant

Predicting Future Oceans.
DOI: https://doi.org/10.1016/B978-0-12-817945-1.00013-7

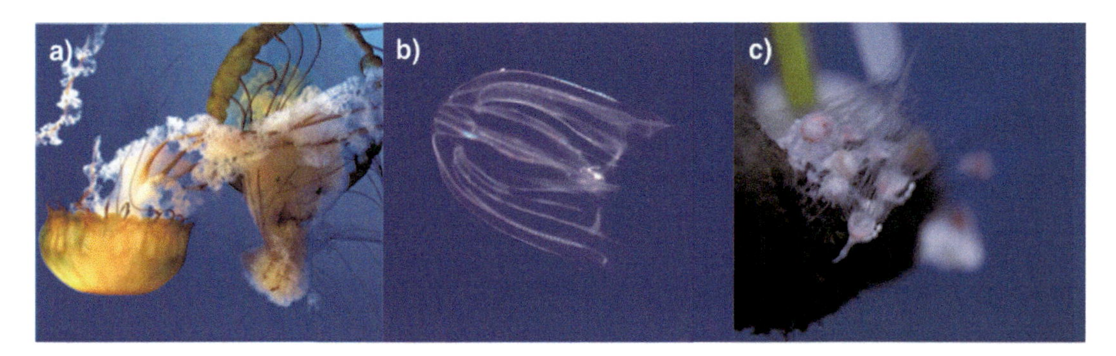

Figure 14.1
(A) *Chrysaora fuscescens*, the sea nettle, is a scyphozoan jellyfish from the phylum Cnidaria. (B) Ctenophores, such as this *Mnemiopsis leidyi*, are typically smaller than cnidarians and can be identified by their rows of cilia. (C) Scyphozoan jellyfish polyps are sessile and hard to locate due to their small size. These *Cassiopea* sp. polyps are ~5 mm in length. *Reproduced with permission from (A) Sam DeLong, (B) Maritime Aquarium at Norwalk, and (C) Dan Bowes.*

jellyfish, *Nemopilema nomurai*, in the Sea of Japan resulted in a $250 million loss of revenue in 2005 and a 10-tonne fishing vessel capsizing in 2009 [11].

Fossil evidence suggests jellyfish have existed since at least 540–500 Mya [12] and have prevailed in oceans since, likely due to a suite of attributes (see below) that make them efficient competitors and survivors in harsh and/or rapidly changing environments. These attributes are inferred to have favored jellyfish survival during periods of low or fluctuating resources and allow for rapid population increase through asexual reproduction. Their resilience to harsh environments, combined with more reported sightings of jellyfish blooms, has led to the perception that jellyfish bloom frequency and magnitudes are increasing worldwide as a result of anthropogenic changes such as overfishing, eutrophication, coastal development, and climate change [13]. Throughout this chapter we discuss the theories behind these perceived increases in jellyfish blooms, present a population modeling approach to assess mechanistic drivers of jellyfish blooms, and explore avenues for future research.

Scyphozoan jellyfish (true jellyfish, hereafter "jellyfish") dominate pelagic cnidarian biomass, are predominantly coastal, and thus make up the majority of jellyfish encountered by humans [14]. They are well documented for their ability to form blooms, as a result of their life cycle that alternates between a sexual medusa stage and an asexual benthic polyp stage [15], with the pelagic medusa stage the one observed by humans (Fig. 14.1). Polyps are a perennial stage that can withstand months of low food conditions [16], allowing jellyfish populations to survive even when recruitment to the medusa stage fails [17]. Despite their importance, very little is known about the polyp stage as they are very small (<5 mm) [18] and are difficult to locate in situ.

Jellyfish have great plasticity in key traits such as feeding and physiology. Experimental work has shown that jellyfish feeding, growth, and reproductive rates will increase with increasing temperatures, food availability (zooplankton biomass), and light exposure [19–21]; temperature has been shown to be a key predictor of the timing and distribution of several jellyfish species in situ (e.g., Refs. [14,22]). They are efficient nonvisual feeders, and as they have similar prey clearance rates as fish [23], they will experience competition release via the overfishing of planktivorous fish [24], and can outcompete fish in highly turbid, eutrophic environments [25]. Overfishing in the Namibian Benguela has resulted in jellyfish biomass now exceeding the biomass of the once abundant sardines (*Sardinops sagax*) and anchovies (*Engraulis encrasicolus*) [10]. Similarly, overfishing in the Black Sea combined with eutrophication and the introduction and spread of the ctenophore, *Mnemiopsis leidyi*, led to a regime shift from a fish-dominated to a ctenophore-dominated ecosystem [24]. Jellyfish can also survive and reproduce in hypoxic conditions, and frequently bloom in areas that experience prolonged hypoxia such as Hiroshima Bay [26] and the Gulf of Mexico [27]. As hypoxic water masses expand under climate change and increased eutrophication, fish populations will experience habitat loss, while jellyfish populations will at least be unaffected and at best will benefit.

The rise of artificial structures through coastal development will increase settlement areas for benthic polyps [28]. Experimentally, polyps have been found to preferentially settle on artificial substrates and field surveys confirm the presence of polyps on artificial substrates such as piers, decks, and marinas [28]. As medusa bloom magnitude is heavily reliant on the success of the polyp population, increasing settlement areas may assist increases in jellyfish blooms in adjacent waters. Correspondingly, the removal of aquaculture rafts, a substrate for polyp attachment, coincided with a decline in medusa densities in a coastal lagoon in Taiwan [29].

Despite the potential for jellyfish populations to benefit and increase in biomass under anthropogenic stressors, it has been difficult to substantiate the hypothesis that jellyfish populations are increasing globally using observations alone, despite unequivocal increases in some areas, such as the Black Sea [30]. This is because jellyfish are difficult to sample accurately with traditional net types as they have fragile bodies and are often damaged when captured [31]; they were historically not recorded, or ignored when sampled [32]. Hence current datasets have limited spatial and temporal coverage, and only consider medusa biomass as in situ observations of polyp communities are rare and difficult to obtain. As jellyfish bloom magnitude is strongly linked to the success and reproductive effort of the polyp community, in order to accurately forecast and predict trends in jellyfish blooms both life history stages need to be considered.

14.2 Modeling jellyfish populations

Population modeling augments existing observational data as it allows us to understand the mechanisms that underlie patterns in the data, interpolate between sparse observations, and—to some degree—explore future scenarios. To more comprehensively understand the drivers of jellyfish blooms and build predictive capacity we have developed a size-structured population model based on the ubiquitous and highly studied scyphozoan jellyfish, *Aurelia* spp. (moon jellyfish) [33], that can be broadly representative of a "general" scyphozoan jellyfish. This model tracks cohorts of individuals from both benthic and pelagic life history stages and uses temperature and/or consumption driven relationships for growth, reproduction, and mortality rates. It is the first size-structured scyphozoan population model that incorporates both life history stages with consumption requirements for somatic growth, reproduction, and mortality. Hence this model can explore population dynamics during the development and demise of a jellyfish bloom across varying environmental conditions and quantitatively test bloom initiation and development hypotheses from the polyp community. Using a combination of this model approach that allows an exploration of polyp dynamics with traditional data-centric approaches will greatly improve how we understand jellyfish bloom dynamics.

14.2.1 Case study: Gulf of Mexico

The Gulf of Mexico is a productive subtropical ecosystem that yields 20% of total US fisheries landings and supports the second largest fishery by volume in North America, the planktivorous Gulf menhaden, *Brevoortia patronus* [34]. It has high biological diversity including among the greatest diversity of pelagic cnidarians in the world with over 115 epipelagic species [27,35]. Forage fish, such as the Gulf menhaden, and jellyfish often overlap in diet, space, and time, yet their contributions to the ecosystem are uneven, as Gulf menhaden have a 10-fold higher reach to footprint ratio than large jellyfish in the Gulf of Mexico [36]. An organism's footprint refers to the amount of energy they consume, and their reach is how much energy they provide to their predators. Jellyfish are voracious feeders, yet are nutritionally poor due to their high water and low carbon content [37] and have few predators in comparison to forage fish. Thus the higher reach to footprint ratios for Gulf menhaden indicates they transfer a greater fraction of energy upward in the food chain to higher trophic levels, whereas jellyfish are akin to a "trophic cul-de-sac," channeling consumed energy away from higher trophic levels. As commercial and recreational fishers target forage fish in most regions over jellyfish, and as both are heavily dependent on primary and secondary production cycles, overfishing of forage fish, or other anthropogenic factors that would benefit jellyfish could result in potential shifts in food web dynamics. Given the diversity and importance of jellyfish in the Gulf of Mexico, understanding the factors influencing bloom dynamics are crucial to better predict future ecological changes.

The moon jellyfish, *Aurelia* spp., blooms regularly in the northern Gulf of Mexico during autumn (October–November) [38]. The medusa stage has been sampled yearly since 1982

[38] making it a valuable case study for model comparison and one of the most high-quality, long-term quantitative jellyfish datasets. During years when there was a cooler than average spring and warmer than average summer and autumn, Robinson and Graham [38] observed higher abundances of jellyfish, leading them to suggest that environmental factors that were affecting polyp and ephyrae production determined the resulting medusa biomass. However, as they had no data on polyp or ephyrae abundance they could not confirm this hypothesis.

Our jellyfish model was forced with local temperature and zooplankton biomass data that were sampled in conjunction with the jellyfish trawls. Temperature influences growth, reproduction, and feeding requirements for the jellyfish, and zooplankton biomass is incorporated into the model as a measure of food supply. Temperature and zooplankton biomass were sufficient drivers to recreate the observed seasonal and interannual population dynamics of *Aurelia* spp. in the Gulf of Mexico during 1982–2007 ($R^2 = 0.7$, $P < .001$; Fig. 14.2A) [33]. Based on model results, several factors compounding across scyphozoan life stages resulted in years with higher medusa biomass, which are discussed below.

An in-depth analysis of population dynamics revealed that higher zooplankton biomass prior to and at the onset of when polyps begin strobilating resulted in higher polyp ingestion rates, allowing for more carbon to be stored for ephyrae production and higher numbers of ephyrae being released. When these optimal environmental conditions continued after the release of ephyrae, the ephyrae were able to ingest more food and thus grow faster, resulting in larger and more successful jellyfish blooms. The higher medusa biomass observed during the 1990s in the Gulf of Mexico was likely a result of earlier polyp reproduction under more favorable environmental conditions, consistent with the hypothesis of Robinson and Graham [38]. Understanding polyp and ephyrae dynamics are crucial as they regulate the success of medusa populations, and need to be integrated into future analyses of jellyfish bloom dynamics.

14.2.2 Case study: global trends in jellyfish blooms

The Gulf of Mexico case study allowed for an in-depth examination of the different mechanisms that drive bloom timing and magnitude. This model was then modified to represent a "generalized" scyphozoan population and run at a global scale to explore global trends in jellyfish blooms. Two meta-analyses have been undertaken exploring trends in global jellyfish observations [32,39] and we use these results to assess our generalized scyphozoan population model. Both meta-analyses used different methods: Condon et al. [39] used only abundance data, whereas Brotz et al. [32] combined abundance data with additional anecdotal sources of data (media articles, interviews, or scientific reports) using a fuzzy logic framework to increase their spatial and temporal coverage. Brotz et al. [32] suggest there have been increases in jellyfish biomass in 62% of the large marine

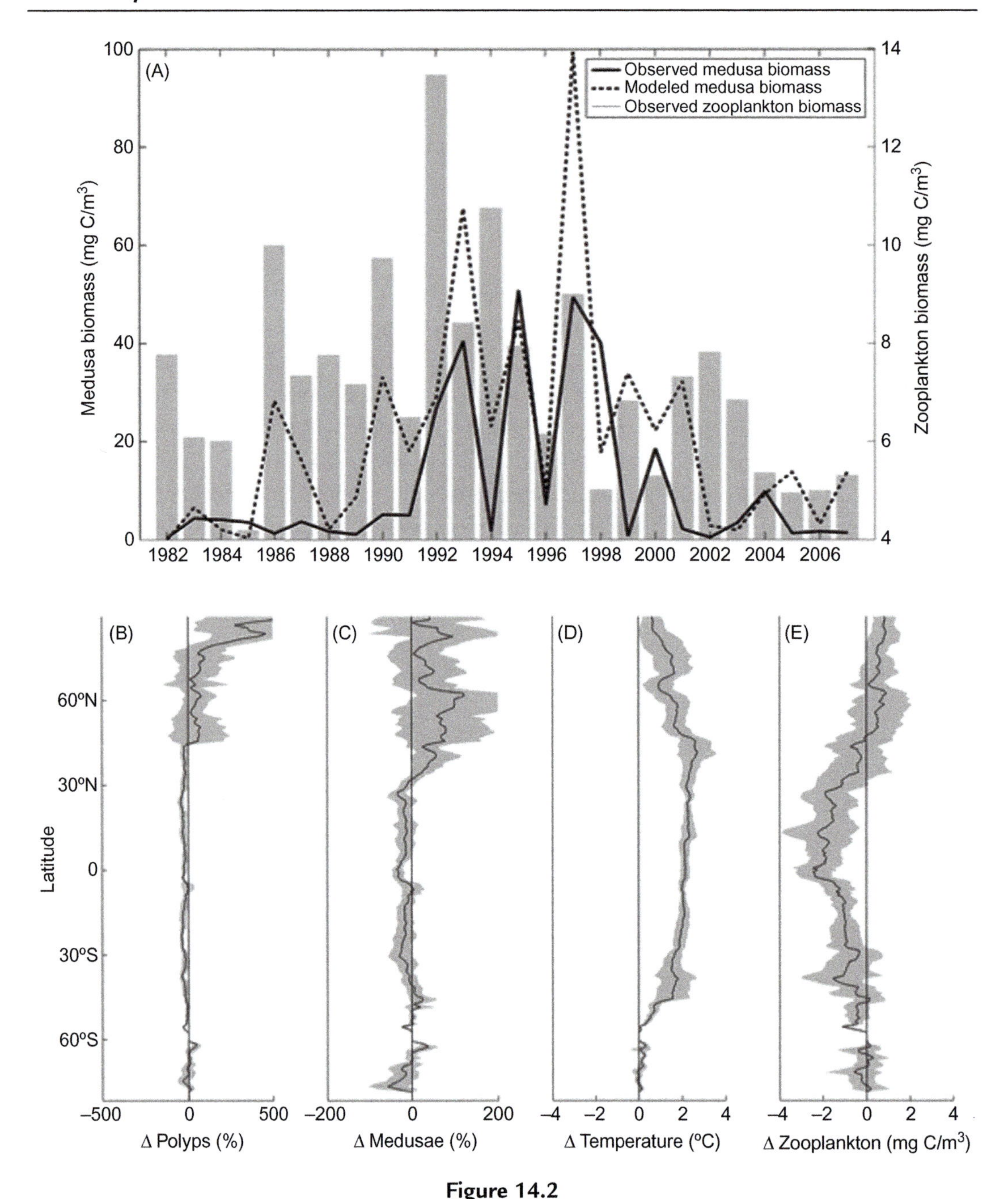

Figure 14.2
(A) Interannual observed medusa biomass (*solid line*), modeled medusa biomass (*dashed line*), and observed zooplankton biomass (*gray columns*) in the Gulf of Mexico. Projected differences in global (B) polyp abundance, (C) medusa biomass, (D) sea surface temperature, and (E) zooplankton biomass from the current period (1951–2000) to the future period (2051–2100) under Representative Concentration Pathway (RCP) 8.5. Values are latitudinal averages (*solid lines*) and standard deviation (*shading*).

ecosystems (LMEs) that were assessed, whereas Condon et al. [39] identified similar areas of increase, yet concluded based on their aggregate index that these increases cannot be substantiated without a longer dataset, instead suggesting that the perceived increases are part of a natural 20-year periodicity. The scyphozoan jellyfish population model was forced with a global mechanistic plankton food web model [40] before performing the same statistical comparisons as in Brotz et al. [32] and Condon et al. [39].

As this work is ongoing, the following results are preliminary. When run at a global scale, spatial and temporal jellyfish bloom dynamics were mostly driven by variations in food supply, whereas the influence of temperature varied regionally. Our model was able to capture the trends observed in the Condon et al. [39] global jellyfish assessment ($r = 0.84$, $P < .001$) and could replicate location-specific trends. The model also matched the long-term linear trends identified by Brotz et al. [32] in the LMEs, where trends are a result of the jellyfish populations responding to natural climatic variability [41]. However, our model could not match observations in LMEs where overfishing, translocation, or eutrophication are believed to be causal links driving changes in the local jellyfish populations [10,41,42]. As the only drivers included in this model are climatic (temperature and zooplankton biomass), it was not expected that our model would be able to recreate the perceived increases in those areas. This highlights the need to incorporate other anthropogenic drivers in future model simulations.

While our model could recreate some of the observed trends from the two meta-analyses, it is important to consider that these datasets are limited in spatial and temporal coverage. The majority of quantitative data is centered in the Northern Hemisphere (87%) and particularly the North Atlantic and Mediterranean (~60%) and from the late 1980s to early 2000s [32,39]. Running our model to include all LMEs from 1948 to 2007, the historical long-term trend in jellyfish biomass suggests there have been no significant increases or decreases in the majority of LMEs (35 of 66) as a result of climatic drivers alone. Significant increases were found only in 10 LMEs, whereas decreases were observed in 21 LMEs. While there are no data to assess our model's ability to predict these trends, and it is likely that other anthropogenic factors that our model has not considered have played a role in several regions, our results indicate that time frames and spatial extent are important when considering whether "global" increases in jellyfish biomass have occurred. This is critical as mis-citation practices in the absence of supporting analysis have caused this theory to be taken as established fact in the media and scientific reporting [43].

Projecting our jellyfish model to 2100 under the high emissions RCP 8.5 scenario [44] suggests that in the absence of other anthropogenic stressors, globally there will be a 7% ± 0.6% increase in medusa biomass. However, our model indicates that dramatic changes in jellyfish populations (> 100% increase) are likely to occur in shallow shelf regions north of 45°N which correspond to latitudes with increases in zooplankton biomass (Fig. 14.2B and E).

As temperature increased across nearly all latitudes this suggests that in the absence of increased zooplankton biomass to support the demands of larger medusa populations, increased temperature alone will not result in more jellyfish. While variations in temperature and zooplankton biomass were sufficient to recreate several trends in jellyfish bloom dynamics, our model missed some localized responses that were due to other anthropogenic factors. To make accurate predictions on how jellyfish blooms are likely to respond in the future, we need to include all potential anthropogenic factors in future simulations, especially those that influence the jellyfish food supply, such as eutrophication (e.g., East China Sea) and overfishing (e.g., Benguela Current, Black Sea).

14.3 Questions, guidance, and directions for future jellyfish research

Whether our oceans are becoming more gelatinous is still up for debate. Are we witnessing a peak in the natural periodicity of jellyfish blooms? Or will our future oceans resemble the renowned fisheries scientist Daniel Pauly's prediction that "My kids will tell their children: eat your jellyfish!" [45], presumably munching on "jellyfish burgers" [46]? Based on our modeling efforts, we suggest that jellyfish blooms are not likely to increase unless there are enough food resources to sustain their populations. Regardless, this debate has increased awareness into a previously understudied field, and we now have a better understanding through observational and modeling studies of the drivers that influence jellyfish bloom dynamics. However, uncertainty still lies in the diversity and interaction of potential drivers, the diversity of jellyfish species, and how to best manage jellyfish populations if they do arise in greater numbers. Jellyfish population models, such as ours, can be used as a framework for developing regional specific predictive models that can be used by managers or stakeholders. To improve predictive capabilities of this and other models, and in order to make informed management decisions, future research should prioritize the issues we have highlighted below.

14.3.1 Increased observational efforts for polyps and medusae

There are no global estimates of polyp abundance or distribution, and to date, there has only been one comprehensive attempt at polyp mapping undertaken on the Korean coastline [47]. While polyp dynamics have been extensively tested in the laboratory, there is a lack of information on the drivers of polyp reproduction in situ which makes it difficult to predict the timing and magnitude of jellyfish blooms. However, polyp mapping is an expensive and timely task as it takes repeated dives by humans or underwater cameras—and this is provided the polyp habitats have first been located. Correspondingly, the lack of medusa observations also makes it difficult to understand drivers and trends in jellyfish blooms. As jellyfish are highly responsive to environmental change, jellyfish monitoring can provide an indicator of ecosystem health.

Recently a cost-effective protocol was developed that can be applied to any trawl-based fishery to sample gelatinous zooplankton, and has been successfully tested in the English Channel and the North Sea [48]. This protocol involves a standardized method for identification and enumeration of gelatinous zooplankton and can be applied with limited training, allowing for the comparison of data across surveys and countries. Additions, such as these that would increase incentives to monitor jellyfish are necessary if we are to include them in ecosystem-based management decisions.

14.3.2 Managing jellyfish populations through fishing or removal

Jellyfish fisheries have existed for hundreds of years, yet in the past decade, demand for jellyfish fisheries has increased in areas where jellyfish are abundant such as Mexico and South Carolina [49]. For example, in 2013 the fishery for the cannonball jellyfish, *Stomolophus meleagris*, became so profitable in Mexico that there were over 1000 boats fishing for jellyfish and the season lasted only 5 days [50]. While establishing jellyfish fisheries provides one way of managing jellyfish biomass, they are not always cost-effective or reliable. Only about 20 of the 1400 species of jellyfish worldwide are preferred for consumption, processing jellyfish is labor intensive and the size of the jellyfish stock can be unpredictable from year to year [49]. To implement any management recommendations we first need to determine what we should manage: the polyps or the medusae [51]? Second, as the combination and scale of drivers of jellyfish blooms vary regionally, and jellyfish can rapidly adapt to changing environmental conditions, management decisions need to be both flexible and localized.

Other methods of controlling jellyfish populations that have been implemented include wire nets that can destroy medusae like an egg slicer [52], a robot that can detect and destroy up to 900 kg of jellyfish biomass per hour [53], and direct removal of polyps through scraping or high-pressure hoses [52]. Polyp removal has been found to be more cost-effective than medusa removal; however, it is not a viable long-term solution as recruitment from remaining polyps, polyps in interconnected areas, or even stray medusae can help recover the jellyfish population [52].

14.3.3 Understanding their trophic role

One compelling reason to avoid complete removal of jellyfish populations is that we do not know enough about their role in the marine ecosystems, and what the effects of removing them are [51]. Jellyfish have some specialist predators and important trophic roles; leatherback turtles swim from Indonesia to California over 2 years to consume jellyfish that bloom off the Californian coast [54] and seabirds track jellyfish swarms to identify where patches of forage fish are located [55]. Although we know that they do not have as large a reach to footprint ratio as forage fish, there have been no assessments yet on how higher trophic predators have adapted to increases in jellyfish populations. Recently, DNA analysis on penguin and albatross scats identified that jellyfish are an important, and previously unknown, component of seabird diets [56,57], indicating the potential for jellyfish to have a higher reach than previously believed. Future studies using techniques designed for gelatinous materials, such as DNA analysis, to determine the role of jellyfish in diets are necessary to determine how ecosystems may respond to increases in jellyfish biomass—are predators consuming jellyfish in the absence of other preferred prey, or do they make up an important component of their diets as they occur in large aggregations and are relatively easy to catch?

With changing ocean dynamics and increased use of ocean habitats by humans, there is a need for continued monitoring of jellyfish blooms and an improved effort to understand their trophic role. While regional increases have occurred, it is important that we do not sensationalize jellyfish blooms by overstating occurrences in the absence of data. However, what the data does indicate is that jellyfish can no longer be ignored, and it is time for jellyfish and other gelatinous zooplankton to be recognized as key members of marine ecosystems.

References

[1] W.M. Graham, F. Pagès, W.M. Hamner, A physical context for gelatinous zooplankton aggregations: a review, Hydrobiologia 451 (1) (2001) 199–212.

[2] D. Pauly, W.M. Graham, S. Libralato, L. Morissette, M.L. Deng Palomares, Jellyfish in ecosystems, online databases, and ecosystem models, Hydrobiologia 616 (2009) 67–85.

[3] J.D.R. Houghton, T.K. Doyle, M.W. Wilson, J. Davenport, G.C. Hays, Jellyfish aggregations and leatherback turtle foraging patterns in a temperate coastal environment, Ecology 87 (8) (2006) 1967–1972.

[4] G.C. Hays, T.K. Doyle, J.D.R. Houghton, A paradigm shift in the trophic importance of jellyfish? Trends Ecol. Evol. 33 (11) (2018) 874–884.

[5] C.P. Lynam, A.S. Brierley, Enhanced survival of 0-group gadoid fish under jellyfish umbrellas, Mar. Biol. 150 (2007) 1397–1401.

[6] M. Omori, E. Nakano, Jellyfish fisheries in southeast Asia, Hydrobiologia 451 (2001) 19–26.

[7] L. Brotz, Jellyfish fisheries—a global assessment, in: D. Pauly, D. Zeller (Eds.), Global Atlas of Marine Fisheries: A Critical Appraisal of Catches and Ecosystem Impacts, Island Press, Washington, DC, 2016.

[8] L. Brotz, D. Pauly, Studying jellyfish fisheries: toward accurate national catch reports and appropriate methods for stock assessments, in: G.L. Mariottini (Ed.), Jellyfish: Ecology, Distribution Patterns and Human Interactions Hauppauge, Nova Publishers, New York, 2017.

[9] J.E. Purcell, S. Uye, W. Lo, Anthropogenic causes of jellyfish blooms and their direct consequences for humans: a review, Mar. Ecol. Prog. Ser. 350 (2007) 153–174.

[10] C.P. Lynam, M.J. Gibbons, Br.E. Axelsen, C.A.J. Sparks, J. Coetzee, B.G. Heywood, et al., Jellyfish overtake fish in a heavily fished ecosystem, Curr. Biol. 16 (13) (2006) R492–R493.

[11] S. Furuya, World worries as jellyfish swarms swell, Nikkei Asian Rev. (2015). Available from: https://asia.nikkei.com/Business/Science/World-worries-as-jellyfish-swarms-swell2.

[12] P. Cartwright, S.L. Halgedahl, J.R. Hendricks, R.D. Jarrad, A.C. Marques, A.G. Collins, et al., Exceptionally preserved jellyfishes from the Middle Cambrian, PLoS One 10 (2007) e1121.

[13] A.J. Richardson, A. Bakun, G.C. Hays, M.J. Gibbons, The jellyfish joyride: causes, consequences and management actions, Trends Ecol. Evol. 24 (6) (2009) 312–322.

[14] C.H. Lucas, D.O.B. Jones, C.J. Hollyhead, R.H. Condon, C.M. Duarte, W.M. Graham, et al., Gelatinous zooplankton biomass in the global oceans: geographic variation and environmental drivers, Global Ecol. Biogeogr. 23 (2014) 701–714.

[15] W.M. Hamner, M.N. Dawson, A review and synthesis on the systematics and evolution of jellyfish blooms: advantageous aggregations and adaptive assemblages, Hydrobiologia 616 (2009) 161–191.
[16] C.H. Lucas, W.M. Graham, C. Widmer, Jellyfish life histories: the role of polyps in forming and maintaining scyphomedusa populations, Adv. Mar. Biol. 63 (2012) 33–196.
[17] S. Willcox, N.A. Moltschaniwskyj, C.M. Crawford, Population dynamics of natural colonies of *Aurelia* sp. scyphistomae in Tasmania, Australia, Mar. Biol. 154 (2008) 661–670.
[18] H. Ishii, T. Watanabe, Experimental study of growth and asexual reproduction in *Aurelia aurita* polyps, Sessile Org. 20 (2) (2003) 69–73.
[19] J.E. Purcell, D. Atienza, V. Fuentes, A. Olariaga, U. Tilves, C. Colahan, et al., Temperature effects on asexual reproduction rates of scyphozoan species from the northwest Mediterranean Sea, Hydrobiologia 690 (2012) 169–180.
[20] J.E. Purcell, R.A. Hoover, N.T. Schwarck, Interannual variation of strobilation by the scyphozoan *Aurelia labiata* in relation to polyp density, temperature, salinity, and light conditions *in situ*, Mar. Ecol. Prog. Ser. 375 (2009) 139–149.
[21] C.H. Lucas, Reproduction and life history strategies of the common jellyfish, *Aurelia aurita*, in relation to its ambient environment, Hydrobiologia 451 (2001) 229–246.
[22] M.B. Decker, C.W. Brown, R.R. Hood, J.E. Purcell, T.F. Gross, J.C. Matanoski, et al., Predicting the distribution of the scyphomedusa *Chrysaora quinquecirrha* in Chesapeake Bay, Mar. Ecol. Prog. Ser. 329 (2007) 99–113.
[23] J. Acuña, A. López-Urrutia, S. Colin, Faking giants: the evolution of high prey clearance rates in jellyfishes, Science 333 (2011) 1627–1629.
[24] G.M. Daskalov, A.N. Grishin, S. Rodionov, V. Mihneva, Trophic cascades triggered by overfishing reveal possible mechanisms of ecosystem regime shifts, Proc. Natl. Acad. Sci. U.S.A. 104 (25) (2007) 10518–10523.
[25] D.L. Aksnes, Evidence for visual constraints in large marine fish stocks, Limnol. Oceanogr. 52 (1) (2007) 198–203.
[26] J. Shoji, T. Kudoh, H. Takatsuji, O. Kawaguchi, A. Kasai, Distribution of moon jellyfish *Aurelia aurita* in relation to summer hypoxia in Hiroshima Bay, Seto Inland Sea, Estuarine, Coastal Shelf Sci. 86 (2010) 485–490.
[27] W.M. Graham, Numerical increases and distributional shifts of *Chrysaora quinquecirrha* (Desor) and *Aurelia aurita* (Linné) (Cnidaria: Scyphozoa) in the northern Gulf of Mexico, Hydrobiologia 451 (2001) 97–111.
[28] C.M. Duarte, K.A. Pitt, C.H. Lucas, J.E. Purcell, S. Uye, K.L. Robinson, et al., Is global ocean sprawl a cause of jellyfish blooms? Front. Ecol. Environ. 11 (2) (2013) 91–97.
[29] W.-T. Lo, J.E. Purcell, J.-J. Hung, H.-M. Su, P.-K. Hsu, Enhancement of jellyfish (Aurelia aurita) populations by extensive aquaculture rafts in a coastal lagoon in Taiwan, ICES J. Mar. Sci. 65 (3) (2008) 453–461.
[30] A.E. Kideys, Fall and rise of the Black Sea ecosystem, Science 297 (2002) 1482–1484.
[31] S.H.D. Haddock, A golden age of gelata: past and future research on planktonic ctenophores and cnidarians, Hydrobiologia 530/531 (2004) 549–556.
[32] L. Brotz, W.W.L. Cheung, K. Kleisner, E. Pakhomov, D. Pauly, Increasing jellyfish populations: trends in Large Marine Ecosystems, Hydrobiologia 690 (2012) 3–20.
[33] N. Henschke, C.A. Stock, J.L. Sarmiento, Modeling population dynamics of scyphozoan jellyfish (*Aurelia aurita*) in the Gulf of Mexico, Mar. Ecol. Prog. Ser. 591 (2018) 167–183.
[34] NMFS. Fisheries of the United States, 2016. U.S. Department of Commerce, NOAA Current Fisheries Statistics No. 2016, 2017.
[35] P.J. Phillips, The Pelagic Cnidaria of the Gulf of Mexico [Dissertation], Texas A & M University, College Station, TX, 1972.
[36] K.L. Robinson, J.J. Ruzicka, M.B. Decker, R.D. Brodeur, F.J. Hernandez, J. Quiñones, et al., Jellyfish, forage fish, and the world's major fisheries, Oceanography 27 (4) (2014) 104–115.
[37] C.H. Lucas, K.A. Pitt, J.E. Purcell, M. Lebrato, R.H. Condon, What's in a jellyfish? Proximate and elemental composition and biometric relationships for use in biogeochemical studies, Ecology 92 (8) (2011) 1704.

[38] K.L. Robinson, W.M. Graham, Long-term change in the abundances of northern Gulf of Mexico scyphomedusae *Chrysaora* sp. and *Aurelia* spp. with links to climate variability, Limnol. Oceanogr. 58 (1) (2013) 235–253.

[39] R.H. Condon, C.M. Duarte, K.A. Pitt, K.L. Robinson, C.H. Lucas, K.R. Sutherland, et al., Recurrent jellyfish blooms are a consequence of global oscillations, Proc. Natl. Acad. Sci. U.S.A. 110 (3) (2013) 1000–1005.

[40] C.A. Stock, J.P. Dunne, J.G. John, Global-scale carbon and energy flows through the marine planktonic food web: an analysis with a coupled physical-biological model, Prog. Oceanogr. 120 (2014) 1–28.

[41] L. Brotz, Changing jellyfish populations: trends in Large Marine Ecosystems. Fisheries Centre Research Report, vol. 19, no. 5, University of British Columbia, Vancouver, Canada, 2011.

[42] U. Niermann, *Mnemiopsis leidyi:* distribution and effect on the Black Sea ecosystem during the first years of invasion in comparison with other gelatinous blooms, in: H. Dumont, T. Shiganova, U. Niermann (Eds.), Aquatic Invasions in the Black, Caspian, and Mediterranean Seas, Kluwer Academic Publishers, Dordrecht, The Netherlands, 2004, pp. 3–31.

[43] M. Sanz-Martin, K.A. Pitt, R.H. Condon, C.H. Lucas, C.N. de Santana, C.M. Duarte, Flawed citation practices facilitate the unsubstantiated perception of a global trend toward increased jellyfish blooms, Global Ecol. Biogeogr. 25 (2016) 1039–1049.

[44] C.A. Stock, J.P. Dunne, J.G. John, Drivers of trophic amplification of ocean productivity trends in a changing climate, Biogeosciences 11 (2014) 7125–7135.

[45] K.R. Weiss, A primeval tide of toxins. LA Times (July 30, 2006).

[46] J. Jacquet, D. Beck, Jellyfish burger wins NSF visualization challenge. Available from: <http://scienceblogs.com/guiltyplanet/2010/02/19/jellyfish-burger-wins-nsf-visu/>, 2010.

[47] J. Chae, B. Kim, G. Sung, W.D. Yoon, Y.N. Kim, I. Hwang, et al., Polyp mapping of *Aurelia* sp. 1 in the southern and western coast of Korea. In: Fifth International Jellyfish Bloom Symposium, May 30th–June 3rd, Barcelona, 2016.

[48] A. Aubert, E. Antajan, C.P. Lynam, S. Pitois, A. Pliru, S. Vaz, et al., No more reason for ignoring gelatinous zooplankton in ecosystem assessment and marine management: concrete cost-effective methodology during routine fishery trawl surveys, Mar. Policy 89 (2018) 100–108.

[49] L. Brotz, A. Schiariti, J. Lopez-Martinez, J. Alvarez-Tello, Y.H.P. Hsieh, R.P. Jones, et al., Jellyfish fisheries in the Americas: origin, state of the art, and perspectives on new fishing grounds, Rev. Fish Biol. Fish. 27 (2017) 1–29.

[50] A. Cisneros-Montemayor, L. Brotz, Jellyfishing in Mexico: the burgers are ready, Sea Around Us Proj. Newsl. (81) (2014).

[51] M.J. Gibbons, F. Boero, L. Brotz, We should not assume that fishing jellyfish will solve our jellyfish problem, ICES J. Mar. Sci. 73 (4) (2016) 1012–1018.

[52] W. Yoon, J. Chae, B.-S. Koh, C. Han, Polyp removal of a bloom forming jellyfish, *Aurelia coerulea*, in Korean waters and its value evaluation, Ocean Sci. J. 53 (2018) 499.

[53] Kelly S. The robot jellyfish shredders. BBC (June 2, 2015).

[54] H. Bailey, S.R. Benson, G.L. Shillinger, S.J. Bograd, P.H. Dutton, S.A. Eckert, et al., Identification of distinct movement patterns in Pacific leatherback turtle populations influenced by ocean conditions, Ecol. Appl. 22 (3) (2012) 735–747.

[55] N.N. Sato, N. Kokubun, T. Yamamoto, Y. Watanuki, A.S. Kitaysky, A. Takahashi, The jellyfish buffet: jellyfish enhance seabird foraging opportunities by concentrating prey, Biol. Lett. 11 (8) (2015) 1–5.

[56] J.C. McInnes, L. Emmerson, C. Southwell, C. Faux, S.N. Jarman, Simultaneous DNA-based diet analysis of breeding, non-breeding and chick Adélie penguins. Royal Society Open, Science 3 (1) (2016) 1–9.

[57] J.C. McInnes, R. Alderman, M. Lea, B. Raymond, B.E. Deagle, R.A. Phillips, et al., High occurrence of jellyfish predation by black-browed and Campbell albatross identified by DNA metabarcoding, Mol. Ecol. 26 (18) (2017) 4831–4845.

CHAPTER 15

Understanding variability in marine fisheries: importance of environmental forcing

Fernando González Taboada

Atmospheric and Oceanic Sciences Program, Princeton University, Princeton, NJ, United States

Chapter Outline

15.1 Overview

Marine capture fisheries contribute to ensure food security, support millions of jobs worldwide and the largest trade network for a food commodity, and form part of the cultural heritage of multiple communities around the world [1]. However, reconciling the sustained provision of fisheries goods and services with the recovery and conservation of healthy marine ecosystems represents a major challenge to ensure human well-being in forthcoming decades [2], and remains a matter of continued debate in the scientific literature (e.g., [3]). As the demand for seafood keeps rising due to increasing human population numbers and socioeconomic development [4], fisheries are more often facing unprecedented changes in physical and chemical conditions associated with climate change [5,6]. Together, these challenges pose an "emerging imperative" to update current management practices toward strategies that anticipate and predict the response of marine fisheries and ecosystems to human interventions in a changing environment [7,8].

Predicting Future Oceans.
DOI: https://doi.org/10.1016/B978-0-12-817945-1.00014-9

The essence and the crux of fisheries management lie in the prevailing conflict among the optimization of fisheries profit, the protection of social values and community welfare, and the preservation of marine ecosystem health [9]. The simultaneous balance of socioeconomical and ecological aspects may be feasible through better management practices [10], and might even pave the way to rebuild fisheries and increase revenue in the long term [11,12]. However, the success of any management strategy rests on its ability to set harvest rates that do not impair future sustainability [13]. Such a premise contrasts with mounting evidence highlighting the importance of environmental constraints on determining the overall abundance, risk of collapse, and rebuilding capacity of marine fisheries (e.g., [14–19]).

Most problems in marine fisheries science require the use of quantitative models to analyze and simulate the dynamics of exploited populations under alternative management scenarios [13,20]. A key aspect to develop, train, and validate these models—and the degree of success in such endeavors—lies in the availability of time series providing accurate information about the abundance and structure of target populations, an account of past changes in fishing effort, and a record of variability in environmental stressors of potential relevance. In practice, however, the latter aspect is often ignored due to the prevailing interest in short-term projections and the difficulties associated with the characterization of reliable relationships between changes in environmental conditions and population dynamics [21,22]. As the oceans warm, become more acidic, and lose oxygen (e.g., [23]), there is no alternative but to account for the impact of environmental variability in fisheries management.

In this chapter, we discuss major patterns of variability observed in time series of the abundance, recruitment, and catch records of commercially exploited marine fish stocks [24–27]. We revisit early classifications of fisheries dynamics [28–31] and provide an updated scheme based on a simple, but biologically motivated, time series analysis. Then we succinctly review some of the hypotheses proposed to explain fluctuations in recruitment success and in adult performance [32–34] and the contribution of fishing to enhance natural variability [27,35–37]. We conclude the chapter highlighting available opportunities to extend current models using ancillary environmental information, the need to pursue ecosystem forecasts [38–40], and how probabilistic methods can be used to integrate different sources of information to anticipate impacts in marine ecosystems [41,42].

15.2 How do fisheries vary in time?

Since its inception as a discipline, fisheries oceanography has tried to explain what factors are behind observed fluctuations in the abundance of species of commercial interest [43]. These fluctuations propagate to fisheries catches and yield, and eventually trigger undesired socioeconomic havoc and environmental degradation. Historically, changes in the abundance of species were commonly ascribed to altered environmental conditions and the internal workings of marine ecosystems [43], always beyond human control and responsibility. Paleoecological analyses corroborate the inherent variability of marine ecosystems [44], as well as analyses of modern time series derived from fisheries catch statistics and abundance estimates [24,25]. Until recently the potential role and far-reaching impacts of fishing and overexploitation were simply overlooked [45,46], even if fishing was identified from the beginning as a major structuring force in marine ecosystems (e.g., [47]).

Nowadays fisheries, and the marine ecosystems they are embedded in, are conceptualized as complex systems where the interaction among environmental variation, human interventions, and the dynamics of marine populations manifest in a variety of behaviors, encompassing endless fluctuations that eventually lead to regime shifts and fisheries collapse [27]. The availability of long time series of catch statistics [48] and catch reconstructions [49], and compilations of the output of quantitative stock assessments [26,50], has allowed the examination of the relative incidence of distinct patterns of variability in fisheries. Both types of information are useful, although the latter complements catch statistics with estimates of stock abundance and recruitment dynamics [51]. Overall these data sources have revealed that dramatic changes and collapses are relatively common in marine capture fisheries [27].

For instance, Mullon et al. [16] estimated that around a quarter of the fisheries included in the FAO database collapsed at some point during the second half of the 20th century, with an almost steady rate of $\sim$60 collapses each 5 years across the 1519 stocks examined. More recent analyses combining stock assessment and life history data confirmed these figures and revealed that overfishing and climate variability are usually associated with fisheries collapse, which is by far more frequent among species with fast dynamics [17,52]. Indeed, greater growth rates convey a greater recovery potential and, on average, many stocks seem able to rebuild in less than two decades according to data from well-managed, temperate fisheries around North America [53,54]. On the other hand, the same analyses reveal that $\sim$30% of these populations fail to recover and remain trapped at very low abundance levels. This state of "impaired recovery" suggests a long-lasting impact of fishing on population structure and life history traits that compromises the recovery potential of multiple fisheries [54,55].

On the other hand, analyses of interannual patterns of variability in fish production and recruitment reveal that most stocks alternate high and low-productivity regimes (i.e., $\sim$70% according to [56,57]). When these shifts in production are examined in detail, they tend to be associated with changes in bottom-up forcing affecting the availability and characteristics of nutrient and food resources that fuel the species targeted by the fisheries (e.g., [58–60]). Ultimately, these shifts reflect the impact of changes in atmospheric and oceanic conditions on mixing and transport that affect local production and retention of planktonic resources and propagules. Altered biogeochemical conditions propagate up through marine food webs and can last for several decades, with archetypal examples like the pseudocyclical Russel cycle in the English Channel [61] and the regime shifts in the North Sea [62], or the multidecadal oscillations in the Pacific [14,63].

15.3 A simple classification

Many of the issues mentioned in the previous section were advanced by the pioneering attempts to classify fisheries according to their variability conducted by Kawasaki [28,29] and Caddy and Gulland [30]. Their approach was further elaborated and formalized by Spencer and Collie [31], who classified 30 selected stocks from 23 commercially exploited invertebrate and fish species into six groups with dynamics ranging from steady-state or low-frequency, smooth trends, to highly irregular and spasmodic fisheries with erratic dynamics, through abundance time series showing strong periodicities. They further show that such variety of dynamics can be emulated with a simple model considering a self-limited species suffering density-dependent predation by a generalist predator in an environment with varying degrees of autocorrelation [31,64].

We revisit here such an approach by analyzing time series of the abundance of 270 commercially exploited marine fish stocks from 132 species. Data were extracted from version 3.0 of the RAM Legacy Stock Recruitment database [26] and updated for stocks included in the ICES Stock Assessment database [65]. Although such choices augmented sample size with respect to previous approaches, it is important to note that our sample is still biased toward economically important and well-studied fisheries from northern hemisphere developed countries. Our analysis followed the basic scheme proposed by Spencer and Collie [31] but differed in some important aspects. We worked with log-transformed data and retained the three basic quantitative descriptors or time series probes used to characterize patterns of variability in fisheries, but we updated the methods to include biological constraints and added a fourth index to the set of probes.

We started calculating (1) Heath's [66] proportional variability index (PV), which is a nonparametric index that ranges from 0 for constant time series to 1 for highly nonlinear sequences, as a synthetic measure of population variability (see also [67]); (2) an estimate of the importance of long-term trends based on the amount of variance explained by the slow varying component of a multiresolution smoother [68]; and (3) the scaled period of fluctuations remaining in the residual time series after subtracting the long-term trend (see below). Finally, (4) we added as a fourth probe the Hurst exponent (H), an index of long-range dependence and short-term predictability, which was calculated after removing a linear trend (i.e., H_{dtr}, to avoid confounding due to periods of exponential increase or decrease; [24]). Values of H close to 1/2 correspond to a random walk; the dynamics are reddened with respect to a white noise process (i.e., variance increases with the length of the series) and feature long excursions above and below from the mean, even when there are no actual trends in the data. As H diverges from 1/2 the behavior becomes more predictable, either because the series has long-memory and perturbations persist for longer periods (i.e., $H \rightarrow 1$), or because the series shows a strong tendency to return to its local mean (i.e., it becomes stationary and less correlated, turning into a white noise process as $H \rightarrow 0$).

The calculation of the long-term trends (2) and the scaled period of oscillations (3) incorporated differences in maturation times (a_{mat}) across species, which were extracted from Thorson et al.'s *FishLife* database [69]. The scale of the kernels used to estimate the large-scale component of the long-term trends was fixed to $3a_{mat}$ to ignore variability below a period of $12a_{mat}$, which we considered an upper threshold for the period of oscillations expected from internal dynamics and species interactions alone [70]. Also to ease interpretation and allow comparisons across species, the actual period of oscillations in the residual time series was divided by a_{mat}, since the scaled period (τ_n) provides key information about population dynamics [70,71]. For values of $\tau_n < 2$, the dynamics are described as "single generation or cohort cycles" and suggest the operation of direct density dependence. When $2 < \tau_n < 4$, the cycles correspond to "delayed-feedback dynamics" that reflect interactions between cohorts in age-structured populations. Finally, $\tau_n > 4$ reveals the presence of specialized, tightly coupled "consumer–resource cycles." However, for generalist species weakly coupled to multiple species of prey, which is a motif that abounds in marine food webs, the dynamics collapse to single-species dynamics with $\tau_n < 4$ [70].

The four time series probes were analyzed using model-based clustering (Fig 15.1) [72,73]. Model selection favored a four mixture component model where differences among groups reflected primarily the importance of long-term trends and the overall variability of population abundance and, secondarily, distinctive patterns in the nature of the dynamics captured by the scaled period and the H. Fig. 15.2 presents example time series for each of the four population variability classes identified. The four groups are defined by the following characteristics:

- A—groups stocks with very irregular dynamics, no apparent trend, and erratic fluctuations with periods that are consistent with delayed-feedback dynamics. In some stocks the oscillations are more pronounced and last longer resembling true consumer–resource cycles ($\tau_n > 4$). Group A comprised only 41 stocks from 27 species from temperate, subpolar, and polar regions. Many of them are small pelagics (herrings, sardines, anchovies, and mackerels), although stocks of demersal species are also frequent (haddocks and hakes).

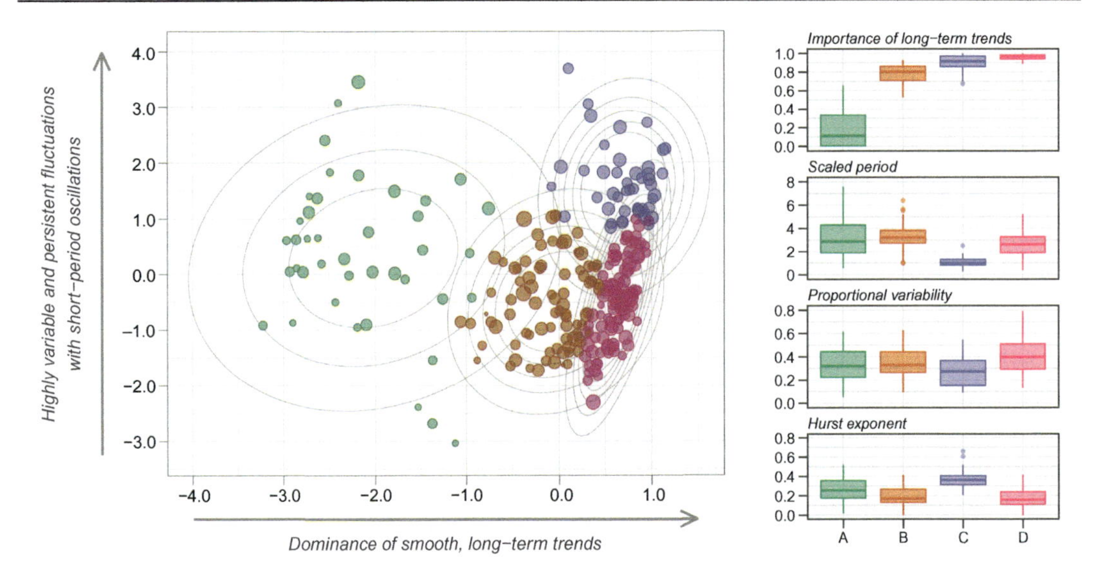

Figure 15.1

Summary of the classification of fish stocks according to the variability in time of their biomass. The classifier considered four different time series probes: (A) Heath's proportional variability index as a synthetic measure of population variability [66,67]; (B) the amount of variance explained by smooth, long-term trends; (C) the period of cyclical components (scaled by the age at maturity of each species) in the residual time series remaining after subtraction of the smooth, long-term trend (B); and (D) the Hurst exponent (H_{dtr}), which is an index of long-range dependence and short-term predictability (calculated after removing the best fit linear trend). The clusters were determined using model-based clustering [72,73]. Time series data was extracted from ICES Stock Assessment database [65] and from version 3.0 of the RAM Legacy Stock Recruitment database [26]. Scaled periods were calculated using age of maturity data from Thorson et al.'s FishLife database [69].

- B—features stocks showing apparent trends accompanied by large fluctuations that resemble delay-feedback dynamics. The group includes 83 stocks from 53 species predominantly distributed over temperate and subpolar waters, although several tropical and subtropical species were included as well. Small and large pelagics, together with flatfishes (flounders and soles), are overrepresented in group B.
- C—includes stocks whose dynamics are dominated by a smooth trend and short-term fluctuations associated with single generation cycles that decay rapidly toward the trend ($\tau_n \approx 1$). Only 46 stocks from 33 species, in most cases temperate, are included in this group. They are mostly slow-growing, long-lived rockfish species.
- D—contains stocks with marked trends which are close to a constant rate increase or, more often, a rapid decrease in abundance. The period of residual oscillations is consistent with delay-feedback dynamics, although regulation is tight and keeps abundance close to the trend. It is the largest group, featuring 100 stocks from 72 species distributed all around the ocean, but mainly in tropical and temperate waters.

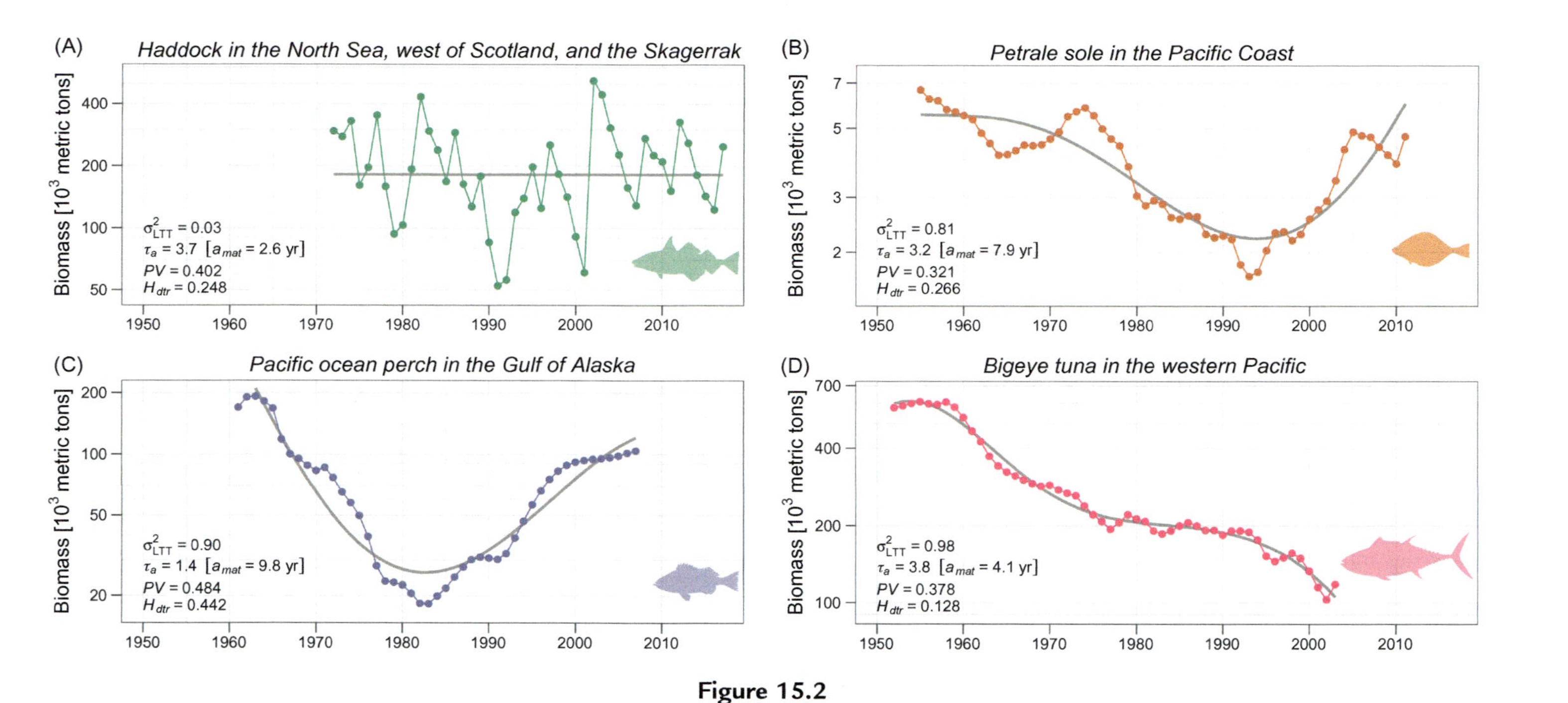

Figure 15.2

Example time series for four selected stocks that illustrate the characteristics of the groups resulting from the classification of fish stocks according to the variability in time of their biomass dynamics (Fig. 15.1). Each panel includes an annual time series of biomass (*colored line and dots*) together with a background, long-term trend (*gray line*); please note the $\log_{10}$ scaling in the ordinate. In the lower left corner of each panel there is a list with the values of the time series probes used in the classification for each particular stock: σ^2 is the amount of variance explained by the long-term trend; τ_a is the normalized period of the oscillations—scaled by the age at maturity of each species, a_{mat}—in the residual time series resulting after subtracting the smooth trend; *PV* is Heath's *PV* [66], which was calculated on raw biomass data; and H_{dtr} is the *H* calculated for linearly detrended biomass [24]. (A) Haddock (*Melanogrammus aeglefinus*) stock in the North Sea, west of Scotland, and the Skagerrak (*had.27.46a20*) has very irregular dynamics with fluctuations that weave around an apparently constant level with a period consistent with delayed-feedback dynamics. (B) Petrale sole (*Eopsetta jordani*) in the Pacific Coast (*PSOLEPCOAST*) also presents large fluctuations, but the dynamics include an apparent trend. (C) Pacific ocean perch (*Sebastes alutus*) in the Gulf of Alaska (*POPERCHGA*) presents a strong trend with irregular fluctuations that seem consistent with single generation cycles. (D) Bigeye tuna (*Thunnus obesus*) in the western Pacific (*BIGEYEWPO*) presents a smooth decline that dominates the dynamics and small amplitude fluctuations of a period consistent again with delayed-feedback dynamics. The code following the name of each stock corresponds to its name in the source database; data for A comes from ICES Stock Assessment database [65], while for stocks B, C, and D, the data was extracted from version 3.0 of the RAM Legacy Stock Recruitment database [26]. Age of maturity was extracted from Thorson et al.'s *FishLife* database [69].

The group included most tunas and billfishes stocks and many demersal species whose dynamics are characterized by a steady decline.

Despite the regularities mentioned, it is important to bear in mind the difficulties associated with this type of analysis (e.g., observation errors and biased sampling of fish stocks, the simple decomposition in long- and short-term components, etc.). However, some generalizations emerge from our simple approach that confirm previous results while incorporating a novel perspective [24,25,28–31]. In general, the abundance of fish stocks features changes of one or two orders of magnitude that, in most cases, are associated with long-term trends ($\sim$85% of the stocks). Fishing is a major candidate to explain these trends, especially in the case of species whose abundance is declining (e.g., group D). However, reddened dynamics with positive temporal autocorrelation prevail across stocks even after removing linear trends (i.e., $H_{dtr} = 0.215[0.057,0.431]90\%$; median and 90% quantile interval, e.g., [24,25]), which suggests the integration of additional sources of external noise. For instance, similar exponent values have been estimated for temperature extremes in the ocean [74].

15.4 Environmental drivers of variability in marine fisheries

Environmental variability affects reproduction and survival of adult fishes, and it is generally assumed to also be the main cause behind the erratic patterns of recruitment observed in marine fisheries [21,57]. Historically, research on the impact of environmental variability on fisheries has focused on population renewal [33,75]. This bias is justified both in theoretical and in empirical grounds, since variability in recruitment rates has been identified as the main process leading to fluctuations in fish abundance [36]. The life cycle of many fishes involves a tiny planktonic stage that is particularly vulnerable to changes in environmental conditions [75,76]. The gamut of potential mechanisms impacting recruitment success was grouped early around three main themes [33,77]: (1) the existence of a critical period determining survival according to a specific set of environmental conditions (calm conditions that concentrate prey [78] or turbulence levels that favor prey encounter [79]) or the ability to exploit seasonal peaks of production (i.e., match–mismatch mechanisms [80]); (2) the dispersal or retention of planktonic eggs and larvae in nursery grounds depending on hydrographic conditions [81]; and more recently (3) the importance of negative interactions with predators and competitors [32].

Adult production and survival are also increasingly recognized as important sources of population variability in fisheries. On some occasions the direct impact of a single environmental factor on adult survival triggers dramatic changes, like the mechanism proposed to explain the decline of cod in the Gulf of Maine due to ocean warming [18]. More often several factors need to be aligned in order to explain fluctuations in fish abundance. In general, mortality due to fishing contributes to increase population variability and the risk of collapse [35,36,82], especially when fishing effort does not take into account changes in environmental conditions [83,84]. Fishing can also prevent species recovery when increased demand rewards fishermen who keep exploiting already depleted resources [37]. The interactions between fishing, environmental variation, and population dynamics hold the key to understand and anticipate unexpected changes.

In the analyses conducted here, single generation cycles turn out to be relatively common ($\tau_n < 2$ for 32%) and suggest the operation of tight density-dependence regulation. Although this type of dynamics cannot generate long-term fluctuations, it reflects a rapid response to fluctuations in resources that erodes short-term predictability. Consumer–resource cycles are far less common but, when present, they manifest as a major source of variability. These longer cycles were found in subpolar and polar stocks, which suggests a link with increased seasonal variation that is consistent with previous findings across systems and taxa [85]. In around half of the stocks, deviations from trends involved short-term oscillations resembling delayed-feedback effects ($2 < \tau_n < 4$ for 55% of the stocks). In age-structured populations with delay feedbacks, external environmental noise can amplify these fluctuations and result in decadal trends under some circumstances by promoting, for instance, recruitment failure at low adult abundances (i.e., the "cultivation–depensation" effect [55]), or changes in recruitment or egg production associated with the amplification of environmental noise (i.e., the "cohort–resonance" effect, [82]). On the other hand, fluctuations in the abundance of generalist predators due to fishing and changes in the environment can promote an entire restructuring of marine ecosystems and delay fisheries recovery via bottom-up effects and trophic cascades [15,86].

15.5 Integration

Taken together, available evidence strongly suggests that environmental variability and fishing pressure are the major drivers behind fluctuations in the abundance of fishing resources. Sometimes the impact of these changes is dramatic and results in the collapse of fish stocks [83,84]. Approaches that attempt to predict the future of fisheries and ignore the effect of environmental variability are likely to fail, at least as much as those ignoring other aspects of the triad formed by economic, social, and ecological aspects of the fisheries [9]. The inclusion of environmental variation into population models to inform management decisions remains anecdotal, especially for short-term forecasts [22]. Looking forward, there are many developments underway bringing dynamic fisheries management to a reality [87]. Advances in the monitoring and modeling of ocean conditions [88], the increasing availability of reliable forecast products and climate change projections [39,89], and the development of methods to integrate different sources of information and propagate uncertainty [90,91] form part of the essential toolbox to advance marine science toward sustainable management.

Studies that have succeeded in revealing reliable relationships between environmental factors and fisheries dynamics share a careful approach to the characterization of environmental variation in relation to the biology of the target species [21,22]. Developments in satellite remote sensing and automatic platforms (floats and gliders) are bringing environmental data of an unprecedented quality, quantity, and nature [92]. These advances bring the opportunity to further extend current approaches to replace strictly empirical relationships with models with a mechanistic foundation. It is time to recognize that temperature often acts as a proxy of other environmental factors with a direct impact on organism performance, including either physiological impacts due to decreases in oxygen concentration and pH, or trophic effects due to shortages in prey availability. New platforms, like bioArgo floats [93], provide for the first time reliable subsurface observations of these biogeochemical variables and provide the opportunity to incorporate realistic constraints on the distribution and performance of fisheries species.

A second development of great utility involves the increasing availability of high-quality forecast products at a range of scales of relevance for marine resource management [38,39], from months to years and decades. Most studies have focused on the use of physical forecasts (i.e., temperature and salinity distributions), and apply them to predict a variety of aspects related to fisheries dynamics, from changes in the spatial distribution of species to changes in phenology and the timing of key events [40]. However, the set of variables of biological relevance that can be reliably assimilated and forecasted is increasing rapidly, including changes in oxygen and pH [94]. These forecasts also allow the development and planning of management strategies beyond usual short-term horizons, which is a key aspect to ensure fisheries sustainability. These developments are fueled by advances in Earth system modeling and by the availability of biogeochemical data to train and develop dynamic prediction models.

Perhaps one of the most important aspects to ensure progress toward dynamic, sustainable management is the need to integrate all the different sources of information to produce useful predictions that incorporate appropriate measurements of uncertainty [90]. In this way probabilistic methods rooted in Bayesian analysis provide the third element of the toolbox. Methods like state–space models have become the standard approach to analyze a variety of problems, from individual growth and movement paths to fisheries stock assessments [95]. These methods present several advantages with respect to competing approaches (e.g., [42]), namely, the ability (1) to propagate uncertainty from observations to parameters' estimates and derived quantities, and (2) to fit complex models taking advantage of the conditional hierarchical structure that naturally emerges among different model components in the Bayesian framework [42,96]. Both advantages bridge the gap between complex and unaligned data sets, noisy biological observations, and the need to improve available theories in marine resource management using mechanistic models [91].

Probabilistic methods have a long tradition in natural resource management [42,96], and indeed, many fishery scientists played a pioneering and leading role in popularizing this type of approach (e.g., [97–100]). Together, developments in Earth observation and modeling, ecological forecasts, and probabilistic modeling pave the way to the development of novel approaches to implement dynamic fisheries management strategies. Looking forward, there are already many "bright spots" on marine resource management [101], we should continue the efforts to integrate knowledge across disciplines in the brave, pioneering tradition of marine science, before it is too late to solve the challenges posed by the environmental changes underway.

Acknowledgments

I would like to thank the Nippon Foundation for support as well as the editors—William Cheung, Yoshitaka Ota, and Andrés Cisneros-Montemayor—for their patience and for the invitation to contribute to this volume. I acknowledge Carlos L. Cáceres, Ricardo González Gil, and Isa Martínez for feedback about early drafts of this manuscript, and Luna González for support. I would also like to thank Ricardo. Anadón, José Luis Acuña, Charles A. Stock, Jorge L. Sarmiento, Colleen Petrik, Kisei Tanaka, Vince Saba, Barb Muhling, Desiree Tommasi, and Jong-Yeon Park for many enriching discussions about plankton, fisheries, oceanography, and society.

References

[1] FAO, The State of World Fisheries and Aquaculture 2018—Meeting the Sustainable Development Goals, Food and Agriculture Organization of the United Nations, Rome, Italy, 2018.
[2] J.F. Caddy, J.C. Seijo, This is more difficult than we thought! The responsibility of scientists, managers and stakeholders to mitigate the unsustainability of marine fisheries, Philos. Trans. R. Soc. B: Biol. Sci. 360 (1453) (2005) 59–75.
[3] A.M. Eikeset, A.B. Mazzarella, B.D. sdóttir, D.H. Klinger, S.A. Levin, E. Rovenskaya, et al., What is blue growth? The semantics of "Sustainable Development" of marine environments, Mar. Policy 87 (2018) 177–179.
[4] G. Merino, M. Barange, J.L. Blanchard, J. Harle, R. Holmes, I. Allen, et al., Can marine fisheries and aquaculture meet fish demand from a growing human population in a changing climate? Global Environ. Change 22 (4) (2012) 795–806.
[5] L.V. Weatherton, A.K. Magnan, A.D. Rogers, U.R. Sumaila, W.W.L. Cheung, Observed and projected impacts of climate change on marine fisheries, aquaculture, coastal tourism, and human health: an update, Front. Mar. Sci. 3 (2016) 48.
[6] M. Barange, T. Bahri, M.C.M. Beveridge, K.L. Cochrane, S. Funge-Smith, F. Poulain, Impacts of Climate Change on Fisheries and Aquaculture: Synthesis of Current Knowledge, Adaptation and Mitigation Options, Rome, Italy, 2018, p. 627.
[7] J.S. Clark, S.R. Carpenter, M. Barber, S. Collins, A. Dobson, J.A. Foley, et al., Ecological forecasts: an emerging imperative, Science 293 (5530) (2001) 657–660.
[8] W.W.L. Cheung, V.W.Y. Lam, J.L. Sarmiento, K. Kearney, R. Watson, D. Pauly, Projecting global marine biodiversity impacts under climate change scenarios, Fish Fish. 10 (3) (2009) 235–251.
[9] A.T. Charles, Fishery conflicts: a unified framework, Mar. Policy 16 (5) (1992) 379–393.
[10] F. Asche, T.M. Garlock, J.L. Anderson, S.R. Bush, M.D. Smith, C.M. Anderson, et al., Three pillars of sustainability in fisheries, Proc. Natl. Acad. Sci. U.S.A. 115 (44) (2018) 11221–11225.

[11] C. Costello, D. Ovando, T. Clavelle, C.K. Strauss, R. Hilborn, M.C. Melnychuk, et al., Global fishery prospects under contrasting management regimes, Proc. Natl. Acad. Sci. U.S.A. 113 (18) (2016) 5125–5129.
[12] W. Bank, The Sunken Billions Revisited: Progress and Challenges in Global Marine Fisheries. World Bank, Washington, DC, 2017. Available from: https://openknowledge.worldbank.org/handle/10986/24056.
[13] C.J. Walters, S.J.D. Martell, Fisheries Ecology and Management, first ed, Princeton University Press, Princeton, NJ, 2004.
[14] F.P. Chavez, J. Ryan, S.E. Lluch-Cota, C.M. Ñiquen, From anchovies to sardines and back: multidecadal change in the Pacific Ocean, Science 299 (5604) (2003) 217–221.
[15] D.Ø. Hjermann, G. Ottersen, N.C. Stenseth, Competition among fishermen and fish causes the collapse of Barents Sea capelin, Proc. Natl. Acad. Sci. U.S.A. 101 (32) (2004) 11679–11684.
[16] C. Mullon, P. Fréon, P. Cury, The dynamics of collapse in world fisheries, Fish Fish. 6 (2) (2005) 111–120.
[17] M.L. Pinsky, O.P. Jensen, D. Ricard, S.R. Palumbi, Unexpected patterns of fisheries collapse in the world's oceans, Proc. Natl. Acad. Sci. U.S.A. 108 (20) (2011) 8317–8322.
[18] A.J. Pershing, M.A. Alexander, C.M. Hernandez, L.A. Kerr, A.L. Bris, K.E. Mills, et al., Slow adaptation in the face of rapid warming leads to collapse of the Gulf of Maine cod fishery, Science 350 (6262) (2015) 809–812.
[19] C.A. Stock, J.G. John, R.R. Rykaczewski, R.G. Asch, W.W.L. Cheung, J.P. Dunne, et al., Reconciling fisheries catch and ocean productivity, Proc. Natl. Acad. Sci. U.S.A. 114 (8) (2017) E1441–E1449.
[20] V. Christensen, J. Maclean (Eds.), Ecosystem Approaches to Fisheries. A Global Perspective. first ed, Cambridge University Press, Cambridge, UK, 2011.
[21] R.A. Myers, When do environment–recruitment correlations work? Rev. Fish Biol. Fish. 8 (3) (1998) 285–305.
[22] M. Skern-Mauritzen, G. Ottersen, N.O. Handegard, G. Huse, G.E. Dingsør, N.C. Stenseth, et al., Ecosystem processes are rarely included in tactical fisheries management, Fish Fish. 17 (1) (2016) 165–175.
[23] E. Howes, F. Joos, M. Eakin, J.P. Gattuso, An updated synthesis of the observed and projected impacts of climate change on the chemical, physical and biological processes in the oceans, Front. Mar. Sci. 2 (2015) 36.
[24] J.M. Halley, K.I. Stergiou, The implications of increasing variability of fish landings, Fish Fish. 6 (3) (2005) 266–276.
[25] H.S. Niwa, Random-walk dynamics of exploited fish populations, ICES J. Mar. Sci. 64 (3) (2007) 496–502.
[26] D. Ricard, C. Minto, O.P. Jensen, J.K. Baum, Examining the knowledge base and status of commercially exploited marine species with the RAM Legacy Stock Assessment Database, Fish Fish. 13 (4) (2012) 380–398.
[27] M.J. Fogarty, R. Gamble, C.T. Perretti, Dynamic complexity in exploited marine ecosystems, Front. Ecol. Evol. 4 (2016) 68.
[28] T. Kawasaki, Fundamental relations among the selections of life history in the marine teleosts, Nippon Suisan Gakkaishi. 46 (3) (1980) 289–293.
[29] T. Kawasaki, Why do some pelagic fishes have wide fluctuations in their numbers? Biological basis of fluctuation from the vewipoint of evolutionary ecology, in: G.D. Sharp, J. Csirke (Eds.), Proceedings of the Expert Consultation to Examine Changes in Abundance and Species Composition of Neritic Fish Resources, No. 291 in FAO Fisheries Report, Rome, Italy, 1983, pp. 1065–1080.
[30] J.F. Caddy, J.A. Gulland, Historical patterns of fish stocks, Mar. Policy 7 (4) (1983) 267–278.
[31] P.D. Spencer, J.S. Collie, Patterns of population variability in marine fish stocks, Fish. Oceanogr. 6 (3) (1997) 188–204.
[32] E.D. Houde, Emerging from Hjort's shadow, J. Northwest Atlant. Fish. Sci. 41 (2008) 53–70.
[33] E.D. Houde, Recruitment variability, Fish Reproductive Biology: Implications for Assessment and Management, second ed, John Wiley & Sons, Ltd, Osney Mead, Oxford, UK, 2016, pp. 98–187.
[34] J.T. Thorson, O.P. Jensen, E.F. Zipkin, How variable is recruitment for exploited marine fishes? A hierarchical model for testing life history theory, Can. J. Fish. Aquat. Sci. 71 (7) (2014) 973–983.

[35] C.N.K. Anderson, C.H. Hsieh, S.A. Sandin, R. Hewitt, A. Hollowed, J. Beddington, et al., Why fishing magnifies fluctuations in fish abundance, Nature 452 (2008) 835–839.
[36] A.O. Shelton, M. Mangel, Fluctuations of fish populations and the magnifying effects of fishing, Proc. Natl. Acad. Sci. U.S.A. 108 (17) (2011) 7075–7080.
[37] J.M. Fryxell, R. Hilborn, C. Bieg, K. Turgeon, A. Caskenette, K.S. McCann, Supply and demand drive a critical transition to dysfunctional fisheries, Proc. Natl. Acad. Sci. U.S.A. 114 (46) (2017) 12333–12337.
[38] A.J. Hobday, C.M. Spillman, J.E. Paige, J.R. Hartog, Seasonal forecasting for decision support in marine fisheries and aquaculture, Fish. Oceanogr. 25 (S1) (2016) 45–56.
[39] D. Tommasi, C.A. Stock, A.J. Hobday, R. Methot, I.C. Kaplan, J.P. Eveson, et al., Managing living marine resources in a dynamic environment: the role of seasonal to decadal climate forecasts, Prog. Oceanogr. 152 (2017) 15–49.
[40] M.R. Payne, A.J. Hobday, B.R. MacKenzie, D. Tommasi, D.P. Dempsey, S.M.M. FÃ¤ssler, et al., Lessons from the first generation of marine ecological forecast products, Front. Mar. Sci. 4 (2017) 289.
[41] J.S. Clark, A.E. Gelfand, A future for models and data in environmental science, Trends Ecol. Evol. 21 (7) (2006) 375–380.
[42] N.T. Hobbs, M.B. Hooten, Bayesian Models. A Statistical Primer for Ecologists, first ed, Princeton University Press, Princeton NJ, 2015.
[43] T.D. Smith, Scaling Fisheries: The Science of Measuring the Effects of Fishing, 1855–1955, first ed, Cambridge University Press, Cambridge, UK, 1994.
[44] T.R. Baumgartner, A. Soutar, V. Ferreira-Bartrina, Reconstruction of the history of Pacific sardine and northern anchovy populations over the past two millennia from sediments of the Santa Barbara basin, California, Calif. Cooperative Ocean. Fish. Invest. Rep. 33 (1992) 24–40.
[45] J. Jackson, J. Jacquet, The shifting baselines syndrome: perception, deception, and the future of our oceans, in: V. Christensen, J. Maclean (Eds.), Ecosystem Approaches to Fisheries. A Global Perspective, first ed, Cambridge University Press, Cambridge, UK, 2011, pp. 128–141.
[46] D.J. McCauley, M.L. Pinsky, S.R. Palumbi, J.A. Estes, F.H. Joyce, R.R. Warner, Marine defaunation: animal loss in the global ocean, Science 347 (2015) 6219.
[47] V. Volterra, Fluctuations in the abundance of a species considered mathematically, Nature 118 (1926) 558–560.
[48] L. Garibaldi, The FAO global capture production database: a six-decade effort to catch the trend, Mar. Policy 36 (3) (2012) 760–768.
[49] D. Zeller, M.L.D. Palomares, A. Tavakolie, M. Ang, D. Belhabib, W.W.L. Cheung, et al., Still catching attention: *Sea Around Us* reconstructed global catch data, their spatial expression and public accessibility, Mar. Policy 70 (2016) 145–152.
[50] R.A. Myers, J. Bridson, N.J. Barrowman, Summary of Worldwide Spawner and Recruitment Data, St. John's, Newfoundland, Canada, 1995, p. 2024.
[51] D. Pauly, R. Hilborn, T.A. Branch, Fisheries: does catch reflect abundance? Nature 494 (2013) 303–306.
[52] M.L. Pinsky, D. Byler, Fishing, fast growth and climate variability increase the risk of collapse, Proc. R. Soc. B: Biol. Sci. 282 (1813) (2015) 20151053.
[53] J.A. Hutchings, J.D. Reynolds, Marine fish population collapses: consequences for recovery and extinction risk, BioScience 54 (4) (2004) 297–309.
[54] J.A. Hutchings, A. Kuparinen, Empirical links between natural mortality and recovery in marine fishes, Proc. R. Soc. B: Biol. Sci. 284 (1856) (2017) 20170693.
[55] C. Walters, J.F. Kitchell, Cultivation/depensation effects on juvenile survival and recruitment: implications for the theory of fishing, Can. J. Fish. Aquat. Sci. 58 (1) (2001) 39–50.
[56] K.A. Vert-pre, R.O. Amoroso, O.P. Jensen, R. Hilborn, Frequency and intensity of productivity regime shifts in marine fish stocks, Proc. Natl. Acad. Sci. U.S.A. 110 (5) (2013) 1779–1784.
[57] C.S. Szuwalski, K.A. Vert-Pre, A.E. Punt, T.A. Branch, R. Hilborn, Examining common assumptions about recruitment: a meta-analysis of recruitment dynamics for worldwide marine fisheries, Fish Fish. 16 (4) (2015) 633–648.

[58] G. Beaugrand, K.M. Brander, J.A. Lindley, S. Souissi, P.C. Reid, Plankton effect on cod recruitment in the North Sea, Nature 426 (2003) 661–664.
[59] W.T. Peterson, F.B. Schwing, A new climate regime in northeast pacific ecosystems, Geophys. Res. Lett. 30 (2003) 1896.
[60] C.T. Perretti, M.J. Fogarty, K.D. Friedland, J.A. Hare, S.M. Lucey, R.S. McBride, et al., Regime shifts in fish recruitment on the Northeast US Continental Shelf, Mar. Ecol. Prog. Ser. 574 (2017) 1–11.
[61] M.C. McManus, P. Licandro, S.H. Coombs, Is the Russell Cycle a true cycle? Multidecadal zooplankton and climate trends in the western English Channel, ICES J. Mar. Sci. 73 (2) (2016) 227–238.
[62] G. Beaugrand, The North Sea regime shift: evidence, causes, mechanisms and consequences, Prog. Oceanogr. 60 (2) (2004) 245–262.
[63] S.R. Hare, N.J. Mantua, Empirical evidence for North Pacific regime shifts in 1977 and 1989, Prog. Oceanogr. 47 (2) (2000) 103–145.
[64] J.H. Steele, E.W. Henderson, Modeling long-term fluctuations in fish stocks, Science 224 (1984) 985–987.
[65] ICES, Stock Assessment Database, Copenhagen, Denmark, ICES, <http://standardgraphs.ices.dk>, 2018 (accessed 01.11.18).
[66] J.P. Heath, Quantifying temporal variability in population abundances, Oikos 115 (3) (2006) 573–581.
[67] J.P. Heath, P. Borowski, Quantifying proportional variability, PLoS One 8 (12) (2014) 1–4.
[68] M.A.R. Ferreira, H.K.H. Lee, Multiscale modeling. A Bayesian perspective, Springer Series in Statistics, first ed, Springer-Verlag, New York, 2007.
[69] J.T. Thorson, S.B. Munch, J.M. Cope, J. Gao, Predicting life history parameters for all fishes worldwide, Ecol. Appl. 27 (8) (2017) 2262–2276.
[70] W.W. Murdoch, B.E. Kendall, R.M. Nisbet, C.J. Briggs, E. McCauley, R. Bolser, Single-species models for many-species food webs, Nature 417 (2002) 541–543.
[71] F. Barraquand, S. Louca, K.C. Abbott, C.A. Cobbold, F. Cordoleani, D.L. DeAngelis, et al., Moving forward in circles: challenges and opportunities in modelling population cycles, Ecol. Lett. 20 (8) (2017) 1074–1092.
[72] C. Fraley, A.E. Raftery, Model-based clustering, discriminant analysis, and density estimation, J. Am. Stat. Assoc. 97 (458) (2002) 611–631.
[73] L. Scrucca, M. Fop, T.B. Murphy, A.E. Raftery, *Mclust 5*: clustering, classification and density estimation using Gaussian finite mixture models, R J. 8 (1) (2016) 205–233. Available from: https://journal.r-project.org/archive/2016/ RJ-2016-021/RJ-2016-021.pdf.
[74] D.A. Vasseur, P. Yodzis, The color of environmental noise, Ecology 85 (4) (2004) 1146–1152.
[75] Fishery science, in: L.A. Fuiman, R.G. Werner (Eds.), Osney Mead, first ed, Blackwell Publishing, Oxford, UK, 2002.
[76] T. Jakobsen, M.J. Fogarty, B.A. Megrey, E. Moksness, Fish Reproductive Biology: Implications for Assessment and Management, second ed, John Wiley & Sons, Ltd, Osney Mead, Oxford, UK, 2016.
[77] J. Hjort, Fluctuations in the great fisheries of northern Europe viewed in the light of biological research, Rapports et Procs-Verbaux des Reunions, Conseil International; pour l'Exploration de la Mer, vol. 20, 1914, pp. 1–228.
[78] R. Lasker, Field criteria for survival of anchovy larvae: the relation between inshore chlorophyll maximum layers and successful first feeding, Fish. Bull. 73 (3) (1975) 453–462.
[79] B.R. MacKenzie, Turbulence, larval fish ecology and fisheries recruitment: a review of field studies, Oceanol. Acta 23 (4) (2000) 357–375.
[80] J.M. Durant, D.Ø. Hjermann, G. Ottersen, N.C. Stenseth, Climate and the match or mismatch between predator requirements and resource availability, Clim. Res. 33 (2007) 271–283.
[81] T.D. Iles, M. Sinclair, Atlantic herring: stock discreteness and abundance, Science 215 (4533) (1982) 627–633.

[82] L.W. Botsford, M.D. Holland, J.C. Field, A. Hastings, Cohort resonance: a significant component of fluctuations in recruitment, egg production, and catch of fished populations, ICES J. Mar. Sci. 71 (8) (2014) 2158–2170.
[83] M. Lindegren, C. Möllmann, A. Nielsen, N.C. Stenseth, Preventing the collapse of the Baltic cod stock through an ecosystem-based management approach, Proc. Natl. Acad. Sci. U.S.A. 106 (34) (2009) 14722–14727.
[84] F. González Taboada, R. Anadón, Determining the causes behind the collapse of a small pelagic fishery using Bayesian population modeling, Ecol. Appl. 26 (3) (2015) 886–898.
[85] B.E. Kendall, J. Prendergast, O.N. Bjørnstad, The macroecology of population dynamics: taxonomic and biogeographic patterns in population cycles, Ecol. Lett. 1 (3) (1998) 160–164.
[86] K. Frank, B. Petrie, J. Fisher, W. Leggett, Transient dynamics of an altered large marine ecosystem, Nature 477 (7362) (2011) 86–89.
[87] S.M. Maxwell, E.L. Hazen, R.L. Lewison, D.C. Dunn, H. Bailey, S.J. Bograd, et al., Dynamic ocean management: defining and conceptualizing real-time management of the ocean, Mar. Policy 58 (2015) 42–50.
[88] J. Morales, V. Stuart, T. Platt, S. Sathyendranath, Handbook of Satellite Remote Sensing Image Interpretation: Applications for Marine Living Resources Conservation and Management, EU PRESPO Project and the International Ocean-Colour Coordinating Group (IOCCG), 2010. Available from: <http://www.ioccg.org/handbook.html>.
[89] C.A. Stock, M.A. Alexander, N.A. Bond, K.M. Brander, W.W.L. Cheung, E.N. Curchitser, et al., On the use of IPCC-class models to assess the impact of climate on Living Marine Resources, Prog. Oceanogr. 88 (1) (2011) 1–27.
[90] W.W.L. Cheung, T.L. Frolicher, R.G. Asch, M.C. Jones, M.L. Pinsky, G. Reygondeau, et al., Building confidence in projections of the responses of living marine resources to climate change, ICES J. Mar. Sci. 73 (5) (2016) 1283–1296.
[91] H.K. Kindsvater, N.K. Dulvy, C. Horswill, M.J. Juan-Jordá, M. Mangel, J. Matthiopoulos, Overcoming the data crisis in biodiversity conservation, Trends Ecol. Evol. 33 (9) (2018) 676–688.
[92] K.S. Johnson, W.M. Berelson, E.S. Boss, Z. Chase, H. Claustre, S.R. Emerson, et al., Observing biogeochemical cycles at global scales with profiling floats and gliders: prospects for a global array, Oceanography 22 (3) (2009) 216–225.
[93] K.S. Johnson, J.N. Plant, L.J. Coletti, H.W. Jannasch, C.M. Sakamoto, S.C. Riser, et al., Biogeochemical sensor performance in the SOCCOM profiling float array, J. Geophys. Res.: Oceans 122 (8) (2017) 6416–6436.
[94] S.A. Siedlecki, I.C. Kaplan, A.J. Hermann, T.T. Nguyen, N.A. Bond, J.A. Newton, et al., Experiments with Seasonal Forecasts of ocean conditions for the Northern region of the California Current upwelling system, Sci. Rep. 6 (2016) 27203.
[95] W.H. Aeberhard, J.M. Flemming, A. Nielsen, Review of state-space models for fisheries science, Annual Rev. Stat. Appl. 5 (1) (2018) 215–235.
[96] J.S. Clark, A. Gelfand (Eds.), Hierarchical Modelling for the Environmental Sciences. Statistical Methods and Applications, Oxford University Press, Oxford, UK, 2006.
[97] J.T. Schnute, A general framework for developing sequential fisheries models, Can. J. Fish. Aquat. Sci. 51 (8) (1994) 1676–1688.
[98] R. Hilborn, M. Mangel, The Ecological Detective. Confronting Models With Data. vol. 28 of Monographs in Population Biology, first ed, Princeton University Press, Princeton, NJ, 1997.
[99] A.E. Punt, R. Hilborn, Fisheries stock assessment and decision analysis: the Bayesian approach, Rev. Fish Biol. Fish. 7 (1) (1997) 35–63.
[100] R.B. Millar, R. Meyer, Non-linear state space modelling of fisheries biomass dynamics by using Metropolis-Hastings within-Gibbs sampling, J. R. Stat. Soc.: Ser. C (Appl. Stat.) 49 (3) (2000) 327–342.
[101] C. Cvitanovic, A.J. Hobday, Building optimism at the environmental science-policy-practice interface through the study of bright spots, Nat. Commun. 9 (2018) 1–5.

CHAPTER 16

Life history of marine fishes and their implications for the future oceans

Colleen M. Petrik[1,2]

[1] Program in Atmospheric and Oceanic Sciences, Princeton University, Princeton, NJ, United States
[2] Department of Oceanography, Texas A&M University, College Station, TX, United States

Chapter Outline

16.1 Life history strategies of fishes

Life history refers to the pattern of survival and reproduction events during the life of an organism. Life history traits include maximum body size, longevity, age at maturity, and fecundity. Life history theory espouses that these traits have been shaped by natural selection to optimize trade-offs related to growth, reproduction, and survival. Thus, organisms that have the same phylogeny share similar traits. Conversely, unrelated organisms occasionally evolve similar traits independently.

Winemiller and Rose [1] evaluated patterns in fish life history traits across species to create an organizing framework. This framework consists of the Equilibrium, Periodic, and Opportunistic (E–P–O) niche scheme, three endpoint strategies of correlated life history traits (Fig. 16.1). Equilibrium species were categorized by moderate to long generation time, low fecundity, and high investment per offspring. Examples of commercial fish species include spiny dogfish and longnose skate. Periodic species were denoted by long generation time, high fecundity, and low investment per offspring—species like Atlantic cod, halibut, and tuna. The Opportunistic strategy consisted of fishes with short generation times, low batch fecundity, and low investment per offspring, such as sardine, anchovy,

Predicting Future Oceans.
DOI: https://doi.org/10.1016/B978-0-12-817945-1.00016-2

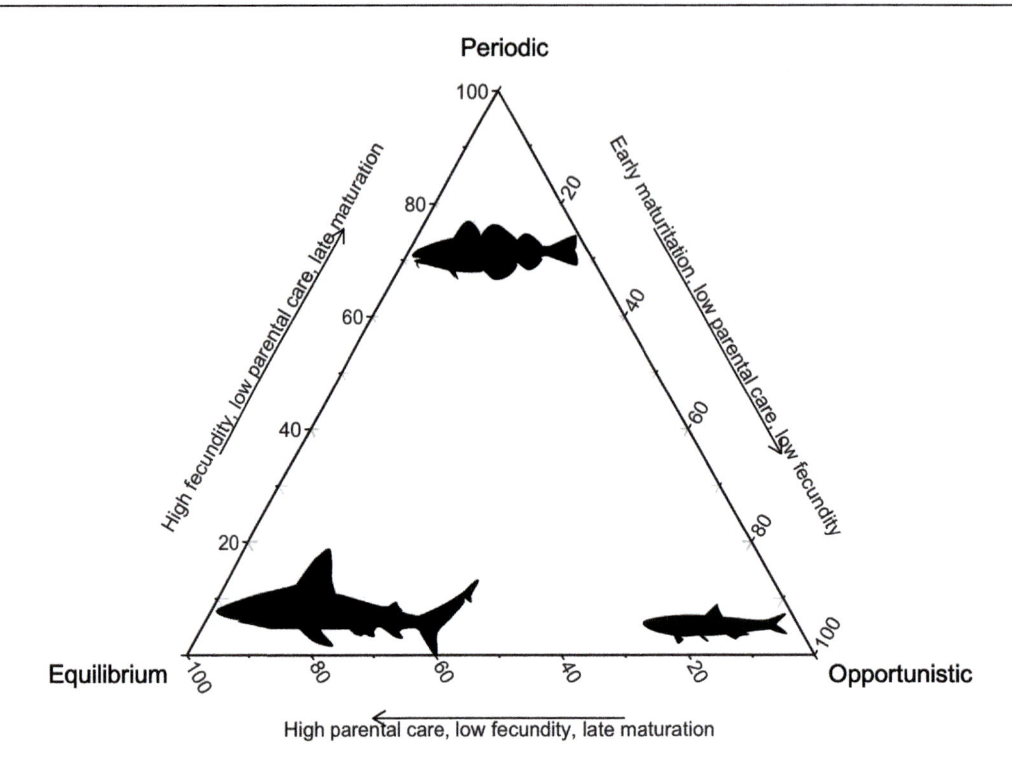

Figure 16.1

The life history endpoint strategies of Winemiller and Rose as a triangular continuum of traits. *Based on K.O. Winemiller, Life history strategies, population regulation, and implications for fisheries management, Can. J. Fish. Aquat. Sci. 62 (2005) 872–885, Figure 1. Fish silhouettes courtesy: The Integration and Application Network, University of Maryland Center for Environmental Science (ian.umces.edu/symbols/).*

and mackerel. These endpoints are a trilateral continuum based on biological constraints and different fish species could theoretically lie anywhere in this space.

The environment has shaped the particular life history traits of any given fish. Because life history traits ultimately determine population dynamics, the E–P–O framework developed by Winemiller and Rose [1] can be used to predict a population's response to different types of disturbances. The E–P–O niche scheme was mapped onto a 2D life history plane by Winemiller [2] that described the gradients of environmental disturbance, variability, and predictability associated with each life history strategy. The Opportunistic group is a colonizing strategy that thrives in environments with high amounts of disturbance and low levels of predictability of resources and mortality [2]. As the scale and predictability of spatiotemporal environmental variability increases, the Periodic life history strategy is favored [2]. Conversely, Equilibrium life history strategies persist in stable environments with high levels of competition and predation [2].

Winemiller and Rose's [1] analysis consisted of 216 North American freshwater and marine fishes, and they noted that marine species tended toward the Periodic endpoint. In this case it seems unlikely that the small differences in traits could be used for understanding population dynamics. The analysis of Winemiller and Rose [1] was extended globally to include marine species of commercially harvested fish stocks [3]. The results show that when examined alone, marine fish species also encapsulate the triangular continuum, but do not fill the space entirely.

16.2 Dominance of large predatory fishes

Life history distinctions may not provide enough information given that the majority of marine fishes fall within the Periodic category, but they differ in their socioeconomic value and ecosystem role. In the marine environment the Opportunists are often planktivorous forage fishes that serve as prey to higher predators, like the Periodic and Equilibrium fishes. These life history strategies can be further divided into various functional groups defined by habitat and feeding preferences. For example, large predatory Periodic fishes can be separated into pelagic and demersal species. Large pelagic fishes inhabit epipelagic and mesopelagic environments, often making large migrations in search of prey [4]. These fishes reach large maximum sizes and often possess homeothermic adaptations to support their higher metabolic rates and fast swimming speeds commensurate with their large migratory abilities [4,5]. Demersal fishes also attain large sizes, but tend to be slower growing and moving compared to the large pelagics [6]. They live near the seafloor and are capable of feeding both on benthic fauna and on pelagic animals [7,8]. In this vein, demersals are more "generalist" predators, while large pelagics are "specialists." These large predatory fishes seldom coexist in the same habitat, with large pelagics dominating in the tropics and subtropics [5] and demersals in temperate and polar environments [9].

Empirical studies have found that net primary productivity (NPP) alone is a weak predictor of regional variations in total fish biomass [10–12], Rather, the production of fish biomass is closely tied to the separation of NPP into pelagic and benthic secondary production and the total amounts of these two types [11,12]. van Denderen et al. [13] expanded this work by hypothesizing that the ratio of the two pathways from NPP to fishes influences which functional type dominates. Indeed the ratio of the fraction of NPP that remained in the pelagic (F_{pelagic}) to the fraction of NPP that was exported to the seafloor (F_{benthic}) explained the majority of the deviance in the relative biomass of large pelagic fish versus demersals in fishery landings [13]. When the amounts of pelagic and benthic resources are similar, the generalist demersals are able to outcompete the large pelagic specialists by feeding on both resource pools while the large pelagics only have access to one. Large pelagics proliferate as the ratio of pelagic to benthic resources increases (Fig. 16.2). Mechanistic food web models have also verified this statistical relationship [13,14].

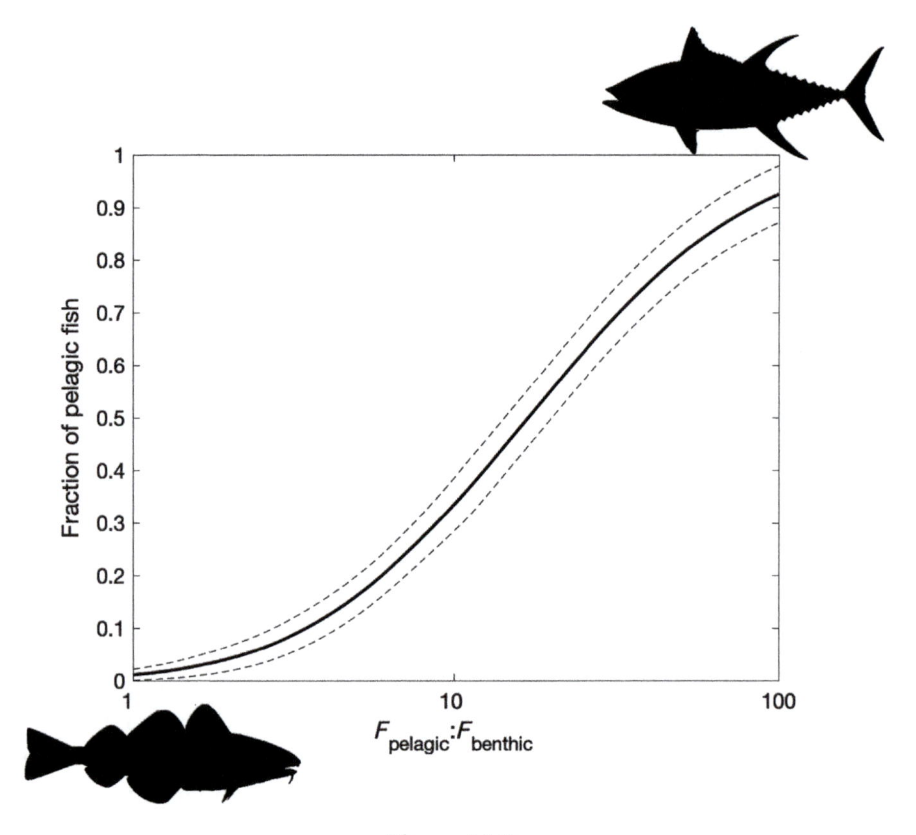

Figure 16.2
Relationship of the fraction of pelagic fish in fishery landings of combined pelagic and demersal fishes with the ratio of pelagic ($F_{pelagic}$) versus benthic ($F_{benthic}$) fractions of net primary production. *Based on D. van Denderen, M. Lindegren, B.R. MacKenzie, R.A. Watson, K.H. Andersen, Global patterns in marine predatory fish, Nat. Ecol. Evol. 2 (2018) 65–70, Figure 3. Fish silhouettes courtesy: The Integration and Application Network, University of Maryland Center for Environmental Science (ian.umces.edu/symbols/).*

This partitioning of production explains the latitudinal patterns in the distributions of large pelagic fish and demersal fish. In oligotrophic, continuously stratified regions like the subtropical gyres, the majority of NPP is recycled within the mixed layer via microbial pathways that support microzooplankton grazers in the pelagic zone. In contrast, the light-limited high latitudes experience strong but short blooms in NPP with high interannual variability. The variability in bloom timing can lead to a mismatch between NPP and the zooplankton grazer population, which has been reduced to low levels via deep winter mixing, resulting in a fraction of ungrazed NPP available for export [15]. Temperature additionally explains the latitudinal patterns in pelagic versus benthic resources, and thus large pelagic versus demersal fish, through its effect of increasing remineralization rates during export, thereby reducing the amount of NPP that reaches the seafloor [16,17].

16.3 Expectations under climate change

It is well known that global air and ocean temperatures are increasing due to anthropogenic greenhouse gas emissions [18]. Higher temperatures favor large pelagics over demersal via two mechanisms: increasing remineralization rates and increasing stratification, both of which act to reduce export. One related measure to export production is the Zooplankton–Phytoplankton Coupling metric, ZPC, of Stock et al. [19], which reflects the proportion of total primary production that is consumed by zooplankton. Global biogeochemistry projections (2051–2100) under a high emission scenario (Representative Concentration Pathway 8.5) compared to mean contemporary simulations (1951–2000) indicate that ZPC generally increases with climate change [19]. These increases are particularly large in mid- and high-latitude regions where climate change leads to shoaling of the winter mixed layer depth [19]. On average the regions with strong increases in ZPC demonstrate increases in zooplankton ingestion rates and decreases in the loss of phytoplankton via aggregation, thus suggesting an increase in the ratio of pelagic resources to benthic resources. Following the results of van Denderen et al. [13] and Petrik et al. [14], these increases in ZPC would lead to a higher fraction of large pelagic fishes in regions that historically have been dominated by demersal predators. There will, of course, be other constraints on the abundance of large pelagics fishes at high latitudes under climate change. These include the effects of temperature on metabolic physiology in addition to whether there is enough total productivity to support higher trophic levels.

Despite their dissimilarities in prey and habitat preferences, large pelagic and demersal fishes exemplify the Periodic life history strategy selected for in seasonally variable environments. It has long been hypothesized by fisheries oceanographers that the survival of larvae and their recruitment to the adult population is influenced by environmental variability (e.g., Ref. [20]). Spawning events should evolve to produce larvae that temporally or spatially match the ideal oceanographic conditions for survival, such as plankton blooms or retentive circulation, however, the variability of these features often leads to failures [21–23]. Thus the long life spans of Periodic fishes are an adaptation to variable recruitment success where longevity serves the purpose of repeated recruitment attempts [24,25] rather than attaining the large sizes of a top predator like the Equilibrium species. Furthermore, their high fecundity also increases the chance that some proportion of offspring will encounter favorable conditions.

However, the E–P–O framework uses mean life history traits for each species, but these traits are plastic and a range of phenotypes may be present within each fish stock [26,27]. There is evidence that fishing pressure and climate change have altered the life history traits of populations, such as age at maturity [28] and maximum size [29]. For example, fishing has depleted many long-lived, large, higher trophic level fishes that have been replaced by faster growing, earlier maturing species of lesser trophic level [30]. Similarly, temperature rise has increased the occurrence of species with Opportunistic and Periodic strategies along European margins [31].

In general, fishing pressure has selected for smaller sizes and earlier maturity ages, which decreases reproductive capacity of population [32]. In addition, fishing truncates the age and size structure of populations [33]. These changes act to increase the influence of recruitment success and environmental conditions on the populations, especially for fishes whose reproductive traits vary with age, such as the Periodic species cod, haddock, plaice, and winter flounder [34]. First, older fishes produce more larvae with increased growth rates and starvation tolerance [33] that may be able to survive in a broader range of environmental conditions. Thus when these spawners are removed from the population, survival of the larvae produced by the younger population is more variable due to the need for meeting a narrower range of environmental conditions each year. Second, spawning time often varies with age [33] such that a population with an intact age structure spreads the risk over a greater period of time so there is a greater probability of larvae temporally matching the ideal conditions for survival. The shifting of populations from "bet-hedging" Periodic strategies to Opportunistic life histories will be further exacerbated by the asymmetric effects of climate change on the timing of spring blooms and spawning. Spawn timing is largely affected by temperature, whereas the spring bloom is affected by additional factors like mixed layer depth, such that climate change could increase the proportion of mismatches. This is particularly significant for populations that spawn in specific geographic locations [35,36].

16.4 Conclusion

Climate change has the potential to drastically alter marine ecosystems [37]. The warming that has occurred over the past several decades produced observable consequences for marine organisms including smaller body sizes, faster maturation rates, vertical displacements, and shifts in geographic ranges [29,38,39]. These changes directly influence predator–prey relationships, including fishing (humans as predators), that depend on size ratios and overlapping habitats. Managing the conservation and sustainable fishing of marine resources on a species-by-species basis under multiple, often interacting, impacts will not be straightforward. Fortunately, life history theory in combination with a trophodynamic understanding of functional types provides frameworks for anticipating the effects of disturbance, including climate change. Thusly these frameworks serve as potentially powerful tools for guiding ecosystem-based fisheries management.

References

[1] K.O. Winemiller, K.A. Rose, Patterns of life-history diversification in North American fishes: implications for population regulation, Can. J. Fish. Aquat. Sci. 49 (1992) 2196–2218.
[2] K.O. Winemiller, Life history strategies, population regulation, and implications for fisheries management, Can. J. Fish. Aquat. Sci. 62 (2005) 872–885.

[3] D.H. Secor, Migration Ecology of Marine Fishes., Johns Hopkins University Press, Baltimore, MD, 2015. 292 pp.
[4] P. Lehodey, I. Senina, R. Murtugudde, A spatial ecosystem and populations dynamics model (SEAPODYM)—modeling of tuna and tuna-like populations, Prog. Oceanogr. 78 (2008) 304–318.
[5] D.G. Boyce, D.P. Tittensor, B. Worm, Effects of temperature on global patterns of tuna and billfish richness, Mar. Ecol. Prog. Ser. 355 (2008) 267–276.
[6] D. Pauly, Tropical fishes: patterns and propensities, J. Fish Biol. 53 (1998) 1–17.
[7] L.P. Garrison, J.S. Link, Dietary guild structure of the fish community in the Northeast United States continental shelf ecosystem, Mar. Ecol. Prog. Ser. 202 (2000) 231–240.
[8] C. Bulman, F. Althaus, X. He, N.J. Bax, A. Williams, Diets and trophic guilds of demersal fishes of the south-eastern Australian shelf, Mar. Freshw. Res. 52 (2001) 537–548.
[9] B. Worm, D.P. Tittensor, Range contraction in large pelagic predators, Proc. Natl. Acad. Sci. U.S.A. 108 (2011) 11942–11947.
[10] J.H. Ryther, Photosynthesis and fish production in the sea, Science 166 (1969) 72–76.
[11] K.D. Friedland, C. Stock, K.F. Drinkwater, J.S. Link, R.T. Leaf, B.V. Shank, et al., Pathways between primary production and fisheries yields of large marine ecosystems, PLoS One 10 (2012) e0133794.
[12] C.A. Stock, J.G. John, R.R. Rykaczewski, R.G. Asch, W.W.L. Cheung, J.P. Dunne, et al., Reconciling fisheries catch and ocean productivity, Proc. Natl. Acad. Sci. U.S.A. 114 (2017) E1441–E1449.
[13] D. van Denderen, M. Lindegren, B.R. MacKenzie, R.A. Watson, K.H. Andersen, Global patterns in marine predatory fish, Nat. Ecol. Evol. 2 (2018) 65–70.
[14] C.M. Petrik, C.A. Stock, K.H. Andersen, P.D. van Denderen, J.R. Watson, Bottom-up drivers of global patterns demersal, forage, and pelagic fishes, Prog. Oceanogr (in press).
[15] M.J. Lutz, K. Caldeira, R.B. Dunbar, M.J. Behrenfeld, Seasonal rhythms of net primary production and particulate organic carbon flux to depth describe the efficiency of biological pump in the global ocean, J. Geophys. Res. 112 (2007) C10011.
[16] L.R. Pomeroy, D.O.N. Deibel, Temperature regulation of bacterial activity during the spring bloom in Newfoundland coastal waters, Science 233 (1986) 359–361.
[17] E.A. Laws, P.G. Falkowski, W.O. Smith, H. Ducklow, J.J. McCarthy, Temperature effects on export production in the open ocean, Global Biogeochem. Cycles 14 (2000) 1231–1246.
[18] IPCC, Climate change 2013: The physical science basis, in: T.F. Stocker, D. Qin, G.-K. Plattner, M. Tignor, S.K. Allen, J. Boschung, et al. (Eds.), Contribution of Working Group I to the Fifth Assessment Report of the Intergovernmental Panel on Climate Change, Cambridge University Press, Cambridge, United Kingdom and New York, 2013. 1535 pp.
[19] C.A. Stock, J.P. Dunne, J.G. John, Drivers of trophic amplification of ocean productivity trends in a changing climate, Biogeosciences 11 (24) (2014) 7125–7135.
[20] R.A. Myers, When do environment-recruit correlations work? Rev. Fish Biol. Fish. 8 (1998) 285–305.
[21] T.D. Iles, M. Sinclair, Atlantic herring: stock discreteness and abundance, Science 215 (1982) 627–633.
[22] D.H. Cushing, The gadoid outburst in the North Sea, Journal du Conseil Permanent International pour l'Exploration de la Mer 41 (1984) 159–166.
[23] D. Hedgecock, Does variance in reproductive success limit effective population sizes of marine organisms, in: A.R. Beaumont (Ed.), Genetics and Evolution of Aquatic Organisms, Chapman & Hall, London, 1994, pp. 122–134.
[24] A. Longhurst, Murphy's law revisited: longevity as a factor in recruitment to fish populations, Fish. Res. 56 (2002) 125–131.
[25] M.S. Love, M. Yoklavich, L.K. Thorsteinson, The Rockfishes of the Northeast Pacific, University of California Press, 2002.
[26] D.A. Roff, The Evolution of Life Histories: Theory and Analysis, Chapman and Hall, New York, 1992.
[27] S.C. Stearns, The Evolution of Life Histories., Oxford University Press, Oxford, 1992.
[28] M. Heino, B. Díaz Pauli, U. Dieckmann, Fisheries-induced evolution, Annu. Rev. Ecol., Evol., Syst. 46 (2015) 461–480.

[29] M. Daufresne, K. Lengfellner, U. Sommer, Global warming benefits the small in aquatic ecosystems, Proc. Natl. Acad. Sci. U.S.A. 106 (2009) 12788–12793.

[30] R.A. Myers, B. Worm, Rapid worldwide depletion of predatory fish communities, Nature 423 (2003) 280–283.

[31] L. Pecuchet, M. Lindegren, M. Hidalgo, M. Delgado, A. Esteban, H.O. Fock, et al., From traits to life-history strategies: deconstructing fish community composition across European seas, Global Ecol. Biogeogr. 26 (2017) 812–822.

[32] T. Rouyer, G. Ottersen, J.M. Durant, M. Hidalgo, D.Ø. Hjermann, J. Persson, et al., Shifting dynamic forces in fish stock fluctuations triggered by age truncation? Global Change Biol. 17 (2011) 3046–3057.

[33] S.A. Berkeley, M.A. Hixon, R.J. Larson, M.S. Love, Fisheries sustainability via protection of age structure and spatial distribution of fish populations, Fisheries 29 (2004) 23–32.

[34] T. Brunel, Age-structure-dependent recruitment: a meta-analysis applied to Northeast Atlantic fish stocks, ICES J. Mar. Sci. 67 (2010) 1921–1930.

[35] R.G. Asch, Climate change and decadal shifts in the phenology of larval fishes in the California Current ecosystem, Proc. Natl. Acad. Sci. U.S.A. 112 (2015) E4065–E4074.

[36] R.G. Asch, B. Erisman, Spawning aggregations act as a bottleneck influencing climate change impacts on a critically endangered reef fish, Divers. Distrib. 24 (2018) 1712–1728.

[37] W.W. Cheung, V.W. Lam, J.L. Sarmiento, K. Kearney, R. Watson, D. Pauly, Projecting global marine biodiversity impacts under climate change scenarios, Fish Fish. 10 (3) (2009) 235–251.

[38] W.W. Cheung, J.L. Sarmiento, J. Dunne, T.L. Frölicher, V.W. Lam, M.D. Palomares, et al., Shrinking of fishes exacerbates impacts of global ocean changes on marine ecosystems, Nat. Clim. Change 3 (3) (2013) 254.

[39] M.L. Pinsky, B. Worm, M.J. Fogarty, J.L. Sarmiento, S.A. Levin, Marine taxa track local climate velocities, Science 341 (6151) (2013) 1239–1242.

SECTION 4

Changing Fisheries and Seafood Supply

CHAPTER 17

Fisheries and seafood security under changing oceans

Elsie M. Sunderland[1,2], Hing Man Chan[3] and William W.L. Cheung[4]

[1]Department of Environmental Health, Harvard T.H. Chan School of Public Health, Boston, MA, United States [2]Harvard John A. Paulson School of Engineering and Applied Sciences, Cambridge, MA, United States [3]Department of Biology, University of Ottawa, ON, Canada [4]Institute for the Oceans and Fisheries, The University of British Columbia, Vancouver, British Columbia, Canada

Chapter Outline

17.1 Introduction

A goal of the Nippon Foundation Nereus Program is to generate knowledge and capacity that supports marine protection policies and a sustainable future ocean (see Chapter 1: Rethinking oceans as coupled human-natural systems). An important benefit of ocean sustainability is the support of long-term meaningful livelihoods and nutritious food for many human populations, particularly for coastal indigenous populations that rely on fish and seafood for trade, food, and the cultural traditions that build and bind their communities [1]. Global change, driven by human emissions of carbon dioxide and other pollutants, such as mercury, and the overexploitation of natural resources, is challenging the capacity of our current and future ocean to provide such benefits [2]. Global change poses a threat to the livelihood, food security, and safety of the human population through multiple components of the coupled human–natural marine system (see Chapter 1: Rethinking oceans as coupled

Predicting Future Oceans.
DOI: https://doi.org/10.1016/B978-0-12-817945-1.00017-4

human-natural systems. Thus, studying these components, their linkages, and their changes under alternative scenarios of human drivers is crucially important to inform evidence-based policy-making in setting targets and implementing interventions for the human dimensions of ocean sustainability [3].

17.2 Fisheries management

The effects of climate change on future fisheries are dependent on the interactions between changing fish stocks and fishing activities, but also the mediation provided by different types of management. Climate change is affecting many aspects of fish stocks and marine ecosystem productivity, including shifting distributions, species assemblages and richness, and the biomass production potential [3]. Some species are particularly sensitive to climate shifts such as temperature increases (e.g., bluefish), while others exhibit less change in foraging territory (see Chapter 20: Climate change adaptation and spatial fisheries management). At a community level, many fisheries are likely to be altered by these northward migrations and redistributions of fish stocks.

Lam (see Chapter 18: Projecting economics of fishing and fishing effort dynamics) explains that these changes in fish stocks affect the revenues, costs and thus the profitability of fishing. Simultaneously, fishers attempt to cope with these climate impacts on their livelihood by changing their fishing activities, for example, by shifting their fishing grounds or targeted species. Moreover, fishing activities are regulated and largely shaped by fisheries management and marine conservation measures. The effectiveness of ocean governance is affected by changing fish stocks and fishing activities, as highlighted by Selden and Pinsky (see Chapter 20: Climate change adaptation and spatial fisheries management). Adaptive co-management efforts and additional precautions in management strategies are needed to promote resilience in these fisheries.

In return, these changes in fisheries and their management will affect many communities and require adaptation to mitigate impacts. One example provided in this section relates to the changing effectiveness of the "Plaice Box" in the North Sea, a special area designed for species conservation, when juvenile fish moved outside of the "Box" due to changes in sea temperature. These types of interactions illustrate the coupled dynamics of climate, fishery adaptation, and conservation measures, as ineffective fisheries management will also affect the long-term benefits of fishers and the fishing sectors. Incorporation of the linkages between changing fish stocks, fishing activities, and fisheries management thus forms an important challenge for "predicting future oceans" to inform fisheries policies in the face of climate change.

Approaches and analytical tools are being developed for the fishing sectors and the management of fisheries to cope and adapt to these climate challenges as highlighted in the contributions of this section. Aspects of fisheries management that show promise in helping fisheries adapt to climate change include economics-based interventions and spatially explicit management measures. It is essential that management strategies for fisheries consider the effects of climate change on sustainable yields, as illustrated by Tanaka (see Chapter 19: Integrating environmental information into stock assessment). Stochasticity in climate can be readily simulated using the best-available models and tools to inform fisheries policy and decision-making, such as setting quotas that account for the effects of environmental changes. Opportunities exist for reducing the risks of climate change on fisheries through the use of informed management. This includes the incorporation of information on changing oceans into scientific assessment for policy making, as shown by Tanaka (see Chapter 19: Integrating environmental information into stock assessment) through case studies in the east coast of North America. This highlights the important role of science in helping fisheries to adapt to climate change impacts.

17.3 Fisheries and food supply

Fisheries support the livelihoods of 10%–12% of the world's population and provide at least 20% of protein to more than 3 billion individuals (see Chapter 18: Projecting economics of fishing and fishing effort dynamics). The chapter by Chen (see Chapter 23: Future global seafood markets) discusses changes that are occurring in the global seafood market as many countries transition from low- to predominantly middle-income status. Seafood is a commodity good and thus local fish landings show relatively small price elasticities with changes in individual income. However, low-income countries tend to consume different fish species and types compared to middle- and high-income countries. Growth in the aquaculture sector has mainly been driven by fish for low-income groups in China. High-income countries tend to consume larger quantities of wild finfish and crustaceans. Global population growth means demand for fisheries products will increase substantially in the future, but most capture fisheries around the world are fully or overexploited [4]. This means growth in aquaculture production, particularly in Asia and Africa, is key to meeting demand for seafood in the future (see Chapter 23: Future global seafood markets).

The chapter by Oyinlola describes how growth in the aquaculture sector is likely to change under a changing climate and as global demand for seafood continues to increase (see Chapter 22: Mariculture: perception and prospects under climate change). Many types of aquaculture are controlled in such a way that there is low environmental risk and little interaction with natural ecosystems. One exception is mariculture that relies on the natural conditions of marine ecosystems to remove waste and also often uses wild capture fish (typically forage fish) as food for farmed species, exerting additional pressure on capture fisheries. Ocean warming may increase the risk of infectious disease spread and changes in physical conditions, and storm frequency may pose an additional threat to the physical infrastructure associated with aquaculture operations. Proper implementation of the ecosystem approach in aquaculture will help address some of these problems.

17.4 Fisheries, food security, and safety

The chapter by Kenny (see Chapter 24: Climate change, contaminants, and country food) describes why Indigenous populations globally are particularly vulnerable to the shifts in marine ecosystems associated with climate change due to their relative high consumption rates of seafood and marine mammals. These populations rely on foods harvested from their local marine environment for key nutrients including high quality protein, vitamins, and essential trace elements that are not easily replaced by market foods. In addition to the nutritional benefits of seafood, harvesting and sharing of these foods has important social and cultural benefits. In Arctic communities, shifts in sea ice distribution are negatively impacting the accessibility of hunters to harvest their traditional foods. Climate change is also exacerbating the accumulation of mercury and other persistent organic contaminants in predatory fish and marine mammals. Relationships between climate, contaminants, and the risks and benefits of the consumption of traditional foods of Indigenous populations are complex. However, maintaining the integrity of fishing resources is essential for protecting the health and well-being of the indigenous peoples living in these communities.

In additional to global environmental change, other dimensions of global change, such as economic development, can also lead to unintended consequences on fisheries and seafood security. Wabnitz (see Chapter 21: Seafood consumption practices under climate change) highlights such potential trade-offs using Palau's ecotourism fisheries and their potential implications for ecosystem health and local food security as a case study. The development of ecotourism, facilitated by globalization, has attracted international tourists and provided an important source of revenues. However, an often overlooked effect of such development is the large increase in consumption of local seafood from these visitors, potentially competing for fish for local consumption. The increased demand for seafood may also put pressure on fish stocks in local waters. As local Palauans are closely linked to fishing as a cultural activity and to seafood as a main source of food, the growing seafood demand to support the blooming ecotourism may impact Palau's culture and food security. However, analyzing economic and ecosystem dynamics, including climate change, can highlight opportunities to resolve such trade-offs, for example, through the promotion of sustainably sourced seafood that does not undermine local food security and the health of the very ecosystems that the sector depends upon.

17.5 Summary

This section highlights the interconnection between global changes such as climate change, globalization of seafood markets, and ecotourism development, and ecosystems, fisheries, and the livelihood and food security and safety of human communities. Global changes render the security of livelihood and food for dependent communities uncertain, and the interconnection between the coupled human and natural marine systems can lead to trade-offs between economic development, human well-being, and environmental health. Nevertheless, the contributions in this section (and indeed throughout this volume) highlight the important role of social and natural sciences in supporting the development of solution options to resolve these global change challenges and the trade-offs associated with different dimensions of sustainable development. It is important to collect local knowledge through participatory research on the perceived changes in the environment and the consequences on their health and livelihood. A good understanding of the capacity and strategies of communities and managers to adapt to the changing environment and seafood supply is also critical to develop local and regional adaptation plans. Strong scientific evidence generated by the integration of research efforts on physical, biological, and social sciences is needed to champion the national and international support required to implement effective interventions for resource management and public health.

References

[1] Millennium Ecosystem Assessment, Ecosystems and Human Well-Being, vol. 5, Island Press, Washington, DC, 2005.

[2] H.-O. Pörtner, D.M. Karl, P.W. Boyd, W. Cheung, S.E. Lluch-Cota, Y. Nojiri, et al., Ocean systems, Climate Change 2014: Impacts, Adaptation, and Vulnerability Part A: Global and Sectoral Aspects Contribution of Working Group II to the Fifth Assessment Report of the Intergovernmental Panel on Climate Change., Cambridge University Press, 2014, pp. 411–484.

[3] W.W.L. Cheung, The future of fishes and fisheries in the changing oceans, J. Fish. Biol. [Internet] 92 (3) (2018) 790–803. Available from: http://doi.wiley.com/10.1111/jfb.13558.

[4] FAO, The State of World Fisheries and Aquaculture 2018 – Meeting the Sustainable Development Goals, Rome, 2018.

CHAPTER 18

Projecting economics of fishing and fishing effort dynamics in the 21st century under climate change

Vicky W.Y. Lam

Nippon Foundation Nereus Program and Changing Ocean Research Unit, Institute for the Oceans and Fisheries, University of British Columbia, Vancouver, BC, Canada

Chapter Outline

18.1 Introduction

Global marine fisheries are important for human well-being, providing food and nutritional security, and jobs and incomes for people living in coastal countries. Global marine fisheries landings are estimated officially at between 80 and 85 million tonnes per year since 1990 [1], with corresponding mean annual gross revenues fluctuating around US$100 billion annually [2]. When accounting for unreported catches, a study estimated annual global catch to be about 130 million tonnes [3]. The global fisheries sector supports the livelihoods of between 660 and 820 million people, directly or indirectly, which is about 10%–12% of the world's population [4] if the dependents of fishers are taken into account. Globally, fisheries and aquaculture also provide more than 3.1 billion people with 20% of their animal protein needs [4] and are a crucial source of micronutrients [5]. However, along with nonclimatic drivers, such as population growth, marine pollution, changes in markets, demographics, and overexploitation, climate change is considered to be one of the major challenges that will significantly shape the future of global fisheries. Several studies suggest that these nonclimatic stresses and changes in management regimes may have a

Predicting Future Oceans.
DOI: https://doi.org/10.1016/B978-0-12-817945-1.00018-6

greater impact on fisheries than climate change in the short term [6], while increasing uncertainty in climate poses a major threat to world fisheries in the long run [7].

Changes in ocean conditions, including temperature, sea ice extent, salinity, pH, oxygen levels, and circulation, lead to shifts in the distribution range of marine species [8–11], changes in primary and secondary productivity, and shifts in timing of biological events [7]. Warmer temperatures may also lead to decreases in maximum body sizes of marine fishes [12]. The combined effects of the predicted distributional shift and changes in ocean productivity under climate change are expected to lead to changes in species composition [13] and hence global redistribution of maximum catch potential (MCP), with projected increases in MCP in high latitudinal regions and decreases in the tropics [14–16]. These changes have large implications for people who depend on fish for food and income, and thus the contribution of fisheries to the global economy [15,17].

In this chapter I will first provide a review of various existing models and studies that have projected the fisheries catch and analyzed the subsequent economic impacts under climate change. Then I will highlight the major components of our latest model for projecting future catch change and subsequent economic impacts under climate change, and describe the ways in which this model improves and is different from the existing ones. Finally, I will further discuss the implications of the economic impacts for various countries under climate change and how the projected fishing effort dynamics may play an important role in planning for climate change-related adaptation measures and solutions.

18.1.1 Existing economic models projecting fisheries catch and economic impact under climate change

The potential impacts of climate change on fish and ecosystems add further pressure to fisheries, which have already been exposed to multiple natural and anthropogenic stressors including extreme weather events, fluctuation in fish stocks, pollution, sedimentation, habitat destruction, shoreline development, coral mining, destructive fishing practices, and overfishing. Evidence of the impact of climate change on fisheries in terms of change in catch and catch composition have already been observed in many different regions [18–20]. Several global-scale studies using different models, which linked the physical and biogeochemical changes to biological to ecological processes, have shown that the global average catch will be adversely impacted under the Intergovernmental Panel on Climate Change (IPCC) Representative Concentration Pathways (RCP) scenarios.

Cheung et al. [14] developed the Dynamic Bioclimate Envelope Model (DBEM) (Fig. 18.1A) which simulates changes in the distributions and relative abundances of the marine species under the climate change scenarios. The DBEM is a mechanistic species distribution model that links prediction of habitat suitability to their spatial and temporal population dynamics and ecophysiology [22,23]. To capture the dynamics and

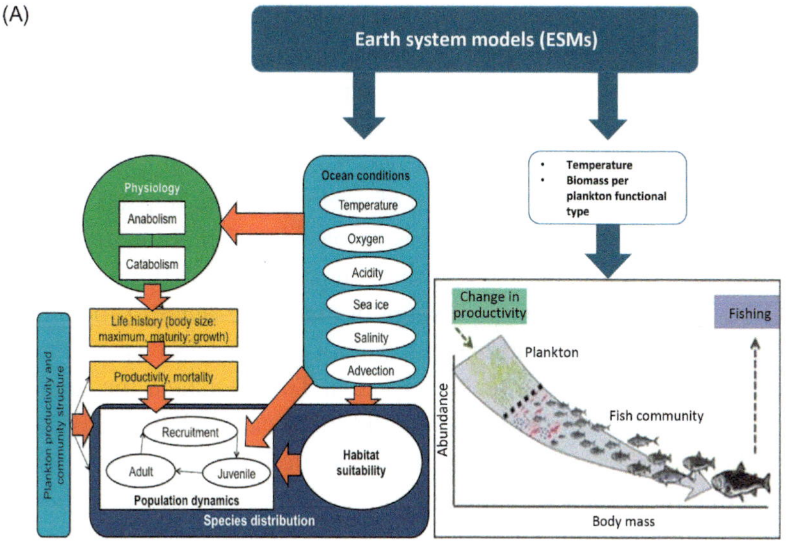

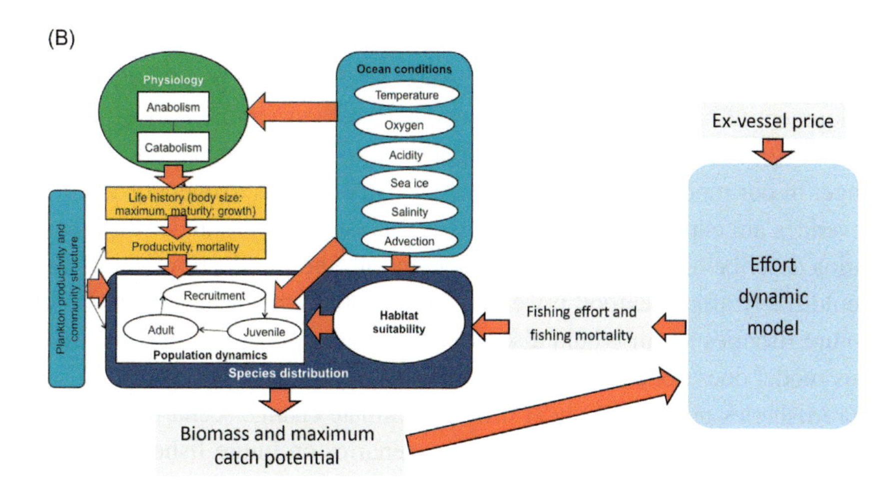

Figure 18.1

(A) Diagram showing the models (dynamic bioclimate envelope model and size-spectrum model) for simulating the future changes in fisheries catch potential under climate change scenarios. (B) Diagram showing the linkage between the DBEM and the effort dynamic model. *DBEM*, Dynamic bioclimate envelope model. *Modified from M. Barange, T. Bahri, M. Beveridge, K. Cochrane, S. Funge-Smith, F. Poulain, Impacts of climate change on fisheries and aquaculture: synthesis of current knowledge, adaptation and mitigation options, in: FAO Fisheries and Aquaculture Technical Paper (FAO) ENG No. 627, 2018 [21].*

interactions of marine food webs, a coupled model that combined a physical–biogeochemical model with a dynamic, size-based ecosystem model (Fig. 18.1A) was developed by Blanchard et al. [16] to project future change in fish biomass and production under climate change in large marine ecosystems. This model was later linked to the model of human response and was used to assess fish production in societies dependent on fisheries under changing climate [15,24]. Understanding the impact of climate change on the socioeconomic aspects through fisheries is crucial for designing socioeconomic policy, food sustainability strategies, adaptation, mitigation, and various fisheries management strategies; however, socioecological linkage assessment is challenging and hence quantitative projection of socioeconomic impacts is usually lacking in the models described above. Other size-based food web models, such as Princeton Ocean ecosystem Model [25] and the composite (hybrid) models, such as, Madingley [26] that projected the future change in marine resources production, have not yet linked to the economic model and hence have not been used to assess the direct economic impacts of climate change on society [27]. To understand the impact of climate change economically, we linked the physical and biological models (DBEM) to the economic parameters to examine the potential economic impact of climate change, focusing on modeling the effects of climate change on revenues through changes in the amount and composition of catches [28].

When modeling and projecting the change in fisheries revenues, it is crucial to understand both the changes in the catch and the price of fish. Price dynamics are affected by the interplay between the supply and demand of seafood products. The preference of consumers and the development of other food supply sectors, such as aquaculture may also affect the future price of seafood and therefore have the potential to alter the economic impact under climate change. In our model, price dynamics are incorporated as exogenous factors and the effects on revenues are explored by conducting an analysis of prices under different scenarios. These scenarios describe how the future development of other production sectors in the economy would likely affect seafood prices. In the meantime the change in the production from the capture fisheries sector would also affect the prices of seafood commodities. However, this model does not consider the price adjustment due to the change in supply from marine capture fisheries production under different climate change scenarios. In this model we explored the effects of different seafood price scenarios on future fisheries' maximum revenue potential (MRP). These price scenarios are based on the results from the United Nations' "Fish to 2020" study [29] that includes (1) baseline; (2) faster aquaculture expansion; (3) slower aquaculture expansion; (4) lower production in China; and (5) fishmeal and fish oil efficiency scenarios [28]. We compare projected MRP from these alternative price scenarios to the "constant price" scenario in which the price stays the same as that in the 2000s. The "constant price" scenario is different from the "baseline" price scenario as the price under the latter scenario changes from 8.7% to 34.8% in the seafood commodity groups.

Specifically, changes in total potential catches may not directly equate to changes in revenues from fisheries. First, climate change may affect catches of species that have different prices in the market. Second, even though potential catches are expected to increase in some countries' Exclusive Economic Zones (EEZs), the fishing sector of these countries may still suffer if they include a substantial distant water fishing fleet that operates in foreign waters that are impacted by climate change. Third, the effects of climate on fishing effort itself, through more extreme weather or changes in vertical fish distribution, may make it more difficult for fishers to access the marine resources.

18.1.2 Incorporating effort dynamics into the model

Previous studies assessing climate change impacts on marine fisheries focused on the impacts from changing ocean conditions on fish distribution, fish production, and the related human well-being. However, these studies often use simple assumptions of fishing effort scenarios without accounting for the effects of changing productivity of living marine resources, fisheries management measures, and the economics of fishing on fishers' behavior and fishing effort [16]. In return, the fishers' decisions and the dynamics of fishing effort would impact the abundance of the fisheries resources. Human activity was incorporated into the Earth system modeling framework by using the BiOeconomic mArine Trophic Size-spectrum (BOATS) model, a bioenergetically constrained macroecological life history fish model that is coupled directly with an economic model [30–32]. In this model the fish catch amount in each grid cell is determined by local biomass density and interactive fishing effort over time. However, this model did not model how the shift in species distribution and composition over time under climate change would impact the spatial distribution and magnitude of fishing costs.

When considering changes in fishing effort dynamics, the projection of future catches and the subsequent fisheries economies may be substantially different from previous studies. Also the projected impact on national economies through fisheries have different implications for different countries. Thus we have recently developed a coupled human–ecosystem model to incorporate the change in fishing effort dynamic into our species distribution model (DBEM). This model is a more holistic approach for projecting future impact of climate change on fisheries by linking the change in fishing effort which is based on the change in catch and profit (and vice versa) to the biological model that projects the change in catch potential (DBEM). Here, profit refers to the financial gains estimated by the difference between the total fishing revenues and the operating fishing costs and subsidies at the end of each fishing season. Based on the dynamic version of the Gordon–Schaefer model [33], fishers seek to maximize their profit and hence the profit from their harvest in each year determines the active fishing effort (in terms of the number of fishing vessels) for the next year after considering the change in biomass of exploited species under changing environmental conditions, operating cost of fishing, costs of purchasing new vessels, and the depreciation cost of existing vessels. If the fishing is profitable, we expect fishing effort to increase over time. Similarly, if profits are negative, fishing effort will decrease (potentially to zero) in the next fishing season. Then, the estimated fishing effort for the next year is fed back to the DBEM by converting it into fishing mortality, which is combined with other biological and environmental variables for projecting next year's biomass and MCP of each marine species. In this grid cell-based model, the total effort allocated to each cell is assumed to be proportional to the sum of biomass, catchability, and the ex-vessel price of the target groups [34]. The operating costs of fishing also change spatially when fish distributions shift, which leads to changes in travel (fuel) costs of fishing vessels.

This model highlights the importance of projecting and applying plausible fishing scenarios in understanding climate change impacts on fisheries. We combined the DBEM with the effort dynamic model for projecting the global MCP under different climate change scenarios (Fig. 18.1B). Under this model, we assume the fisheries are under the open access system, which implies that the incentives for the fishers to allocate their effort are based only on the profit from their catch. Although there is government intervention and control in many countries' EEZs, open access conditions are still dominant in the high seas and many countries [35]. The open access management is still the dominant approach in many fisheries and hence leads to the collapse of many fisheries of the world [35,36]. Our model has the potential to incorporate different management scenarios, for example, imposing Marine Protected Areas by modifying several parameters in the model, such as imposing a cap on the effort, a cap on the catch amount, and modifying the catchability of different species, to project the change in MCP under climate change scenarios. Results from the recent study would have great implications on the design of fisheries' management plans for adapting to climate change.

18.1.3 Potential socioeconomic impacts

Globally, our study revealed that the marine fisheries revenues is projected to be negatively impacted in 89% of the world's fishing countries under the IPCC RCP 8.5 scenario in the 2050s relative to the current status [28] (Fig. 18.2). Assuming constant price, global MCP is projected to decrease globally by 7.7% (± 4.4%) by 2050 (average between 2041 and 2060) relative to 2000 (average between 1991 and 2010) under the business-as-usual scenario (RCP 8.5). In contrast, global fisheries revenue (or landed value at—MRP) is projected to decrease by 10.4% (± 4.2%), that is, about 35% more than the impact on MCP (Table 18.1) on a global scale. With an estimated total MRP of US$100 billion, the variation in projected change in MRP between different Earth System Models (ESMs) ranges from US$6 to 15 billion, which seems small in one particular sector but may amplify when the economic impact of fisheries-dependent (direct and indirect) sectors are considered [37]. The projected changes in the MCP in a country's EEZs do not directly translate to the change in the fisheries revenues of a country as the catch composition changes and vessels of some countries do not only fish in their own EEZs, but also in the high seas and other countries' EEZs. The implications of the impact on the fisheries sector under climate changes vary among different regions and countries, with the extent being the greatest in the coastal

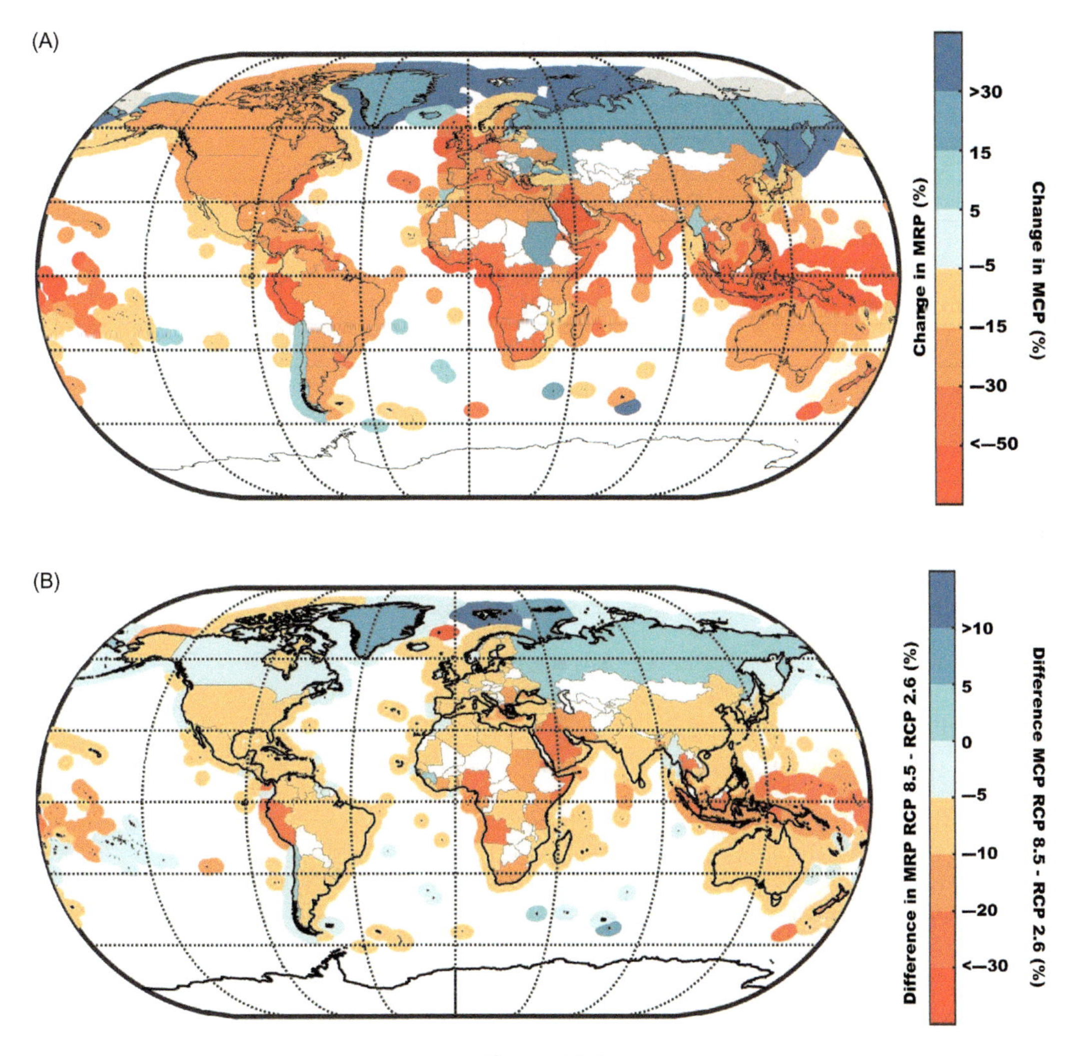

Figure 18.2

Impacts of climate change on MCP and MRP by the 2050s (average between 2041 and 2060) relative to the 2000s (average between 1991 and 2010): (A) mean percentage change in projected MCP of 280 EEZs and mean percentage change in projected MRP of 192 fishing nations in the 2050s relative to the level in the 2000s under RCP 8.5 scenario; (B) differences in percentage change in MCP and MRP between RCP 8.5 and RCP 2.6 scenarios in the 2050s. *EEZs*, Exclusive economic zones; *MCP*, maximum catch potential; *MRP*, maximum revenue potential; *RCP*, Representative Concentration Pathways.

low-income food-deficit countries including small island countries, African countries, and tropical Asian countries [28] (Fig. 18.3). These least developed countries usually rely heavily on fish and fisheries as a major source of otherwise limited animal protein, nutritional needs, and income and job opportunities. Therefore negative impacts on the catch and total fisheries

Table 18.1: Projected percentage change in global maximum catch potential (MCP) and fisheries maximum revenue potential (MRP) in the 2050s from the current status under different climate change scenarios.

	Model uncertainty				
	Percentage change in maximum catch potential				
	GFDL	IPSL	MPI	Mean	Standard deviation
RCP 2.6	− 1.66	− 8.49	− 2.03	− 4.06	3.84
RCP 8.5	− 4.44	− 12.66	− 6.02	− 7.71	4.36
	Percentage change in fisheries maximum revenue potential				
	GFDL	IPSL	MPI	Mean	Standard deviation
RCP 2.6	− 5.07	− 11.15	− 5.12	− 7.11	3.50
RCP 8.5	− 6.88	− 15.03	− 9.21	− 10.37	4.20

To explore the effects of different ESMs on the results, we use outputs from three ESMs that are available for the Coupled Models Intercomparison Project Phase 5 (CMIP5): the Geophysical Fluid Dynamics Laboratory Earth System Model 2M (GFDLESM2M), the Institute Pierre Simon Laplace Coupled Model Version Five-Medium Resolution (IPSL) (IPSL-CM5-MR), and Max Planck Institute for Meteorology Earth System Model (MPI-ESM-MR). *GFDL,* GFDLESM2M; *IPSL,* IPSL-CM5-MR; *MPI,* MPI-ESM-MR; *RCP,* Representative Concentration Pathways.

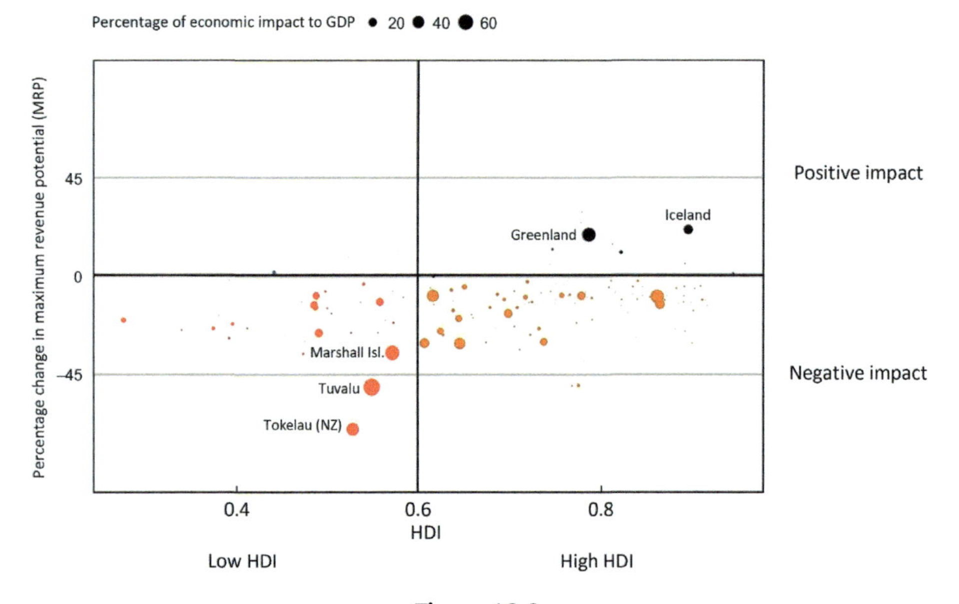

Figure 18.3

Percentage change in fisheries' MRP is mapped against HDI of countries. The bigger the size of the bubble the larger the percentage of economic impact of the fisheries sector to the total GDP. *GDP,* Gross domestic product; *HDI,* human development index; *MRP,* maximum revenue potential.

revenues obtained by these countries may have greater implications for their jobs, economies, food, and nutritional security than the impacts on high Human Development Index countries [5,38]. These least developed countries are also highly vulnerable to the impact of climate change but have relatively low capacity to adapt to rapid change [39,40].

Table 18.2: Percentage change in fisheries maximum revenue potential (MRP) in the 2050s relative to the 2000s under different price scenarios and projections.

	Percentage change in fisheries MRP in the 2050s relative to the 2000s							
	Constant price	Baseline	Faster aquaculture expansion	Lower China production	Fish mean and fish oil efficiency	Slower aquaculture expansion	Mean	Standard deviation
RCP 8.5	− 6.88	12.78	− 14.77	13.77	10.75	39.58	9.20	18.93
RCP 2.6	5.07	14.88	− 13.34	15.88	12.81	42.36	11.25	19.37

RCP, Representative Concentration Pathways.

Our projected MRP are most sensitive (i.e., with the largest changes from the "constant price" scenario) to the "slower aquaculture expansion" scenario. The increase in prices of all seafood commodity groups leads to an increase in MRP of approximately seven and nine times the magnitude of change under the "constant price" scenario for RCP 8.5 and RCP 2.6 scenarios, respectively (Table 18.2). Since global seafood production from aquaculture has been increasing at an annual rate of 8% in the past decade [41], this scenario is unlikely to happen based on the current trend [41]. Only two of six price scenarios show a negative impact on MRP (Table 18.2). Under the "faster aquaculture expansion" scenario, seafood prices of two seafood commodity groups (i.e., low-value food fish and mollusks) decrease, resulting in a more substantial decrease (more than double) in fisheries MRP relative to the "constant prices" scenario (Table 18.2). The majority of price scenarios assume an increase in price by 7%–57% by 2050 relative to 2000 that compensates for the projected decrease in catch and results in a projected increase in fisheries MRP even under the high-emission scenario (11%–40%).

Climate change may affect the subsequent benefits to indirect sectors (secondary and ancillary) that are related to fisheries, for example, fish canning, processing industries, and boat repair. Women are heavily engaged in the fishing industry, both fishing and gleaning directly and in the post-harvesting sectors, including selling and processing fish at fish landing sites and local markets. Women often work in canning, processing, and other ancillary economic activities related to marine resources, and are also involved in book-keeping operations and in taking care of the home while fishers are away. However, women in the post-harvest sector are usually in a disadvantaged position because of the marginalization of many fishing communities and their lack of access to credits and capacity development. Gender equity and empowering women's participation in climate change discourse and actions have been included in the Gender Action Plan of the United Nations Framework Convention on Climate Change [42]. However, the impacts of climate change on women in fisheries-related sectors requires much more attention and research focus.

Climate change may also worsen climate-related shocks and stresses, and hence is an obstacle for the progress of poverty reduction [43]. Hence, climate change adaptation and mitigation strategies must be human-centered with an emphasis on the need for poverty eradication, food security, empowering local stakeholders, and addressing power imbalances and inequity in fishing and aquaculture communities [42]. Also, more effort will be needed to understand the effect of climate change on the dynamics of fish prices and costs of fishing. Both of these changes would lead to the adjustment of the fishing effort and hence intensify or lessen the overcapacity issue. Studies have attempted to project how fishers may respond to changes in fish distribution and abundance by incorporating different management systems [30,44]. These studies found that fisheries did not simply follow the fish but are both driven and constrained by multiple other factors. Although a recent study showed that improved fisheries management can counteract the adverse impacts of climate change on fisheries [45], the understanding of the impacts of climate change on the effectiveness of management and trade practices [15] is still not adequate and will need to be further explored.

References

[1] FAO, The State of World Fisheries and Aquaculture 2016, FAO, Rome, 2016.

[2] W. Swartz, R. Sumaila, R. Watson, Global ex-vessel fish price database revisited: a new approach for estimating 'missing' prices, Environ. Resour. Econ. 56 (4) (2013) 467–480.

[3] D. Pauly, D. Zeller, Catch reconstructions reveal that global marine fisheries catches are higher than reported and declining, Nat. Commun. 7 (2016) 10244.

[4] FAO, The State of World Fisheries and Aquaculture, FAO, Rome, 2014.

[5] C.D. Golden, E.H. Allison, W.W.L. Cheung, M.M. Dey, B.S. Halpern, D.J. Mccauley, et al., Nutrition: fall in fish catch threatens human health, Nature 534 (2016) 317–320.

[6] A. Eide, Economic impacts of global warming: the case of the Barents Sea fisheries, Nat. Resour. Model. 20 (2007) 199–221.

[7] H.-O. Pörtner, D.M. Karl, P.W. Boyd, W. Cheung, S. Lluch-Cota, Y. Nojiri, et al., Ocean systems, in: Climate Change 2014: Impacts, Adaptation, and Vulnerability. Part A: Global and Sectoral Aspects. Contribution of Working Group II to the Fifth Assessment Report of the Intergovernmental Panel on Climate Change, 2014, pp. 411–484.

[8] W.W. Cheung, V.W. Lam, J.L. Sarmiento, K. Kearney, R. Watson, D. Pauly, Projecting global marine biodiversity impacts under climate change scenarios, Fish Fish. 10 (2009) 235–251.

[9] R.A. Clark, C.H. Fox, D. Viner, M. Livermore, North Sea cod and climate change – modelling the effects of temperature on population dynamics, Global Change Biol. 9 (2003) 1669–1680.

[10] M.L. Pinsky, B. Worm, M.J. Fogarty, J.L. Sarmiento, S.A. Levin, Marine taxa track local climate velocities, Science 341 (2013) 1239–1242.

[11] G. Rose, On distributional responses of North Atlantic fish to climate change, ICES J. Mar. Sci. 62 (2005) 1360–1374.

[12] W.W. Cheung, J.L. Sarmiento, J. Dunne, T.L. Frölicher, V.W. Lam, M.D. Palomares, et al., Shrinking of fishes exacerbates impacts of global ocean changes on marine ecosystems, Nat. Clim. Change 3 (2013) 254–258.

[13] G. Beaugrand, M. Edwards, V. Raybaud, E. Goberville, R.R. Kirby, Future vulnerability of marine biodiversity compared with contemporary and past changes, Nat. Clim. Change 5 (2015) 695–701.

[14] W.W. Cheung, V.W. Lam, J.L. Sarmiento, K. Kearney, R. Watson, D. Zeller, et al., Large-scale redistribution of maximum fisheries catch potential in the global ocean under climate change, Global Change Biol. 16 (2010) 24–35.

[15] M. Barange, G. Merino, J. Blanchard, J. Scholtens, J. Harle, E. Allison, et al., Impacts of climate change on marine ecosystem production in societies dependent on fisheries, Nat. Clim. Change 4 (2014) 211–216.

[16] J.L. Blanchard, S. Jennings, R. Holmes, J. Harle, G. Merino, J.I. Allen, et al., Potential consequences of climate change for primary production and fish production in large marine ecosystems, Philos. Trans. R. Soc. B 367 (2012) 2979–2989.

[17] U.R. Sumaila, W.W. Cheung, V.W. Lam, D. Pauly, S. Herrick, Climate change impacts on the biophysics and economics of world fisheries, Nat. Clim. Change 1 (2011) 449–456.

[18] P.R. Last, W.T. White, D.C. Gledhill, A.J. Hobday, R. Brown, G.J. Edgar, et al., Long-term shifts in abundance and distribution of a temperate fish fauna: a response to climate change and fishing practices, Global Ecol. Biogeogr. 20 (2011) 58–72.

[19] J.A. Nye, J.S. Link, J.A. Hare, W.J. Overholtz, Changing spatial distribution of fish stocks in relation to climate and population size on the Northeast United States continental shelf, Mar. Ecol. Prog. Ser. 393 (2009) 111–129.

[20] M.L. Pinsky, M. Fogarty, Lagged social-ecological responses to climate and range shifts in fisheries, Clim. Change 115 (2012) 883–891.

[21] M. Barange, T. Bahri, M. Beveridge, K. Cochrane, S. Funge-Smith, F. Poulain, Impacts of climate change on fisheries and aquaculture: synthesis of current knowledge, adaptation and mitigation options, FAO Fisheries and Aquaculture Technical Paper (FAO), 2018.

[22] W.W.L. Cheung, M.C. Jones, G. Reygondeau, C.A. Stock, V.W.Y. Lam, T.L. Frolicher, Structural uncertainty in projecting global fisheries catches under climate change, Ecol. Model. 325 (2016) 57–66.

[23] W.W. Cheung, J. Dunne, J.L. Sarmiento, D. Pauly, Integrating ecophysiology and plankton dynamics into projected maximum fisheries catch potential under climate change in the Northeast Atlantic, ICES J. Mar. Sci. 68 (2011) 1008–1018.

[24] G. Merino, M. Barange, J.L. Blanchard, J. Harle, R. Holmes, I. Allen, et al., Can marine fisheries and aquaculture meet fish demand from a growing human population in a changing climate? Global Environ. Change 22 (2012) 795–806.

[25] J.R. Watson, C.A. Stock, J.L. Sarmiento, Exploring the role of movement in determining the global distribution of marine biomass using a coupled hydrodynamic–size-based ecosystem model, Prog. Oceanogr. 138 (2015) 521–532.

[26] M.B. Harfoot, T. Newbold, D.P. Tittensor, S. Emmott, J. Hutton, V. Lyutsarev, et al., Emergent global patterns of ecosystem structure and function from a mechanistic general ecosystem model, PLoS Biol. 12 (2014) e1001841.

[27] D.P. Tittensor, T.D. Eddy, H.K. Lotze, E.D. Galbraith, W. Cheung, M. Barange, et al., A protocol for the intercomparison of marine fishery and ecosystem models: Fish-MIP v1. 0, Geosci. Model Dev. 11 (4) (2018) 1421–1442.

[28] V.W.Y. Lam, W.W.L. Cheung, G. Reygondeau, U.R. Sumaila, Projected change in global fisheries revenues under climate change, Sci. Rep. 6 (2016) 32607.

[29] C.L. Delgado, Fish to 2020: Supply and Demand in Changing Global Markets, WorldFish, 2003.

[30] E. Galbraith, D. Carozza, D. Bianchi, A coupled human-Earth model perspective on long-term trends in the global marine fishery, Nat. Commun. 8 (2017) 14884.

[31] D.A. Carozza, D. Bianchi, E.D. Galbraith, The ecological module of BOATS-1.0: a bioenergetically constrained model of marine upper trophic levels suitable for studies of fisheries and ocean biogeochemistry, Geosci. Model Dev. 9 (2016) 1545–1565.

[32] D.A. Carozza, D. Bianchi, E.D. Galbraith, Formulation, general features and global calibration of a bioenergetically-constrained fishery model, PLoS One 12 (2017) e0169763.

[33] M.B. Schaefer, Some considerations of population dynamics and economics in relation to the management of the commercial marine fisheries, J. Fish. Res. Board Can. 14 (1957) 669–681.
[34] T.C. Tai, T. Cashion, V.W. Lam, W. Swartz, U.R. Sumaila, Ex-vessel fish price database: disaggregating prices for low-priced species from reduction fisheries, Front. Mar. Sci. 4 (2017) 363.
[35] B. Worm, E.B. Barbier, N. Beaumont, J.E. Duffy, C. Folke, B.S. Halpern, et al., Impacts of biodiversity loss on ocean ecosystem services, Science 314 (5800) (2006) 787–790.
[36] C. Costello, J. Lynham, S.E Lester, S.D. Gaines, Economic incentives and global fisheries sustainability, Annu. Rev. Resour. Econ. 2 (1) (2010) 299–318.
[37] V.W.Y. Lam, W.W. L. Cheung, U.R. Sumaila, Marine capture fisheries in the Arctic: winners or losers under climate change and ocean acidification? Fish Fish. 17 (2) (2014) 335–357.
[38] U. Srinivasan, W. Cheung, R. Watson, U. Sumaila, Food security implications of global marine catch losses due to overfishing, J. Bioecon. 12 (2010) 183–200.
[39] E.H. Allison, A.L. Perry, M.C. Badjeck, W. Neil Adger, K. Brown, D. Conway, et al., Vulnerability of national economies to the impacts of climate change on fisheries, Fish Fish. 10 (2009) 173–196.
[40] R. Blasiak, J. Spijkers, K. Tokunaga, J. Pittman, N. Yagi, H. Österblom, Climate change and marine fisheries: least developed countries top global index of vulnerability, PLoS One 12 (2017) e0179632.
[41] S. Ponte, I. Kelling, K.S. Jespersen, F. Kruijssen, The blue revolution in Asia: upgrading and governance in aquaculture value chains, World Dev. 64 (2014) 52–64.
[42] D.C. Kalikoski, S. Jentoft, A. Charles, D. Salazar Herrera, K. Cook, C. Béné, et al., Understanding the Impacts of Climate Change for Fisheries and Aquaculture: Applying a Poverty Lens, FAO, 2018.
[43] S. Hallegatte, M. Bangalore, L. Bonzanigo, M. Fay, T. Kane, U. Narloch, et al., Shock Waves: Managing the Impacts of Climate Change on Poverty, The World Bank, 2015.
[44] A.C. Haynie, L. Pfeiffer, Why economics matters for understanding the effects of climate change on fisheries, ICES J. Mar. Sci. 69 (2012) 1160–1167.
[45] S.D. Gaines, C. Costello, B. Owashi, T. Mangin, J. Bone, J.G. Molinos, et al., Improved fisheries management could offset many negative effects of climate change, Sci. Adv. 4 (2018) eaao1378.

CHAPTER 19

Integrating environmental information into stock assessment models for fisheries management

Kisei R. Tanaka

Postdoctoral Research Fellow, Atmospheric and Oceanic Sciences Program, Princeton University, Princeton, NJ, United States

Chapter Outline

19.1 The role of stock assessment models in fisheries management

Sustainable exploitation of commercial fisheries is incumbent on a predictable relationship between stock size and both short- and long-term fishing efforts. The overarching goal is to maintain resource status at a level that can allow long-term economic and societal gains [1]. Furthermore, fisheries managers generally assume that the production of the fishery can be optimized and sustainable if the targeted stock abundance is maintained above a specific threshold (that is, a harvest control rule is specified) [2,3]. In many cases management and advice of commercial fish stocks is based on single-species stock assessment models that track population dynamics through time [4]. The primary purposes of these single-species stock assessment models are to (1) estimate current and historical biomass/abundance, (2) analyze trends in stock removal by fishing, (3) evaluate stock status relative to reference points, and (4) provide the scientific basis for the establishment of harvest control rules.

Predicting Future Oceans.
DOI: https://doi.org/10.1016/B978-0-12-817945-1.00021-6

Fitting stock assessment models to data, fishery scientists and resource managers aim to address two fundamental management questions: how many fish are in a given area? and how many fish can be sustainably exploited?

In a typical single-species stock assessment framework (i.e., production and stage-structured), a population dynamics model is fitted to available data such as abundance indices and age/size compositions from scientific surveys and commercial catch data to estimate population status and productivity. Variants of the following basic population dynamics equation are used in most of the single-species stock assessment models:

$$B_{t+1} = B_t + (R_t + G_t) - (F_t - M_t)$$

where biomass (B) in the next time step ($t + 1$; season or year) is determined by a combination of the following processes: recruitment (R), growth (G), natural mortality (M), and fishing mortality (F) [1,2,5,6]. Model parameters are estimated by finding the values that maximize the likelihood of observed data (i.e., maximum likelihood estimation). Key model outputs usually include estimates of stock abundance/biomass, recruitment, and fishing mortality that are generally used to derive biological reference points that can help fishery managers to make informed decisions. Single-species stock assessment models can be broadly categorized as (1) surplus production (or biomass dynamics) models, (2) age-structured models, and (3) size (length or stage) structured models. Many stock assessment models are age-structured, while size-structured models are often applied to species that are hard to age (e.g., crustaceans) [7].

Stock assessment models have changed dramatically over the past decades as methodologies and computing power have evolved. This has allowed for enhanced mechanistic understanding of population dynamics. Present day stock assessment models routinely integrate a wide range of data sources within the same model. This type of model is called an integrated stock assessment model and incorporates data comprising fishing effort and landings or data derived from scientific surveys [8]. Integrated stock assessment models typically include two submodeling components: (1) a process model that can project changes in population dynamics over the study period given a set of life history parameters (e.g., growth and mortality) and prespecified condition (e.g., annual exploitation rate); and (2) an observation model that can link observations and predictions by specifying probability distribution for each data source. A fully integrated stock assessment process (1) generally considers many sources of data; (2) estimates stock biomass and status relative to target, threshold, and/or limit reference points; and (3) assists in developing harvest control rules that utilize the outcomes of the stock assessments. To this date, integrated stock assessment models have played a central role in providing a scientific basis for informed assessment and management of fishery resources [1,7,9].

19.2 Equilibrium paradigm assumptions under changing climate

> Since fishing pressure can be managed but the environment cannot, the default assumption in fisheries models has been to assume that the changes are due to fishing pressure. Thus, we use models without systematic environmental changes and leave the challenge of realistically considering environmental change for the next generation of ecological detectives.
>
> ***Hilborn and Mangel [10].***

Society has recognized the impact of dynamic environmental conditions on fisheries production, with rising concerns over climate change. Climate-driven changes in the marine environment have caused increases in water temperature as well as changes in the timing of the spring transition, seasonal sea ice retreat, salinity, pH levels, ocean upwelling, and circulation patterns [11]. All these climate-driven changes can impact exploited fish populations. It has been long accepted that such bottom-up forcing can play a significant role in annual variability in the population dynamics of exploited marine fish species (e.g., Refs. [12–16]). Many commercial stocks have already shown both gradual and abrupt responses to climate change and variability [17–19]. Some recent examples can be found in crustacean and small pelagic fisheries [17,20–22]. These bottom-up changes alter the functioning and structure of the marine ecosystem, posing a growing concern for commercial fisheries. Climate variability challenges the equilibrium assumptions underlying the population dynamics models used in stock assessment and introduces substantial uncertainty into fisheries management.

The traditional, yet predominant, single-species stock assessment model has been based on the equilibrium paradigm, which assumes that (1) nature exists in balance in general and the underlying structure of the ecosystem remains unchanged on a scale of decades; (2) variability in key population parameters (growth, natural mortality, and recruitment) are centered on a stationary mean and average abundance of species is roughly constant over time; and finally (3) the productivity of a stock is closely linked to the stock abundance and can be controlled through regulating the harvest rate [1,4]. Traditional stock assessment models generally do not explicitly account for a stochastically changing environment and assume that fish stocks are driven by density-dependent processes (e.g., surplus production models: [1,2,23–25]). For example, Ricker's stock–recruitment relationship [23,26] can incorporate environmental variables; however, recruitment is often modeled as a function of spawning stock size even though the year-to-year stock–recruitment relationship can show significant variability [27].

Under the assumptions of the equilibrium paradigm, any inaccuracies in stock prediction as well as unexpected fluctuations in population status were assumed to be the results of stochastic variability in the environment (i.e., environmental and/or demographic noise;

[1,2,4,28]). However, changes in the structure and functioning of marine ecosystems may increase or decrease the carrying capacity of the environment to support an exploited fish stock over time. Changes in carrying capacity will also increase the uncertainty in the dynamics of fish stocks as they experience changes in productivity, survival, recruitment, migration behaviors, trophic interactions, and habitat availability [29]. Climate-driven changes in stock productivity and resilience will influence the capacity of stocks to endure historical levels of fishing mortality [30]. This can lead to accepted management benchmarks to maintain maximum sustainable yields (MSY) in a static environment no longer being applicable and a harvesting rate no longer being sustainable [31,32]. The missing environmental information in traditional fisheries assessment models has led to inaccurate estimations of the status of exploited stocks and has contributed to the inadvertent depletion of many stocks [33,34]. Most stock assessment models to date continue to assume constant environmental conditions and do not explicitly account for the impact of environmental stochasticity [4]. Recent literature has highlighted the need to expand existing stock assessment models to incorporate the role of dynamic environmental conditions as a key step toward adaptive fishery management in a changing environment [35–38].

19.3 Incorporating environmental variability into stock assessment models

The characterization of the climate-driven marine ecosystem changes, as well as modeling of its effect on exploited commercial fish stocks, has become the central research topic within the coupled climate–fisheries discipline [39]. As contemporary management advice is dominated by single-species stock assessments, such models are becoming the preferred platform to incorporate environmental factors, albeit few results have been translated into management decisions [28]. Over the last decade efforts have been made to link environmental variability to fishery population dynamics and a number of stock assessment programs have been revised to incorporate environmental effects. These efforts range from understanding how climate change can influence stock monitoring programs to how climate change can affect fishery population dynamics and production.

One of the primary ways that climate change can affect fish population dynamics is by altering seasonal and annual recruitment dynamics (e.g., recruit survivorship). Therefore the typical approach to incorporate environmental signals [e.g., sea surface temperature (SST) anomalies] into stock assessment models is to add a parameter to the Ricker or Beverton–Holt stock–recruitment function to allow the estimated recruitment to deviate from the mean levels [1]. These explicit incorporations of environmental variables into stock assessment models allow stock assessment modelers to analyze whether changes in recruitment are due to internal factors (e.g., spawning stock biomass) or external factors (i.e., environment-driven changes in recruit survivorship). There are only a few examples where the successful incorporation of environmental variables into stock assessment models has influenced management decisions. For example, Jacobson and McCall [40] incorporated SST to explain variability in the stock–recruitment relationship of sardines in the California current region. The study led to the development of an environmentally explicit harvest control rule with a precautionary decision criterion to reduce catches under cold SST regimes. Churchill [41] used the correlation between recruitment estimates and a wind index related to larval retention of Atlantic cod. Kienzle et al. [42,43] recently improved the stock assessment of brown tiger prawn in Moreton Bay by incorporating a temperature-induced physiological change in catchability, weekly temperature measurements, and variable recruitment timing into the population dynamic model.

However, in practice only few stock assessments assume environment-driven recruitment due to a lack of reliable relationships between environmental and recruitment dynamics (e.g., Refs. [44–46]). Myers' [45] meta-analysis found that only a few environment–recruitment correlations held up when reevaluated with more data. Later analyses on the US Pacific sardine stock questioned the robustness of this indicator scrutinized the relationship [47,48], and a more appropriate measure of SST was introduced and adapted based on the findings by Lindegren and Checkley [49]. For Atlantic cod, Hare et al. [50] revised the wind–recruitment correlation obtained by Churchill et al. [41] with updated recruitment estimates, and found that the correlation was much less significant. The failure of these relationships to withstand the test of time is likely due to various confounding and unaccounted factors in the empirical environment–fishery relationships as well as the analyses not capturing the fundamental processes that govern fisheries dynamics [45,51]. Potential sources of these "breaking" environment–population dynamics relationships are (1) nonlinearity, (2) multidimensionality, (3) direct and indirect effects, (4) temporal lags, (5) spatial considerations, (6) effect of population structure, and (7) spurious regressions [28]. These indicate that either the environment–population relationship is real but masked by other factors (1–3), the environment–population link is not real and is a product of limited sample size or autocorrelated time series (4–7), or a functional relationship may be understood over a given range but if the range changes the relationship may go away (5–7) [52–54]. These frequent documentations of "breaking relationships" suggest that incorporation of environment–recruitment relationships in stock assessment requires a clear mechanistic understanding as well as modeled relationships cross-validated with new data [28,45].

Aside from explicitly incorporating environmental variability in the stock assessment models, efforts are made to account for the impact of climate change on the quality of fishery-independent surveys [55–57]. In many fisheries, abundance indices (catch-per-unit-effort) derived from fishery-independent surveys play a key role in the assessment and management process [58]. For example, in an integrated stock assessment model, surveys are used as a proxy for abundance and usually a catchability coefficient is estimated for the entire time series. The ability of these indices to capture stock dynamics will be relatively unbiased under stable climate conditions (i.e., time-constant catchability). However, climate change and variability can influence spatial and temporal availability of suitable habitats and the timing of migrations of commercial fish stocks [30,59]. This will affect the effectiveness of surveys based on spatiotemporally fixed grids and may lead to a potential disparity in the spatial extent and timing of surveys as indices for some stocks may no longer track changes in abundance accurately (i.e., time-varying catchability; [55,60,61]). For example, changes in water temperature have been shown to influence the effectiveness of bottom trawl surveys for butterfish in the northwest Atlantic [62]. Furthermore, studies have shown that climate-driven changes in habitat suitability index (HSI) can affect underlying stock production (e.g., [21]). Similar problems likely exist for fishery-dependent indices of abundance [63–65]. Incorporating habitat data to inform and improve stock assessments has advanced with the development of species distribution models. For the northwest Atlantic butterfish stock assessment, a thermal HSI model was used to quantify the magnitude of overlap between the amount of habitat sampled versus the total amount of habitat [55]. The quantified overlap between surveyed versus total habitat was used to constrain habitat availability estimates in the stock assessment models of some commercial stocks (butterfish, scup, and bluefish).

19.4 Case study of American lobster in the Gulf of Maine

The American lobster (*Homarus americanus*) fishery in the Gulf of Maine supports one of the highest grossing wild-caught fisheries in the northern hemisphere, recognized as the most productive lobster fishery in the world [63,66]. The 2016 ex-vessel value of the US American lobster fishery exceeded US$669 million and represented an historic high [67]. American lobsters are ectothermic and experience strong bottom-up control (e.g., availability of climatically suitable habitat) throughout their range and life stages [63,68]. The Gulf of Maine is a part of the Northwest Atlantic Shelf Large Marine Ecosystem [69] that is experiencing strong climate-driven changes. The Gulf of Maine SST has increased 0.03°C per year since 1982 [70] and bottom temperature has shown a similar increasing trend [71], which is a growing concern for its socioeconomically important fisheries. Both mesoscale climate change and variability will challenge equilibrium assumptions underlying the population dynamics and biological reference points of the lobster stocks (ASMFC 2015), and introduce substantial uncertainty into management of the fishery [21,22,66,68,72].

It was hypothesized that the American lobster recruitment dynamics were driven by climate variability, therefore incorporating environmental signals can potentially improve

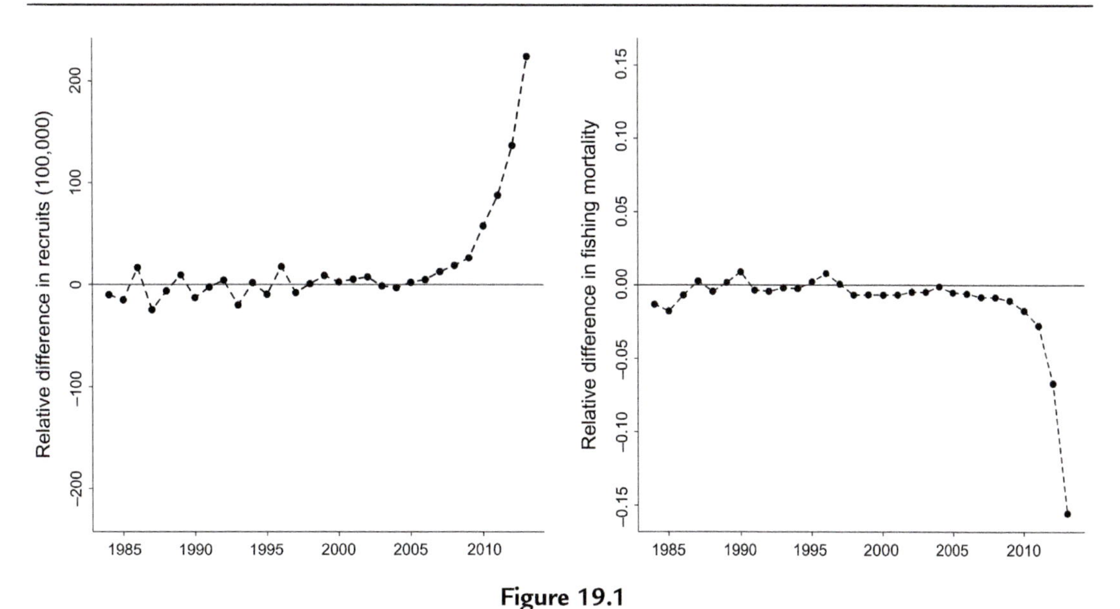

Figure 19.1
Relative difference in estimated American lobster annual recruitment (left) and fishing mortality (right) with and without environmentally informed recruitment dynamics.

recruitment estimates. Tanaka [73,74] developed a model-based framework that can incorporate climate/environmental variability into single-species stock assessments. The framework consists of the two submodeling components: (1) an empirical bioclimate envelope model that quantifies the spatiotemporal variability of lobster recruit HSI due to bottom temperature and salinity; and (2) an integrated size-structured population dynamics model that incorporates environmental effects, HSI, to inform recruitment dynamics. Within this proposed modeling framework, changes in annual median lobster recruitment HSI over 30 years were treated as an index of climate-driven environmental variability, which was assumed to have influenced American lobster recruitment dynamics during 1984–2013. The performance of the assessment model with an environmentally explicit recruitment function is evaluated by comparing relevant assessment outputs such as recruitment, annual fishing mortality, and magnitude of retrospective biases. The environmentally informed lobster assessment model (1) estimated higher recruitment and lower fishing mortality in the late 2000s and early 2010s (Fig. 19.1) and (2) showed reduced retrospective patterns and improved model fit (Fig. 19.2).

This analysis indicates that climate-driven changes in lobster HSI contributed to increased lobster recruitment and present potential improvement to the species' assessment. The proposed model-based framework can improve our understanding of environment-driven marine ecological processes and our ability to assess the status of exploited fishery

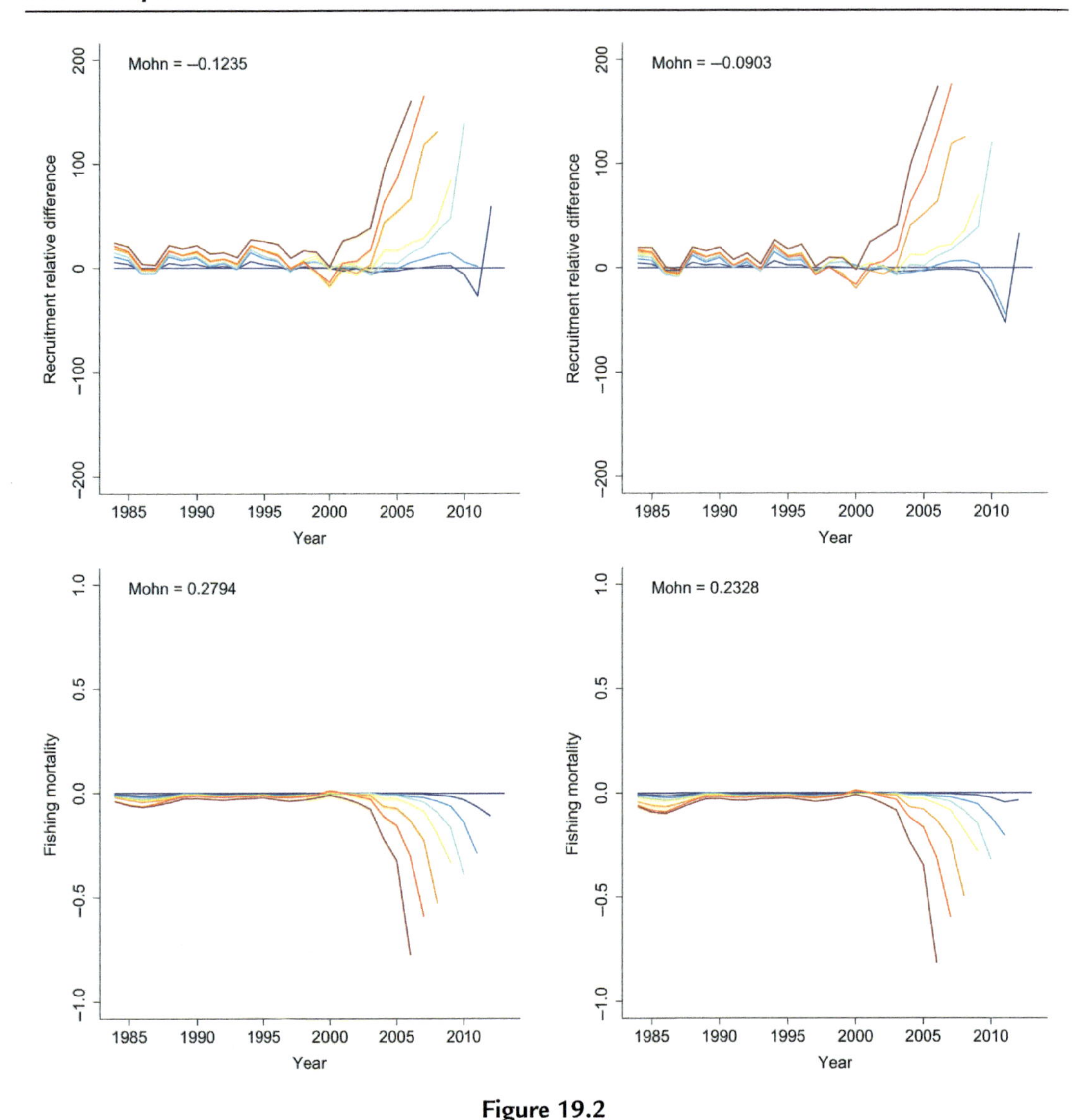

Figure 19.2

Relative difference in retrospective analysis of recruitment (top panels) and fishing mortality (bottom panels) estimates to terminal year 2006 of American lobster based on the model without (left panels) and with (right panels) environmentally informed recruitment dynamics. Mohn's coefficient shows the magnitude of retrospective bias, where the value is zero when the peeled outputs match exactly with full time series output (i.e., lower values indicate a better model performance).

resources, which can potentially enhance our adaptive management capacity in a changing environment. The framework can improve the assessment of the US American lobster stocks and is extendable to other commercial fish stocks that are impacted by environmental change.

19.5 Conclusion

With an increasing number of studies reporting the impact of climate change on marine ecosystems [75], the focus of research is no longer whether, but how stocks are affected, and what can science do to adapt to the foreseeable changes? Assessment of commercial fisheries is becoming increasingly complex and uncertain as ocean conditions change. Management based on MSY estimated from stock assessment models that assume equilibrium conditions is difficult given the dynamic nature of the marine environment. For example, some stock assessment models may not have the capacity to deal with a spatially explicit stock structure (e.g., index of abundance declining in one area but increasing in adjacent areas). If the models do not account for changes in fishing effort and species movement between different locations, estimates of stock status and productivity are likely biased (see recent findings and discussions from Atlantic bluefin tuna [76–78]).

Significant improvements have been made to stock assessment models but predicting changes in population dynamics with precision and accuracy is not always feasible in a complex marine environment. Successful examples of incorporating environmental signals in stock assessments can be found in environmentally explicit single-species stock–recruitment models to highly complex ecosystem models [39,79–81]. However, only a fraction (<2%) of stock assessments/tactical fisheries management accounts for climate and/or ecosystem factors [4]. This is partially due to (1) inadequate representation of climate–fishery processes as relationships between fish distributions, recruitment or productivity, and environment are not straightforward; and (2) the criteria for incorporating an environmental index into a stock assessment or management decision is high as fishery managers generally favor methodological consistency in stock assessments over innovative approaches for reaching agreements [82].

It is also not always feasible or sensible to try and include environmental variables, as every model is subject to the classic dilemma of never satisfying (1) realism (the qualitative aspect), (2) precision (the quantitative aspect), and (3) generality (the aspect of the universality of applicability) at the same time [83]. Therefore, when analyzing stock dynamics and the surrounding ecosystem, stock assessment scientists and fishery managers face a trade-off between these features. This is also the trade-off between the systematic and the statistical error, where the former and latter lead to decreasing and increasing model complexity, respectively. A simple generalized model has low data requirements but may fail to capture key processes, while a more complex realistic model can be data intensive and have limited applicability.

It is important that the stock assessment scientists aim for an appropriate level of model complexity that can describe key processes with a satisfactory statistical performance [84]. In general, the incorporation of environmental factors increases model complexity with more parameters. This makes a model become more susceptible to higher statistical error [28]. Despite the trade-off, the incorporation of environmental data in stock assessment models has progressed and it is expected to play a significant role for future assessment of marine resources. These efforts will be a key step toward implementing adaptive ecosystem-based fisheries management which will be central to sustainable management.

Acknowledgments

The author thanks Dr. Samuel Truesdell, Dr. Fernando González Taboada, Dr. Charles Stock, Mackenzie Mazur, and Hsiao-Yun Chang for their generous support, ideas, and perspectives.

References

[1] R. Hilborn, C.J. Walters, Quantitative Fisheries Stock Assessment: Choice, Dynamics and Uncertainty, Springer Science & Business Media, 1992, p. 570.
[2] M. Haddon, Modelling and Quantitative Methods in Fisheries, second ed., CRC Press, Boca Raton, FL, 2011, 452 pp.
[3] K.A. Vert-pre, R.O. Amoroso, O.P. Jensen, R. Hilborn, Frequency and intensity of productivity regime shifts in marine fish stocks, Proc. Natl. Acad. Sci. U.S.A. 110 (5) (2013) 1779–1784.
[4] M. Skern-Mauritzen, G. Ottersen, N.O. Handegard, G. Huse, G.E. Dingsør, N.C. Stenseth, et al., Ecosystem processes are rarely included in tactical fisheries management, Fish Fish. (2015) 165–175.
[5] E.S. Russell, Some theoretical consideration on the overfishing problem, ICES J. Mar. Sci. 6 (1) (1931) 3–20.
[6] T.J. Quinn, R.B. Deriso, Quantitative Fish Dynamics, Oxford University Press, 1999, 560 pp.
[7] A.E. Punt, T. Huang, M.N. Maunder, Review of integrated size-structured models for stock assessment of hard-to-age crustacean and mollusc species, ICES J. Mar. Sci. 69 (2013) 516–528.
[8] D. Fournier, C.P. Archibald, A general theory for analyzing catch at age data, Can. J. Fish Aquat. Sci. 39 (8) (1982) 1195–1207.
[9] R. Hilborn, The state of the art in stock assessment: where we are and where we are going, Sci. Mar. 67 (1) (2003) 15–20.
[10] R. Hilborn, M. Mangel, The Ecological Detective: Confronting Models With Data, Princeton University Press, 1997, 315 pp.
[11] M. Lindegren, K. Brander, Adapting fisheries and their management to climate change: a review of concepts, tools, frameworks, and current progress toward implementation, Rev. Fish. Sci. Aquac. 8249 (2018) 1–16.
[12] F.P. Chavez, J. Ryan, S.E. Lluch-Cota, M. Ñiquen, From anchovies to sardines and back: multidecadal change in the Pacific Ocean, Science 292 (5515) (2003) 217–221.
[13] D.H. Cushing, Climate and Fisheries, Academic Press, 1982, 373 pp.
[14] D.M. Ware, G.A. McFarlane, Climate induced changes in hake abundance and pelagic community interactions in the Vancouver Island upwelling system, in: R.J. Beamish (Ed.), Climate Change and Northern Fish Populations Issue 121 of Canadian Special Publication of Fisheries and Aquatic Sciences, Can. Spec. Publ. Fish. Aquat. Sci. 1995, pp. 509–521.
[15] J. Hjort, Fluctuations in the great fisheries of northern Europe viewed in the light of biological research, Conseil Permanent International Pour L'exploration De La Mer, Rapports et Procès-Verbaux des Réunions, Copenhagen, Denmark, 1914, 225 pp.
[16] E.A. Logerwell, N. Mantua, P.W. Lawson, R.C. Francis, V.N. Agostini, Tracking environmental processes in the coastal zone for understanding and predicting Oregon coho (*Oncorhynchus kisutch*) marine survival, Fish. Oceanogr. 12 (6) (2003) 554–568.

[17] P. Lehodey, J. Alheit, M. Barange, T. Baumgartner, G. Beaugrand, K. Drinkwater, et al., Climate variability, fish, and fisheries, J. Clim. 19 (20) (2006) 5009–5030.

[18] M.L. Pinsky, B. Worm, M.J. Fogarty, J.L. Sarmiento, S.A. Levin, Marine taxa track local climate velocities, Science 341 (6151) (2013) 1239–1242.

[19] J.A. Hare, W.E. Morrison, M.W. Nelson, M.M. Stachura, E.J. Teeters, R.B. Griffis, et al., A vulnerability assessment of fish and invertebrates to climate change on the Northeast U.S. continental shelf, PLoS One 11 (2) (2016) e0146756.

[20] B.P. Finney, J. Alheit, K.-C. Emeis, D.B. Field, D. Gutiérrez, U. Struck, Paleoecological studies on variability in marine fish populations: a long-term perspective on the impacts of climatic change on marine ecosystems, J. Mar. Syst. 79 (3–4) (2010) 316–326.

[21] K.E. Mills, A.J. Pershing, C.J. Brown, Y. Chen, F.-S. Chiang, D.S. Holland, et al., Fisheries management in a changing climate: lessons from the 2012 ocean heat wave in the Northwest Atlantic, Oceanography 26 (2) (2013) 191–195.

[22] K.R. Tanaka, J.-H. Chang, Y. Xue, Z. Li, L. Jacobson, Y. Chen, Mesoscale climatic impacts on the distribution of *Homarus americanus* in the US inshore Gulf of Maine, Can. J. Fish. Aquat. Sci. (2018) 1–58.

[23] W.E. Ricker, Handbook of computations for biological statistics of fish populations, Can. Fish. Res. Board Bull. 119, 1958, p. 300.

[24] M.B. Schaefer, Some aspects of the dynamics of populations important to the management of the commercial marine fisheries, Inter-Am. Trop. Tuna Commun. Bull. 1 (2) (1954) 23–56.

[25] M.B. Schaefer, Some considerations of population dynamics and economics in relation to the management of the commercial marine fisheries, J. Fish. Res. Board Canada 14 (5) (1957) 669–681.

[26] W.E. Ricker, Stock and recruitment, J. Fish. Res. Board Canada 11 (5) (1954) 559–623.

[27] J.A. Koslow, Fecundity and the stock–recruitment relationship, Can. J. Fish. Aquat. Sci. 49 (2) (1992) 210–217.

[28] F. Keyl, M. Wolff, Environmental variability and fisheries: what can models do? Rev. Fish. Biol. Fish. 18 (3) (2008) 273–299.

[29] J.E. Overland, J. Alheit, A. Bakun, J.W. Hurrell, D.L. Mackas, A.J. Miller, Climate controls on marine ecosystems and fish populations, J. Mar. Syst. 79 (3–4) (2010) 305–315.

[30] National Marine Fisheries Service (NMFS), Marine Fisheries Habitat Assessment Improvement plan, Report of the National Marine Fisheries Service Habitat Assessment Improvement Plan Team. NOAA Tech. Memo, NMFS-F/SPO-108, U.S. Department of Commerce, Silver Spring, MD, 2010, p. 115.

[31] K. Brander, Impacts of climate change on fisheries, J. Mar. Syst. 79 (3–4) (2010) 389–402.

[32] A. McIlgorm, S. Hanna, G. Knapp, P. Le Floc'H, F. Millerd, M. Pan, How will climate change alter fishery governance? Insights from seven international case studies, Mar. Policy 34 (1) (2010) 170–177.

[33] K. Brander, Impacts of climate change on marine ecosystems and fisheries, J. Mar. Biol. Assoc. India 51 (1) (2009) 1–13.

[34] A.D. Rijnsdorp, M.A. Peck, G.H. Engelhard, C. Mollmann, J.K. Pinnegar, Resolving the effect of climate change on fish populations, ICES J. Mar. Sci. 66 (7) (2009) 1570–1583.

[35] M.L. Pinsky, N.J. Mantua, Emerging adaptation approaches for climate-ready fisheries management, Oceanography 27 (4) (2014) 146–159.

[36] D. Tommasi, C. Stock, A. Hobday, R. Methot, I. Kaplan, P. Eveson, et al., Managing living marine resources in a dynamic environment: the role of seasonal to decadal climate forecasts, Prog. Oceanogr. 152 (2017) 15–49.

[37] S.R. Sagarese, M.D. Bryan, J.F. Walter, M. Schirripa, A. Grüss, M. Karnauskas, Incorporating ecosystem considerations within the stock synthesis integrated assessment model for Gulf of Mexico Red Grouper (*Epinephelus morio*), in: Sedar42-Rw-01, North Charleston, SC, 2015, p. 27.

[38] J. Cao, Y. Chen, R.A. Richards, Part 1—Improving assessment of Pandalus stocks using a seasonal, size-structured assessment model with environmental variables. Model description and application, Can. J. Fish. Aquat. Sci. 14 (2016) 1–14.

[39] J. Link, Ecosystem-Based Fisheries Management: Confronting Tradeoffs, Cambridge University Press, 2010, 224 pp.
[40] L.D. Jacobson, A.D. MacCall, Stock-recruitment models for Pacific sardine (*Sardinops sagax*), Can. J. Fish. Aquat. Sci. 52 (3) (1995) 566–577.
[41] J.H. Churchill, J. Runge, C. Chen, Processes controlling retention of spring-spawned Atlantic cod (*Gadus morhua*) in the western Gulf of Maine and their relationship to an index of recruitment success, Fish. Oceanogr. 20 (1) (2011) 32–46.
[42] M. Kienzle, A.J. Courtney, M.F. O'Neill, Environmental and fishing effects on the dynamics of brown tiger prawn (*Penaeus esculentus*) in Moreton Bay (Australia), Fish. Res. 155 (2014) 138–148.
[43] M. Kienzle, D. Sterling, S. Zhou, Y.-G. Wang, Maximum likelihood estimation of natural mortality and quantification of temperature effects on catchability of brown tiger prawn (*Penaeus esculentus*) in Moreton Bay (Australia) using logbook data, Ecol. Modell. 322 (2016) 1–9.
[44] Ø. Ulltang, Stock assessment and biological knowledge: can prediction uncertainly be reduced? ICES J. Mar. Sci. 53 (4) (1996) 659–675.
[45] R.A. Myers, When do environment—recruitment correlations work ? Rev. Fish. Biol. Fish. 8 (1998) 285–305.
[46] P. Lehodey, I. Senina, R. Murtugudde, A spatial ecosystem and populations dynamics model (SEAPODYM)—modeling of tuna and tuna-like populations, Prog. Oceanogr. 78 (4) (2008) 304–318.
[47] S. McClatchie, R. Goericke, G. Auad, K. Hill, Re-assessment of the stock–recruit and temperature–recruit relationships for Pacific sardine (*Sardinops sagax*), Can. J. Fish. Aquat. Sci. 67 (11) (2010) 1782–1790.
[48] L.D. Jacobson, S. Mcclatchie, Comment on temperature-dependent stock–recruit modeling for Pacific sardine (*Sardinops sagax*) in Jacobson and MacCall (1995), McClatchie et al. (2010), and Lindegren and Checkley (2013), Can. J. Fish. Aquat. Sci. 70 (2013) 1566–1569.
[49] M. Lindegren, D.M. Checkley, Temperature dependence of Pacific sardine (*Sardinops sagax*) recruitment in the California Current Ecosystem revisited and revised, Can. J. Fish. Aquat. Sci. 70 (2) (2013) 245–252.
[50] J.A. Hare, E.N. Brooks, M.C. Palmer, J.H. Churchill, Re-evaluating the effect of wind on recruitment in Gulf of Maine Atlantic Cod (*Gadus morhua*) using an environmentally-explicit stock recruitment model, Fish. Oceanogr. 24 (1) (2015) 90–105.
[51] E.R. Deyle, M. Fogarty, C. Hsieh, L. Kaufman, A.D. MacCall, S.B. Munch, et al., Predicting climate effects on Pacific sardine, Proc. Natl. Acad. Sci. U.S.A. 110 (16) (2013) 6430–6435.
[52] B.J. Pyper, R.M. Peterman, Comparison of methods to account for autocorrelation in correlation analyses of fish data, Can. J. Fish. Aquat. Sci. 55 (9) (1998) 2127–2140.
[53] C.W.J. Granger, N. Hyung, Y. Jeon, Spurious regressions with stationary series, Appl. Econ. 33 (7) (2001) 899–904.
[54] J.-H. Chang, Y. Chen, W. Halteman, C. Wilson, Roles of spatial scale in quantifying stock–recruitment relationships for American lobsters in the inshore Gulf of Maine, Can. J. Fish. Aquat. Sci. 73 (2016) 885–909.
[55] Northeast Fisheries Science Center, Butterfish Stock Assessment for 2014, Northeast Fisheries Science Center, Woods Hole, MA, 2014.
[56] J. Cao, J.T. Thorson, R.A. Richards, Y. Chen, Spatio-temporal index standardization improves the stock assessment of northern shrimp in the Gulf of Maine, Can. J. Fish. Aquat. Sci. 74 (11) (2017) 1781–1793.
[57] A.L. Perry, P.J. Low, J.R. Ellis, J.D. Reynolds, Climate change and distribution shifts in marine fishes, Science 308 (5730) (2005) 1912–1915.
[58] National Research Council, Improving Fish Stock Assessments, National Academies Press, Washington, DC, 1998, 188 pp.
[59] G.H. Engelhard, D.A. Righton, J.K. Pinnegar, Climate change and fishing: a century of shifting distribution in North Sea cod, Global Change Biol. 20 (2014) 2473–2483.

[60] A.Z. Horodysky, S.J. Cooke, R.W. Brill, Physiology in the service of fisheries science: why thinking mechanistically matters, Rev. Fish. Biol. Fish. 25 (3) (2015) 425–447.

[61] B. Muhling, M. Lindegren, L. Worsøe Clausen, A. Hobday, P. Lehodey, Impacts of climate change on pelagic fish and fisheries, in: B.F. Phillips, M. Pérez-Ramírez (Eds.), Climate Change Impacts on Fisheries and Aquaculture: A Global Analysis, first ed., John Wiley & Sons Ltd, 2018, pp. 771–814.

[62] J. Kohut, L. Palamara, E. Curchitser, M.J. Oliver, M. Breece, J. Manderson, et al., Toward dynamic marine spatial planning tools: can we inform fisheries stock assessments by using dynamic habitat models informed by the Integrated Ocean Observing System (IOOS)? Oceans—St John's, IEEE, St. John's, NL, Canada, 2014, pp. 1–7.

[63] Atlantic States Marine Fisheries Commission (ASMFC), American Lobster Benchmark Stock Assessment and Peer Review Report, Atlantic States Marine Fisheries Commission (ASMFC), Woods Hole, MA, 2015.

[64] State of Maine Department of Marine Resources (Maine DMR), Maine Ventless Trap Lobster Monitoring Survey, Augusta, ME, 2006, p. 4.

[65] State of Maine Department of Marine Resources (Maine DMR), The Sea Sampling Program: DMR Lobster Research, Monitoring, and Assessment Program, 2014.

[66] R.S. Steneck, R.A. Wahle, American lobster dynamics in a brave new ocean, Can. J. Fish. Aquat. Sci. 70 (2013) 1612–1624.

[67] Atlantic Coastal Cooperative Statistics Program (ACCSP), Data warehouse. URL: <http://www.accsp.org/data-warehouse> (cited June 1, 2017). Available from: <http://www.accsp.org/data-warehouse>, 2017 (accessed 01.06.17).

[68] N. Caputi, S. Lestang, S. de, Flusher, R.A. Wahle, The impact of climate change on exploited lobster stocks, in: B.F. Phillips (Ed.), Lobsters: Biology, Management, Aquaculture & Fisheries, second ed., John Wiley & Sons, Ltd, 2013, pp. 84–112.

[69] D.W. Townsend, A.C. Thomas, L.M. Mayer, M.A. Thomas, J.A. Quinlan, Oceanography of the northwest Atlantic continental shelf (1, W), in: A.R. Robinson, K.H. Brink (Eds.), The Sea: The Global Coastal Ocean: Interdisciplinary Regional Studies and Syntheses, Harvard University Press, Cambridge, MA, 2006, pp. 119–168.

[70] A.J. Pershing, M.A. Alexander, C.M. Hernandez, L.A. Kerr, A. Le Bris, K.E. Mills, et al., Slow adaptation in the face of rapid warming leads to collapse of the Gulf of Maine cod fishery, Science 350 (6262) (2015) 809–812.

[71] K.M. Kleisner, M.J. Fogarty, S. McGee, A. Barnett, P. Fratantoni, J. Greene, et al., The effects of sub-regional climate velocity on the distribution and spatial extent of marine species assemblages, PLoS One 11 (2) (2016) e0149220.

[72] A. Le Bris, K.E. Mills, R.A. Wahle, Y. Chen, M.A. Alexander, A.J. Allyn, et al., Climate vulnerability and resilience in the most valuable North American fishery, Proc. Natl. Acad. Sci. U.S.A. (21) (2018) 201711122.

[73] K.R. Tanaka, Incorporating Environmental Variability Into Assessment and Management of American Lobster (*Homarus americanus*), University of Maine, 2018.

[74] K.R. Tanaka, J. Cao, B.V. Shank, S.B. Truesdell, M. Mazur, L. Xu, et al., A model-based approach to incorporate environmental variability into assessment of a climatically-influenced commercial fishery: a case study with the American lobster fishery in the Gulf of Maine and Georges Bank. ICES J Mar Sci. (fsz024), 2019, 1–13.

[75] K. Brander, Climate and current anthropogenic impacts on fisheries, Clim. Change 119 (1) (2013) 9–21.

[76] W. Golet, N. Record, S. Lehuta, M. Lutcavage, B. Galuardi, A. Cooper, et al., The paradox of the pelagics: why bluefin tuna can go hungry in a sea of plenty, Mar. Ecol. Prog. Ser. 527 (2015) 181–192.

[77] A.S.M. Vanderlaan, A.R. Hanke, J. Chassé, J.D. Neilson, Environmental influences on Atlantic bluefin tuna (*Thunnus thynnus*) catch per unit effort in the southern Gulf of St. Lawrence, Fish. Oceanogr. 23 (1) (2014) 83–100.

[78] ICCAT, Report of the 2014 Atlantic Bluefin Tuna Stock Assessment Session, ICCAT, Madrid, Spain, 2014.

[79] H. Townsend, K. Aydin, K. Holsman, C. Harvey, I. Kaplan, E. Hazen, et al. Report of the 4th National Ecosystem Modeling Workshop (NEMoW 4): Using Ecosystem Models to Evaluate Inevitable Trade-offs: NOAA Technical Memorandum NMFS-F/SPO-173, Silver Spring, MD, 2017.

[80] C.A. Stock, M.A. Alexander, N.A. Bond, K.M. Brander, W.W.L. Cheung, E.N. Curchitser, et al., On the use of IPCC-class models to assess the impact of climate on Living Marine Resources, Prog. Oceanogr. 88 (1–4) (2011) 1–27.

[81] F.G. Taboada, R. Anadón, Determining the causes behind the collapse of a small pelagic fishery using Bayesian population modeling, Ecol. Appl. 26 (3) (2016) 886–898.

[82] R. Hilborn, The evolution of quantitative marine fisheries management 1985-2010, Nat. Resour. Model. 25 (1) (2012) 122–144.

[83] R. Levins, The strategy of model building in population ecology, Am. Sci. 54 (4) (1966) 421–431.

[84] C.J. Walters, S.J.D. Martell, Fisheries Ecology and Management, Princeton University Press, 2004, 399 pp.

CHAPTER 20

Climate change adaptation and spatial fisheries management

Rebecca Selden and Malin Pinsky

Ecology, Evolution, and Natural Resources, Rutgers University, New Brunswick, NJ, United States

Chapter Outline

20.1 Past, present, and future shifts in species distributions and catch potential

This is a fascinating and eye-opening time to be observing ocean life. The ocean has few barriers to dispersal, which means that marine animals often colonize new areas quickly when conditions become suitable. Similarly, marine species tolerate relatively narrow ranges of environmental conditions, so even small changes in their environment can switch conditions from suitable to not, or vice versa. With anthropogenic climate change and the increasing rapidity of habitat transformation [1], the suitability of environmental conditions for marine species is now changing rapidly.

The historical record provides insights into the rapid rate at which marine species respond to environmental change. Atlantic cod abundance, for example, increased rapidly in west Greenland during a warm period from 1925 to 1935, sparking a new international fishery 1000 km north of previous locations [2]. Cod abundance declined and the fishery disappeared just as quickly when conditions cooled. While this older example reveals rapid responses to climate variability, recent shifts in species distributions reveal a strong signature of anthropogenic climate change and have been largely toward the poles [3]. On average, the leading (a.k.a. the poleward) edges of marine species distributions have been moving into new territory at 72 ± 14 km/decade, which is an order of magnitude faster than

Predicting Future Oceans.
DOI: https://doi.org/10.1016/B978-0-12-817945-1.00023-X

similar observations on land [3]. At the other edge of species distributions, marine species' trailing (a.k.a. equatorward) edges are twice as likely to have retracted to higher latitudes as are terrestrial species' range edges [4].

Shifts in geographic distributions are anything but uniform across species, however, with some species shifting north and others south, some fast and others slow or not at all. A sizeable fraction of this variation across species can be explained by the rate and direction of local temperature change [5]. Temperature varies across space as a gradient, so any given temperature can be drawn as a line on a map, called an isotherm. As temperatures warm or cool through time, the isotherm moves across the map. The rate and direction at which the isotherm moves is called climate velocity. Climate velocities can be complex, creating a spatial mosaic of different directions and rates that influence species distributions. Marine species in general have not lagged behind climate velocities, though such lags have become apparent on land [5].

These responses mean that many marine species are presently found in areas in which they were historically rare, and are rare in areas where they were historically abundant. Black sea bass, for example, was centered off Virginia in the late 1960s, but is now (2010s) centered off New Jersey, 250 km further north. Off the west coast of North America in 2013–15, an exceptionally large marine heat wave ("the Blob") allowed many species to move dramatically further north, including market squid breeding in Alaska [6]. Market squid are typically found in southern and central California. Climate projections suggest that marine heat waves like this will become dramatically more common in the future [7]. To keep track of these changes, online resources like RedMap (http://www.redmap.org.au/) and OceanAdapt (http://oceanadapt.rutgers.edu) are now available to document changes in species distributions as they occur.

Future projections of climate change under a range of greenhouse gas emissions scenarios help us understand potential future shifts in species distributions. Modeling efforts suggest that poleward range boundaries for marine fishes are likely to shift poleward by 50 km/decade [8], with various efforts now projecting distributions for nearly 13,000 marine species [9] or examining patterns at finer spatial scales [10]. A common theme tends to be that species richness will decline in the tropics as species shift poleward, while new combinations of species will be created in temperate zones. Areas with weak latitudinal gradients in temperature, such as the west coast of North America, may experience shifts in species distributions over 1000 km [10].

The mechanisms that underlie shifting species distributions, however, are diverse. At the poleward edge, a species range can expand through adult movement or the dispersal of larvae by ocean currents. Bluefish, for example, respond quickly to interannual variation in temperature, likely through behavioral decisions about how much to migrate along the east coast of the United States [11]. At the equatorward trailing edge, population persistence depends primarily on survival rather than dispersal, and long-lived species can persist for years after conditions are unsuitable for growth or reproduction. Within a species range, the distribution of biomass (often measured with biomass centroids) can be affected by spatial differences in population growth rates, independently from any changes in range edges.

From a fisheries perspective, the important link is between species distribution and opportunities to catch a species. While ecological surveys and models quantify the average probability of occurrence or the average density of biomass across a reasonably large area or time span (e.g., 0.25° × 0.25° grids over a year), fishers do not catch averages. Successful fishing is about targeting the precise location (sometimes down to the meter) and time (down to the minute or hour) when fish will be most abundant. Fish that school and aggregate can remain locally abundant and available to fisheries even as regional abundance declines [12]. These processes can disconnect large-scale changes in abundance from the perception of change at local scales. However, as will be explained in more detail below, changes in species distribution have already had dramatic impacts on fishing opportunities and fisher behavior in many cases.

20.2 Impacts on fishing communities

As climate change causes shifts in both the spatial distribution and productivity of species, the fishing communities that rely upon these resources must adapt or cope in order to maintain their livelihoods. Synchronous changes in both biomass and species distributions can lead to differential gains and losses in resource access across fisheries ports [13]. As the abundance of species change within fishing grounds, fishers are faced with the decision about whether to shift where they are fishing to follow species they traditionally targeted or shift what they are fishing to take advantage of new species. While fishers are adept at adapting to variability in conditions, permanent changes in availability of target species can alter the long-term viability of fishing communities. These choices are further influenced by the management context (e.g., rules and regulations), economics (costs and prices), and social dynamics of the fishery (e.g., mobility or alternative career options).

Our research on fishing communities in the northeast United States suggests that following the fish is the exception rather than the rule, and communities are more likely to change their catch composition [14]. Management plays a significant role in the viability of these strategies in the short and long term. For some species, it is not possible to follow the fish into new areas in the near term due to area-based prohibitions of gear types [15], resulting in a lag between poleward shifts in species distributions and the redistribution of landings. However, the long-term viability of switching to some target species may also be constrained by management structures that allocate quotas based on historical landings of a species. For example, trawl vessels based in New Jersey have increased their landings of fluke and scup over time, but the quota allocated to states further north is set based on their proportion of landings in the 1980s, which was quite low [14]. As such, vessels, communities, and states that currently target a diverse suite of species may be more resilient to future changes. In general, portfolio theory suggests that increasing catch diversity stabilizes catches under variable conditions due to complementarity in species responses [16]. Indeed communities in Alaska with higher portfolio diversities were better able to buffer the impact of both market and regime shifts [17].

As variability increases with climate change, diversification is likely to be increasingly important for fisheries resilience [18]. However, as permits become increasingly specialized for individual species, diversifying catch in response to changing availability can be difficult. In these cases, the absence of management for a new target species may facilitate adaptation by allowing fishers to take advantage of an emerging species. However, unregulated fishing on emerging fisheries can represent a risk to the long-term sustainability of the stock. Fisheries managers in the mid-Atlantic instituted emergency management measures to constrain the expansion of the emerging fishery for blueline tilefish after landings increased 20-fold over a 3-year period, and the species is now formally included in the fisheries management plan for golden tilefish [19]. Future projections of species ranges, adaptive management frameworks, monitoring of (currently) nontarget species, quota trading methods, and strengthened collaboration among adjacent management organizations can help avoid similar surprises in the future.

20.3 Challenges for spatial fisheries management

Shifts in species distributions can create many challenges for fisheries management, from scientific issues like understanding how many fish are in the water, to political questions of allocations and perceptions of fairness or equity. Changes in distribution can lead to changes in the stock structure, including altering the spatial footprint of populations, merging previous stocks into one stock, or splitting previous single stocks into many [20]. Many stock assessments use statistical surveys as an index of biomass, and shifts in distribution may move a stock partially outside the survey footprint such that the survey index is no longer correlated in the same way to population biomass.

Many fisheries also use spatial closed areas or spatial management measures. Shifts in species distributions can reduce the effectiveness of these areas by moving the species outside the boundaries of the management zone. For example, the Plaice Box in the southern North Sea was designed to protect juvenile plaice from fishing. As temperatures warmed, the juvenile plaice moved deeper, a shift that took them outside the Plaice Box and again exposed them to fishing effort [21].

For various historical reasons, many fisheries also allocate fishery access spatially, a system which becomes challenging when species shift their distribution. In the summer flounder fishery on the east coast of the United States, for example, each state is allocated a fraction of the overall quota for the season and the fractions do not change through time. Northward shifts in summer flounder, however, mean that summer flounder are now more abundant in states with small allocations, which has created conflict and friction among stakeholders. As of 2018 New York state had filed a legal petition to increase their allocation, based in part on the argument that the mismatch between allocations and species distribution was unfair to fishers in their state (https://www.dec.ny.gov/docs/fish_marine_pdf/sumfldpetforrulemaking.pdf). Such conflict has the potential to be quite widespread. The Common Fisheries Policy for the European Union, for example, uses a conceptually similar approach to allocate fishing opportunities among countries.

More drastically, shifts in species distributions can also alter which stakeholders have access to a fishery [22]. If new stakeholders gain access and do not have preagreed mechanisms for coordinating their catches with other stakeholders, a race to fish can ensue in which stakeholders attempt to harvest the fish before others do. Northeast Atlantic mackerel shifted into Icelandic waters in 2007 and sparked a race between Iceland, the European Union, and other stakeholders that led to the overfishing of mackerel, then later spread into a trade war. Even a decade later, Iceland and the European Union had not reached an agreement on cooperative management measures. The conditions for similar conflicts are set to appear all over the world as species distributions shift into the waters of at least 70 new nations [22].

20.4 Tools for climate-ready management in the context of shifting species distributions

With a few exceptions, current fisheries management approaches do not account for long-term climate change [23]. Instead management strategies are designed to be robust to stochasticity around a stationary average. When change is instead directional, management based on static fishery reference points can be biased [24] and greater precaution may be needed. Incorporating information on climate change into harvest control rules results in better outcomes in both yield and species biomass, but the largest gains are simply from improving management now under current conditions [25].

One of the major sticking points in the context of shifting species distributions is how to allocate quotas among different management units, where such approaches are used. Options include mechanisms to allow quota sharing between states or management regions, or to adjust allocations based on a combination of historical landings and current distributions or on the degree of utilization of current quotas—both of which are being considered by the Atlantic States Marine Fisheries Commission in the United States [26]. While requiring a more fundamental transition toward ecosystem-based fisheries management, shifting from quotas set on a species basis to that for an area would potentially streamline management and allow fishers to adjust their catch composition based on availability [27]. In the northwest Atlantic, delineations are being considered based on ecological production units at the subnational level [28], but these types of mixed species fishery "portfolios" could also serve the basis for setting total allowable catch (TAC) among adjacent coastal nations [29].

Individual rights-based approaches like catch shares may serve not only as a useful model for these types of area-based approaches but also as a viable tool in its own right to prepare for climate-ready management. Multispecies individual catch shares allow fishers to trade quotas among species at prespecified rates (1:1 or with specific species conversions) [30]. These types of species exchanges can result in catch exceeding the desired TAC [31], but this can be reduced by restricting the amount of the quota that can be converted. In Iceland, where species exchanges are common, owners cannot convert more than 5% of their annual catch entitlements, and no more than 2% can be converted into any one species [30]. Real-time updating of catches with web-based tools has provided additional safeguards and reduced administrative costs. While not included in any of the current multispecies programs that allow transferable quotas, additional limits on conversion could potentially include species-specific information about susceptibility to overfishing to further reduce the risk of overexploitation.

In general, the increasing availability of data at fine temporal and spatial scales increases the potential to implement dynamic ocean management that is responsive to environmental changes. Dynamic ocean management uses near real-time data on physical, biological, and socioeconomic conditions to inform spatial uses of the ocean [32]. When applied to bycatch reduction, efficiency gains are clear: local closures and rules to move on from areas with high bycatch result in less foregone catch than traditional static area-based closures [33]. However, the degree to which dynamic ocean management provides benefits over coarser management frameworks in the context of climate change remains to be tested.

Adaptive co-management is a bottom-up approach that may help achieve climate-ready fisheries, particularly in areas with limited resources. Government agencies set performance standards to ensure compliance with laws, and cooperative entities based on partnerships between government, fishing groups, and nongovernmental organizations would be responsible for data collection and testing out new ideas [34]. As is done in many countries around the world, such efforts could be funded by cooperatives that invest a portion of their profits in sustainability and adaptation. Through this flexibility, adaptive co-management may provide a self-sustaining mechanism for climate change adaptation.

As the oceans move increasingly into no-analog conditions with future warming, designing management to cope with these changes may seem daunting. As highlighted above, many of the management frameworks currently in use may be well suited for the task. While finding the management strategy that is most robust to climate change may require iteration, failure to begin the process most surely results in greater long-term risk to the sustainability of valuable marine resources.

References

[1] D.J. McCauley, M.L. Pinsky, S.R. Palumbi, J.A. Estes, F.H. Joyce, R.R. Warner, Marine defaunation: animal loss in the global ocean, Science 347 (2015) 1255641. Available from: https://doi.org/10.1126/science.1255641.

[2] K.M. Brander, Global fish production and climate change, Proc. Natl. Acad. Sci. U.S.A. 104 (2007) 19709–19714.

[3] E.S. Poloczanska, C.J. Brown, W.J. Sydeman, W. Kiessling, D.S. Schoeman, P.J. Moore, et al., Global imprint of climate change on marine life, Nat. Clim. Change 3 (2013) 919–925. Available from: https://doi.org/10.1038/nclimate1958.

[4] M.L. Pinsky, A.M. Eikeset, D.J. McCauley, J.L. Payne, J.M. Sunday, Greater vulnerability to warming of marine versus terrestrial ectotherms. Nature, 569 (7754) (2019) 108–111. Available from: https://www.nature.com/articles/s41586-019-1132-4.

[5] M.L. Pinsky, B. Worm, M.J. Fogarty, J.L. Sarmiento, S.A. Levin, Marine taxa track local climate velocities, Science 341 (2013) 1239–1242. Available from: https://doi.org/10.1126/science.1239352.

[6] L. Cavole, A. Demko, R. Diner, A. Giddings, I. Koester, C. Pagniello, et al., Biological impacts of the 2013–2015 warm-water anomaly in the Northeast Pacific: winners, losers, and the future, Oceanography 29 (2016). Available from: https://doi.org/10.5670/oceanog.2016.32.

[7] T.L. Frölicher, E.M. Fischer, N. Gruber, Marine heatwaves under global warming, Nature 560 (2018) 360–364. Available from: https://doi.org/10.1038/s41586-018-0383-9.

[8] W.W.L. Cheung, V.W.Y. Lam, J.L. Sarmiento, K. Kearney, R. Watson, D. Pauly, Projecting global marine biodiversity impacts under climate change scenarios, Fish Fish. 10 (2009) 235–251.

[9] J.G. Molinos, B.S. Halpern, D.S. Schoeman, C.J. Brown, W. Kiessling, P.J. Moore, et al., Climate velocity and the future global redistribution of marine biodiversity, Nat. Clim. Change 6 (2015) 83–88. Available from: https://doi.org/10.1038/nclimate2769.

[10] J.W. Morley, R.L. Selden, R.J. Latour, T.L. Frölicher, R.J. Seagraves, M.L. Pinsky, Projecting shifts in thermal habitat for 686 species on the North American continental shelf, PLoS One 13 (2018) e0196127. Available from: https://doi.org/10.1371/journal.pone.0196127.

[11] J.W. Morley, R.D. Batt, M.L. Pinsky, Marine assemblages respond rapidly to winter climate variability, Global Change Biol 23 (2017) 2590–2601. Available from: https://doi.org/10.1111/gcb.13578.

[12] M.G. Burgess, C.J. Costello, A. Fredston-Hermann, M.L. Pinsky, S.D. Gaines, D. Tilman, et al., Range contraction enables harvesting to extinction, Proc. Natl. Acad. Sci. U.S.A. 114 (2017) 3945–3950. Available from: https://doi.org/10.1073/pnas.1607551114.

[13] R.L. Selden, J.T. Thorson, J.F. Samhouri, S. Brodie, G. Carroll, E. Willis-Norton, et al., Adapting to change? Availability of fish to west coast communities. ICES J. Mar. Sci. (submitted).

[14] Eva A. Papaioannou, R. Selden, K. St. Martin, J. Olson, J. Schenkel, B. McCay et al., Not all those who wander are lost– Responses of fishers' communities to shifts in the distribution and abundance of fish. Front. Mar. Sci. (submitted).

[15] M.L. Pinsky, M. Fogarty, Lagged social-ecological responses to climate and range shifts in fisheries, Clim. Change 115 (2012) 883–891. Available from: https://doi.org/10.1007/s10584-012-0599-x.

[16] D.E. Schindler, R.W. Hilborn, B. Chasco, C.P. Boatright, T.P. Quinn, La Rogers, et al., Population diversity and the portfolio effect in an exploited species, Nature 465 (2010) 609–612. Available from: https://doi.org/10.1038/nature09060.

[17] T.J. Cline, D.E. Schindler, R.W. Hilborn, Fisheries portfolio diversification and turnover buffer Alaskan fishing communities from abrupt resource and market changes, Nat. Commun. 8 (2017) 14042. Available from: https://doi.org/10.1038/ncomms14042.

[18] L.E. Dee, S.J. Miller, L.E. Peavey, D. Bradley, R.R. Gentry, R. Startz, et al.,). Functional diversity of catch mitigates negative effects of temperature variability on fisheries yields, Proc. R. Soc. B: Biol. Sci. 283 (2016) 20161435. Available from: https://doi.org/10.1098/rspb.2016.1435.

[19] GARFO, NOAA Fisheries Announces Final Blueline Tilefish Amendment to the Golden Tilefish Fishery Management Plan. <https://www.greateratlantic.fisheries.noaa.gov/mediacenter/2017/11/14_blueline_tilefish_amendment.html>, 2017.

[20] J.S. Link, J.A. Nye, J.A. Hare, Guidelines for incorporating fish distribution shifts into a fisheries management context, Fish Fish. 12 (2011) 461–469. Available from: https://doi.org/10.1111/j.1467-2979.2010.00398.x.

[21] O.A. van Keeken, M. van Hoppe, R.E. Grift, A.D. Rijnsdorp, Changes in the spatial distribution of North Sea plaice (*Pleuronectes platessa*) and implications for fisheries management, J. Sea Res. 57 (2007) 187–197. Available from: https://doi.org/10.1016/j.seares.2006.09.002.

[22] M.L. Pinsky, G. Reygondeau, R. Caddell, J. Palacios-Abrantes, J. Spijkers, W.W.L. Cheung, Preparing ocean governance for species on the move, Science 360 (2018) 1189–1191. Available from: https://doi.org/10.1126/science.aat2360.

[23] M.L. Pinsky, N.J. Mantua, Emerging adaption approaches for climate ready fisheries management, Oceanography 27 (2014) 146–159.

[24] C.S. Szuwalski, A.B. Hollowed, Climate change and non-stationary population processes in fisheries management, ICES J. Mar. Sci.: Journal du Conseil 73 (2016) 1297–1305. Available from: https://doi.org/10.1093/icesjms/fsv229.

[25] S.D. Gaines, C. Costello, B. Owashi, T. Mangin, J. Bone, J.G. Molinos, et al., Improved fisheries management could offset many negative effects of climate change, Sci. Adv. 4 (2018) eaao1378. Available from: https://doi.org/10.1126/sciadv.aao1378.

[26] R.E. Beal, Adapting fisheries management to changes in species abundance and distribution resulting from climate change, ASMFC Fish. Focus 27 (2018) 3.

[27] M. Fogarty, The art of ecosystem-based fishery management, Can. J. Fish. Aquat. Sci. 490 (2014) 479–490.

[28] P. Pepin, J. Higdon, M. Koen-Alonso, M. Fogarty, N. Ollerhead, Application of ecoregion analysis to the identification of Ecosystem Production Units (EPUs) in the NAFO Convention Area. SCR 14/069, in: NAFO Scientific Council Research Documents, 2014, pp. 1–13.

[29] M. Burden, K. Kleisner, J. Landman, E. Priddle, K. Ryan, Climate-Related Impacts on Fisheries Management and Governance in the North East Atlantic, Environmental Defense Fund, London, 2017. URL https://www.edf.org/sites/default/files/documents/climate-impacts-fisheries-NE-Atlantic_0.pdf.

[30] J.N. Sanchirico, D. Holland, K. Quigley, M. Fina, Catch-quota balancing in multispecies individual fishing quotas, Mar. Policy 30 (2006) 767–785. Available from: https://doi.org/10.1016/j.marpol.2006.02.002.

[31] P.J. Woods, D.S. Holland, G. Marteinsdóttir, A.E. Punt, How a catch–quota balancing system can go wrong: an evaluation of the species quota transformation provisions in the Icelandic multispecies demersal fishery, ICES J. Mar. Sci. 72 (2015) 1257–1277. Available from: https://doi.org/10.1093/icesjms/fsv001.

[32] R.L. Lewison, A.J. Hobday, S.M. Maxwell, E. Hazen, J.R. Hartog, D.C. Dunn, et al., Dynamic ocean management: identifying the critical ingredients of dynamic approaches to ocean resource management, Bioscience 65 (2015) 486–498. Available from: https://doi.org/10.1093/biosci/biv018.

[33] D.C. Dunn, S.M. Maxwell, A.M. Boustany, P.N. Halpin, Dynamic ocean management increases the efficiency and efficacy of fisheries management, Proc. Natl. Acad. Sci. U.S.A. 113 (2016) 201513626. Available from: https://doi.org/10.1073/pnas.1513626113.

[34] J.R. Wilson, S. Lomonico, D. Bradley, L. Sievanen, T. Dempsey, M. Bell, et al., Adaptive comanagement to achieve climate-ready fisheries, Conserv. Lett. (2018) 1–7. Available from: https://doi.org/10.1111/conl.12452.

CHAPTER 21

Adapting tourist seafood consumption practices in Pacific Islands to climate change

Colette C.C. Wabnitz

Institute for the Oceans and Fisheries, University of British Columbia, Vancouver, BC, Canada

Chapter Outline

> Sustainability of tourism in general fully depends on the sustainability of the resource they are exploiting [...]. There has to be more obligation on the part of tourism ventures to protect the assets providing them with a tourist attraction.
>
> —*Ian Campbell, WWF.*

21.1 Introduction

A recent report concluded that by the year 2040 tourism could bring an additional US$1.89 billion in revenue and 127,600 jobs to Pacific Island countries and territories (PICTs) through an additional 1 million tourists [1]. The report highlights four strategies to ensure sustainable tourism development in the region: improving international transport connections; attracting high-end tourists; improving public sector engagement; and strengthening linkages between tourism and local economies. Part of the latter should be a focus on sustainable seafood provision with a reduction in reef fish intake, as reef fish are critical to local food security and their populations and habitats have registered significant

Predicting Future Oceans.
DOI: https://doi.org/10.1016/B978-0-12-817945-1.00020-4

declines and suffered widespread degradation in health, respectively [2], and stand to be severely impacted by climate change [3].

For approximately half of PICTs, tourism is the leading export earner [4,5]. Pacific Islands are a popular destination, with arrivals having increased at an average rate of around 4.9% over the last decade [6] and the region remaining above the world average per capita [7]. Therefore, it is not surprising that governments throughout the region see the industry as a means of potentially achieving sustainable development. Indeed, tourism is deemed to represent a unique opportunity because factors that characterize PICTs and are seen as barriers to other forms of economic growth—small and dispersed populations, small land areas, and remoteness from markets—can actually be advantageous in a tourism context [8]. At a regional level, total contributions from the tourism sector account for 7% of Gross Domestic Product (GDP) and 6% of formal employment [9]; country-level estimates can be much higher (Table 21.1). Despite the devastating impacts of tropical cyclone Winston in 2015, Fiji registered a record number of visitor arrivals in 2016, with earnings up by 2.7% to total US$1.6 billion compared to the previous year, and accounting for over 13% of GDP [10]. The ocean, the Pacific region's predominant resource, is what attracts most visitors. In Fiji, swimming was one of the primary activities for more than 75% of tourists surveyed, with over 50% reporting a variety of other water- or beach-related activities [11]. Shark diving in Palau, one of the main attractions for tourists visiting the island country, was estimated to generate US$1.2 million and US$1.5 million annually, in salaries and tax income alone, respectively [12,13]. Killing the 100 sharks that are known to interact with divers at popular dive sites for sale, on the other hand, would bring in about US$10,800 [12,13]. In Fiji, the Shark Reef Marine Reserve, is a self-sustaining and profitable project where local villages exchanged their traditional fishing rights for the protection of a stretch of reef and income generated through diver user fees [14]. Shark diving in the country as a whole was estimated by Vianna et al. [15] to contribute US$42.2 million annually to Fiji's economy.

While tourism development can significantly contribute to traditional economies and provide a key source of employment in many coastal communities [16,17], it also presents a number of important challenges—notably through increased coastal development, associated sedimentation, increased waste generation [18–20], and pressure on natural resources, among other impacts [21–23]. In Palau, for instance, large groups of often inexperienced divers have raised concerns about impacts on the reef through overcrowding and poor diver behavior (e.g., coral holding or kicking, resting on corals, etc.) contributing to coral reef decline [24–26]. Throughout the region, rapid development and high visitation rates have also led to important social and cultural impacts, such as the erosion of traditional customs and community well-being [27,28].

Table 21.1: Pacific Island countries and territories overseas visitor numbers (2015–17), land area (km^2), Exclusive Economic Zone (EEZ, '000 km^2), midyear population estimate, estimates of the tourism's sector direct and total GDP contribution, as well as direct and total employment contribution.

Country/Territories	Overseas visitors – Arrival numbers 2015		2016		2017		Midyear population estimate	Direct GDP contribution (%) 2017	Total GDP contribution (%) 2017	Direct employment contribution (%) 2017	Total employment contribution (%) 2017	Land area (km^2)	EEZ ('000 km^2)
American Samoa	44,045		38,285		42,316		56,700	*	*	*	*	199	390
Cook Islands	125,130	p	146,473	p	161,362	p	15,200	*	*	*	*	237	1830
Federated States of Micronesia	31,200		29,600		nc		105,300	*	*	*	*	701	2980
Fiji Islands	868,596	p	962,952	p	1,027,309	p	888,400	14.4	40.3	13	36.5	18,333	1290
Guam	1,409,050		1,535,518		1,543,990		172,400	*	*	*	*	541	218
Kiribati	2095		4029		6600		120,100	8.7	20.9	6.9	17.3	811	3550
Marshall Islands	6452		5336		...		55,500	*	*	*	*	181	2,131
Nauru	...		2991		...		11,000	—	—	—	—	21	320
Niue	8206		8920		11,556p		1520	*	*	*	*	259	390
Northern Mariana Islands (CNMI)	478,922		501,489		621,846p		56,200	*	*	*	*	457	1823
New Caledonia	558,075		625,139		613,975		285,500	*	*	*	*	18,576	1740
Palau	168,764		146,629		122,050		17,900	*	*	*	*	444	629
Papua New Guinea	198,685		197,632		...		8,558,800	0.7	1.8	0.5	1.6	462,840	3120
Pitcairn	nc		nc		nc		49	—	—	—	—	47	800
French Polynesia	239,077		241,339		254,358		277,100	*	*	*	*	3521	5030
Samoa	139,043		146,089		157,515		196,700	*	*	*	*	2934	120
Solomon Islands	19,524		23,194		25,709		682,500	4.3	10.4	3.7	9.1	28,230	1340
Tokelau	120		81		70		1400	—	—	—	—	12	290
Tonga	72,553		86,662		86,579		100,300	6.8	18.2	7	19.6	749	700
Tuvalu	2541		2609		2560		10,200	*	*	*	*	26	900
Vanuatu	287,423		351,599		332,659		304,500	18.2	46.1	14.4	39.3	12,281	680
Wallis and Futuna	nc		nc		nc		11,700	—	—	—	—	142	300
Other oceanic states*	4,659,501	—	5,056,566	—	4,377,052			12.7	33.4	17.4	41.5		

Notes: (1) Overseas visitors include sea and air arrivals. It includes excursionists/same-day-visitors, that is, visitors from cruise ships as well. (2) 'nc' means data have not been compiled. (3) '...' means data may have been compiled, but not made available to SPC. (4) 'p' means provisional. (5) 'r' means revised. (6) The Aviation Desk of Wallis and Futuna who collects arrival data does not make a difference between local and foreign travelers. Therefore, there are no statistics available on visitor arrivals. (7) Data for American Samoa reflects tourist (not visitor) arrivals from 2005 to 2012.
(8) Data for French Polynesia reflects tourist (and not) visitor arrivals from 2005 to 2014. (9) Data for GDP and employment contributions from www.wttc.org/. (10) '*' denotes countries for which statistics were aggregated by WTTC.

Source: *The Pacific Community (SPC) 2018, except where indicated.*

While a number of studies have focused on the tourism industry's contribution to a decline in the health of marine ecosystems through diving, coastal development as well as sewage, tourists' impact on ecosystems through the consumption of mostly reef-sourced seafood in restaurants generally has been overlooked (but see [29–31]). Food is an important part of the tourist experience and the use of local food can be a key component of developing visitors' experience and creating a sense of authenticity [32]. Food and beverage consumption also represents a significant part of tourist expenditure. In Fiji, for instance, roughly 20% of expenses are on food. In the case of agriculture, tourism, if well-managed, creates a demand that can be an effective tool to grow and sustain the sector, reduce economic leakage[1] [33,34]—a key impediment to the sustainable development of the industry [35]—and improve the distribution of economic benefits to rural communities, as well as contribute to the ethos of sustainable tourism [36–38]. For most small Pacific islands this will be difficult, as the proportion of arable land is limited, and transport infrastructure as well as agrotechnology lags behind. Moreover, any expansion of agricultural production will need to secure the health and nutrition of rural producers first, whilst also protecting the country's unique and valuable biodiversity (e.g., not occur at the expense of forest loss). In the case of seafood, the vast majority of, if not all, fish and invertebrates harvested to supply the tourism industry are wild-caught. Tourism, through sport fishing and the consumption of local seafood by visitors, represents a powerful driver of fishery resources exploitation in many developing nations. Tourists will usually or always order seafood when dining out, and in many places are actively encouraged to catch it themselves. Supplying this demand also tends to skew fishing pressure toward higher trophic level reef fish [39] or important grazers, which are generally already overexploited, expensive seafood such as crustaceans and mollusks [40,41], or undersized fish that fit best on a plate.

21.2 Case study: Palau

The Republic of Palau is an archipelago consisting of over 700 islands (only 12 of which are inhabited) stretching over 700 km, about 750 km east of the Philippines and 1300 km southwest of Guam. Palau is recognized as a world-class diving location, consistently ranking as one of the top destinations for divers [42,43] (Fig. 21.1), so much so that tourism numbers have exceeded local residents by a factor of between 5 and 9 over the past 12 years (Table 21.1). It is particularly popular for the abundance of large pelagic fish, especially sharks [13].

A recent study explored the current and future impacts of tourists on reefs, based on existing visitation rates, currently proposed government tourism and conservation

[1] Revenue generated by tourism that is lost to other countries' economies. In the context of food, for example, preparing dishes based predominantly on imported foods results in a loss of foreign exchange earnings.

Figure 21.1
Tourists and manta rays. *Image by Peter Mumby.*

strategies, and projected climate change impacts on local ecosystems [30]. Tourist impacts such as diving and snorkeling as well as reef fish consumption were included in a social–ecological model (Fig. 21.2) and included the following assumptions (see [30] for full details):

1. Approximately 50% of tourists visiting Palau will dive or snorkel.
2. The average reef fish consumption for tourists was derived by taking the average per person consumption of fresh fish in tourists' countries of origin, weighting it by the proportion of arrivals from each region into Palau in 2015, further scaling it by the average number of days spent in Palau [12] and assuming, based on local experience, that 80% of meals consisted of reef fish.
3. Estimated impacts of divers on corals derived from the literature.
4. Published regional projections of climate change impacts on reef fish and their reef environment.

The study's results showed that aside from climate change, fishing of reef fish for both tourist and local consumption has the greatest impacts on Palau's ecosystems. Given the importance of the tourism sector from a revenue generating perspective, management efforts targeted at reducing reef fish consumption and maintaining high ecotourist visitation rates to Palau should be prioritized. Such high ecotourist visitation rates may also maximize employment given the number of services they rely upon (e.g., diving, snorkeling, water-based activities, tours). High visitation rates, through the levying of green fees ($100 per person), would also generate funds that could be spent on effective management,

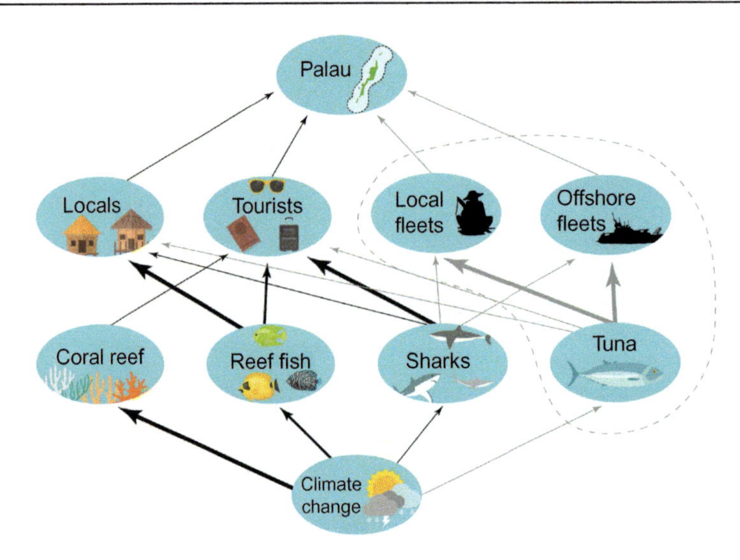

Figure 21.2
Conceptual model framework, including physical (climate change), ecological (coral reef, reef fish, sharks), and human (local population, tourists, fishing fleets) sectors. Components encircled by a gray dotted line (tuna stocks, local fishing fleets, local- and foreign-based offshore fleets) are considered conceptually, but not included in the model. The thickness of the arrows illustrates the importance of individual impacts or contributions of one component (on) to another.

monitoring, and enforcement of local marine reserves. This would in turn act as positive reinforcement for ecotourism development, as healthy reefs, and particularly large fish, are key factors in determining the attractiveness of a site to divers and their willingness to pay higher fees [44,45]. Key means to reduce pressure on local reef fish populations include the adequate enforcement of existing (and future) Marine Protected Areas (MPAs), regular monitoring of stocks, sustainable means of harvest, and a shift in consumption to more pelagic species - including tuna, wahoo and dolphinfish. The local Palauan government is leading the way with a recent Presidential directive indicating that "all government food service systems must serve pelagic fish such as tuna, with reef fish no longer allowed to be served at any government event or by any government-backed foodservice system. [46]" Exports of reef products for consumption outside of Palau should also be more strictly regulated. Given the importance of reef fish in traditional Palauan diets and culture, it is important that any changes are considerate of social ties, nutritional security and traditional customs.

A number of initiatives are currently under way between the government of Palau and researchers, both from Palau and at international institutions, to develop a portfolio of policy and management options addressing the above to support food security and sustainable development—including tourism and marine resource use—in the context of the newly designated National Marine Sanctuary [47,48].

21.3 Tourism and sustainable seafood

As highlighted in the Palau case study, climate change represents a significant underlying challenge, and is expected to continue to affect not only marine environments [49,50], but also local islanders' food security, culture, and livelihoods. Coastal fisheries of most PICTs will not meet their food security needs by 2030 due to population growth, overfishing, reduced productivity because of climate change, and inadequate national distribution networks [51]. Coastal regions stand to suffer the most important losses, with significant declines expected in the abundance of nearshore fisheries species due to degradation of coral reefs and other essential habitats, such as mangroves and seagrass beds. Climate change will have significant (direct and indirect) impacts on tourism [52,53] and the sector recognizes the need to address this challenge now [4,54]. From a fisheries perspective, while modeling studies suggest that strong local mitigation measures can substantially reduce projected declines [3], strong participatory and culturally appropriate resource management can help support adaptation measures that protect local communities' food security and livelihoods. One such measure is to transfer some level of consumption from reef fish onto tuna and other pelagic fish, particularly for tourists.

A survey conducted in Sanibel Island, southwest Florida, found that a large proportion of tourists rarely or never knew which seafood they would order before dining at a restaurant [55]. They would therefore likely be receptive to seafood recommendations from restaurant staff or seafood specials. Similar surveys could be developed in other locations and sustainability initiatives designed around educating local restaurant management about local sustainable options as well as encourage staff to advertise only sustainable options as daily specials, for example.

A number of developments around the world highlight that the tourism industry is joining in on the increasing interest in sustainable food systems and cuisine. In the Maldives, the Soneva Resorts group executive and the sustainability team developed a list of seafood to avoid and conducted associated training to ensure the culinary team is aware of the list and understands its importance [56]. In 2010, Marriott was the first large global hotel chain to launch a sustainable seafood program, FutureFish, designed to support the company's hotels to source, cook, and serve sustainable fish [56]. In August 2014, Hyatt Hotels announced the first phase of a long-term seafood sustainability strategy in partnership with WWF during which the group will work toward responsibly sourcing more than 50% of its inventory by 2018 [56]. On Zanzibar in Tanzania, the Chumbe Island Coral Park implemented a sustainable seafood buying policy to ensure that they would purchase only sustainably sourced marine products from local fishermen, and avoid species that are overfished or have been caught using destructive fishing techniques [57]. Including fewer reef fish species and a majority of pelagic fish as part of the options tourists can choose from would support sustainable resource management, local communities' food security and livelihoods, as well as responsible tourism development.

21.4 Conclusion

The role of tourism in fisheries' removals through seafood consumption has been overlooked, and is a particular concern in islands with high and increasing visitation rates. The tourism industry can and should take advantage of the increasing interest in sustainable food systems and cuisine by promoting sustainably sourced seafood that does not undermine local food security and the health of the very ecosystems that the sector depends upon. Seafood for tourists should predominantly consist of tuna and other pelagic fish. In addition to supporting the ecological sustainability of marine resources supplied to tourists, efforts should be made to ensure that the tourism-fishing sector contributes equal income opportunities among seafood trade stakeholder groups. Tourism-led demand for seafood should also not undermine local food security either through a decline in availability of popular or culturally important fish, or through increases in prices, which would see local residents purchase cheaper and potentially less nutritious food options.

According to current scenarios [58], increases in sea surface temperatures of between 1°C and 3°C by 2100 are projected to significantly impact ecosystems in the tropical Pacific [3,59]. Thus the conservation of coastal ecosystem biodiversity, resilience, functions, benefits, and services has gained in value and urgency. Combined with sustainable agriculture where possible, an emphasis on locally sourced sustainable products (over imports) will not only support the resilience of reefs into the future and the food security of local communities, but also reduce countries' economic leakage and carbon footprint.

While this chapter focuses on PICTs, the suggestions outlined here are very much applicable to other Small Islands Developing States that derive significant benefits from tourism and strive for sustainable and equitable ocean stewardship in a changing climate.

Acknowledgments

Many thanks are due to my Nereus Program colleagues for the many interesting discussions and enriching opportunities over the past few years. I also wish to acknowledge the Nippon Foundation for their support of such an incredible and creative network. Many thanks also to Andrés Cisneros-Montemayor for constructive comments on early drafts of this chapter and colleagues throughout the Pacific region - notably at SPC - for their efforts, support and inspiration.

References

[1] The World Bank Group, Pacific Possible, Tourism, 2016.

[2] C. Moritz, J. Vii, W. Lee Long, J. Tamelander, A. Thomassin, S. Planes (Eds.), Status and Trends of Coral Reefs of the Pacific, Global Coral Reef Monitoring Network, 2018.

[3] R.G. Asch, W.W.L. Cheung, G. Reygondeau, Future marine ecosystem drivers, biodiversity, and fisheries maximum catch potential in Pacific Island countries and territories under climate change, Mar. Policy 88 (2018) 285–294.

[4] S. Becken, J.E. Hay, Tourism and Climate Change: Risks and Opportunities, Channel View Publications, 2007.

[5] Commonwealth of Australia, Pacific Economic Survey: Engaging With the World, Australian Agency for International Development, Canberra, Australia, 2009.

[6] J.M. Cheer, S. Pratt, D. Tolkach, A. Bailey, S. Taumoepeau, A. Movono, Tourism in Pacific island countries: a status quo round-up, Asia Pac. Policy Stud. (2018). Available from: https://doi.org/10.1002/app5.250.

[7] World Travel and Tourism Council, Travel and tourism, Economic Impact 2018, World, 2018.

[8] UNWTO, Challenges and Opportunities for Tourism Development in Small Island Developing States, UNWTO, Madrid, Spain, 2012.

[9] H. Seidel, P.N. Lal, Economic Value of the Pacific Ocean to the Pacific Island Countries and Territories., IUCN, Gland, Switzerland, 2010.

[10] South Pacific Tourism Organisation, Annual Review of Visitor Arrivals in Pacific Island Countries, 2017.

[11] M. Verdone, A. Seidl, Fishing and Tourism in the Fijian Economy., IUCN, Gland, Switzerland, 2012.

[12] G. Vianna, M. Meekan, D. Pannell, S. Marsh, J. Meeuwig, Wanted Dead or Alive? The Relative Value of Reef Sharks as a Fishery and an Ecotourism Asset in Palau, Australian Institute of Marine Science and University of Western Australia, Perth, 2010.

[13] G.M.S. Vianna, M.G. Meekan, D.J. Pannell, S.P. Marsh, J.J. Meeuwig, Socio-economic value and community benefits from shark-diving tourism in Palau: a sustainable use of reef shark populations, Biol. Conserv. 145 (1) (2012) 267–277.

[14] J.M. Brunnschweiler, The Shark Reef Marine Reserve: a marine tourism project in Fiji involving local communities, J. Sustain. Tour. 18 (1) (2010) 29–42.

[15] G.M.S. Vianna, J.J. Meeuwig, D. Pannell, H. Sykes, M.G. Meekan, The Socioeconomic Value of the Shark-Diving Industry in Fiji, University of Western Australia, Perth, Australia, 2011.

[16] J. Cinner, Coral reef livelihoods, Curr. Opin. Environ. Sustain. 7 (2014) 65–71.

[17] M. Honey, Ecotourism and Sustainable Development: Who Owns Paradise? Island Press, 1999.

[18] Y. Golbuu, A. Bauman, J. Kuartei, S. Victor, The state of coral reef ecosystems of Palau, in: J.E. Waddell (Ed), The State of Coral Reef Ecosystems of the United States and Pacific Freely Associated States: 2005, NOAA Technical Memorandum NOS NCCOS 11, 2005, pp. 488–507.

[19] S. Gossling, P. Peeters, C.M. Hall, J.P. Ceron, G. Dubois, L. Lehmann, et al., Tourism and water use: Supply, demand, and security. An international review, Tour. Manage. 33 (1) (2012) 1–15.

[20] A. DeGeorges, T.J. Goreau, B. Reilly, Land-sourced pollution with an emphasis on domestic sewage: lessons from the Caribbean and implications for coastal development on Indian Ocean and Pacific Coral reefs, Sustainability 2 (9) (2010) 2919.

[21] J. Davenport, J.L. Davenport, The impact of tourism and personal leisure transport on coastal environments: a review, Estuarine Coastal Shelf Sci. 67 (1–2) (2006) 280–292.

[22] G. Musa, Sipadan: a SCUBA-diving paradise: an analysis of tourism impact, diver satisfaction and tourism management, Tour. Geograph. 4 (2) (2002) 195–209.

[23] B. Garrod, S. Gössling, New Frontiers in Marine Tourism: Diving Experiences, Sustainability, Management, Elsevier, Amsterdam, The Netherlands, 2008.

[24] Anon, Palau Islands Have Been Inundated With Chinese Tourists. Available from: <http://www.news.com.au/travel/world-travel/pacific/palau-islands-have-been-inundated-with-chinese-tourists/news-story/75a4d19601a930e431298983a2b28937>, 2015.

[25] C. Poonian, P.Z.R. Davis, C.K. McNaughton, Impacts of recreational divers on Palauan Coral reefs and options for management, Pac. Sci. 64 (4) (2010) 557–565.

[26] E.I. Otto, M. Gouezo, S. Koshiba, G. Mereb, R. Jonathan, D. Olsudong, et al., Impact of snorkelers on shallow coral reef communities in Palau, in: PICRC Technical Report 16-15, Palau International Coral Reef Center, Palau, 2015.

[27] E. Agyeiwaah, B. McKercher, W. Suntikul, Identifying core indicators of sustainable tourism: a path forward? Tour. Manage. Perspect. 24 (2017) 26–33.

[28] A. Movono, H. Dahles, S. Becken, Fijian culture and the environment: a focus on the ecological and social interconnectedness of tourism development, J. Sustain. Tour. 26 (3) (2018) 451–469.

[29] N.S. Smith, D. Zeller, Unreported catch and tourist demand on local fisheries of small island states: the case of The Bahamas, 1950-2010, Fish. Bull. 114 (1) (2016) 117–131.

[30] C.C.C. Wabnitz, A.M. Cisneros-Montemayor, Q. Hanich, Y. Ota, Ecotourism, climate change and reef fish consumption in Palau: benefits, trade-offs and adaptation strategies, Mar. Policy 88 (2018) 323–332.

[31] C. Birkeland, Working with, not against, coral-reef fisheries, Coral Reefs 36 (1) (2017) 1–11.

[32] G. Richards, Gastronomy: an essential ingredient in tourism production and consumption, in: A. Hjalager, G. Richards (Eds.), Tourism and Gastronomy., Routledge, London, UK, 2002, pp. 1–20.

[33] R. Scheyvens, M. Russell, Tourism and poverty alleviation in Fiji: comparing the impacts of small- and large-scale tourism enterprises, J. Sustain. Tour. 20 (3) (2012) 417–436.

[34] S. Pratt, Same, same but different: perceptions of south pacific destinations among Australian travelers, J. Travel. Tour. Mark. 30 (6) (2013) 595–609.

[35] S. Pratt, The economic impact of tourism in SIDS, Ann. Tour. Res. 52 (2015) 148–160.

[36] R. Torres, Linkages between tourism and agriculture in Mexico, Ann. Tour. Res. 30 (3) (2003) 546–566.

[37] T. Berno, Sustainability on a Plate: Linking Agriculture and Food in the Fiji Islands Tourism Industry, in: R.M. Torres, J.H. Momsen (Eds.), Tourism and Agriculture: New Geographies of Consumption, Production and Rural Restructuring. London, UK and New York, 2011, pp. 87–103.

[38] H. McBain, Caribbean tourism and agriculture: linking to enhance development and competitiveness, No. 2, United Nations Publications, ECLAC Studies and Perspectives Series, The Caribbean, New York, 2007.

[39] G.J. Rodrigues, S. Villasante, Disentangling seafood value chains: tourism and the local market driving small-scale fisheries, Mar. Policy 74 (2016) 33–42.

[40] M. Thyresson, B. Crona, M. Nyström, M. de la Torre-Castro, N. Jiddawi, Tracing value chains to understand effects of trade on coral reef fish in Zanzibar, Tanzania, Mar. Policy 38 (2013) 246–256.

[41] B. Crona, M. Nystrom, C. Folke, N. Jiddawi, Middlemen, a critical social-ecological link in coastal communities of Kenya and Zanzibar, Mar. Policy 34 (4) (2010) 761–771.

[42] SportDiver, 50 Best Dive Sites in the World. Available from: <https://www.sportdiver.com/photos/planets-50-greatest-dives>, 2013.

[43] Scuba Travel, Top 10 Dives: Best Diving in the World 2018. Available from: <https://www.scubatravel.co.uk/topdives.html>, 2018.

[44] H. Koike, A. Friedlander, K. Oleson, S. Koshiba, K. Polloi, Final report on diver's perception survey for Palau's Kemedukl and Maml, in: Project: Stock Assessment for Humphead Wrasse and Humphead Parrotfish. PICRC Technical Report 14-02, 2014.

[45] D.A. Gill, P.W. Schuhmann, H.A. Oxenford, Recreational diver preferences for reef fish attributes: economic implications of future change, Ecol. Econ. 111 (2015) 48–57.

[46] https://www.seafoodsource.com/news/environment-sustainability/tuna-will-take-center-stage-on-palau-s-menu.

[47] Palau National Marine Sanctuary Act, 2015.

[48] T. Remengesau, Healthy Oceans and Seas: A Way Forward, Address to the United Nations, 2014, p. 4.

[49] M.S. Pratchett, A.S. Hoey, S.K. Wilson, Reef degradation and the loss of critical ecosystem goods and services provided by coral reef fishes, Curr. Opin. Environ. Sustain. 7 (2014) 37–43.

[50] M.S. Pratchett, P.L. Munday, S.K. Wilson, Effects of climate-induced coral bleaching on coral-reef fishes: ecological and economic consequences, Oceanogr. Mar. Biol.: Annu. Rev. 46 (2008) 251–296.

[51] Q. Hanich, C.C.C. Wabnitz, Y. Ota, M. Amos, C. Donato-Hunt, A. Hunt, Small-scale fisheries under climate change in the Pacific Islands region, Mar. Policy 88 (2018) 279–284.

[52] R.B. Richardson, K. Witkowski, Economic vulnerability to climate change for tourism-dependent nations, Tour. Anal. 15 (2010) 315–330.

[53] L.M. Klint, T. DeLacy, S. Filep, Climate change and island tourism, in: S. Pratt, D. Harrison (Eds.), Tourism in Pacific Islands: Current Issues and Future Challenges, Routledge, New York, 2015, pp. 257–277.

[54] G. Sem, R. Moore, The Impact of Climate Change on the Development Prospects of the Least Developed Countries and Small Island Developing States., United Nations Office of the High Representative for the Least Developed Countries, Landlocked Developing Countries and Small Island Developing States, New York, 2009.

[55] L. Biery, The sustainable special: can restaurants encourage sustainable seafood consumption? Sea Around Us Project Newsl. 63 (2011) 1–2.

[56] International Tourism Partnerships, ITP Presents... Sustainable Seafood in Hotels. Available from: <https://www.tourismpartnership.org/blog/itp-presents-sustainable-seafood-in-hotels-3/>, 2015.

[57] Chumbe Island Team, Sustainable Seafood Guide 2017. Available from: <http://www.chumbeisland.com/fileadmin/downloads/Textlink-pdfs/CHICOP_seafood_guide_2017.pdf>, 2017.

[58] IPCC, Special Report on Emission Scenarios, Cambridge University Press, UK, 2000.

[59] J.D. Bell, C. Reid, M.J. Batty, P. Lehodey, L. Rodwell, A.J. Hobday, et al., Effects of climate change on oceanic fisheries in the tropical Pacific: implications for economic development and food security, Clim. Change 119 (1) (2013) 199–212.

CHAPTER 22

Mariculture: perception and prospects under climate change

Muhammed A. Oyinlola

Changing Ocean Research Unit Institute for the Oceans and Fisheries The University of British Columbia, Vancouver, BC, Canada

Chapter Outline

22.1 Background

The world's population will reach an estimated 9.7 billion people by 2050, increasing global food demand [1,2]. Meeting the fundamental food demand for everyone in the world is a crucial challenge to sustainable development [3]. Some studies point out that the world will need 70%–100% more food by 2050 [4,5]. However, the ability to meet food demand for the increasing population will depend on the efficiency of food production systems, the sustainability of production methods, and the equality in the distribution of food across the world's populations [2].

Predicting Future Oceans.
DOI: https://doi.org/10.1016/B978-0-12-817945-1.00019-8

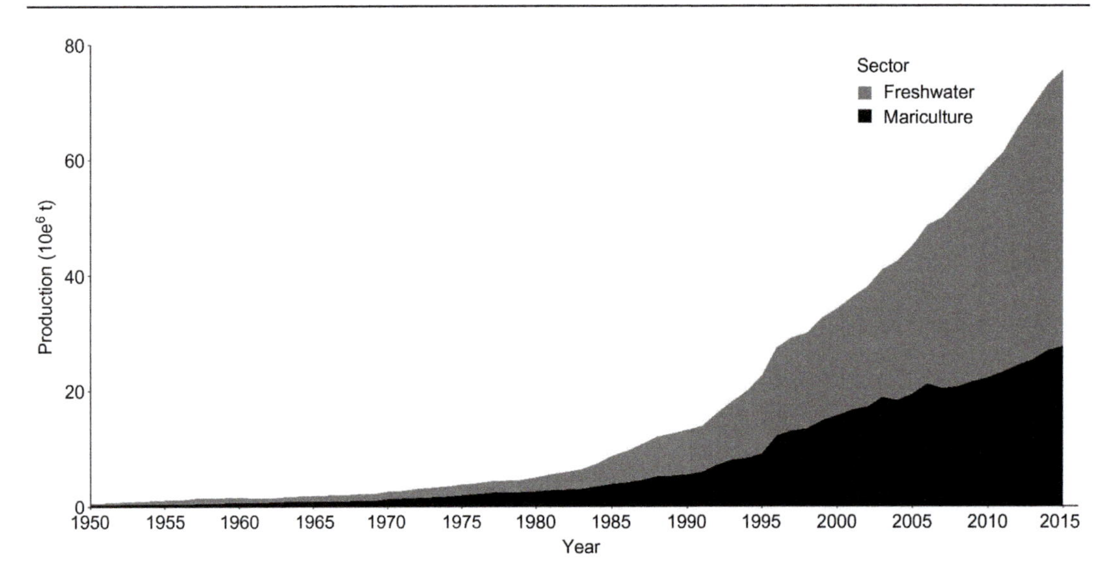

Figure 22.1

Global trends in aquaculture production (1950 - 2015) from freshwater aquaculture and mariculture. Data from Sea Around Us (www.seaaroundus.org) and FAO (http://www.fao.org/fishery/statistics/en).

Aquaculture has played a significant role in filling the gap between food fish demand and supply from capture fisheries, with both freshwater and marine aquaculture (including brackish environments) providing around 80 million tonnes of food fish (excluding aquatic plants) for consumption worldwide in 2016 [6] (Fig. 22.1). The global production of food fish from aquaculture has multiplied during the past four decades from 2.57 million tonnes in 1970 to 80 million tonnes in 2016 [6], at an average annual rate of 5.8% (2001–16).

Despite aquaculture's contribution to global food security, the food system has received a negative image driven by the concerns associated with a few commodities like salmon and shrimp, and practices which affect the whole aquaculture industry [7]. These prejudices leave out other farmed species and aquaculture systems that are practiced, especially in developing countries where aquaculture's contribution to food security and livelihoods is enormous. Aquaculture could assist in increasing seafood supply to offset declining supply from capture fisheries, especially under the effects of climate change with the allied negative consequences on food security. Nevertheless, climate change might threaten aquaculture's contribution to food security. This chapter aims to provide basic aquaculture concepts and types, and the possible impacts of climate change on global future seafood supply from marine aquaculture.

22.2 Aquaculture definitions

Multiple definitions for aquaculture exist, and this has affected the overall understanding of its principles and practice especially among the general public who liken it to capture fisheries regarding food hunting rather than a food production system [7]. However, aquaculture is distinct because of stock ownership and deliberate intervention in the production cycle [8].

The United States National Aquaculture Act of 1980 defined aquaculture as:

> The propagation and rearing of aquatic organisms in controlled or selected environments for any commercial, recreational or public purpose [9].

The above definition failed to embrace the understanding of "rearing" in context. In 1988, the Food and Agriculture Organization of the United Nations introduced a definition of aquaculture to reduce the confusion between capture fisheries and ocean ranching (ocean ranching is a type of fish farming in which juvenile fish are released into the ocean to grow unprotected and unassisted to be subsequently harvested [10]).

This definition is accepted as the modern definition of aquaculture:

> The farming of aquatic organisms including fish, molluscs, crustaceans and aquatic plants with some sort of intervention in the rearing process to enhance production, such as regular stocking, feeding, protection from predators [11].

However, this definition lacks the current practice of controlling physiochemical water parameters to make the rearing intervention suitable for the aquatic animal to grow and develop in a stress-free environment. Hence, for this book aquaculture will be defined as:

> The farming of aquatic organisms including finfishes, crustaceans, molluscs, amphibians, and other aquatic animals in a controlled environment (freshwater, marine, or brackish) with some intervention in the rearing process to enhance production, such as regular stocking, feeding, protection from predators.

22.3 Aquaculture farming systems and practice

22.3.1 Extensive aquaculture system

An extensive aquaculture system is a production system that relies on the natural primary productivity of the water body for farmed species nourishment, growth, and development. Principally, herbivore and omnivore species are often farmed in an extensive system. These species could be freshwater fish or brackish species, such as some Carp (*Cyprinus* spp.), catfish (*Clarias* spp.), shrimp (*Penaeus* spp.), mullet (*Mugil* spp.), and marine species such as bivalve mussels.

The following features characterize this system of production [12,13]:

1. low abiotic and biotic control (e.g., environments, nutrition, predators, competitors, and diseases vectors);
2. low stocking density with low production yield, <1000 kg/ha/year;

3. high water usage, 2100 L/kg;
4. no feeding, dependence on natural food organisms;
5. polyculture—two or more fish species or aquatic organisms farmed together; and
6. high dependence on water quality of natural water bodies (e.g., bays, rivers, dams).

22.3.2 Intensive aquaculture system

An intensive aquaculture system involves high intervention in the farming or rearing procedure, ranging from feeding, solid waste management, to environmental control. The benefits of the intensive farming system are centered on a high return on investment because of the high inputs required and the ability to increase the scale of production by decreasing the cost per unit of output. All aquatic organisms could be farmed with this method.

The following are the characteristic features of this method [12,13]:

1. high abiotic and biotic control (e.g., environmental parameters—water quality, nutrition, predators, competitors, and disease vectors);
2. high stocking density with high production yield c.1,340,000 kg/ha/year;
3. low water usage 50 L/kg;
4. high quantity feed requirement;
5. monoculture practice only; and
6. high technology involvement.

22.3.3 Semi-intensive aquaculture system

A distinctive definition for "semi-intensive farming system" has been complicated over the years as the level of supplementary aquafeeds and fertilization (nutrient enrichment) of the water body is used to characterize the system [14]. The operation may involve the addition of supplemental aquafeeds or other inputs like aeration and waste management equipment. However, farmed species still partially rely on the primary productivity of the water body for nourishment and optimal water parameters for growth and development.

22.3.4 Open aquaculture system

An open aquaculture system is found in freshwater lakes and rivers, coastal waters, and offshore. In most cases a net, cage, or line/rope is submerged in the aquatic ecosystem, which allows free exchange of water from the surrounding environment. The open aquaculture system is considered a high environmental risk method because of its direct contact with the ocean ecosystem. A lot of environmental issues in aquaculture today are related to open systems [15,16]. Most commercial open systems have been mainly limited to the farming of high economic value finfish like bivalves, salmon, cobia, trout, seabass, and sea bream [17]. Indeed, major mariculture activities are practiced with this method.

22.3.5 Closed aquaculture systems

Closed aquaculture systems (CAS) are any system of aquatic organism production that creates a controlled interface between the farm organisms and the natural environment [18]. Aquatic organisms are typically raised in tanks where they are fed, respire, and excrete, with a highly sophisticated waste management procedure which allows water to pass through some compartments for solid waste removal, and biological filtration thereby makes the water able to be used several times before being discarded. A typical example of CAS is the recirculating aquaculture system (RAS). Examples of species currently farmed in RAS include *Salmo salar* (Atlantic salmon), *Rachycentron canadum* (Cobia), *Clarias gariepinus* (Catfish), *Oreochromis niloticus* (Tilapia), and *Dicentrarchus labrax* (European bass).

22.3.6 Semiclosed aquaculture systems

Contrary to open and CAS, a semiclosed aquaculture system is a land-based aquaculture system, in which water is exchanged between the aquaculture holding embankment (ponds, tanks, etc.) and the natural water environment. This system does have a water inlet that allows the inflow of fresh (unused) water into the ponds and outflow of wastewater into the natural water environment. A typical example of this system is a raceway.

22.3.7 Integrated aquaculture systems

An integrated aquaculture system is the combined production system of two or more aquatic organisms with or without other agriculture/livestock farming operations in a given space. The rationale for integrated practices is to use the by-products/wastes from one subsystem as a valuable input to another subsystem. An example of such a system is integrated multitrophic aquaculture (IMTA) systems, where lower trophic level species, such as seaweed and bivalves, are farmed with finfish, thereby improving inorganic and organic nutrient waste recycling. This can reduce the environmental footprint of mariculture farms [19], although large commercial IMTA farms are still not common [20].

22.3.8 Capture-based aquaculture

The term capture-based aquaculture (CBA) first emerged in 2004 [21]. Since then, the farming of aquatic organisms without a controlled breeding process for the production of seed has been termed CBA, and it is defined as:

> The practice of collecting seed material from early life stages to adults from wild and its subsequent on growing in captivity to marketable size, using aquaculture techniques [21].

The CBA industry is a diverse worldwide industry because of factors such as little technical input, the limited biological and economic feasibility of producing seeds, and wild seeds usually being of high quality [22]. The development of the industry is solely driven by the market demand for some high-value species whose life cycles currently cannot be enclosed on a commercial scale [21]. Some examples of the species harvested as wild seeds or spats include shrimps (*Penaeidae*), tunas (*Thunnus* spp.), and groupers (*Epinephelus* spp.).

22.3.9 Closed life cycle aquaculture or hatchery-based aquaculture

The production of seeds from hatcheries through the manipulation of adult maturation and reproduction and larval or juvenile rearing is called hatchery-based aquaculture [23]. The process of artificial propagation of seeds largely depends on the aquatic organisms to be farmed. However, inducement of the brood stocks (hypophyzation) with or without hormone treatment is the most common method. Artificial fertilization, incubation of eggs, and subsequent rearing of larvae would follow. A large number of farmed species are reared in closed life cycle aquaculture; examples include *S. salar* (Atlantic salmon), *R. canadum* (Cobia), *C. gariepinus* (catfish), and *Chanos chanos* (milkfish).

22.3.10 Production based on the aquatic environment

Mostly defined by the level of salinity and the species to be farmed, an aquaculture operation or practice can take place in all aquatic ecosystems. Freshwater aquaculture, also called inland aquaculture, is the farming of aquatic organisms like molluscs, crustaceans, aquatic plants, and fish in water not exceeding 0.5 PSU. Freshwater aquaculture contributed 51.4 million tonnes to the total aquatic food production in 2016 [6]. The subsector makes a substantial contribution to the supply of affordable protein food, particularly in developing countries. Species such as *Tilapia* spp. and catfishes are farmed in a freshwater environment.

Brackish water farming is the rearing of aquatic organisms where the end product is raised in brackish water, such as estuaries, coves, bays, lagoons, and fjords, in which the salinity generally fluctuates between 0.5 PSU and full strength seawater [11]. Brackish water supports a large number of shrimps and prawn [*Fenneropenaeus indicus* (Indian prawn), *Penaeus monodon* (giant tiger prawn), *Litopenaeus vannamei* (whiteleg shrimp), etc.] and finfish [*Anguilla* (short-finned eel), *Oncorhynchus mykiss* (rainbow trout), *O. niloticus* (Nile tilapia)]. However, most production is recorded on aquaculture databases under mariculture or freshwater depending on the level of salinity required for the stress-free growth and development of the farmed species.

Generally, aquaculture in marine and brackish environments is called mariculture. This is defined as the rearing of the end product taking place in seawater, such as fjords, inshore and open waters, and inland seas or an inland facility (RAS) in which the salinity generally exceeds 20 PSU [11]. The subsector produces a large proportion of carnivorous species and bivalves which have high economic value and contributed about 28.7 million tonnes to total global production from farmed aquatic food for human consumption in 2016 [6].

22.4 Mariculture, environmental sustainability, and food security

Mariculture is a specialized subsector of aquaculture which involves the farming of marine organisms. It is not clear if seafood production from capture fisheries will increase. However, mariculture could boost seafood supply, especially if practiced sustainably.

Over the last decade, the contribution of mariculture to global aquaculture production has increased markedly from 14% in 2000 to 35.9% in 2016 [6]. The *Sea Around Us* global mariculture database (www.searoundus.org) estimated production of 5 million tonnes in 1990; this production increased to 21 million tonnes by the end of 2010. The increased production has been attributed to the growing demand for seafood (marine and brackish waters) in developed countries [24,25], which focuses on high-value omnivorous, carnivorous finfish, crustacean species, and marine bivalves [25].

Despite the interest in expanding mariculture to increase seafood production, concerns about its environmental impact [8] and the availability of suitable space [26] are also growing. Issues related to the environment, such as organic farm waste from mariculture activities [27] can lead to sediment organic enrichment and eutrophication [28,29], which subsequently can change the physiochemical properties and microflora biodiversity of benthic sediments in and around mariculture areas [27,29]. Also fishmeal and oil, which are mainly obtained from small pelagic forage fishes, are essential components of aquafeed, particularly for omnivorous and carnivorous farmed species [30]. Thus the demand for fishmeal and oil exerts pressure on wild fish stocks. Other potential environmental concerns from mariculture include pollution derived from the application of pharmaceuticals, antibiotics, and heavy metal contamination [27,28], alien species introduction, genetic interactions, disease transfer [31], and the destruction of coastal habitats such as mangroves [32].

Various attempts are being made to estimate the potentially suitable area for mariculture production [20,33,34] amidst other ocean activities (e.g., energy production, shipping, marine protection, and fishing). Findings show that the global area with a suitable environment for mariculture is much larger than the area where mariculture is currently operating [20,33]. Using species distribution models, Oyinlola et al. [20] estimated the total suitable mariculture area for the 102 species to be 72 million km^2. Sixty-six million km^2 of this area is suitable for finfish, 39 million km^2 for crustaceans, and 31 million km^2 for shellfish (Fig. 22.2). This prediction included areas currently used for mariculture purposes. Also, the study shows areas with the highest mariculture potential in regions such as the southwestern Atlantic coast (30–45 species richness) and West Africa

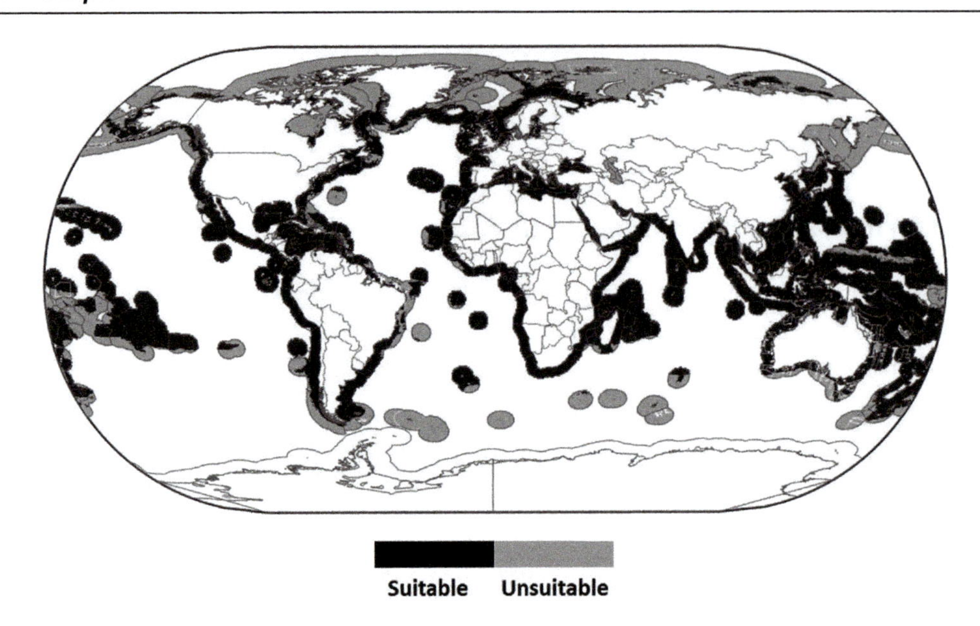

Figure 22.2
Suitable areas for mariculture, including fish and invertebrate species. Data from Oyinlola et al. [20]

(35–40 species richness). Other notable areas with high mariculture species richness include the Gulf of Mexico, the Caribbean Sea, the East China Sea, the Yellow Sea, the Sea of Japan, and the Banda Sea off the coast of Timor-Leste. However, despite results showing high species richness for these potential mariculture areas, most of the species currently are not being farmed in these regions.

Further expansion of mariculture can contribute to solving food security challenges [35–37] and the development of producing countries through an increase in employment and livelihood opportunities, especially in regions with large suitable marine areas [38]. Currently, there are over 112 countries and territories in which mariculture contributes to the gross domestic product [39], and more than half of these countries are developing nations. The future sustainability of mariculture can have significant effects on these countries.

22.5 Mariculture, climate change, and future seafood supply

Climate change is expected to impact aquaculture [2], although the extent, types, and distribution of impacts are not well understood. Most mariculture organisms are ectotherms, and their metabolism is affected by temperature [40]. Thus the environment directly affects their growth rate and reproduction. Species' environmental preferences and tolerances also restrict the area where particular species can be farmed [41].

Consequently, changes in environmental conditions under climate change will affect the suitability of the environment for farming different species. Ocean acidification could also have an impact on the growth and survival of shellfish, particularly in their early life stages [42,43]. Other elements of climate change, such as sea level rise, saltwater intrusion, and ocean productivity, will also affect the suitability of coastal and ocean areas for mariculture [44].

Climate change will also affect mariculture indirectly through the reduction in fishmeal and fish oil supplies, disease, and socioeconomic changes. Finfish and crustacean mariculture is currently dependent on wild capture fisheries, of which the abundance and biomass will be affected by climate change, overfishing, and other human stressors [45,46]. Although fishmeal and oil as feed ingredients can be restricted for use in higher value starter (fry), finisher, and brood stock feed only [47], the demand for nutrients from wild capture fisheries from a rapidly growing mariculture is still a concern. Efforts have been made for more than a decade to find a replacement for fishmeal and oil using plant sources [48–51].

One of the challenges to the sustainability of mariculture is the impact of infectious disease on farm production [52]. Climate change may also increase the succession of mariculture disease due to environmental stress and virulent pathogens [53]. Disease outbreaks have been considered a significant constraint to mariculture development in many countries [54,55]. Global economic losses from disease outbreaks for aquaculture have been estimated to be about US$6 billion per year since 1990, with the shrimp industry having lost about US$10 billion from 1990 to 2014 [56]. Although vaccines [57] and antibiotics have been developed and applied to control diseases in aquaculture [54], infectious diseases of farmed species have not been eliminated and continue to be a major uncertainty in aquaculture production [58].

Disease problems in mariculture may be exacerbated by extreme weather events and other ocean changes associated with climate change [59]. Ocean warming, sea level rise, and changes in salinity may affect disease occurrence in marine systems [60]. Warming is suggested to affect early life stages (egg and larvae) of pathogens [61]. Also, high stocking density of farms, disease control capacity, and the degree of globalization of the farmed species may increase the exposure of farmed organisms to disease because of the cooperating industry integrated business models, making mariculture more vulnerable to climate change impacts [62].

22.6 Conclusion

Societal perceptions and misunderstanding of aquaculture could be enhanced with public education and continuous review of industry practices. Nevertheless, constructive public scrutiny has helped and prompted the sector to improve, especially in reducing its environmental impact and increasing quality and food safety [7]. With climate change impacts, adaptation measures could be employed to minimize the adverse effects of climate change on the future seafood supply. With regards to site selection and suitability of farm locations, efforts toward proper implementation of the ecosystem approach to aquaculture [63] with the consideration of climate change could be a vital adaptation tool. Better biosecurity procedures should also be encouraged among farmers to reduce the disease susceptibility of stressed farmed species due to ocean warming. Detailed feed optimal ratio procedures could be employed to reduce the amount of fishmeal needed to achieve farmed species' best growth performance [64]. Other measures including the use of farming systems, such as IMTA, will assist in reducing the negative impact on food and nutritional security. This approach could improve water quality, reduce the stress of farm species, reduce bioremediation needs and environmental impacts, and improve the social acceptability of aquaculture.

References

[1] A.J. McMichael, Impact of climatic and other environmental changes on food production and population health in the coming decades, Proc. Nutr. Soc. 60 (02) (2007) 195–201.

[2] S.S.D. Silva, D. Soto, Climate change and aquaculture: potential impacts, adaptation and mitigation, in: FAO Fisheries and Aquaculture Technical Paper No 530, Rome, 2009, No. 530, pp. 151–212.

[3] UN, World Economic and Social Survey 2013 Sustainable Development Challenges, Department of Economic and Social Affairs, 2013.

[4] N. Alexandratos, J. Bruins, World Agriculture Towards 2030:2050, ESA Work, 2012:Pap, 3.

[5] D. Buncombe, B.D. Ian Cruet, M.G. Jim Dun well, J. Jones, J. Pretty, W. Sutherland, et al., Reaping the Benefits: Science and the Sustainable Intensification of Global Agriculture, The Royal Society Chicago, 2009.

[6] FAO, The State of World Fisheries and Aquaculture 2018—Meeting the Sustainable Development Goals, Rome, 2018. Licence: CC BY-NC-SA 3.0 IGO.

[7] K. Bacher, Perceptions and misconceptions of aquaculture: a global overview, GLOBEFISH Research Programme 120 (I) (2015).

[8] R.L. Naylor, R.J. Goldburg, J.H. Primavera, N. Kautsky, M.C. Beveridge, J. Clay, et al., Effect of aquaculture on world fish supplies, Nature 405 (6790) (2000) 1017–1024.

[9] T. Banerjee, The national aquaculture act of 1980, Editorial Page 6 (1) (1981) 18.

[10] A. Isaksson, Salmon ranching: a world review, Aquaculture 75 (1–2) (1988) 1–33.

[11] FAO, CWP handbook of fishery statistical standards, Section J: Aquaculture, CWP data collection, in: FAO Fisheries and Aquaculture Department [online] Rome Updated 10 January 2002 [cited June 26, 2015] <http://wwwfaoorg/fishery/cwp/handbook/J/en>, 2002.

[12] M. Timmons, J. Ebeling, Recirculating aquaculture, in: NRAC Publication No. 01-007, Cayuga Aqua Ventures, Ithaca, NY, 2007.

[13] V. Crespi, A. Coche, Glossary of Aquaculture, Food and Agriculture Organization of the United Nations, 2008.

[14] H. Nilsson, K. Wetengere, Adoption and viability criteria for semi-intensive fish farming: a socio-economic study in Ruvuma and Mbeya regions in Tanzania, in: ALCOM/FAO Field Document No. 28, ALCOM, Harare, Zimbabwe, 1994.

[15] M.S. Islam, Nitrogen and phosphorus budget in coastal and marine cage aquaculture and impacts of effluent loading on ecosystem: review and analysis towards model development, Mar. Pollut. Bull. 50 (1) (2005) 48–61.
[16] Y. Olsen, L. Olsen, K. Tsukamoto (Eds.), Environmental impact of aquaculture on coastal planktonic ecosystems, in: Fisheries for Global Welfare and Environment Memorial Book of the 5th World Fisheries Congress 2008, TERRAPUB, Tokyo, Japan, 2008.
[17] M. Halwart, D. Soto, J.R. Arthur, Cage Aquaculture: Regional Reviews and Global Overview, Food & Agriculture Org. 2007.
[18] CAAR, Global assessment of closed system aquaculture, in: Prepared for the David Suzuli Foundation and the Geogia Strait Alliance, 2007.
[19] M. Troell, A. Joyce, T. Chopin, A. Neori, A.H. Buschmann, J.-G. Fang, Ecological engineering in aquaculture—potential for integrated multi-trophic aquaculture (IMTA) in marine offshore systems, Aquaculture 297 (1–4) (2009) 1–9.
[20] M.A. Oyinlola, G. Reygondeau, C.C.C. Wabnitz, M. Troell, W.W.L. Cheung, Global estimation of areas with suitable environmental conditions for mariculture species, PLoS One 13 (1) (2018) e0191086.
[21] F. Ottolenghi, C. Silvestri, P. Giordano, A. Lovatelli, M.B. New, Capture-Based Aquaculture: The Fattening of Eels, Groupers, Tunas and Yellowtails, FAO, 2004.
[22] O. Hermansen, B. Dreyer, Capture-Based Aquaculture-Sustainable Value Adding to Capture Fisheries? 2008.
[23] Y. Sadovy de Mitcheson, M. Liu, Environmental and biodiversity impacts of capture-based aquaculture, in: FAO Fisheries Technical Paper (FAO), 2008.
[24] B. Campbell, D. Pauly, Mariculture: A global analysis of production trends since 1950, Mar. Policy 39 (2013) 94–100.
[25] FAO, The State of World Fisheries and Aquaculture, FAO Fisheries and Aquaculture Department, Rome, 2014, p. 243.
[26] J. Hofherr, F. Natale, P. Trujillo, Is lack of space a limiting factor for the development of aquaculture in EU coastal areas? Ocean Coastal Manage. 116 (2015) 27–36.
[27] A.H. Buschmann, F. Cabello, K. Young, J. Carvajal, D.A. Varela, L. Henríquez, Salmon aquaculture and coastal ecosystem health in Chile: analysis of regulations, environmental impacts and bioremediation systems, Ocean Coastal Manage. 52 (5) (2009) 243–249.
[28] J. Li, F. Li, S. Yu, S. Qin, G. Wang, Impacts of mariculture on the diversity of bacterial communities within intertidal sediments in the Northeast of China, Microb. Ecol. 66 (4) (2013) 861–870.
[29] S. Mirto, S. Bianchelli, C. Gambi, M. Krzelj, A. Pusceddu, M. Scopa, et al., Fish-farm impact on metazoan meiofauna in the Mediterranean Sea: analysis of regional vs. habitat effects, Mar. Environ. Res. 69 (1) (2010) 38–47.
[30] M. Metian, A.G.J. Tacon, Fishing for feed or fishing for food: increasing global competition for small pelagic forage fish, AMBIO 38 (6) (2009) 294–302.
[31] K. Grigorakis, G. Rigos, Aquaculture effects on environmental and public welfare—the case of Mediterranean mariculture, Chemosphere 85 (6) (2011) 899–919.
[32] M.A. Rimmer, K. Sugama, D. Rakhmawati, R. Rofiq, R.H. Habgood, A review and SWOT analysis of aquaculture development in Indonesia, Rev. Aquacult. 5 (4) (2013) 255–279.
[33] R.R. Gentry, H.E. Froehlich, D. Grimm, P. Kareiva, M. Parke, M. Rust, et al., Mapping the global potential for marine aquaculture, Nat. Ecol. Evol. 1 (9) (2017) 1317–1324.
[34] J. McDaid Kapetsky, J. Aguilar-Manjarrez, J. Jenness, A. Dean, A. Salim, A Global Assessment of Offshore Mariculture Potential From a Spatial Perspective, FAO, Roma, Italia, 2013.
[35] A.G. Tacon, Increasing the Contribution of Aquaculture for Food Security and Poverty Alleviation, 2001.
[36] M. Ahmed, M.H. Lorica, Improving developing country food security through aquaculture development—lessons from Asia, Food Policy 27 (2) (2002) 125–141.
[37] O.R. Liu, R. Molina, M. Wilson, B.S. Halpern, Global opportunities for mariculture development to promote human nutrition, PeerJ 6 (2018) e4733.

[38] R. Subasinghe, D. Soto, J. Jia, Global aquaculture and its role in sustainable development, Rev. Aquacult. 1 (1) (2009) 2–9.
[39] B.M.M. Campbell A Global Analysis of Historical and Projected Mariculture Production Trends, 2011, pp. 1950–2030.
[40] L. Williams, A. Rota, Impact of climate change on fisheries and aquaculture in the developing world and opportunities for adaptation, in: Fisheries Thematic Paper: Tool for Project Design [Online], (2010). Available from: <http://www.ifad.org/lrkm/pub/fisheries.pdf.2010>.
[41] W.W. Cheung, R. Watson, D. Pauly, Signature of ocean warming in global fisheries catch, Nature 497 (7449) (2013) 365–368.
[42] F. Gazeau, J.P. Gattuso, C. Dawber, A.E. Pronker, F. Peene, J. Peene, et al., Effect of ocean acidification on the early life stages of the blue mussel *Mytilus edulis*, Biogeosciences 7 (7) (2010) 2051–2060.
[43] K.J. Kroeker, R.L. Kordas, R. Crim, I.E. Hendriks, L. Ramajo, G.S. Singh, et al., Impacts of ocean acidification on marine organisms: quantifying sensitivities and interaction with warming, Glob. Change Biol. 19 (6) (2013) 1884–1896.
[44] K. Cochrane, C. De Young, D. Soto, T. Bahri, Climate change implications for fisheries and aquaculture, in: FAO Fisheries and Aquaculture Technical Paper, 530, 2009, p. 212.
[45] A. Schmittner, Decline of the marine ecosystem caused by a reduction in the Atlantic overturning circulation, Nature 434 (7033) (2005) 628–633.
[46] J.R. Porter, L. Xie, A. J. Challinor, K. Cochrane, S.M. Howden, M.M Iqbal, et al., Food security and food production systems. (2014) 485–533.
[47] A.G.J. Tacon, M. Metian, Global overview on the use of fish meal and fish oil in industrially compounded aquafeeds: Trends and future prospects, Aquaculture 285 (1–4) (2008) 146–158.
[48] R.W. Hardy, Utilization of plant proteins in fish diets: effects of global demand and supplies of fishmeal, Aquacult. Res. 41 (5) (2010) 770–776.
[49] P. Gómez-Requeni, M. Mingarro, J. Calduch-Giner, F. Médale, S. Martin, D. Houlihan, et al., Protein growth performance, amino acid utilisation and somatotropic axis responsiveness to fish meal replacement by plant protein sources in gilthead sea bream (*Sparus aurata*), Aquaculture 232 (1) (2004) 493–510.
[50] A.N. Lunger, S. Craig, E. McLean, Replacement of fish meal in cobia (*Rachycentron canadum*) diets using an organically certified protein, Aquaculture 257 (1) (2006) 393–399.
[51] V. Patil, T. Källqvist, E. Olsen, G. Vogt, H.R. Gislerød, Fatty acid composition of 12 microalgae for possible use in aquaculture feed, Aquacult. Int. 15 (1) (2006) 1–9.
[52] T.L.F. Leung, A.E. Bates, N. Dulvy, More rapid and severe disease outbreaks for aquaculture at the tropics: implications for food security, J. Appl. Ecol. 50 (1) (2013) 215–222.
[53] R. Callaway, A.P. Shinn, S.E. Grenfell, J.E. Bron, G. Burnell, E.J. Cook, et al., Review of climate change impacts on marine aquaculture in the UK and Ireland, Aquat. Conserv.: Mar. Freshw. Ecosyst. 22 (3) (2012) 389–421.
[54] T. Defoirdt, P. Sorgeloos, P. Bossier, Alternatives to antibiotics for the control of bacterial disease in aquaculture, Curr. Opin. Microbiol. 14 (3) (2011) 251–258.
[55] M.G. Bondad-Reantaso, R.P. Subasinghe, J.R. Arthur, K. Ogawa, S. Chinabut, R. Adlard, et al., Disease and health management in Asian aquaculture, Vet. Parasitol. 132 (3–4) (2005) 249–272.
[56] W. Bank, Reducing Disease Risk in Aquaculture. Agriculture and Environmental Services Discussion Paper 09, World Bank Group, Washington, DC <https://gaallianceorg/wp-content/uploads/2015/02/wb-riskinaquaculture1pdf>, 2014.
[57] J.C. Leong, J.L. Fryer, Viral vaccines for aquaculture, Annu. Rev. Fish Diseas. 3 (1993) 225–240.
[58] H. McCallum, D. Harvell, A. Dobson, Rates of spread of marine pathogens, Ecol. Lett. 6 (12) (2003) 1062–1067.
[59] A. Karvonen, P. Rintamaki, J. Jokela, E.T. Valtonen, Increasing water temperature and disease risks in aquatic systems: climate change increases the risk of some, but not all, diseases, Int. J. Parasitol. 40 (13) (2010) 1483–1488.

[60] C.D. Harvell, C.E. Mitchell, J.R. Ward, S. Altizer, A.P. Dobson, R.S. Ostfeld, et al., Climate warming and disease risks for terrestrial and marine biota, Science 296 (5576) (2002) 2158–2162.

[61] A. Stien, P.A. Bjørn, P.A. Heuch, D.A. Elston, Population Dynamics of Salmon Lice *Lepeophtheirus salmonis* on Atlantic Salmon and Sea Trout, 2005.

[62] D.J. Marcogliese, Implications of climate change for parasitism of animals in the aquatic environment, Can. J. Zool. 79 (8) (2001) 1331–1352.

[63] D. Soto, J. Aguilar-Manjarrez, C. Brugère, D. Angel, C. Bailey, K. Black, et al., Applying an ecosystem-based approach to aquaculture: principles, scales and some management measures, Build. Ecosyst. Approach Aquacult. 11 (2008) 14.

[64] W.W.L. Cheung, M.A. Oyinlola, Vulnerability of flatfish and their fisheries to climate change, J. Sea Res. 140 (2018) 1–10.

CHAPTER 23

The big picture: future global seafood markets

Oai Li Chen[1,2]

[1]*Nippon Foundation Nereus Program, University of British Columbia, Vancouver, BC, Canada*
[2]*Changing Ocean Research Unit, University of British Columbia, Vancouver, BC, Canada*

Chapter Outline

The world has changed. Today, more than half the world's population is in the "middle class" [1]. Using the World Bank's classification of economies into four income groups[1]: low income, lower middle income, upper middle income, and high income [2], Fig. 23.1 illustrates the transitioning of the world's economies from low-income to higher income levels over the period between 1993 and 2013. About half of global economic demand is generated from household consumption, and the rapid growth of the middle class, most of which is taking place in Asia, will have significant global and local economic effects [3]. The world's population is projected to increase by 2.4–9.7 billion people in 2050 [4]. Understanding where most people live now and in the future can help us understand the world better and make better decisions. For instance, "Where will future seafood markets be?"

23.1 Fish is a necessity food item

The basic idea of income elasticity of demand is to measure how the demand for a good changes relative to a change in income, with other factors remaining the same. The higher the income elasticity, the more sensitive the demand for a good is to changes in real

[1] Calculated using the World Bank Atlas Method, the current low income, lower middle income, upper middle income, and high-income economies are defined as GNI per capita of $995 or less; between $996 and $3895; between $3895 and $12,055; and $12,056 or more in 2017, respectively [13].

Predicting Future Oceans.
DOI: https://doi.org/10.1016/B978-0-12-817945-1.00022-8

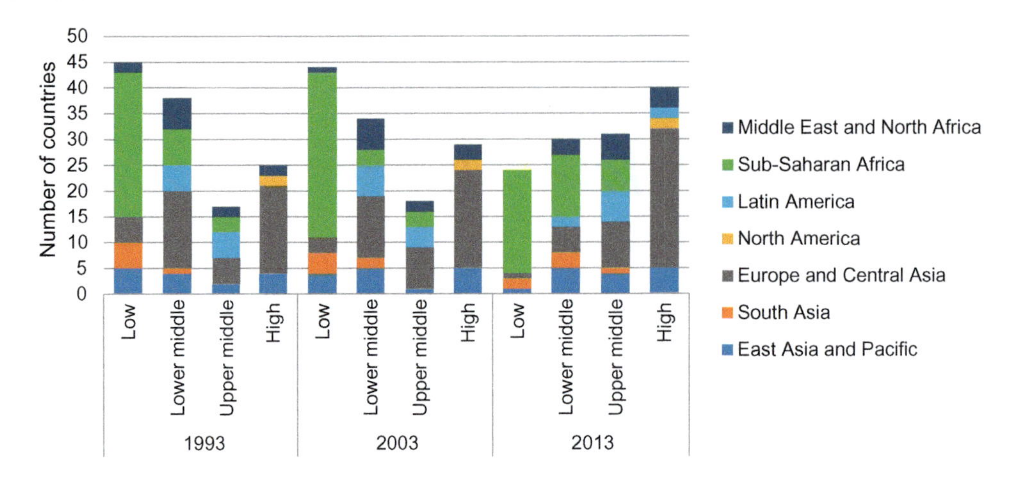

Figure 23.1

World's income over the period 1993–2013. *Author's illustration based on estimates from The World Bank, World Development Indicators, Income group classification [Data file]. Retrieved from: <https://datahelpdesk.worldbank.org/knowledgebase/articles/906519-world-bank-country-and-lending-groups>, 2018.*

income. According to the international cross-country food consumption patterns studies using 1996 and 2005 International Comparison Program[2] (ICP) data [5], all food categories, including fish, are considered to be necessities (income elasticity is less than 1), not a luxury item in household expenditure. This indicates that when real income increases by 1%, demand for seafood will increase but by less than 1%. Fig. 23.2 illustrates the estimated income elasticity of demand for different food categories including fish (finfish and shellfish) of 144 countries. Overall, lower income countries are more responsive to changes in real income than wealthier countries. For instance, with a 1% increase in real income, Congo (low-income) will increase their demand for fish by 0.8%, while United States (high-income) will only add 0.26% more fish to their plate. Lower income countries also spend a greater portion of their budget on lower value (in terms of price) food, such as cereals compared to higher income countries. Therefore, as their real income increases, they exhibit greater demand for higher-value food items, such as fish, meat, and dairy. Conversely, bigger cutbacks are also made on these higher-value items when real income falls.

23.2 Current global seafood markets

An increasing proportion of the world's population falls in the middle of the income distribution [1,2], suggesting an increasing demand for seafood. Next we will explore the

[2] ICP data consist of comparative price data and expenditure values of countries' GDP. They are used to produce robust purchasing power parities (PPPs) which are vital for undertaking cross-country analysis [5].

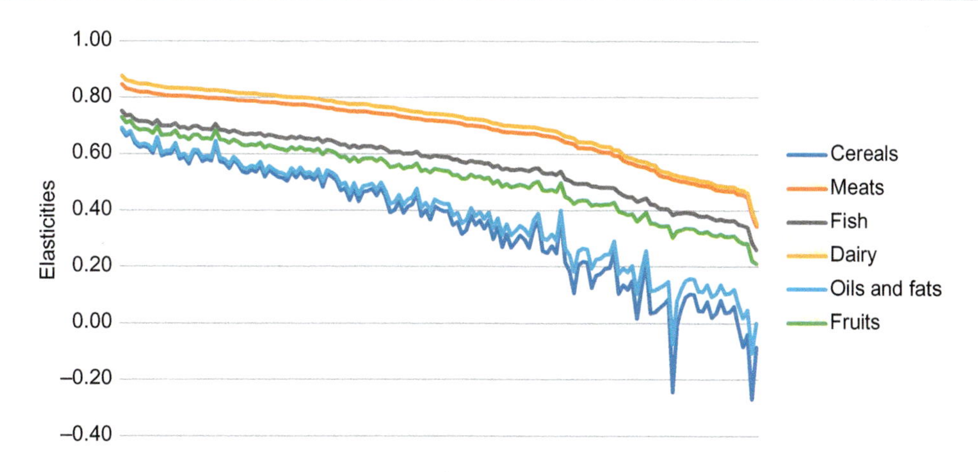

Figure 23.2

Income elasticity for different food categories.

The income countries are sorted by income level from the lowest income country on the left to the highest income country at the right of the figure. *Author's illustration based on estimates from A. Muhammad, J.L. Seale, B. Meade, A. Regmi, International Evidence on Food Consumption Patterns: An Update Using 2005 International Comparison ProgramData (March 1, 2011), USDA-ERS Technical Bulletin No. 1929, 2011. using International Comparison 2005 data.*

fish consumption per capita patterns across income groups between the period of 1993–2013, in 125 countries (Fig. 23.1).

These countries cover the world's 10 largest countries (population) as well as the 32 leading economies by 2050: G7 (the United States, Japan, Germany, the United Kingdom, France, Italy, and Canada); advanced economies (Australia, South Korea, Spain); E7—largest emerging economies (China, India, Brazil, Russia, Indonesia, Mexico, and Turkey); other G20 economies (Argentina, South Africa, and Saudi Arabia); potential fast-growing frontier economics (Vietnam and Nigeria); Poland (largest economy in Central and Eastern Europe); and Malaysia (fast-growing medium-sized economy). Collectively these economies account for 84% of the world's GDP [6]. In addition, they also account for 92%, 65%, and 67% of the global supply of aquaculture, inland capture, and marine capture, respectively. The remaining 93 countries included in the study capture 5%, 24%, and 26% of the aquaculture, inland capture, and marine capture, respectively [7].

The calculated fish consumption per capita in different income groups over the last three decades are presented in Fig. 23.3. Overall it shows that the per capita fish consumption across income groups exhibits similar patterns over time, with low-income countries consuming less fish per capita compared to the high-income countries. The obvious change observed during this time horizon is a change in the composition of the fish consumption. This change is attributed to the development of the aquaculture sector, mostly taking place

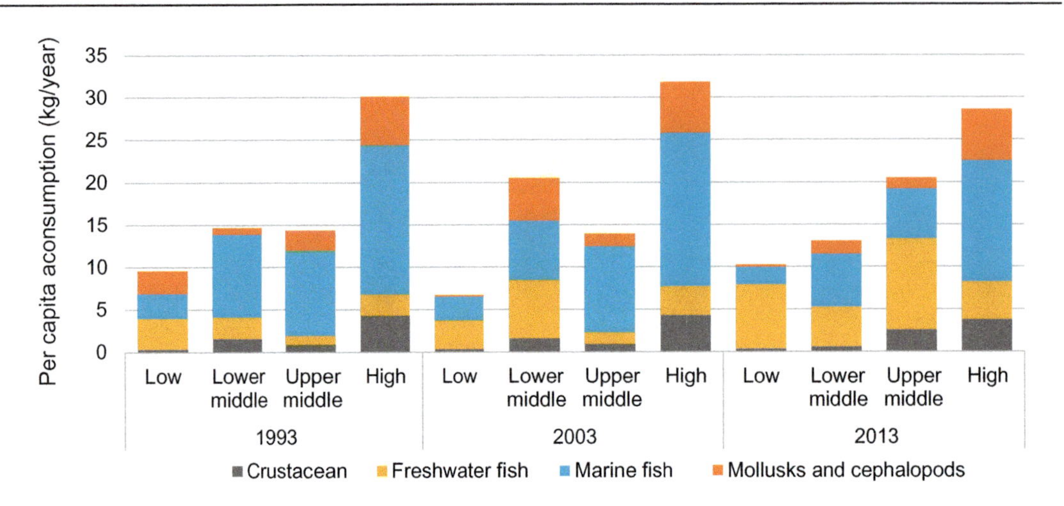

Figure 23.3

Per capita apparent seafood consumption across income groups. *Author's calculation based on per capita apparent consumption from FAO, FishStat J database. <http://www.fao.org/fishery/statistics/software/fishstatj/en>, 2017; and income group data from The World Bank, World Development Indicators, Income group classification [Data file]. Retrieved from: <https://datahelpdesk.worldbank.org/knowledgebase/articles/906519-world-bank-country-and-lending-groups>, 2018. Data are 3-year averages.*

in Asia. The growth in the aquaculture sector, mostly pro-poor (low-cost) production focus [8–11], managed to cater to the demand of lower income groups, and therefore sparked a tremendous increase in per capita consumption of freshwater fish in lower income groups. Overall high-income countries consumed more high-value fish commodities such as marine fish and crustaceans. It also indicates a lower marine fish consumption over the decades, mostly driven by a decline in, or stagnating, marine fish supply, coupled with increasing population over the same period.

As expected, Asia's fish market doubled between 1993 and 2013, driven by growth in income, a higher population, and the development of the aquaculture sector. The East Asian, Pacific, and South Asian markets represent 70% of the global seafood market and have remained the top markets for the past two decades. The Sub-Saharan African market has also shown a tremendous expansion, more than doubling during the past two decades, and has surpassed the market in North America (Fig. 23.4). Such growth was driven mostly by Nigeria's growing population, which is currently the seventh largest in the world, coupled with the transition to the lower–middle income level in 2013 [2].

23.3 Future global seafood markets

The world's population is projected to reach 9.7 billion people in 2050. Most growth is expected in India, Nigeria, Pakistan, the Democratic Republic of the Congo, Ethiopia, the

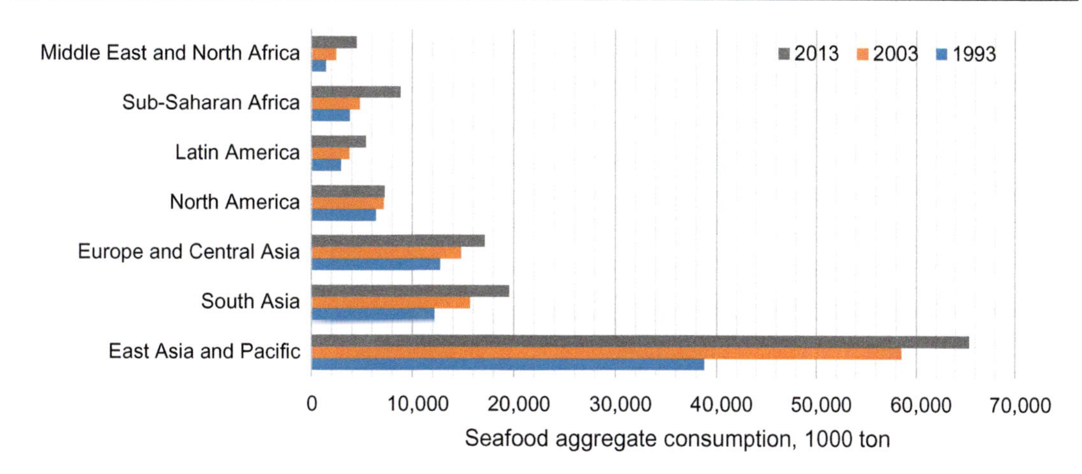

Figure 23.4

Aggregate total apparent seafood consumption across regions. *Author's calculation based on per capita apparent consumption from FAO, FishStat J database, <http://www.fao.org/fishery/statistics/software/fishstatj/en>, 2017 and income group data from The World Bank, World Development Indicators, Income group classification [Data file]. Retrieved from: <https://datahelpdesk.worldbank.org/knowledgebase/articles/906519-world-bank-country-and-lending-groups>, 2018a. Data are 3-year averages.*

United Republic of Tanzania, the United States, Indonesia, and Uganda. The countries with largest populations are projected to be China (1.34 billion), India (1.65 billion), Indonesia (321 million), Nigeria (403 million), Pakistan (305 million), and the United States (387 million) [4].

The leading 32 economies would be home for 6.7 billion people. Furthermore, PwC [6] project average real GDP per capita PPP to grow in all the leading economics during 2014–50, with faster growth, at 5% per annum, in newly emerging economics (Vietnam, Nigeria), with a slower growth, 3%–5%, in established emerging economics such as China. Advanced economies are projected to grow at 1.5%–2.5% per annum.

Based on these two key drivers—population and income growth—I employed median consumption/income per capita PPP to explore the potential change in the world marketplace into 2050. Given that fish is a necessity food item, median consumption/income per capita PPP conveys a better picture of typical individual material well-being in a country than an average GDP per capita PPP, which is inevitably skewed to the right [12]. Fig. 23.5 illustrates that if the United Nation's population growth projections [4] are accurate, coupled with the projected [6] and recent 1994–2013 [13] growth in real income per capita, the world's economic drive will likely to shift to Asia, Latin America, and Sub-Saharan Africa.

Capture fisheries output reached its potential in the 1990s [14]. If the current world average fish consumption per capita remains unchanged at 19 kg per annum [15], the aquaculture

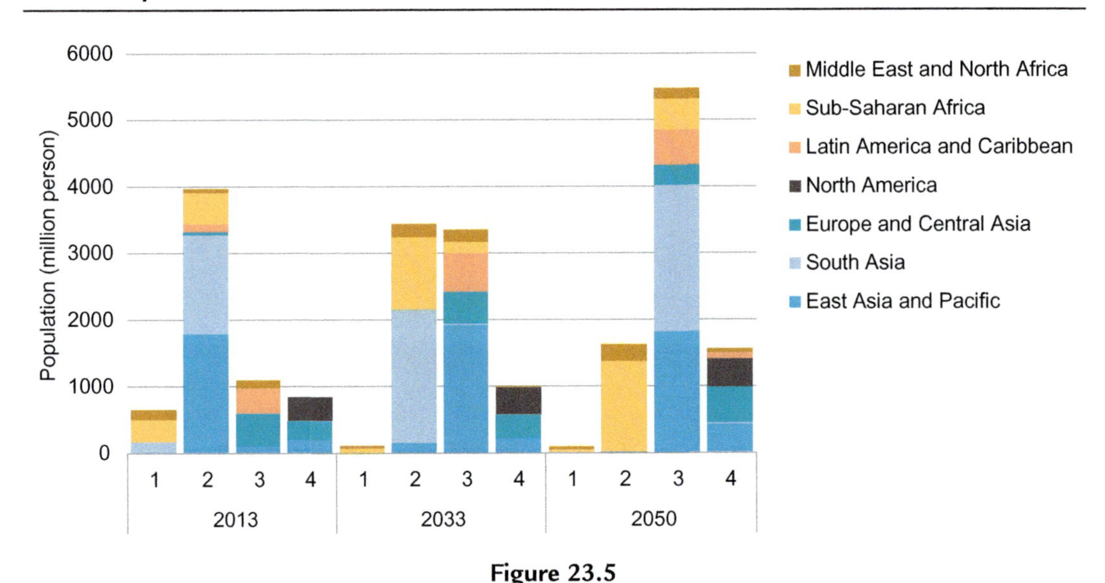

Figure 23.5

World's consumption/income over the period 1993–2050.
1, 2, 3, and 4 denote low-income, lower–middle income, upper–middle income, high-income groups, respectively. 1, 2, 3, and 4 are defined as real median consumption/income per capita PPP less than $2/day, between $2 and $8/day, between $8 and $32/day, and $32 or more/day, respectively.
The author's calculation is based on population projection data from The World Bank [4], median consumption/income per capita PPP [12], annual average growth of GDP per capita PPP of 32 leading economies projected by PwC [6], and annual average growth of GDP per capita PPP (covering period between 1994 and 2013) of other economies are calculated using the average real GDP per capita PPP from World Bank [13]. All estimates are 3-year averages.

sector needs to grow at around 2.6% per annum to be able to cater to the increasing demand over the period between 2013 and 2050. This is equivalent to 1.6 times more than what the world aquaculture sector is currently producing (54 billion tons). Yet aquaculture growth has shown a diminishing growth rate over the last few decades: 11% per annum in the 1980s, 9.6% per annum in the 1990s, 6% per annum in the 2000s, and recent growth at 5% per annum from 2011 to 2015. It shows a consistent slower growth by 2% per annum each decade [7]. Therefore a sustained development in the aquaculture sector is vital to meet the growing market demand in the future. Currently, the Asia-Pacific region produces 90% of the world aquaculture outputs, making them a potential market hub in the future [7].

Fish is the world's most traded food commodity [16–17]. The three largest trading regions, based on total export and import values from 2006–17, are Europe and Central Europe (35%), East Asia and Pacific (29%), and North America (11%). East Asia and the Pacific's trade value almost doubled, faster than the world's average growth, over the past decade, with half of it traded within the region; 36% was equally shared by Europe, Central Asia, and North America, and 14% with other regions [18]. The past decade also highlights the expansion of trade relations with non-Western markets, at a very fast pace, including Latin America and the Caribbean (8% per annum), South Asia (8% per annum), Sub-Saharan Africa (11% per annum), and trading partners within the region (6% per annum) [18]. With greater spending power over time, these relations will likely continue to grow.

23.4 Conclusion

Global wealth will continue to increase gradually in the future, with most people likely falling into the upper middle sector of the income distribution by 2050. Most of the population will reside on the Asian continent and Sub-Saharan Africa. Subsequently, this indicates that their consumption power is expected to increase their seafood market demand. Being the leading aquaculture producers, seafood consumers, and seafood traders, the future of the seafood market in the Asia-Pacific region will be extremely dynamic.

References

[1] H. Kharas, The unprecedented expansion of the global middle class: an update, in: Global Economy and Development Working Paper 100, Brookings Institution, 2017.
[2] The World Bank, World Development Indicators, Income Group Classification [Data file]. Retrieved from: <https://datahelpdesk.worldbank.org/knowledgebase/articles/906519-world-bank-country-and-lending-groups>, 2018.
[3] OECD, Perspectives on Global Development: Social Cohesion in a Shifting World, OECD Publishing, 2012, 263 pp.
[4] The World Bank, World Development Indicators, Projected Population [Data file]. Retrieved from: < http://databank.worldbank.org/data/home.aspx >, 2018.
[5] A. Muhammad, J.L. Seale, B. Meade, A. Regmi, International Evidence on Food Consumption Patterns: An Update Using 2005 International Comparison ProgramData (March 1, 2011), in: USDA-ERS Technical Bulletin No. 1929, 2011.
[6] PwC, The World in 2050: Will the Shift in Global Economic Power Continues? PwC, 2015, 263 pp.
[7] FAO, FishStat J Database, <http://www.fao.org/fishery/statistics/software/fishstatj/en>, 2017.
[8] B. Belton, S.R. Bush, C.L. David, Not just for the wealthy: rethinking farmed fish consumption in the global south, Global Food Secur. 16 (2018) 85–92.
[9] M.M. Dey, M.F. Alam, M.L. Bose, Demand for aquaculture development: perspectives from Bangladesh for improved planning, Rev. Aquacult. 2 (2010) 16–32. Available from: https://doi.org/10.1111/j.1753-5131.2010.01020.x.
[10] A.B.M.M. Haque, M.M. Dey, Impacts of community-based fish culture in seasonal floodplains on income, food security and employment in Bangladesh, Food Secur. (2016). Available from: https://doi.org/10.1007/s12571-016-0629-z.
[11] K.A. Toufique, B. Belton, Is Aquaculture pro-poor? Empirical evidence of impacts on fish consumption in Bangladesh, World Dev. 64 (2014) 609–620.
[12] N. Birdsall, C.J. Meyer, The median is the message: a good enough measure of material wellbeing and shared development progress, Global Policy 6 (2015) 343–357. Available from: https://doi.org/10.1111/1758-5899.12239.
[13] The World Bank, World Development Indicators. GDP Per Capita, PPP (constant 2011 international $) [Data file]. Retrieved from: http://databank.worldbank.org/data/home.aspx, 2018.

[14] D. Pauly, D. Zeller, Catch reconstructions reveal that global marine fisheries catches are higher than reported and declining, Nat. Commun. (2016). Available from: https://doi.org/10.1038/ncomms10244.
[15] FAO, Food Balance Sheets, Food and Agricultural Organization of the United Nations, Rome, 2018.
[16] F.R. Asche, M.F. Bellemare, C. Roheim, M.D. Smith, T. Tveteras, Fair enough? Food security and the international trade of seafood, World Dev. 67 (2015) 151–160.
[17] S. Tveterås, F. Asche, M.F. Bellemare, M.D. Smith, A.G. Guttormsen, A. Lem, K. Lien, S. Vannuccini, Fish Is Food - The FAO's Fish Price Index, PLoS ONE 7 (5) (2012) e36731. Available from: https://doi.org/10.1371/journal.pone.0036731.
[18] United Nations. UN Comtrade Database. Available from: <http://comtrade.un.org/>, 2018.

CHAPTER 24

Climate change, contaminants, and country food: collaborating with communities to promote food security in the Arctic

Tiff-Annie Kenny

Department of Biology, University of Ottawa, Ottawa, ON, Canada

Chapter Outline

24.1 Seafood, food security, and public health

Changing ocean conditions associated with global environmental change are projected to impact marine primary productivity, species distributions, and abundance, with repercussions for fisheries yields [1]. For seafood-dependent peoples, these changes may have significant economic [2], food security [3,4], and public health impacts [5]. Seafood is a rich source of protein, micronutrients, and polyunsaturated fatty acids—the cardiovascular health and cognitive development benefits of which are extensively reported [6,7]—that is indispensable for the food and nutrition security of millions of people around the globe [8,9].

Predicting Future Oceans.
DOI: https://doi.org/10.1016/B978-0-12-817945-1.00024-1

Climate-related declines in fisheries yields are projected to render over 10% of the global population vulnerable to malnutrition [5]. This raises significant health equity concerns, as the majority of countries which are highly seafood-dependent are also low-income, food-deficient, countries [10]. Furthermore, the adverse human health impacts of changing oceans may be experienced more prominently in some populations—including among coastal Indigenous Peoples, where seafood consumption markedly exceeds national averages [11] and among whom significant disparities in socioeconomic status and health already exist [12]. While seafood declines are of concern to public health (through the potential attenuation of nutrient intake and the lack of available and affordable nutrient-rich alternatives for some populations), seafood-rich diets are associated with elevated exposures to environmental contaminants, such as mercury (a potent neurotoxin), with coastal/small-island communities and Inuit experiencing the highest exposures [13]. Importantly the harvest, sharing, and consumption of locally harvested foods by Indigenous Peoples through identity-linked social practices is involved in the maintenance and perpetuation of collective ways of being (e.g., intergenerational transfer of knowledge, principles of sharing/reciprocity, and community relations) [14]. Thus the potential impact of changing oceans (including climate change and the contamination of marine food webs) and marine harvests extends from physical health outcomes, and includes consequences for cultural, psychosocial, mental, and spiritual dimensions of health and well-being.

While collective action on the global scale is required to mitigate global ocean change, anticipating and proactively mitigating the negative impacts of these changes at local scales may attenuate their negative consequences for human health. Arctic Indigenous communities are on the forefront of global ocean change. The Arctic, among the most ecologically sensitive regions on Earth to climate variability [15], has experienced global warming more rapidly than elsewhere on the planet [16]. Climate change is a serious threat to Arctic biodiversity [17] and Arctic marine ecosystems are expected to undergo profound changes in the coming decades due to sea ice decline, ocean acidification, and related ecological changes [18,19]. Inuit rely on the harvest of marine (as well as terrestrial and freshwater) animals (termed "country foods") for various facets of life and community [20,21]. Marine country foods are exceptionally rich in polyunsaturated fatty acids [22] and micronutrients [23,24]. Contemporary Inuit diets involving high consumption of these foods are associated with positive physical health outcomes, such as favorable blood lipid profiles [25,26] and lower risk of coronary heart disease [27]. Environmental contaminants in Arctic marine biota [28], however, may attenuate their positive health benefits [29] and present a food safety concern [30]. Nevertheless, given their rich nutrient profile [31–33], the lack of affordable healthful dietary alternatives in Arctic communities [34], and critically the importance of marine country foods to Inuit culture, identity, and social relations [35,36], human health and well-being in the Arctic remain inextricably linked to the status of Arctic marine ecosystems and species [37,38].

This chapter presents a multidisciplinary synthesis of the current state and future outlook of various socioeconomic and environmental drivers of change as they relate to Arctic marine ecosystems, country food harvest, and food security for Inuit in Canada. The importance of terrestrial species, such as caribou (*Rangifer tarandus*) for Inuit food and nutrition security, though beyond the scope of the present work, is recognized [39]. Interconnectivity between terrestrial and aquatic (marine water and freshwater) food productions systems [40] (e.g., dramatic declines of key terrestrial species, like caribou [41], may be a driver of demand for marine species) emphasizes the need for holistic, systems-based (food system and ecosystem) approaches to food security and fisheries/wildlife management [42].

24.2 Changing Arctic communities

With a median age of 22 years, the Inuit population in Canada is very young (relative to 40 years for the general Canadian population) [43] and increasingly urbanized [44]. The majority of the nearly 60,000 Inuit people in Canada reside in coastal communities across the Inuit Nunangat—the traditional Inuit homeland, which encompasses 50% of the Canadian coastline [45–47]. Distinct geographical, cultural, and sociopolitical histories have resulted in different traditions, economies, and sociopolitical organizations between regions and communities. In general, communities across Inuit Nunangat are small (i.e., two thirds have a total population inferior to 1000 people), remote (i.e., lack year-round road access), and remain defined by a close relationship to Arctic ecosystems and sea ice conditions (i.e., for travel and hunting) for various facets of life. Strong traditions of country food sharing and reciprocity characterize Inuit culture [48,49], with the vast majority of Inuit households reporting sharing country food with others [50].

During the last several decades, Inuit in Canada have endured significant lifestyle changes, involving settlement into permanent communities, the proliferation of physical infrastructure (e.g., transportation and telecommunication), and the development of wage economies. During this period, considerable declines in the harvest of country food species have been documented [51], paralleled by the reduced consumption of country foods, particularly among younger generations [52]. Elucidating the cause of these declines is multifaceted and complex, with factors such as the escalating costs of harvest, and changes to subsistence economies (among many other social, economic, and ecological factors) believed to play a role [51]. The contemporary economy in Inuit communities is characterized by complex interplay between subsistence and wage-based activities, with households balancing time and resources derived from each [53,54]. While elements of a cash-based economy are often required to support harvest (i.e., purchase of harvesting equipment, gas, and supplies), employment in the wage economy may also limit the time individuals can devote to harvesting activities [55].

Despite these changes, the majority (70%) of Inuit adults take part in harvesting activities and report that at least half of the meat and fish consumed in their household is country food [50,56]. Inuit continue to harvest and consume a rich diversity of approximately 200 species of local fauna, including tens of (marine and terrestrial) mammals, roughly 50 species of fish/seafood, and over 100 plants and avian species—with caribou (*R. tarandus*), ringed seal (*Pusa hispida*), beluga whales (*Delphinapterus leucas*), and Arctic char (*Salvelinus alpinus*) among the most frequently harvested and consumed across the Arctic [20,33,57,58]. The species harvested, the parts consumed, and the preparation/preservation methods, are subject to considerable regional variation, based on climate, ecology, species distance of wildlife populations to communities, and other social factors [59,60].

Acknowledging the mixed nature of the economy, significant socioeconomic disparities persist across Inuit Nunangat. Less than half of Inuit in Canada (45%) report having a high school diploma (relative to 86% of the non-Indigenous population), while an income gap of almost $70,000 ($23,500 vs $92,000) exists between Inuit and non-Indigenous people in Inuit Nunangat [56]. Socioeconomic disparities and political marginalization are articulated in significant health disparities among Inuit, relative to the non-Indigenous population in Canada [61,62]. Age-standardized mortality rates due to major chronic diseases in Inuit Regions of the Canadian Arctic are considerably higher than Canada as a whole (235.5 compared to 167.1 per 100,000, between 2004 and 2008) [63].

Conceptualizations of Inuit health which emphasize biomedical statistics on health deficits, however, fail to capture Inuit conceptualizations of health [64]. Holistic approaches that recognize the social determinants of Inuit health, including culture and language, self-determination, housing, and food security (among others) are integral to understanding and addressing health. Moreover, for Inuit, health, wellness, and cultural identity are embedded in ideologies of the food system—including the relationship between humans and nonhuman animals [21,65]. Inuit Qaujimajatuqangit—Inuit culture, knowledge, laws, principles, values, beliefs, and attitudes—defines Inuit social organization and cultural ecology and articulates extensive multigenerational knowledge of Arctic environments, ecosystems, and human–animal relationships, increasingly included within Arctic environmental and harvest research and policy [66,67].

24.2.1 Food systems and food security

Within the food system, health and social inequities are expressed in extreme disparities in food security experienced across Inuit Nunangat. Over 60% of Inuit households in Canada experienced food insecurity in 2007–08 [66], approximately eight times the rate among Canadian households [67]. Food insecurity among Inuit is associated with disturbed eating patterns, reduced diet quality, and increased susceptibility to chronic and infectious disease [68,69]. Despite the existence of a federal food subsidy program, the price of market foods (particularly nutritious perishable foods such as fresh produce) in Inuit communities of northern Canada is extremely high [34]. For example, fresh produce can be between 52% (apples) and 303% (celery) higher in the Arctic communities than elsewhere in Canada [34]. The average cost to feed a family of four (e.g., the Revised Northern Food Basket) is estimated to cost over twice as much in Arctic communities, relative to the capital city of Ottawa (CAD 410 vs CAD 192 per week) [34]. For several decades, Inuit have indicated that they cannot afford to purchase sufficient food to meet their family's needs [70].

For many Arctic species in the Inuit food system, there is no nutritionally or culturally equivalent substitute. Thus, factors that promote the transition away from country food consumption are of significant concern to human health. In 1987 country foods provided up to almost half (43.5% of total diet energy) of the total diet for Inuit men [32]. Market foods, though largely unavailable to Inuit during the first half of the 20th century (with the exception of a few staples, such as flour) now constitute approximately three-quarters of the total diet [33]. Subject to different social, economic, and environmental/ecological conditions, country food consumption varies significantly between generations, and across age. Older adults consume significantly more country foods and a greater proportion of their total diet as country food than do younger adults (over twice as much) [33]. While most Inuit adults (~75%) now prefer to eat a mixed diet of both country food and market food [71–73], greater reliance on the latter has led to poor dietary quality [68,74] and is linked to increased incidence of obesity and unhealthy body weight [52,65].

Changing demographic structures, dietary patterns, and community/household economies have repercussions for current and future country food need and harvest levels, harvest sustainability and community well-being (e.g., food sharing), and public health (nutrition, food security, and exposure to environmental contaminants). While predicting future eating habits is extremely challenging, cohort modeling approaches provide a useful tool to forecast future demand for seafood in societies experiencing profound demographic and structural change in consumer preferences [75]. Diet modeling approaches also provide a means to engage with Indigenous communities and to translate nutrient requirements and food safety concerns into culturally appropriate public health recommendations embedded in community-specific food realities (e.g., food availability, prices, contaminant levels, and dietary preferences) [76].

24.3 Climate change and country foods

24.3.1 Changing Arctic marine environments and ecosystems

Arctic sea ice (extent and thickness) has declined sharply (at a rate of 10% per decade) in recent decades [77], while Arctic sea temperatures have increased rapidly [78]. Diminishing sea ice is expected to contribute to Arctic temperature amplification, exacerbate heating of the surface waters, and reinforce reductions in ice/snow cover [79]. Projections from climate models suggest that the Arctic could be largely free of sea ice by the late 2030s [18]. Most Arctic species, including marine mammals, are narrowly distributed, highly specialized, and highly sensitive to changes in sea ice conditions [80]. Diminishing sea ice is also contributing to ocean acidification in Arctic marine waters, among the world's most sensitive in terms of acidification response to increasing levels of CO_2 [19]. Changing climatic and biophysical conditions resulting from sea ice decline and acidification (coupled with changing conditions in sub-Arctic oceans and ecosystems) are driving changes in the productivity, range (poleward range shift of sub-Arctic species), distribution, and migration of Arctic marine biota—from planktonic communities to large marine mammals.

24.3.2 Inuit subsistence harvest

Inuit have witnessed climate-related changes on both the biotic (e.g., abundance, distribution, and migration of wildlife populations) and abiotic (e.g., changing sea ice conditions) dimensions of their local food systems that may affect harvest, food security, and dietary quality [37,81]. Most changes are related to changing weather (e.g., more variable and unpredictable weather, stronger winds and changes in prevailing wind direction, increased rainfall, decreased snowfall, fewer extreme cold temperatures, and more extreme warm summer temperatures), ice and hydrologic systems (e.g., longer ice-free season, thinner ice, earlier snow melt, lower freshwater levels, increased coastal erosion, decreasing lake and stream levels), and animal populations (e.g., decreased wildlife health/quality/abundance, changing migration patterns, and the appearance of new wildlife and plant species) [82,83].

Inuit harvesters have reported that changing sea ice conditions are already contributing safety hazards and risks on ice and impacting access to country foods [84]. Importantly these changes interact with other social, economic, and political factors. For example, changing sea ice conditions may affect the accessibility of hunting areas which may require longer travel distances for harvesters; however, harvesters may lack the time availability (due to the employment and the development of a wage economy in the North) or resources to adapt to these changes [85]. Likewise, declining species abundance in the Arctic has the potential to introduce new harvest quotas and tighten existing ones further. In addition to constraining access to country foods and potentially affecting diet quality [39], harvest quotas may restrict the flexibility with which hunters can respond to changing accessibility of hunting areas and abundance of animals [86]. Disruptions to harvest may also interrupt intergenerational knowledge exchange, which has direct implications for how communities interact with the impacts of climate change in the future (i.e., dynamic vulnerability [87]). Moreover, these emerging conservation issues may perpetuate sociopolitical challenges related to Arctic wildlife management. Despite considerable advancements in co-management (now a collective responsibility administered between Indigenous and public governments, at various levels across the Canadian Arctic) Arctic wildlife management has historically been a site of contention and cultural resistance [85].

On the other hand, climate-related shifts in species distributions and increased marine productivity could signify the development of emerging Arctic fisheries (e.g., Atlantic cod and Greenland halibut) [88]. These ecological changes have been paralleled by significant fisheries policy development—including the first territorial fishery strategy formulated in the North [89]. Indeed, commercial fisheries are rapidly expanding in the Arctic; in less than a decade (2006–14), fisheries value in the Nunavut territory has more than doubled (from CAD 38 million to CAD 86 million) [89]. Interactions between commercial and subsistence marine production systems, as well as the respective policies/systems of wildlife co-management and economic (fisheries) development in the Canadian Arctic have received limited attention to date. Greater knowledge of the various dynamics driving demand and change in marine resource use, is necessary to understand how communities respond to change and how policy and management may interact.

While it is well established that climate change is having a significant impact on Arctic marine environments and ecosystems, given the magnitude and diversity of changes observed, it is challenging to predict the cumulative effect of climate change on harvest, food security, and health. Furthermore, climate change impacts are differentially experienced within communities and households. This includes impacts mediated through gender, as gender results in different pathways through which changing climatic conditions affect people locally [90,91]. Moreover, the potential food security and human health impacts of Arctic marine ecosystem change are mediated through local adaptation strategies (e.g., how hunters substitute one species for another [92]) which are often community and household specific.

24.4 Contaminants and country foods

Although generally distant from industrial activities, studies over the last several decades have documented high levels of environmental contaminants [e.g., mercury and persistent organic pollutants (POPs)] in Arctic biota from long-range atmospheric and aquatic transport and bioaccumulation and biomagnification through marine food webs [30,93]. As many high-trophic level marine animals (e.g., ringed seal, beluga whales) are staples in the diet of many Inuit communities, exposure to environmental contaminants from country food consumption presents human health risks [30,93].

International treaties (e.g., the Stockholm Convention on POPs and the Minamata Convention on Mercury) have been adopted to restrict or eliminate the production and release of these toxic compounds. As a result, concentrations of many legacy POPs in

Canadian Arctic biota have declined significantly and are expected to further decline [94]. For mercury, however, there has been no consistently documented trend of reducing levels in Arctic biota [95] and it is expected that without improved pollution abatement, levels of mercury in Arctic biota will be substantially higher in the coming decades [96]. Emerging compounds and new local sources of anthropogenic pollution (e.g., increased shipping) present concern for the future Arctic Ocean. Furthermore, there are complex climate-related influences on physiochemical process and food web structures that may affect contaminant cycles and human exposures. For example, climate-related sea ice decline may volatilize some compounds—including some POPs, which have increased in atmospheric concentrations at certain Arctic locations [97].

While it is extremely challenging to predict how changing pollution, climate, and food webs could impact future human exposures and health in the Arctic, mechanistic dynamic models of chemical emissions, contaminant fate, transport, and bioaccumulation across food webs are powerful tools to explore exposure in subsistence-based populations, including the Arctic [98] and the Faroe Islands [99]. Exposure models, however, are highly sensitive to changes in intakes and patterns of country food consumption [100], such that reductions in human exposures for some compounds (including mercury) are believed to derive from declining trends in country food consumption, rather than changing contaminant concentrations in Arctic biota.

24.4.1 Human health risks of contaminants

Biomonitoring is key to understanding spatially and/or temporally variable human exposure to contaminants [101]. Although the majority of the Inuit population is below Health Canada guideline levels for heavy metals, mercury, and POPs, the body burden of these contaminants often exceeds exposures observed in the general population of Canada [102]. For example, average mercury blood concentration among Inuit women (18–45 years) in Nunavut was approximately eight times higher than the female Canadian national average, although still below the 8 ppb population guideline [103]. Nevertheless, a significant proportion of Inuit from Nunavik (eastern Arctic), including women of childbearing age (where human health risks from environmental contaminants are pronounced) have exhibited blood mercury levels that exceed the health guidelines [104]. It is well established that low-level prenatal exposure to neurotoxins such as mercury from maternal seafood consumption may place the fetus at increased risk of adverse neurodevelopmental outcomes [105].

The presence of contaminants in country food represents a particular case for risk management, as country foods are beyond the safety monitoring, policy, and legislation, of the industrial agrifood industry [106]. Though local consumption advisories may be issued for species or animal parts (e.g., ringed seal liver), ultimately the decision to consume country foods resides with consumers. Community perceptions and responses to contaminants have been varied, and the perceived risk of contaminants, as well as cultural resistance to contaminant discourse [107,108], may determine, more practicably than scientific food safety characterizations, the extent to which the environmental contaminants influence food choices and harvest. It is important to stress that threats from environmental contamination must be weighed judiciously against the many health and social benefits that are derived from the harvesting, consumption, and sharing of country foods [109].

24.5 Collaborating with communities to improve food security

The confluence of changing socioeconomic, climatic, and environmental (biophysical, ecological, toxicological) factors renders it highly difficult to predict how changes to Arctic marine environments and ecosystems, presented in the subsections above, will interact and affect food security and public health for Inuit in Canada. Furthermore, climate change and environmental contaminants are among a multiplicity of other stressors (e.g., oil and gas activities, mining, tourism, shipping, fisheries) that confront contemporary Arctic ecosystems and communities. As change continues to accelerate in the Arctic, there is a need to co-produce information with communities to anticipate and adapt to the impacts of ocean change on food security and health.

From a human health perspective, adaptation focuses mainly on developing preventive efforts and early-warning systems through monitoring, modeling, and scenarios [110]. Scenarios and models—from dynamic models of marine ecosystems and fisheries [88,111,112], to mechanistic models of contaminant fate, accumulation, and human exposures [98], to public health and economic models of human diets and seafood demand [75,76]—offer tools to examine and visualize possible futures and impacts of the world's changing oceans on public health. Models and scenarios, however, embody international epistemologies and ontologies of resource use and time [110] that may be incongruous with cultural conceptions of the future, food security, and human–animal relationalities in Indigenous communities.

In adaptation planning, community health must be viewed beyond the traditional biomedical model focused on mitigating the prevalence of disease. This comprises holistic approaches that recognize the social determinants of Indigenous health, including the importance of culture and language, self-determination, and the health of local environments and ecosystems (among others). The "food system" paradigm, for example, provides a holistic framework for analyzing the multifaceted relationships between the changing Arctic Ocean, subsistence harvest, and human health [113]. While ideological dichotomies, institutional structures, and the lack of common data platforms for oceanography, marine ecology, subsistence harvest, community economics, and public health research have traditionally hindered efforts to derive holistic understandings of changing oceans and human health, the equitable inclusion of local knowledge and ways of knowing are critical to developing culturally relevant information that can be translated to action (policy and programming) at the local level. These understandings must be co-produced through participatory, community-based, research processes that involve all partners equitably. In this capacity, the conceptual, methodological, and governance foundation of future oceans research involving coastal Indigenous communities must be premised upon the recognition and enactment of respect, reciprocity, and relationality, as well as the commitment to developing and maintaining equitable and mutually beneficial research collaborations that are inclusive and respectful of the knowledge, methodologies, protocols, worldviews, ontologies, and political systems of the participating communities.

References

[1] W.W.L. Cheung, V.W.Y. Lam, J.L. Sarmiento, K. Kearney, R. Watson, D. Zeller, et al., Large-scale redistribution of maximum fisheries catch potential in the global ocean under climate change, Global Change Biol. 16 (2010) 24–35. Available from: https://doi.org/10.1111/j.1365-2486.2009.01995.x.

[2] V.W.Y. Lam, W.W.L. Cheung, G. Reygondeau, U.R. Sumaila, Projected change in global fisheries revenues under climate change, Sci. Rep. 6 (2016). Available from: https://doi.org/10.1038/srep32607.

[3] S. Hughes, A. Yau, L. Max, N. Petrovic, F. Davenport, M. Marshall, et al., A framework to assess national level vulnerability from the perspective of food security: the case of coral reef fisheries, Environ. Sci. Policy 23 (2012) 95–108. Available from: https://doi.org/10.1016/j.envsci.2012.07.012.

[4] J.C. Rice, S.M. Garcia, Fisheries, food security, climate change, and biodiversity: characteristics of the sector and perspectives on emerging issues, ICES J. Mar. Sci. 68 (2011) 1343–1353. Available from: https://doi.org/10.1093/icesjms/fsr041.

[5] C. Golden, E.H. Allison, W.W.L. Cheung, M.M. Dey, B.S. Halpern, D.J. McCauley, et al., Fall in fish catch threatens human health, Nature 534 (2016) 317–320.

[6] D. Mozaffarian, J.H.Y. Wu, Omega-3 fatty acids and cardiovascular disease: effects on risk factors, molecular pathways, and clinical events, J. Am. Coll. Cardiol. 58 (2011) 2047–2067. Available from: https://doi.org/10.1016/j.jacc.2011.06.063.

[7] N. Siriwardhana, N.S. Kalupahana, N. Moustaid-Moussa, Health benefits of n-3 polyunsaturated fatty acids: eicosapentaenoic acid and docosahexaenoic acid, Adv. Food Nutr. Res. 65 (2012) 211–222. Available from: https://doi.org/10.1016/B978-0-12-416003-3.00013-5.

[8] C. Béné, M. Barange, R. Subasinghe, P. Pinstrup-Andersen, G. Merino, G.-I. Hemre, et al., Feeding 9 billion by 2050—putting fish back on the menu, Food Sec. 7 (2015) 261–274. Available from: https://doi.org/10.1007/s12571-015-0427-z.

[9] HLPE, Sustainable Fisheries and Aquaculture for Food Security and Nutrition. A Report by the High Level Panel of Experts on food Security and Nutrition of the Committee on World Food Security, FAO, Rome, 2014.

[10] N. Kawarazuka, The Contribution of Fish Intake, Aquaculture, and Small-Scale Fisheries to Improving Nutrition: A Literature Review, Penang, Malaysia, 2010.

[11] A.M. Cisneros-Montemayor, D. Pauly, L.V. Weatherdon, Y. Ota, A global estimate of seafood consumption by coastal Indigenous peoples, PLoS One 11 (2016). Available from: https://doi.org/10.1371/journal.pone.0166681.

[12] I. Anderson, B. Robson, M. Connolly, F. Al-Yaman, E. Bjertness, A. King, et al., Indigenous and tribal peoples' health (The Lancet–Lowitja Institute Global Collaboration): a population study, Lancet 388 (2016) 131–157.

[13] N. Basu, M. Horvat, D.C. Evers, I. Zastenskaya, P. Weihe, J. Tempowski, et al., Review of mercury biomarkers in human populations worldwide between 2000 and 2018, Environ. Health Perspect. 126 (2018) 106001. Available from: https://doi.org/10.1289/EHP3904.

[14] H.V. Kuhnlein, B. Erasmus, D. Spigelski, Indigenous Peoples' Food Systems: The Many Dimensions of Culture, Diversity and Environment for Nutrition and Health, Food and Agriculture Organization of the United Nations, Centre for Indigenous Peoples' Nutrition and Environment, Rome, Italy, 2009.

[15] A.W.R. Seddon, M. Macias-Fauria, P.R. Long, D. Benz, K.J. Willis, Sensitivity of global terrestrial ecosystems to climate variability, Nature 531 (2016) 229–232. Available from: https://doi.org/10.1038/nature16986.

[16] IPCC, Climate Change 2013: The Physical Science Basis, Contribution of Working Group I to the Fifth Assessment Report of the Intergovernmental Panel on Climate Change, Cambridge University Press, Cambridge, United Kingdom and New York, 2014. <https://doi.org/10.1017/CBO9781107415324>.

[17] E. Post, M.C. Forchhammer, M.S. Bret-Harte, T.V. Callaghan, T.R. Christensen, B. Elberling, et al., Ecological dynamics across the Arctic associated with recent climate change, Science 325 (2009) 1355–1358. Available from: https://doi.org/10.1126/science.1173113.

[18] M. Wang, J.E. Overland, A sea ice free summer Arctic within 30 years? Geophys. Res. Lett. 36 (2009). Available from: https://doi.org/10.1029/2009GL037820. n/a–n/a.

[19] AMAP, Arctic Ocean Acidification Assessment: Summary for Policy-Makers, Arctic Monitoring and Assessment Programme (AMAP), Oslo, Norway, 2013.

[20] T.A. Kenny, H.M. Chan, Estimating wildlife harvest based on reported consumption by Inuit in the Canadian Arctic, Arctic 70 (2017) 1–12. Available from: https://doi.org/10.14430/arctic4625.

[21] K. Borré, Seal blood, Inuit blood, and diet: a biocultural model of physiology and cultural identity, Med. Anthropol. Q. 5 (1991) 48–62. Available from: https://doi.org/10.2307/648960.

[22] M. Lucas, F. Proust, C. Blanchet, A. Ferland, S. Déry, B. Abdous, et al., Is marine mammal fat or fish intake most strongly associated with omega-3 blood levels among the Nunavik Inuit? Prostaglandins Leukot. Essent. Fatty Acids 83 (2010) 143–150.

[23] J.R. Geraci, T.G. Smith, Vitamin C in the diet of Inuit hunters from Holman, Northwest Territories, Arctic 32 (1979) 135–139. Available from: https://doi.org/10.14430/arctic2611.

[24] H.V. Kuhnlein, V. Barthet, A. Farren, E. Falahi, D. Leggee, O. Receveur, et al., Vitamins A, D, and E in Canadian Arctic traditional food and adult diets, J. Food Compos. Anal. 19 (2006) 495–506. Available from: https://doi.org/10.1016/j.jfca.2005.02.007.

[25] Y.E. Zhou, S. Kubow, G.M. Egeland, Highly unsaturated n-3 fatty acids status of Canadian Inuit: International Polar Year Inuit Health Survey, 2007-2008, IJCH 70 (2011) 498–510.

[26] X.F. Hu, K. Singh, T.A. Kenny, H.M. Chan, Prevalence of heart attack and stroke and associated risk factors among Inuit in Canada: a comparison with the general Canadian population, Int. J. Hyg. Environ. Health (2018) 1–8. Available from: https://doi.org/10.1016/j.ijheh.2018.12.003.

[27] X.F. Hu, T.A. Kenny, H.M. Chan, Inuit country food diet pattern is associated with lower risk of coronary heart disease, J. Acad. Nutr. Diet. 118 (2018) 1237–1248.e1. Available from: https://doi.org/10.1016/j.jand.2018.02.004.

[28] B.M. Braune, P.M. Outridge, A.T. Fisk, D.C.G. Muir, P.A. Helm, K. Hobbs, et al., Persistent organic pollutants and mercury in marine biota of the Canadian Arctic: an overview of spatial and temporal trends, Sci. Total Environ. 351-352 (2005) 4–56. Available from: https://doi.org/10.1016/j.scitotenv.2004.10.034.

[29] X.F. Hu, B.D. Laird, H.M. Chan, Mercury diminishes the cardiovascular protective effect of omega-3 polyunsaturated fatty acids in the modern diet of Inuit in Canada, Environ. Res. 152 (2017) 470–477. Available from: https://doi.org/10.1016/j.envres.2016.06.001.

[30] J. Van Oostdam, S.G. Donaldson, M. Feeley, D. Arnold, P. Ayotte, G. Bondy, et al., Human health implications of environmental contaminants in Arctic Canada: a review, Sci. Total Environ. 351 (2005) 165–246. Available from: https://doi.org/10.1016/j.scitotenv.2005.03.034.

[31] D. Kinloch, H.V. Kuhnlein, D. Muir, Inuit foods and diet: a preliminary assessment of benefits and risks, Sci. Total Environ. 122 (1992) 247–278. Available from: https://doi.org/10.1016/0048-9697(92)90249-R.

[32] H.V. Kuhnlein, Benefits and risks of traditional food for Indigenous peoples: focus on dietary intakes of Arctic men, Can. J. Physiol. Pharmacol. 73 (1995) 765–771. Available from: https://doi.org/10.1139/y95-102.

[33] T.A. Kenny, X.F. Hu, H.V. Kuhnlein, S.D. Wesche, H.M. Chan, Dietary sources of energy and nutrients in the contemporary diet of Inuit adults: results from the 2007–08 Inuit Health Survey, Public Health Nutr. 21 (2018) 1319–1331. Available from: https://doi.org/10.1017/S1368980017003810.

[34] T.A. Kenny, M. Fillion, J. MacLean, S.D. Wesche, H.M. Chan, Calories are cheap, nutrients are expensive—the challenge of healthy living in Arctic communities, Food Policy 80 (2018) 39–54. Available from: https://doi.org/10.1016/j.foodpol.2018.08.006.

[35] R.G. Condon, P. Collings, G.W. Wenzel, The best part of life: subsistence hunting, ethnicity, and economic adaptation among young adult Inuit males, Arctic 48 (1995) 31–46.

[36] M.T. Harder, G.W. Wenzel, Inuit subsistence, social economy and food security in Clyde River, Nunavut, Arctic 65 (2012) 305–318. Available from: https://doi.org/10.14430/arctic4218.

[37] S.D. Wesche, H.M. Chan, Adapting to the impacts of climate change on food security among Inuit in the Western Canadian Arctic, EcoHealth 7 (2010) 361–373. Available from: https://doi.org/10.1007/s10393-010-0344-8.

[38] T.L. Nancarrow, H.M. Chan, Observations of environmental changes and potential dietary impacts in two communities in Nunavut, Canada, Rural Remote Health 10 (2010) 1370.

[39] T.A. Kenny, M. Fillion, S. Simpkin, S.D. Wesche, H.M. Chan, Caribou (*Rangifer tarandus*) and Inuit nutrition security in Canada, EcoHealth 15 (2018) 590–607. Available from: https://doi.org/10.1007/s10393-018-1348-z.

[40] R.S. Cottrell, A. Fleming, E.A. Fulton, K.L. Nash, R.A. Watson, J.L. Blanchard, Considering land-sea interactions and trade-offs for food and biodiversity, Global Change Biol. 24 (2017) 580–596. Available from: https://doi.org/10.1111/gcb.13873.

[41] L.S. Vors, M.S. Boyce, Global declines of caribou and reindeer, Global Change Biol. 15 (2009) 2626–2633. Available from: https://doi.org/10.1111/j.1365-2486.2009.01974.x.

[42] P. Loring, C. Gerlach, Food security and conservation of Yukon River Salmon: are we asking too much of the Yukon River? Sustainability 2 (2010) 2965–2987. Available from: https://doi.org/10.3390/su2092965.

[43] Statistics Canada, Aboriginal Statistics at a Glance—Median Age of Population, 1–1. <https://www150.statcan.gc.ca/n1/pub/89-645-x/2010001/median-age-eng.htm>, 2015 (accessed 23.10.18).

[44] M. Morris, A statistical portrait of Inuit with a focus on increasing urbanization: implications for policy and further research, Aboriginal Policy Stud. 5 (2016). Available from: https://doi.org/10.5663/aps.v5i2.27045.

[45] Statistics Canada, 2006 Census: Aboriginal Peoples in Canada in 2006: Inuit, Métis and First Nations, 2006 Census: Inuit, 2018, pp. 1–3.

[46] Statistics Canada, Aboriginal Peoples in Canada: First Nations People, Métis and Inuit, 2016, pp. 1–13.

[47] ITK, Health Indicators of Inuit Nunangat Within the Canadian Context: 1994-1998 and 1999-2003, Inuit Tapiriit Kanatami (ITK). <https://www.itk.ca/publication/health-indicators-inuit-nunangat-within-canadian-context>, 2010.

[48] G.W. Wenzel, Sharing, money, and modern Inuit subsistence : obligation and reciprocity at Clyde River, Nunavut, Senri Ethnol. Stud. 53 (2000) 61–85.

[49] P. Collings, M.G. Marten, T. Pearce, A.G. Young, Country food sharing networks, household structure, and implications for understanding food insecurity in Arctic Canada, Ecol. Food Nutr. 55 (2016) 30–49. Available from: https://doi.org/10.1080/03670244.2015.1072812.

[50] ITK, Inuit Statistical Profile 2008, Inuit Tapiriit Kanatami. <https://www.itk.ca/wp-content/uploads/2016/07/InuitStatisticalProfile2008_0.pdf>, 2008.

[51] G.W. Wenzel, J. Dolan, C. Brown, Wild resources, harvest data and food security in Nunavut's Qikiqtaaluk Region: a diachronic analysis, Arctic 69 (2016) 147–159. Available from: https://doi.org/10.14430/arctic4562.

[52] N. Sheikh, G.M. Egeland, L. Johnson-Down, H.V. Kuhnlein, Changing dietary patterns and body mass index over time in Canadian Inuit communities, IJCH 70 (2011) 511–519. Available from: https://doi.org/10.3402/ijch.v70i5.17863.

[53] D.C. Natcher, Subsistence and the social economy of Canada's Aboriginal North, North. Rev. 30 (2009) 83–98.
[54] P.J. Usher, Evaluating country food in the northern native economy, Arctic 29 (1976) 105–120. Available from: https://doi.org/10.14430/arctic2795.
[55] Council of Canadian Academies, Aboriginal Food Security in Northern Canada: An Assessment of the State of Knowledge, Council of Canadian Academies. <http://www.scienceadvice.ca/uploads/eng/assessments%20and%20publications%20and%20news%20releases/food%20security/foodsecurity_fullreporten.pdf>, 2014.
[56] ITK, Inuit Statistical Profile 2018, Inuit Tapiriit Kanatami. <https://www.itk.ca/2018-inuit-statistical-profile/>, 2018.
[57] H. Priest, P.J. Usher, The Nunavut Wildlife Harvest Study, Nunavut Wildlife Management Board, Iqaluit, NU, 2004.
[58] P.J. Usher, M.A. Wendt, Inuvialuit Harvest Study: Statistical Assessment of the Harvest Survey Data Base 1988–1996, Inuvik, NWT, 1999.
[59] F. Boas, The Central Eskimo, University of Nebraska Press, Lincoln, Nebraska, 1964.
[60] L.R. Binford, Nunamiut Ethnoarchaeology, Academic Press, New York, 1978.
[61] S. Chatwood, P. Bjerregaard, T.K. Young, Global health—a circumpolar perspective, Am. J. Public Health 102 (2012) 1246–1249. Available from: https://doi.org/10.2105/AJPH.2011.300584.
[62] P. Bjerregaard, T.K. Young, E. Dewailly, S.O.E. Ebbesson, Indigenous health in the Arctic: an overview of the circumpolar Inuit population, Scand. J. Public Health 32 (2004) 390–395. Available from: https://doi.org/10.1080/14034940410028398.
[63] Statistics Canada, Aboriginal Peoples Survey (APS), pp. 1–8. <http://www23.statcan.gc.ca/imdb/p2SV.pl?Function = getMainChange&Id = 109115>, 2012 (accessed 28.11.16).
[64] ITK, Social Determinants of Inuit Health in Canada, Inuit Tapiriit Kanatami (ITK). <https://www.itk.ca/wp-content/uploads/2016/07/ITK_Social_Determinants_Report.pdf>, 2014.
[65] H.V. Kuhnlein, O. Receveur, R. Soueida, G.M. Egeland, Arctic Indigenous peoples experience the nutrition transition with changing dietary patterns and obesity, J. Nutr. 134 (2004) 1447–1453.
[66] R. Rosol, C. Huet, M. Wood, C. Lennie, G. Osborne, G.M. Egeland, Prevalence of affirmative responses to questions of food insecurity: International Polar Year Inuit Health Survey, 2007-2008, IJCH 70 (2011) 488–497. Available from: https://doi.org/10.3402/ijch.v70i5.17862.
[67] Health Canada, Household Food Insecurity in Canada in 2007-2008: Key Statistics and Graphics—Food and Nutrition Surveillance—Health Canada, pp. 1–9. <http://www.hc-sc.gc.ca/fn-an/surveill/nutrition/commun/insecurit/key-stats-cles-2007-2008-eng.php>, 2012.
[68] G.M. Egeland, L. Johnson-Down, Z.R. Cao, N. Sheikh, H.A. Weiler, Food insecurity and nutrition transition combine to affect nutrient intakes in Canadian Arctic communities, J. Nutr. 141 (2011) 1746–1753. Available from: https://doi.org/10.3945/jn.111.139006.
[69] J.A. Jamieson, H.A. Weiler, H.V. Kuhnlein, G.M. Egeland, Traditional food intake is correlated with iron stores in Canadian Inuit men, J. Nutr. 142 (2012) 764–770. Available from: https://doi.org/10.3945/jn.111.140475.
[70] J. Lambden, O. Receveur, J. Marshall, H.V. Kuhnlein, Traditional and market food access in Arctic Canada is affected by economic factors, IJCH 65 (2006) 331–340. Available from: https://doi.org/10.3402/ijch.v65i4.18117.
[71] G.M. Egeland, Inuit Health Survey 2007-2008: Inuvialuit Settlement Region, Centre for Indigenous Peoples' Nutrition and Environment, Montreal, QC, 2010.
[72] G.M. Egeland, Inuit Health Survey 2007–2008: Nunatsiavut, Centre for Indigenous Peoples' Nutrition and Environment (CINE), Montreal, QC, 2010.
[73] G.M. Egeland, Inuit Health Survey 2007–2008: Nunavut, Centre for Indigenous Peoples' Nutrition and Environment (CINE), Montreal, QC, 2010.
[74] E. Erber, B.N. Hopping, L. Beck, T. Sheehy, E. De Roose, S. Sharma, Assessment of dietary adequacy in a remote Inuvialuit population, J. Hum. Nutr. Diet. 23 (2010) 35–42. Available from: https://doi.org/10.1111/j.1365-277X.2010.01098.x.

[75] H. Mori, D.L. Clason, A cohort approach for predicting future eating habits: the case of at-home consumption of fresh fish and meat in an aging Japanese society, Int. Food Agribusiness Manage. Rev. 7 (2004) 22–41.

[76] N. Willows, L. Johnson-Down, T.A. Kenny, H.M. Chan, M. Batal, Modelling optimal diets for quality and cost: examples from Inuit and First Nations communities in Canada, Appl. Physiol. Nutr. Metab. (2018). Available from: https://doi.org/10.1139/apnm-2018-0624.

[77] J.C. Comiso, C.L. Parkinson, R. Gersten, L. Stock, Accelerated decline in the Arctic sea ice cover, Geophys. Res. Lett. 35 (2008) 413. Available from: https://doi.org/10.1029/2007GL031972.

[78] M. Steele, W. Ermold, J. Zhang, Arctic Ocean surface warming trends over the past 100 years, Geophys. Res. Lett. 35 (2008) 67. Available from: https://doi.org/10.1029/2007GL031651.

[79] J.A. Screen, I. Simmonds, The central role of diminishing sea ice in recent Arctic temperature amplification, Nature 464 (2010) 1334–1337. Available from: https://doi.org/10.1038/nature09051.

[80] K.L. Laidre, I. Stirling, L.F. Lowry, Ø. Wiig, M.P. Heide-Jørgensen, S.H. Ferguson, Quantifying the sensitivity of Arctic marine mammals to climate-induced habitat change, Ecol. Appl. 18 (2008) S97–S125. Available from: https://doi.org/10.1890/06-0546.1.

[81] T.L. Nancarrow, H.M. Chan, A. Ing, H.V. Kuhnlein, Climate change impacts on dietary nutrient status of Inuit in Nunavut, Canada, FASEB J. 22 (2008). Suppl. 1096–1097.

[82] S. Nickels, C. Furgal, M. Buell, H. Moquin, Unikkaaqatigiit: putting the human face on climate change—perspectives from Inuit in Canada, Inuit Tapiriit Kanatami, Nasivvik Centre for Inuit Health and Changing Environments (Université Laval), Ajunnginiq Centre (National Aboriginal Health Organization), 2006.

[83] A. Downing, A. Cuerrier, A synthesis of the impacts of climate change on the First Nations and Inuit of Canada, Indian J. Tradit. Knowl. 10 (2011) 57–70.

[84] J.D. Ford, B. Smit, J. Wandel, M. Allurut, Climate change in the Arctic: current and future vulnerability in two Inuit communities in Canada, Geogr. J. 174 (2008) 45–62. Available from: https://doi.org/10.1111/j.1475-4959.2007.00249.x.

[85] T.K. Suluk, S.L. Blakney, Land claims and resistance to the management of harvester activities in Nunavut, Arctic (2008). Available from: https://doi.org/10.2307/40513357.

[86] J.D. Ford, B. Smit, J. Wandel, J. MacDonald, Vulnerability to climate change in Igloolik, Nunavut: what we can learn from the past and present, Polar Rec. 42 (2006) 127–138. Available from: https://doi.org/10.1017/S0032247406005122.

[87] J.W. Handmer, S. Dovers, T.E. Downing, Societal vulnerability to climate change and variability, Mitig. Adapt. Strateg. Global Change 4 (1999) 267–281. Available from: https://doi.org/10.1023/A:1009611621048.

[88] V.W.Y. Lam, W.W.L. Cheung, U.R. Sumaila, Marine capture fisheries in the Arctic: winners or losers under climate change and ocean acidification? Fish Fish. 17 (2016) 335–357. Available from: https://doi.org/10.1111/faf.12106.

[89] Government of Nunavut, Nunavut Fisheries Strategy 2016-2020, Department of Environment, Fisheries and Sealing Division, 2016.

[90] M.C. Beaumier, J.D. Ford, Food insecurity among Inuit women exacerbated by socio-economic stresses and climate change, Can. J. Public Health 101 (2010) 196–201.

[91] A. Bunce, J.D. Ford, S. Harper, V. Edge, IHACC Research Team, Vulnerability and adaptive capacity of Inuit women to climate change: a case study from Iqaluit, Nunavut, Nat. Hazards 16 (2016) 268. Available from: https://doi.org/10.1007/s11069-016-2398-6.

[92] W.D. Hansen, T.J. Brinkman, F.S. Chapin III, C. Brown, Meeting indigenous subsistence needs: the case for prey switching in rural Alaska, Hum. Dimens. Wildl. 18 (2013) 109–123. Available from: https://doi.org/10.1080/10871209.2012.719172.

[93] T. Bidleman, R.W. Macdonald, J. Stow, Canadian Arctic Contaminants Assessment Report II: Sources, Occurrence, Trends and Pathways in the Physical Environment, Indian and Northern Affairs, 2003.

[94] S.G. Donaldson, J. Van Oostdam, C. Tikhonov, M. Feeley, B. Armstrong, P. Ayotte, et al., Environmental contaminants and human health in the Canadian Arctic, Sci. Total Environ. 408 (2010) 5165–5234. Available from: https://doi.org/10.1016/j.scitotenv.2010.04.059.

[95] B. Braune, J. Chételat, M. Amyot, T. Brown, M. Clayden, M. Evans, et al., Mercury in the marine environment of the Canadian Arctic: review of recent findings, Sci. Total Environ. 509–510 (2015) 67–90. Available from: https://doi.org/10.1016/j.scitotenv.2014.05.133.
[96] UNEP, Global Mercury Assessment 2013, United Nations (UN), 2015.
[97] J. Ma, H. Hung, C. Tian, R. Kallenborn, Revolatilization of persistent organic pollutants in the Arctic induced by climate change, Nat. Clim. Change 1 (2011) 255–260. Available from: https://doi.org/10.1038/nclimate1167.
[98] F. Wania, M.J. Binnington, M.S. Curren, Mechanistic modeling of persistent organic pollutant exposure among indigenous Arctic populations: motivations, challenges, and benefits, Environ. Rev. 25 (2017) 396–407. Available from: https://doi.org/10.1139/er-2017-0010.
[99] S. Booth, D. Zeller, Mercury, food webs, and marine mammals: implications of diet and climate change for human health, Environ. Health Perspect. 113 (2005) 521–526. Available from: https://doi.org/10.1289/ehp.7603.
[100] M.J. Binnington, M.S. Curren, C.L. Quinn, J.M. Armitage, J.A. Arnot, H.M. Chan, et al., Mechanistic polychlorinated biphenyl exposure modeling of mothers in the Canadian Arctic: the challenge of reliably establishing dietary composition, Environ. Int. 92–93 (2016) 256–268. Available from: https://doi.org/10.1016/j.envint.2016.04.011.
[101] AMAP, AMAP Assessment 2015: Human Health in the Arctic, Arctic Monitoring and Assessment Programme (AMAP), Oslo, Norway, 2018.
[102] H.M. Chan, C. Kim, K. Khoday, O. Receveur, H.V. Kuhnlein, Assessment of dietary exposure to trace metals in Baffin Inuit food, Environ. Health Perspect. 103 (1995) 740–746. Available from: https://doi.org/10.1289/ehp.95103740.
[103] H.M. Chan, Inuit Health Survey 2007-2008: Contaminant Assessment in Nunavut, 2012.
[104] AMAP, Arctic Pollution 2011, Arctic Monitoring and Assessment Program (AMAP), 2011.
[105] G. Muckle, E. Dewailly, P. Ayotte, Prenatal exposure of Canadian children to polychlorinated biphenyls and mercury, Can. J. Public Health 89 (Suppl. 1) (1998). S20–S25, 22–27.
[106] H.V. Kuhnlein, H.M. Chan, Environment and contaminants in traditional food systems of northern Indigenous peoples, Annu. Rev. Nutr. 20 (2000) 595–626. Available from: https://doi.org/10.1146/annurev.nutr.20.1.595.
[107] K.A. Friendship, C. Furgal, The role of Indigenous knowledge in environmental health risk management in Yukon, Canada, IJCH 71 (2012) 75. Available from: https://doi.org/10.3402/ijch.v71i0.19003.
[108] J.D. O'Neil, B. Elias, A. Yassi, Poisoned food: cultural resistance to the contaminants discourse in Nunavik, Arctic Anthropol. 34 (1997) 29–40.
[109] H.V. Kuhnlein, H.M. Chan, O. Receveur, D. Muir, R. Soueida, Arctic Indigenous women consume greater than acceptable levels of organochlorines, J. Nutr. 125 (1995) 2501–2510.
[110] D.C. Natcher, O. Huntington, H. Huntington, F.S. Chapin, S.F. Trainor, L. DeWilde, Notions of time and sentience: methodological considerations for Arctic climate change research, Arctic Anthropol. 44 (2007) 113–126. Available from: https://doi.org/10.1353/arc.2011.0099.
[111] C. Hoover, M. Bailey, J. Higdon, S.H. Ferguson, Estimating the economic value of narwhal and beluga hunts in Hudson Bay, Nunavut, Arctic (2013). Available from: https://doi.org/10.2307/23594602.
[112] P.M. Suprenand, C.H. Ainsworth, C. Hoover, Ecosystem Model of the Entire Beaufort Sea Marine Ecosystem: A Temporal Tool for Assessing Food-Web Structure and Marine Animal Populations From 1970 to 2014, 2018. Marine Science Faculty Publications. 261. Available online at: https://scholarcommons.usf.edu/msc_facpub/261.
[113] T.A. Kenny, The Inuit Food System: Ecological, Economic, and Environmental Dimensions of the Nutrition Transition, University of Ottawa, 2017.

SECTION 5

Changing Social World of the Oceans

CHAPTER 25

The changing social world of the oceans

Larry B. Crowder[1] and Wilf Swartz[2]

[1]Edward F. Ricketts Provostial Professor of Marine Ecology and Conservation at Hopkins Marine Station, Stanford University, CA, United states [2]Nippon Foundation Nereus Program, Institute for Oceans and Fisheries, University of British Columbia, Vancouver, BC, Canada

Chapter Outline

Until recently, sustainable development was perceived as an essentially environmental issue that attempts to integrate environmental concerns into economic decision-making. The emphasis of natural resource management, including marine fisheries, was on understanding the environmental response to exploitation pressure and designing policies that align economic incentives so that desirable environmental outcomes, that is, sustainable resource use, can be attained. Yet, despite considerable advancements in our understanding of natural systems, aided by improvements in our capacity to observe and process significant volumes of environmental data, natural resource challenges remain. We now recognize that these sustainability challenges are embedded in coupled social–ecological systems [1]. These systems comprise complex networks of interactions that straddle natural and human communities. Thus, the scientific understanding required to address such systems and to comprehensively predict their response to global environmental changes must also straddle natural and social sciences.

In the first contribution to this section, Oestreich et al. (see Chapter 26: Impact of environmental change on small-scale fishing communities) unpack the theoretical issues surrounding such systems, particularly around the concepts of resilience and adaptive capacity in communities faced with the challenges of global environmental change. Drawing from ecological theories, these concepts have been used extensively in assessments of community responses, but the utility of such frameworks is defined by specific dynamics of each community and their local social contexts. Oestreich et al. develop an analytical framework that explicitly accounts for a variety of local social factors in understanding how communities respond to climate-driven stressors. By using interdisciplinary approaches, such a framework will facilitate understanding the "why" of community response.

Predicting Future Oceans.
DOI: https://doi.org/10.1016/B978-0-12-817945-1.00025-3

Seary's description of mangrove fishing communities from Cambodia (see Chapter 27: The future of mangrove fishing communities) and Mason's call for the integration of fishers' perspectives and values (see Chapter 28: Ocean policy on the water), in that sense, represent excellent case studies in accounting for local social contexts in assessing community's response to climate-driven environmental changes. Seary's insights on the adaptive capacity of these fishing communities, particularly concerning the role of family members that engage in non-fishing occupations while maintaining family fishing needs, challenge the conventional notion of a linear relationship between population growth and exploitation pressure. Her observation that "mangrove fishing" varies considerably from community to community lends further support to the complexity and diversity of these social–ecological systems and the importance of local context in making management decisions. Meanwhile, Mason's contribution demonstrates how environmental policies can be more effective and equitable by considering individual determinants of adaptive capacity, such as individual's agency, worldview, and values, including the cultural factors and power dynamics that shape them. Mason argues for greater community participation in academic research and in policy development, and puts forth research approaches founded upon building trust and listening to community members, allowing them space to share their knowledge, perceptions, concerns, and values as equal partners in knowledge production.

Co-creation of knowledge through dialogues that explicitly recognize the value of both scientific and traditional knowledge of local communities as equal is also the central theme of Vierros and Ota's contribution (see Chapter 29: Integration of traditional knowledge in policy). Through their discussion on the role of traditional knowledge of the Pacific island communities, Vierros and Ota reveal that the interactions between peoples and their oceans are not limited to fisheries but are embedded in social organization and cultural identity that are practiced and developed through traditional knowledge. Tenure systems governed by traditional knowledge do not treat nature and culture separately. Such governance systems are capable of developing robust adaptation strategies for global environmental changes, but only if local sociocultural contexts are fully appreciated and the traditional knowledge that emerged from generations of local social–ecological interactions are used.

For Indigenous peoples, their local cultural links with the environment are particularly salient as these relationships underpin their identity as distinct Peoples. Cisneros-Montemayor and Ota, recognizing the importance of fishing as a fundamental aspect of self-identity among coastal Indigenous peoples, put forth an approach to connecting the knowledge and perspectives of these peoples to national and international ocean policy initiatives with their analysis of seafood consumption by Indigenous peoples (see Chapter 30: Coastal indigenous peoples in global ocean governance). Their study is the first of its kind to be undertaken on a global scale, and their analyses of around 1900 coastal Indigenous communities, representing almost 30 million people, further stress that the full inclusion of Indigenous peoples into ocean policies, including recognition of their rights to access fisheries resources and the continuation of cultural practices, is imperative.

The final two contributions of this section by Teh et al. (see Chapter 31: The relevance of human rights to socially responsible seafood) and Swartz (see Chapter 32: Corporate social responsibility in the global seafood industry) examine how these socioecological considerations could be integrated in national and international seafood industries to ensure that the sector is not only environmentally sustainable but also socially responsible. Teh et al. identify three core aspects of social responsibility as the protection of rights and access, the recognition of equality and equitable opportunities, and the improvements in food and livelihood security. While these components are enshrined in various international instruments, their application and implementation with respect to fisheries issues remain limited, both conceptually and in practice. Swartz focuses on the self-regulation by private seafood firms. Known as Corporate Social Responsibility, such approaches demand that private firms voluntarily commit to codes of conduct that exceed compliance with existing laws and regulations. While both public and private mechanisms described by Teh et al. and Swartz could be harnessed to promote a more socially responsible seafood sector, concerns remain regarding their capacity to ensure that local social contexts, which are crucial to protecting the rights of all individuals involved, are respected. In this regard, Swartz's concludes that a more conciliatory, cooperative approach throughout seafood supply chains may help integrate local concerns and contexts, where socially responsible business practices are promoted not through adhesion to universal best practice standards but through active engagement with each partner throughout the supply chain.

In summary, all contributions in this section call for more explicit recognition of the essential role of local social contexts in understanding the complex social–ecological system that is ocean governance. While effective international and national frameworks for integrating social considerations must be present, ocean sustainability can only be attained through a greater understanding of the unique local social and cultural relationships that each coastal community has with its surrounding ocean, gained through direct engagement with the people of these communities.

Reference

[1] E. Ostrom, A general framework for analyzing sustainability of social-ecological systems, Science 325 (5939) (2009) 419–422.

CHAPTER 26

The impact of environmental change on small-scale fishing communities: moving beyond adaptive capacity to community response

William K. Oestreich[1], Timothy H. Frawley[1], Elizabeth J. Mansfield[1], Kristen M. Green[2], Stephanie J. Green[1,3,4], Josheena Naggea[2], Jennifer C. Selgrath[1], Shannon S. Swanson[2], Jose Urteaga[2], Timothy D. White[1] and Larry B. Crowder[1]

[1]*Hopkins Marine Station, Stanford University, Pacific Grove, CA, United States* [2]*School of Earth, Energy, and Environmental Sciences, Stanford University, Stanford, CA, United States*
[3]*Center for Ocean Solutions, Stanford University, Pacific Grove, CA, United States*
[4]*Department of Biological Sciences, University of Alberta, Edmonton, AB, Canada*

Chapter Outline

26.1 Introduction

Already threatened by overfishing, habitat loss, and inadequate systems of governance [1], the shocks and stressors associated with global environmental change represent a significant and daunting challenge for fishing communities worldwide [2–4]. Previous research on the

Predicting Future Oceans.
DOI: https://doi.org/10.1016/B978-0-12-817945-1.00027-7

responses of fisheries to climate change has focused on the geographic shifts of industrial fisheries in response to warming waters and changes in the abundance and distribution of target species. For example, industrial fisheries have displayed movements of 10s–1000s of kilometers in avoidance of warming waters [5–8].

Yet small-scale fisheries (SSFs), which employ over 90% of the world's capture fishers and provide livelihoods and food security for many millions of individuals around the globe [9], may be even more vulnerable to environmental change than their industrial counterparts (see special issue of Marine Policy, Volume 88, Feb 2018). Most SSFs lack the capacity for significant geographic redistribution, as they utilize small vessels with limited range and rely upon geographic features that are accessible by foot or small vessels (i.e., reefs, beaches, estuaries, etc.) within close proximity to the coastal communities they inhabit. Furthermore, SSFs in island, low-elevation, and ice-edge systems are disproportionately vulnerable to sea-level rise [10], which threatens to inundate coastal habitats as well as cause significant (if not catastrophic) damage to local homes and infrastructure. Climate adaptation strategies have been proposed for industrial fisheries [3], but adaptation of SSFs to climate change and other environmental stressors has received substantially less research attention (Marine Policy, Volume 88, Feb 2018). Even within the scholarship surrounding adaptation, much of the work has focused on hypothetical adaptive capacity indices [11], rather than the observed response of SSF communities to climate variability and/or change. Here, following a brief review of resilience and adaptive capacity theory (Section 26.2), we present an analytical framework for assessing observable responses and mediating attributes of SSF communities to environmental stressors (Section 26.3). We intentionally avoid abstract indices of resilience or adaptive capacity [11] and focus upon how communities empirically respond to those environmental perturbations whose frequency and intensity is expected to increase alongside climate change.

26.2 Resilience and adaptive capacity: theory and limitations

Social–ecological systems (SES), including SSFs, are comprised of complex networks of interactions. The idea of resilience is central to understanding the complexities of such systems and the human and natural communities which comprise them. Resilience has its origins in ecological theory, which focused on the ability of an ecological system to absorb change and disturbance while maintaining interactions and structures found within the system [12]. In the years since its introduction, the concept of resilience has expanded to include human contexts [13,14] and evolved to encompass many definitions. While social scientists and development scholars have adopted resilience theory to describe the ability of human communities to "withstand external shocks to their social infrastructure, such as environmental variability or social, economic, and political upheaval" [13], more recently

SES scholars have expanded the framework to investigate human–environment interactions and the ability of coupled systems to persist and develop with change [15,16].

Assessment of adaptive capacity, or "the ability of systems, institutions, humans, and other organisms to adjust to potential damage, take advantage of opportunities, or to respond to the consequences" [17] relies heavily on concepts from resilience theory [11]. The main bridging concept that integrates adaptive capacity and resilience is vulnerability [18,19], which in the social context can be broken down further into components of exposure to perturbations, the sensitivity of the system to this exposure, and a system's capacity to adapt [20]. Seminal theoretical contributions to the field have described the dimensions, indicators [21], and/or determinants [22] of adaptive capacity, emphasizing factors such as resource availability and distribution, human and social capital (i.e., education or property rights), technology availability, decision-making authority or agency [23], and institutional design [24]. Most recently, Cinner et al. [25] synthesized much of this work to outline five interlinked domains (assets, flexibility, social organization, learning, and agency) across which investment and efforts may be distributed in order to build adaptive capacity of tropical coastal communities.

Ultimately the utility of such frameworks and concepts is determined by the context of scale [11]. Studies conducted at broad spatial scale may be helpful in developing generalizable approaches to address problems in regions where empirical studies are scarce. For example, Berkes and Seixas [26] apply broad concepts of resilience theory to the assessment of lagoon fisheries from around the globe, extracting "potential surrogates" of resilience (i.e., proxy variables for assessing resilience) from their analysis. On the other hand, long-term studies conducted over limited spatial scales can provide nuance by revealing important information about the contextual and developmental factors influencing resilience and adaptive capacity within specific socioecological contexts. Cinner et al. [27] provide an example of this approach, using nine indicators to assess how the adaptive capacity of Kenyan SSF communities varied over a 4-year period.

As more researchers explore adaptive capacity, they have introduced new methods and theories. The field is becoming increasingly fragmented, however, with many case studies and few empirically validated or widely accepted definitions, theories, or methods [11,28]. Most studies on adaptive capacity have sought to identify the characteristics and conditions that determine a group's capacity to adapt (i.e., the "determinants" of adaptive capacity), but measuring or evaluating these determinants and linking them to specific goals and processes remains a challenge. When adaptive capacity is defined as a latent characteristic, an act in the future, there is no physical outcome to be measured [28]. In practice, adaptive capacity is likely to vary depending on the changes occurring, the systems with which they interact, and the specific objectives that are being pursued. To date little work examines the observable response of communities to change and the factors mediating these responses.

Monitoring adaptive response and considering how adaptive capacity is impacted as systems react to change may allow a deeper understanding of feedbacks, trade-offs, and potential improvements to techniques for assessing and building adaptive capacity [27]. But few post-assessment evaluations exist to quantify how social factors match with specific community responses to perturbations. Below we present an analytical approach to systematically investigate how the characteristics of coastal communities and the resource systems in which they are embedded mediate response to different kinds of environmental perturbations.

26.3 An analytical framework for community response and mediating factors

Here, we propose a framework for analyzing SSF community responses to climate change-related stressors (e.g., warming events, hypoxic outbreaks, more intense storms, coastal flooding, shifts in fish distribution). Our framework evolved from an extensive review of the existing literature describing the responses of fishing communities to environmental stressors. We identify both climate change-related stressors and community characteristics that affect community response. Our framework aims to categorize responses observed across communities as adaptive, reactive, or coping [46], and to decipher whether potential mediating factors amplify or dampen these responses (Fig. 26.1). Our framework allows

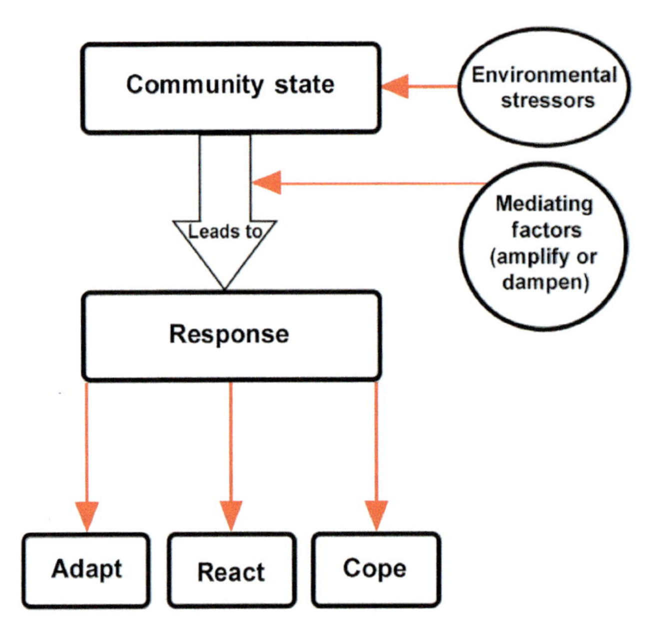

Figure 26.1
Framework for assessing observations of community response to environmental stressors as amplified or dampened by mediating factors.

for the analysis of observed responses (i.e., studies describing a climate-related perturbation, community response(s), and mediating community attributes), and may be supplemented and/or improved as adaptation initiatives progress worldwide and additional insight is accrued.

26.3.1 Climate-driven stressors

Although the full spectrum of climate-driven stressors remains unknown, we can learn about how communities will respond to climate-related threats by evaluating community responses to existing biophysical change. SSF communities throughout history have faced a plethora of threats, ranging from acute stressors (e.g., typhoons; [29]) to slow-accumulating stressors (e.g., chronic overfishing; [30]). The ability of SSF communities to respond and adapt to changing environmental conditions is a fundamental part of the culture and ethos that has enabled their long-term persistence. However, as the rate of environmental change accelerates, coastal communities may be unable to keep pace with increasing exposure to biophysical stressors. Emerging stressors will become more frequent and increasingly severe and may amplify impacts of existing stressors [31]. These stressors may fall into one of several categories: weather and climate, ocean conditions, and biological stressors. Examples of "weather and climate" stressors include sea level rise, long-term temperature changes, extreme temperatures, and changing rainfall patterns. "Ocean conditions" stressors include perturbations such as ocean acidification, ice loss, hypoxia, and ocean warming. Stressors in the "biological" category include species range shifts, smaller individuals, and disease. To understand how such emerging environmental perturbations challenge SSF communities, we identify three components of any given environmental stressor that could initiate the cascade of responses: what the stressor challenging the community is, what the impacts of that stressor are, and the specific groups of people affected by these impacts.

26.3.2 Mediating factors

Researchers have identified a host of social factors both theoretically and empirically linked to a community's response to environmental stress. These factors include access to assets; learning and knowledge; governance and institutions; and diversity and flexibility (Table 26.1). Factors that characterize access to assets include physical assets and infrastructure at the community and household level (e.g., housing, roads, vehicles, boats, fishing gear; [32]), access to technology [35], nonphysical assets affecting social interactions and status such as authority and education [24,33], and financial assets including capital, credit, markets, or aid [34,37,44]. Factors that characterize access to learning and knowledge include the extent to which local ecological knowledge (LEK) is maintained within the community [39], access to a diversity of knowledge sources [24], and capacity for local institutions to foster community members' learning and ability to

Table 26.1: Factors with potential to mediate community responses, sorted into four domains: diversity and flexibility, access to assets, learning and knowledge, and governance and institutions.

Domain	Factor	Description	Sources
Access to assets	Community infrastructure	Physical infrastructure in place that effects responses or variability to change, quality, isolation, presence, transportation, connectivity, etc.	[32]
	Household material assets	E.g., boat, house, specific gear, etc.	[11]
	Authority	Provision of accepted or legitimate forms of power; whether or not institutional rules are embedded in constitutional laws	[24,33]
	Human resources	Availability of expertise, knowledge and human labor	[24,34]
	Financial resources	Availability of financial resources to support policy measures and financial incentives	[24,34,35]
	Natural capital	Access to stocks of natural resources	[36]
	Access to credit	Financial support that is repaid	[11,27,32]
	Access to aid	Financial or material support that is not repaid	[32]
	Access to markets	Access to and/or diversification of markets (local, regional, or international)	[37]
	Access to information	Spaces and platforms for learning; can be achieved through multiple paths, e.g., formal education, informal education, media sources, electronic resources, etc.	[24]
	Fishing gear	Type of technology used for marine fishery resource extraction	[38]
	Technology use	Access to technologies that may directly or indirectly enable many of the adaptive strategies identified as possible in the management of climate change (e.g., warning systems, protective structures, crop breeding and irrigation, settlement and relocation or redesign, flood control measures)	[35]
Learning and knowledge	LEK	Influence of or access to LEK. "LEK is often considered a unique source of information in remote areas, far from research centers, where local ecological and social systems are poorly understood"	[39]
	Learning capacity	Presence of institutional patterns that promote mutual respect and trust; ability of institutional patterns to learn from past experiences and improve their routines; evidence of changes in assumptions underlying institutional patterns; institutional openness towards uncertainties; institutional provision of monitoring and evaluation processes of policy experiences	[24]
	Diversity of knowledge sources	Diversity of knowledge and information sources; intergenerational learning capacity	[11]
	Risk perception	Recognition of causality and human agency, perception of risk, knowledge of disturbances	[11]

(Continued)

Table 26.1: (Continued)

Domain	Factor	Description	Sources
Governance and institutions	Trust	Levels of trust, social capital, and networks (Whitney); presence of institutional patterns that promote mutual respect and trust (Gupta)	[11,24]
	Leadership	Presence of local environmental institutions and strength of social norms, accountability of managers and governance bodies; can fall under three categories: visionary, entrepreneurial, or collaborative	[11,24]
	Gender	Gender and race relations; gender of individuals in a community affects their ability to adapt or respond to a certain stressor or change, women and men are responsible for certain tasks, or more able to adapt to change, education of women can be another role, and traditional gender roles can play into the response	[11,40]
	Social capital	Levels of trust, social capital, and networks; describes relationships of trust, reciprocity, and exchange; the evolution of common rules; and the role of networks. How do individuals use their relationship to others for their own and the collective good? Are there connections or relationships that affect how communities respond?	[11,41]
	Regulations	Quality of governance and leadership in environmental policies and agencies; accountability of managers and governance bodies; active risk management and adaptive governance processes	[11]
	Stakeholder engagement	Levels of participation and quality of decision-making processes; implies a process of active participation of diverse stakeholders working together in concert to develop a unified proposal or common focus (in terms of visions, objectives, points of view, and concerted action)	[11,42]
Diversity and flexibility	Livelihood diversity	Options for altering one's livelihood within fishing portfolios or outside the fishing sector entirely	[11,43]
	Occupational mobility	Options for changing one's occupation completely	[11,27]
	Geographic isolation	Ability or willingness to move in times of uncertainty; includes elements of place attachment	[11]
	Room for autonomous change	Also can be described as agency; has three key attributes: (1) continuous access to information, (2) the ability to act according to a plan, and (3) the capacity to improvise	[24]

Brief descriptions of each factor are included along with sources for more detailed information. *LEK*, Local ecological knowledge.

perceive risk [11]. The effect of governance and institutions depends on a broad suite of factors that include levels of trust and engagement among community members—elements that reflect social capital held by individuals and across social networks in the community [11,24,42,45]. Leadership shown by local institutions and community members is highly related to the ability of communities to respond through its effect on regulations, social norms, and government accountability [11,24]. The domain of diversity and flexibility extends across many elements of communities and their members' lives. The effect of diversity and flexibility across this spectrum can be characterized by livelihood diversity and occupational mobility [11], geographic flexibility or isolation (the ability or willingness to relocate; [11]), and the concept of agency, or room for autonomous change [24]. While a full accounting of metrics is beyond the scope of this chapter, each of these factors and literature that can provide insights into their measurement are found in Table 26.1.

26.3.3 Response framework—adapt, react, or cope

Impacted communities can respond to stressors in a variety of ways, but any response will set communities on pathways that can lead to resilient futures or that can leave communities highly vulnerable to future threats. To understand the spectrum of community responses, we draw on a conceptual framework that divides responses into three categories: adapt, react, and cope (Fig. 26.1) [46]. Communities that exhibit adaptive responses take a proactive and anticipatory approach to change. Adaptive responses of communities are based on knowledge or experience of past threats, and these responses are structured to anticipate future change. These adaptive responses allow communities to ensure that that their SES will be resilient to stressors that occur in the future. Anticipated stressors may be general or specifically defined [29]. For example, communities may prepare in advance for poor fishing conditions during future hypoxia events by diversifying livelihoods in anticipation. By contrast, communities that exhibit a reactive response to change take actions that are unplanned and occur in response to impacts. Because reactive responses tend to address symptoms rather than drivers of impacts, reactive responses may impede the capacity of communities to adapt to future change. Switching target species in response to a range shift of a community's primary target species during El Niño conditions, for example, is a reactive response. Coping occurs when communities passively accept the consequences of an event; communities do not change their behavior or the system to alter the outcome of the current stressor or to reduce the potential impacts from any future stressors. Examples include cases in which communities "wait it out" rather than adaptively or reactively responding. The adapt, react, cope framework [46] is useful to assess responses to existing threats, and the community response to a particular threat may foster a more robust response to future threats. But with climate change, the novel threats and enhanced intensity of perturbations may exceed adaptive responses generated by single threats or those previously experienced. How previous community responses might enable responses to novel threats or new interacting threats remains an open question.

26.3.4 Dampen or amplify?

Each of the mediating factors described above may dampen or amplify the three potential response types (adapt, react, cope) of a community faced with anthropogenic environmental stressors. For example, increased access to physical, financial, and social resources may amplify adaptive responses, and thus reduce the tendency for communities to react or to cope, because these assets support the development and maintenance of alternative livelihoods. Similar relationships may exist for communities that have access to extensive knowledge and learning opportunities. High levels of trust and connectivity among community members may support adaptive responses by widely disseminating information about potential future risks and opportunities for mitigation from key knowledge holders throughout the community. Weak local institutions that are unable to influence social norms or regulations are unlikely to foster accountability by governing bodies and individuals to adaptively manage risks to resources, and thus may result in reactive or coping strategies. Adoption of coping strategies may also be enhanced by geographic isolation, as adaptive or even reactive relocation may not be feasible. Similar inflexibility in occupation may also result in lower likelihood of adaptive response.

26.3.5 Example framework application

In an example of a study which has documented the adaptive response of SSF communities impacted by an environmental stressor, Coulthard [38] describes the responses of artisanal fishing communities on the lagoon of Pulicat Lake in Southeast India to the adverse impacts of weakening monsoon seasons. The study discusses the responses of two villages on the lagoon: the traditionally higher-caste Nadoor kuppam and the traditionally lower-caste Dhonirevu. These communities, targeting primarily white (*Fenneropenaeus indicus*) and tiger (*Penaeus monodon*) prawns, experienced declining catch (and subsequent lower income and food security) as a result of increasing salinity and temperature during weak monsoon seasons. This environmental stressor, categorized as an acute, climate and weather stressor (as described in Section 26.3.1) elicited three responses across two distinct communities: (1) waiting it out (Nadoor kuppam); (2) occupational multiplicity outside of fishing-related livelihoods (Dhonirevu); and (3) using different SSF gear (Dhonirevu). Under the framework described here, response (1) is classified as a coping response, whereas responses (2) and (3) are classified as adaptive responses. In this case study, all three responses were mediated primarily by factors in the domains of "diversity and flexibility" and "access to assets" (Table 26.1). The coping response of Nadoor kuppam was amplified by their access to financial resources (i.e., family members with jobs and income from fish trade, due to social status) that allowed for "waiting it out" (i.e., continuing standard fishing practices even with low catch, while relying upon another income source). Their lack of willingness to diversify their occupations (beyond fishing-related livelihoods) or fishing gear (largely due to their higher-caste status, according to the author) dampened the likelihood of reactive or adaptive responses. Both adaptive responses by Dhonirevu were amplified by factors under the domains of diversity and flexibility, specifically occupational mobility and livelihood diversity. The gear shift in response (3) from regulated stake nets to smaller, unregulated handheld cast and gill nets, was also amplified by access to such gear, a factor in the domain of access to assets. This example categorization of observed responses and mediating factors, in combination with categorization of observed responses found elsewhere in the literature, allows for an analytical approach to understanding the community characteristics driving specific response types of SSF communities to climate-driven stressors.

26.4 Conclusions

Here we provide a framework for the assessment and monitoring of observed community responses to climate-related stressors and their links to a variety of social factors. We are currently implementing this framework using observations of SSF community response to climatic stress events from the existing literature, as in the example found in Section 26.3.5. Our analysis will highlight the relative importance and interdependence of social factors in eliciting adaptive, reactive, or coping responses by communities. Yet to continue to test and apply this framework, we will need research teams to collect data on stressors, responses, and the factors that mediate them simultaneously in SES. Currently, few studies include the appropriate suite of factors to fully evaluate community response to climate perturbations. Such interdisciplinary studies will facilitate understanding the "why" of community response. To facilitate adaptation at the local level, resilience science must appraise the developmental and contextual processes by which communities negotiate adversity and respond to change [47]. The ongoing application of our framework to existing observations is a significant step in this direction, and can help communities dependent on marine resources with identifying, developing, or strengthening factors linked to community responses to best prepare them for future environmental variation.

References

[1] E.M. Finkbeiner, N.J. Bennett, T.H. Frawley, J.G. Mason, D.K. Briscoe, C.M. Brooks, et al., Reconstructing overfishing: moving beyond Malthus for effective and equitable solutions, Fish Fish. 18 (6) (2017) 1180–1191.

[2] W.W. Cheung, R.D. Brodeur, T.A. Okey, D. Pauly, Projecting future changes in distributions of pelagic fish species of Northeast Pacific shelf seas, Prog. Oceanogr. 130 (2015) 19–31.

[3] D.D. Miller, Y. Ota, U.R. Sumaila, A.M. Cisneros-Montemayor, W.W.L. Cheung, Adaptation strategies to climate change in marine systems, Global Change Biol. (2017). Available from: https://doi.org/10.1111/gcb.13829.

[4] E.H. Allison, A.L. Perry, M.C. Badjeck, W. Neil Adger, K. Brown, D. Conway, et al., Vulnerability of national economies to the impacts of climate change on fisheries, Fish Fish. 10 (2) (2009) 173–196.

[5] D.A. Kroodsma, J. Mayorga, T. Hochberg, N.A. Miller, K. Boerder, F. Ferretti, et al., Tracking the global footprint of fisheries, Science 359 (2018) 904–908.

[6] P. Lehodey, M. Bertignac, J. Hampton, A. Lewis, J. Picaut, El Niño Southern Oscillation and tuna in the western Pacific, Nature 389 (6652) (1997) 715.

[7] P. Michael, C. Wilcox, G. Tuck, A. Hobday, P. Strutton, Japanese and Taiwanese pelagic longline fleet dynamics and the impacts of climate change in the southern Indian Ocean, Deep-Sea Res. II 140 (2017) 242–250.
[8] J.W. Morley, R.L. Selden, R.J. Latour, T.L. Frolicker, R.J. Seagraves, M.L. Pinsky, Projecting shifts in thermal habitat for 686 species on the North American continental shelf, PLoS One 13 (5) (2018) e0196127.
[9] F. Berkes, Managing Small-Scale Fisheries: Alternative Directions and Methods, IDRC, 2001.
[10] E. Le Cornu, A.N. Doerr, E.M. Finkbeiner, D. Gourlie, L.B. Crowder, Spatial management in small-scale fisheries: a potential approach for climate change adaptation in Pacific Islands, Mar. Policy 88 (2018) 350–358.
[11] C.K. Whitney, N.J. Bennett, N.C. Ban, E.H. Allison, D. Armitage, J.L. Blythe, et al., Adaptive capacity: from assessment to action in coastal social-ecological systems, Ecol. Soc. 22 (2) (2017) 22.
[12] C.S. Holling, Resilience and stability of ecological systems, Annu. Rev. Ecol. Syst. 4 (1973) 1–23.
[13] W.N. Adger, Social and ecological resilience: are they related? Prog. Hum. Geogr. 24 (3) (2000) 347–364.
[14] F. Berkes, C. Folke, J. Colding (Eds.), Linking Social and Ecological Systems: Management Practices and Social Mechanisms for Building Resilience, Cambridge University Press, New York, 1998.
[15] C. Folke, Resilience: the emergence of a perspective for social–ecological systems analyses, Global Environ. Change 16 (3) (2006) 253–267.
[16] B. Walker, C.S. Holling, S.R. Carpenter, A. Kinzig, Resilience, adaptability, and transformability in social-ecological systems, Ecol. Soc. 9 (2) (2004) 5.
[17] Intergovernmental Panel on Climate Change (IPCC), Annex II: glossary, in: K.J. Mach, S. Planton, C. von Stechow (Eds.), Climate Change 2014: Synthesis Report. Contribution of Working Groups I, II and III to the Fifth Assessment Report of the Intergovernmental Panel on Climate Change, IPPC, Geneva, Switzerland, 2014, pp. 117–130 [core writing team, R.K. Pachauri and L.A. Meyer, editors].
[18] W.N. Adger, Vulnerability, Global Environ. Change 16 (3) (2006) 268–281.
[19] J.J. McCarthy, O.F. Canziani, N.A. Leary, D.J. Dokken, K.S. White (Eds.), Climate Change 2001: Impacts, Adaptation, and Vulnerability: Contribution of Working Group II to the Third Assessment Report of the Intergovernmental Panel on Climate Change, vol. 2, Cambridge University Press, 2001.
[20] G.C. Gallopín, Linkages between vulnerability, resilience, and adaptive capacity, Global Environ. Change 16 (3) (2006) 293–303.
[21] J. Hinkel, "Indicators of vulnerability and adaptive capacity": towards a clarification of the science–policy interface, Global Environ. Change 21 (1) (2011) 98–208.
[22] N. Brooks, W.N. Adger, P.M. Kelly, The determinants of vulnerability and adaptive capacity at the national level and the implications for adaptation, Global Environ. Change 15 (2) (2005) 151–163.
[23] G. Yohe, R.S. Tol, Indicators for social and economic coping capacity—moving toward a working definition of adaptive capacity, Global Environ. Change 12 (1) (2002) 25–40.
[24] J. Gupta, C. Termeer, J. Klostermann, S. Meijerink, M. van den Brink, P. Jong, et al., The adaptive capacity wheel: a method to assess the inherent characteristics of institutions to enable the adaptive capacity of society, Environ. Sci. Policy 13 (6) (2010) 459–471.
[25] J.E. Cinner, W.N. Adger, E.H. Allison, M.L. Barnes, K. Brown, P.J. Cohen, et al., Building adaptive capacity to climate change in tropical coastal communities, Nat. Clim. Change 8 (2) (2018) 117–123.
[26] F. Berkes, C.S. Seixas, Building resilience in lagoon social–ecological systems: a local-level perspective, Ecosystems 8 (8) (2005) 967–974.
[27] J.E. Cinner, C. Huchery, C.C. Hicks, T.M. Daw, N. Marshall, A. Wamukota, et al., Changes in adaptive capacity of Kenyan fishing communities, Nat. Clim. Change 5 (9) (2015) 872.
[28] A.R. Siders. Adaptive Capacity to Climate Change (Doctoral dissertation), 2018. Retrieved from <https://purl.stanford.edu/zq078jz9132>.
[29] B. Walker, D. Salt, Resilience Thinking: Sustaining Ecosystems and People in a Changing World, Island Press, 2006.

[30] J.B. Jackson, M.X. Kirby, W.H. Berger, K.A. Bjorndal, L.W. Botsford, B.J. Bourque, et al., Historical overfishing and the recent collapse of coastal ecosystems, Science 293 (5530) (2001) 629–637.
[31] M.C. Badjeck, E.H. Allison, A.S. Halls, N.K. Dulvy, Impacts of climate variability and change on fishery-based livelihoods, Mar. Policy 34 (3) (2010) 375–383.
[32] N. Brooks, W.N. Adger, G. Bentham, M. Agnew, S. Eriksen, New Indicators of Vulnerability and Adaptive Capacity. *Technical Report*, Tyndall Centre for Climate Change Research, Norwich, 2005.
[33] F. Biermann, 'Earth system governance' as a crosscutting theme of global change research, Global Environ. Change 17 (3-4) (2007) 326–337.
[34] R. Nelson, P. Kokic, S. Crimp, P. Martin, H. Meinke, S.M. Howden, et al., The vulnerability of Australian rural communities to climate variability and change: Part II—Integrating impacts with adaptive capacity, Environ. Sci. Policy 13 (1) (2010) 18–27.
[35] B. Smit, O. Pilifosova, Adaptation to climate change in the context of sustainable development and equity, in: J.J. McCarthy, O. Canziani, N.A. Leary, D.J. Dokken, K.S. White (Eds.), Climate Change 2001: Impacts, Adaptation and Vulnerability. IPCC Working Group II, Cambridge University Press, Cambridge, 2001, pp. 877–912.
[36] B. Walker, L. Gunderson, A. Kinzig, C. Folke, S. Carpenter, L. Schultz, A handful of heuristics and some propositions for understanding resilience in social-ecological systems, Ecol. Soc. 11 (1) (2006) 13.
[37] F. Poulain, A. Himes-Cornell, C. Shelton, Methods and tools for climate change adaptation in fisheries and aquaculture, in: M. Barange, T. Bahri, M.C.M. Beveridge, K.L. Cochrane, S. Funge-Smith, F. Poulain (Eds.), Impacts of Climate Change on Fisheries and Aquaculture: Synthesis of Current Knowledge, Adaptation and Mitigation Options. FAO Fisheries and Aquaculture Technical Paper No. 627, FAO, Rome, 2018.
[38] S. Coulthard, Adapting to environmental change in artisanal fisheries—insights from a South Indian Lagoon, Global Environ. Change 18 (3) (2008) 479–489.
[39] L.C. Gerhardinger, E.A. Godoy, P.J. Jones, Local ecological knowledge and the management of marine protected areas in Brazil, Ocean Coastal Manage. 52 (3-4) (2009) 154–165.
[40] Haan et al., Chronic vulnerability to food insecurity in Kenya—2001, in: A World Food Programme Pilot Study for Improving Vulnerability Analysis, 2001.
[41] W.N. Adger, Social capital, collective action, and adaptation to climate change, Der klimawandel, VS Verlag für Sozialwissenschaften, 2010, pp. 327–345.
[42] M. Bouamrane, M. Spierenburg, A. Agrawal, A. Boureima, M.C. Cormier-Salem, M. Etienne, et al., Stakeholder engagement and biodiversity conservation challenges in social-ecological systems: some insights from biosphere reserves in western Africa and France, Ecol. Soc. 21 (4) (2016). Available from: https://doi.org/10.5751/ES-08812-210425.
[43] W.N. Adger, T.P. Hughes, C. Folke, S.R. Carpenter, J. Rockström, Social-ecological resilience to coastal disasters, Science 309 (5737) (2005) 1036–1039.
[44] R. Mendelsohn, W.D. Nordhaus, The impact of global warming on agriculture: a Ricardian analysis: reply, Am. Econ. Rev. 89 (4) (1999) 1046–1048.
[45] W.N. Adger, Social aspects of adaptive capacity, Climate Change, Adaptive Capacity and Development, Imperial College Press, 2003, pp. 29–49.
[46] N.J. Bennett, P. Dearden, G. Murray, A. Kadfak, The capacity to adapt? Communities in a changing climate, environment, and economy on the northern Andaman coast of Thailand, Ecol. Soc. 19 (2) (2014) 5.
[47] S. Coulthard, Can we be both resilient and well, and what choices do people have? Incorporating agency into the resilience debate from a fisheries perspective, Ecol. Soc. 17 (1) (2012) 4.

CHAPTER 27

The future of mangrove fishing communities

Rachel Seary[1,2]

[1]Cambridge Coastal Research Unit, Department of Geography, University of Cambridge, Cambridge, United Kingdom [2]UN Environment World Conservation Monitoring Centre, Cambridge, United Kingdom

Chapter Outline

27.1 An introduction to mangrove fishing issues

Management of mangrove fisheries is rarely the target of coastal fisheries or forest management, which therefore often fails to acknowledge the multidimensionality and context dependency of mangrove fishing. Mangrove fishery research itself has even failed to demonstrate this complexity, with mangrove fishery valuation studies focusing typically on one sector, gear, or target (Seary, unpublished PhD thesis). This could lead to the interests of many fishing communities who rely on mangrove fishery livelihoods being invisible or underrepresented in management measures. Mangrove use by fishers is highly context dependent, therefore impacts of future change are also likely to be context dependent and influenced by local cultural values, governance, and ability to adapt. Compounded by uncertain responses to future climatic change, there is a large question mark over the future of mangrove fishery community linkages and a number of knowledge gaps to address. This chapter will summarize the knowledge on likely influencing factors on future mangrove fishery links and the social aspects which may play into future mangrove fishery community linkages. It will also highlight gaps in research required to

Predicting Future Oceans.
DOI: https://doi.org/10.1016/B978-0-12-817945-1.00036-8

ultimately understand how mangrove fishery communities could be impacted by future environmental change.

Mangrove forests occur in the coastal saline or brackish environment of 105 tropical and subtropical nations, covering an estimated 8,349,500 ha worldwide [1]. Mangroves provide a number of vital ecosystem services, particularly to coastal communities, and have most recently been valued at almost US$200,000 ha/year [2]. These services include, but are not limited to, coastal protection from extreme weather events, climate mitigation through the efficient storage of carbon, and the provision of forest and fisheries products [3–6]. This role in fisheries production provides a vital link between mangrove forests and coastal fishing communities, whose livelihoods can be very closely tied to the mangrove for both income and subsistence. We acknowledge mangrove fisheries to include any activity in which collection or culture of fish or invertebrate species is benefited by the presence of mangrove forests. This can include products collected directly in the mangrove or mangrove associates caught offshore, as well as those obtained through mariculture activities. Mangrove fishing can provide both income and subsistence as well as back-up or last resort occupations to fishing communities, making them important even where mangrove fishing is not the dominant fishing industry in a local area.

Mangrove extent has been shown as a good predictor of fisheries productivity and has been positively correlated with local fisheries catches worldwide [5]. Mangroves are thought to enhance fisheries through two main mechanisms of food and shelter provision [7]. Biodiversity of fish and invertebrates within mangroves make them important locations for direct capture fisheries, as is widely documented for prawn, crab, and oyster fisheries [8–10]. However, one of the major roles of mangroves in enhancing fisheries production is through providing nursery habitat for juvenile fish and invertebrates [7]. Mangrove forests provide ideal refugial space for prey and young fish through the complexity of above-ground root structures which allow them to avoid predation and invest more time in feeding [7,11]. Ontogenetic (life history) migration of mangrove-dwelling juvenile fish recruiting to adjacent reef or offshore habitats means that adult fish stocks outside of the mangrove forest are also replenished [7,12]. Mangrove forests are also highly productive environments and their leaves and woody matter make up an important part of the detrital food web, also being directly consumed by commercially valuable detritivores, such as mangrove crab [7]. Offshore and adjacent habitats are also thought to benefit from nutrient outwelling from mangrove systems [11].

Fishing by coastal communities can therefore benefit from mangrove forest presence both directly and indirectly, exhibiting complex and variable connections to the mangrove between and within fishing communities. Sustained provision of this ecosystem service which supports mangrove fishing communities is however uncertain due to the current global decline in mangrove extent.

27.2 How do human impacts threaten mangrove fishing?

An estimated 2.3 million km^2 of mangrove forest has been lost worldwide between 2000 and 2012 [13].While mangrove loss has slowed in recent years from an estimated 1% per year in 1990–2000, to between 0.16% and 0.39% per year from 2000 to 2012, considerably widespread and disproportionate damage and loss continue [1]. Whilst some countries have achieved stable mangrove forest extent, others such as countries in the SE Asia region continue to lose between 3.58% and 8.08% per year [1]. Mangrove loss and associated changes to ecosystem service provision will therefore be disproportionately felt by coastal communities. Also depending on the levels of dependence on mangrove products, mangrove loss will likely impact many developing countries the worst [14].

Destruction of mangrove forests for alternate human uses with perceived higher revenue, such as aquaculture ponds, rice agriculture, and oil palm plantations are some of the leading causes of mangrove loss [15]. Disproportionate loss can also somewhat be reflective of varying national levels of pressure and mitigation by policy and management regarding mangrove use [14]. Research on mangrove fishery value lags behind its coastal habitat counterparts, meaning that even baseline data on mangrove fishing is often lacking. Limited understanding of what a mangrove fishery can encompass, together with undervaluation and underrepresentation of the importance of mangrove value to varying groups in a local context means that mangrove fishing is rarely a management target.

As mangrove area is often linked positively to fish abundance, biodiversity, and local fisheries production, the loss of mangrove area through human-induced deforestation or degradation is expected to have negative impacts on fishable biomass [16]. The human activities which lead to mangrove loss and degradation are well studied (Table 27.1), however less focus has been placed into investigating the implications that mangrove loss has on fisheries and therefore the communities who rely upon them.

Conversion for aquaculture is one of the leading causes of mangrove loss globally. In SE Asia aquaculture is responsible for 30% of mangrove forest loss [15]. Conversion activities for aquaculture and agriculture are conducted on the premise of larger profits that can be gained than available from the system in its natural state. However, failed inactive aquaculture ponds are common in SE Asia as are stories of fisheries losses as a result. Conversion for shrimp aquaculture in Southern Thailand, which was destroying 30 km of mangrove annually between 1990 and 1993, was estimated to have cost as much as US $408,000 in welfare losses from artisanal fisheries (consumer or producer surplus) [19]. Economic losses resulting from deforestation were reportedly more impactful for shellfish products than demersal fisheries [19]. The unsustainable nature of shrimp ponds, which often fail within 5–6 years of establishment, mean that more destruction continues to make room for new ponds, although now at a lesser rate than before [19]. While conversion for

Table 27.1: Description of the human activities which impact mangrove forests and their various influences on mangrove condition and therefore fish habitat.

Activity	Impact
Aquaculture [15,17–19]	Immediate forest/habitat loss Groundwater withdrawal Alterations to river flow Postcollapse coastal erosion, saltwater intrusion, and flooding
Agriculture (rice, oil palm, and sugarcane) [15,20]	Immediate forest/habitat loss or density decrease Flooding and coastal erosion Fish kills caused by water pollution
Coastal development [21,22]	Immediate forest/habitat loss from city expansion Sewage overflows Mangrove water contamination from developments Fish habitat loss from building of bridges and levees Alteration of water flows
Fishing [21–24]	Habitat loss or stress from damaging practices Unsustainable cutting for construction of gear Unsustainable cutting for oyster harvest Decreased fishing productivity caused by overfishing Introduction of invasive species from ballast waters
Damming [21,25]	Alteration of tidal cycles Can lead to decrease in fish abundance in mangroves and reduced catches of estuarine-dependent species
Unsustainable logging [21,26]	Immediate forest loss Reduction of mangrove tree biodiversity Lost blue carbon stocks
Tourism [22,27]	Additional infrastructure to accommodate tourists (see coastal development) Solid waste pollution

aquaculture slows, oil palm production, promoted by governments as an attractive livelihood alternative [15], accelerates and represents the next new threat to mangrove destruction.

While mangrove fishery importance continues to be undervalued, conservation rather than conversion is unlikely to come out on top of cost–benefit analyses used in land-use change decisions. Better understanding of the impacts caused by human-induced deforestation on mangrove fishery productivity is necessary in order to comprehend the consequences for those reliant on mangroves for subsistence and economy. The first step necessary in such pursuit is investigating the full complexity of what mangrove fisheries can encompass and therefore benefit to the whole community in each local context.

Fishing itself, however, is also a threat to mangrove fishery production. Fisheries in the Sundarbans mangrove forest, Bangladesh, are an example of observed catch declines following ongoing fish, crab, and prawn overexploitation, together with destruction of mangrove areas for shrimp aquaculture ponds [28]. The socioeconomic drivers of fishing effort in mangrove fisheries and the impacts of overfishing should therefore not be excluded as a driver of fishable biomass in mangrove ecosystems. Management of mangroves which focusses on control of fishing however is often lacking enforcement, or is absent entirely. Conversely, efforts by communities and governments to replant lost mangroves are widespread and have been successful in restoring mangroves to their original extent. Unfortunately, restoration is not always done well and often fails as a result of replanting where the reasons for original degradation have not been addressed [29]. Mangroves can regrow naturally where conditions are appropriate, however often human actions have degraded the environment to an extent that mangroves cannot regrow naturally and often require expensive interventions [29].

Whether restored mangroves function equally to natural mangroves in enhancing fish productivity and therefore fisheries is understudied. Further, biodiversity of fauna in mangroves is rarely included as an indicator of whether mangrove restoration or management has been successful. Some evidence however does support that mangrove restoration encourages the recolonization of faunal biodiversity in mangroves [30]. It should be noted that some species are thought to be more sensitive to mangrove destruction than others, such as herbivorous crabs and mollusks [30]. Indicators of community composition of mangrove fauna should therefore also be used to understand whether a restored mangrove is successfully functioning or providing the ecosystem services required. This is particularly important for fisheries communities who may rely on particular species or groups for income or subsistence. Mangrove restoration done well, however, may offer sustained function of mangrove fishery production for fishing communities in future.

27.3 Will future climate change influence mangrove fishery productivity?

Destruction for human uses continues to be the largest issue threatening mangrove extent, however future environmental change may also influence the extent, distribution, and functioning of mangrove forests. The impact on mangroves of various aspects of climate change such as sea-level rise (SLR), increased storminess and changes to temperature, precipitation, atmospheric CO_2, and ocean circulation have been investigated [31–33]. All of these components of climate change are expected to alter to some degree the productivity and respiration of mangroves and their associated biological community, along with their links with adjacent systems [34]. Less research however has investigated how these changes to mangrove forests will influence fisheries that depend upon them. Nevertheless, the climate-induced changes in marine ecosystems and resident fish populations are very likely to force changes in fisheries productivity and impact fishing community livelihoods [35].

SLR is projected to be the greatest climate change-related risk to mangrove forests [31–33,36–38]. The decline or degradation of mangrove area driven by SLR could impact mangrove fisheries production through the reduction of critical coastal habitat for fish and invertebrate species [35]. The ability of mangrove areas to keep up with changing sea level through surface elevation will be determined by regional vulnerability [32,36]. Vulnerability will be determined by a range of factors which include forest species composition, geomorphic setting, lateral accommodation space, sediment supply, and tidal range [32,38–40].

A combination of these factors render some areas, such as the low-lying Pacific Islands, particularly vulnerable to SLR [31,33]. It has been estimated that 69% of Indo-Pacific mangrove areas studied are not keeping pace with SLR and are largely controlled by the availability of suspended matter [33]. The impacts of SLR are often compounded by limitations placed on sediment supply by human activity [41,42]. Conversely, sedimentation due to land subsidence or sediment supply increases due to human land use changes can also have erosional effects on mangrove extent [41]. Altered water turbidity as a result of these changes to sediment supply also has potential to change habitat suitability for fish [41]. Other climate-related variability and change, such as temperature, precipitation, atmospheric CO_2, and ocean circulation, although expected to have some influence on mangrove ecosystem function, are not well studied and even less so is the subsequent impact on mangrove fisheries. Global increases in temperature, atmospheric CO_2, and precipitation have the potential to increase mangrove productivity and provoke expansion into higher latitudinal ranges [32,38,43]. The predicted increased frequency and intensity of storm events are expected to have mixed impacts on mangrove forests, through loss of mangrove area through defoliation, tree mortality, and alterations to soil elevation or conversely increased resilience to SLR through increased allochthonous sediment input (sediment originating elsewhere) [32,38,44]. Human responses to climate change, for example, mitigation actions, such as construction of sea walls, water irrigation activities or managed retreat, may also have impacts on mangrove response to climate change, which may indirectly impact upon mangrove fisheries productivity [32].

There are also aspects of climate change which may impact fish and invertebrates directly. Global catch potential and revenue from fisheries are projected to decrease by 7.7% and 10.4%, respectively, by 2050 relative to 2000 under high CO_2 climate scenarios, and are predicted to have the worst impacts on the countries with the least adaptive capacity [45]. Further, marine fish are most at risk of climate change impacts in the tropics (where most mangroves are distributed) due to low tolerance to warming [46]. No research has investigated how this will affect mangrove-associated fish specifically. For coral reef fisheries, however, studies have suggested temperature-induced coral mortality can cause collapses in coastal fisheries which will impact coral reef fishing communities, the effects of which can be first felt several years after the habitat destruction [47]. These delayed impacts are generated first from the loss of complex structures supporting small class coral reef fish, which in turn reduces recruitment of adult fish into the fishable stock [47]. The removal of mangrove habitat through climate change impacts, and therefore refugial space for fish of commercial or subsistence value could have similar impacts on adult stock recruitment and therefore fisheries production as is seen for coral reefs, but this impact requires investigation. Connectivity between habitats is also an important factor influencing the nursery function [48]. Climate change-induced fragmentation or losses of other habitats such as coral reefs could also therefore have implications for mangrove fisheries.

Adaptation and diversification by fishing communities are often discussed as solutions to dealing with climate change-related shifts in fish productivity. However, there are also many social and political factors to be considered which may influence future mangrove fishery community function under future environmental change.

27.4 What social and political factors will influence continued mangrove fishing by communities?

Fishing effort within mangrove forests can be largely driven by a number of socioeconomic variables, such as the proximity, size, and demand of human populations, the cultural, political, and economic conditions of the local population, access to alternative livelihoods, and varying levels of fisheries management [7]. Rapid population growth in coastal areas can indeed influence mangrove fishing effort as fishing and related activities are a dominant source of income for the majority of people living in close proximity to mangrove forests [22,49]. Growth of coastal fishing populations does not however always equate to greater fishing effort, nor greater mangrove fishery production.

Fishing production in mangroves does not always increase linearly with effort, as fishing itself can cause degradation of mangroves which can hinder mangrove fishery production [7]. Diversification into new fishing activities or nonfishing occupations is often promoted as the solution to adapting to negative future changes in such cases of mangrove forests and fisheries degradation. There are however a number of factors which influence an individual's, household's, or fishing community's ability to adapt [50].

In coral reef fisheries the adaptive capacity has been linked to the number of family members available to take up new occupations, and the number of different activities conducted per household [47]. This was also observed in mangrove fishing communities in Koh Kong Province, Cambodia, where households with greater numbers of family members were able to diversify into new activities, such as mariculture or non-fishing occupations, while maintaining family fishing needs (Seary, unpublished PhD thesis). Larger households as a result of population growth will therefore not be necessarily correlated to increased mangrove fishing effort. Further, the future of mangrove fishery communities may see mangroves being used in more diverse ways, particularly with an increase in mariculture of mangrove-associated species in recent years.

Adaptive capacity, or entry to new activities, may also be linked to wealth. Research on the east coast of India suggested that local communities living around mangroves had a positive attitude towards mangrove conservation and use of alternative resources; however, those who were too poor to afford these alternative resources admitted that they would continue to take mangrove products even from protected areas [51]. Management of mangrove resources should therefore take into account the socioeconomic drivers influencing the interactions between communities and mangrove forests on the land–sea interface.

27.5 Conclusions

Competing uses for mangrove areas, including fishing but particularly activities which directly damage mangroves, threaten to weaken, change, or force adaptation to links between mangroves and fishing. Mitigation against damaging impacts to mangroves through management and conservation efforts will help sustain mangrove fishing communities. Better information about the connection between mangrove habitat (including restored mangroves), fisheries production, and fishing communities is necessary in order to encourage this.

While mangrove habitat changes with future environmental degradation and climate change, adaptation by mangrove fishing communities will be a large social driver of sustained mangrove fishing livelihoods. Management of mangrove uses, particularly where uses are varied and complex, will have great influence over how different groups of a mangrove fishing community will be able to adapt to future changes. As mangrove fishing activities vary considerably from community to community, their dependence on mangroves and therefore response to future change will also vary by local context. Management must therefore also focus on the local scale in order to meet the demands of such complex systems. More research which demonstrates this complexity in mangrove fishing on the local scale is therefore a priority in finding management solutions that can be tailored to the needs of local mangrove fishing communities.

References

[1] S.E. Hamilton, D. Casey, Creation of a high spatio-temporal resolution global database of continuous mangrove forest cover for the 21st century (CGMFC-21), Global Ecol. Biogeogr. 25 (6) (2016) 729–738.

[2] R. Costanza, R. De Groot, P. Sutton, S. Van Der Ploeg, S.J. Anderson, I. Kubiszewski, et al., Changes in the global value of ecosystem services, Global Environ. Change 26 (2014) 152–158. Available from: https://doi.org/10.1016/j.gloenvcha.2014.04.002.

[3] A.L. McIvor, I. Möller, T. Spencer, M. Spalding, Reduction of wind and swell waves by mangroves. Natural Coastal Protection Series: Report 1. Cambridge Coastal Research Unit Working Paper 40, 2012, (November 2015). Available from: http://www.naturalcoastalprotection.org/documents/reduction-of-wind-and-swell-wavesby-mangroves.

[4] M.D. Spalding, S. Ruffo, C. Lacambra, I. Meliane, L. Zeitlin, C.C. Shepard, et al., Ocean & coastal management the role of ecosystems in coastal protection: adapting to climate change and coastal hazards, Ocean Coast Manage 90 (2014) 50–57. Available from: https://doi.org/10.1016/j.ocecoaman.2013.09.007.
[5] M. Carrasquilla-Henao, F. Juanes, Mangroves enhance local fisheries catches: a global meta-analysis, Fish Fish. 18 (1) (2017) 79–93.
[6] S. Bouillon, A.V. Borges, E. Castañeda-Moya, K. Diele, T. Dittmar, N.C. Duke, et al., Mangrove production and carbon sinks: a revision of global budget estimates, Global Biogeochem. Cycles 22 (2) (2008) 1–12.
[7] A.J. Hutchison, M. Spalding, P. Ermgassen, The role of mangroves in fisheries enhancement, Nat. Conserv. Wetl. Int. 54 (2014) (November 2015) Available from: http://preventionweb.net/go/40622.
[8] S.Y. Lee, Relationship between mangrove abundance and tropical prawn production: a re-evaluation, Mar. Biol. 145 (5) (2004) 943–949.
[9] N.R. Loneragan, N. Ahmad Adnan, R.M. Connolly, F.J. Manson, Prawn landings and their relationship with the extent of mangroves and shallow waters in western peninsular Malaysia, Estuar. Coast Shelf Sci. 63 (1–2) (2005) 187–200.
[10] F.J. Manson, N.R. Loneragan, B.D. Harch, G.A. Skilleter, L. Williams, A broad-scale analysis of links between coastal fisheries production and mangrove extent: a case-study for northeastern Australia, Fish Res. 74 (1–3) (2005) 69–85.
[11] V.C. Chong, Mangroves-fisheries linkages – the Malaysian perspective, Bull. Mar. Sci. 80 (3) (2007) 755–772.
[12] I.A. Kimirei, I. Nagelkerken, Y.D. Mgaya, C.M. Huijbers, The mangrove nursery paradigm revisited: otolith stable isotopes support nursery-to-reef movements by Indo-Pacific fishes, PLoS One 8 (6) (2013) e66320.
[13] M.C. Hansen, P.V. Potapov, R. Moore, M. Hancher, S.A. Turubanova, A. Tyukavina, et al., High-resolution global maps of 21st-century forest cover change, Science 850 (2013) 2011–2014.
[14] H. Van Lavieren, M. Spalding, Securing the future of mangroves, in: Policy Brief, UN Univ., 2012, 53. Available from: <http://www.ganadapt.org/files/Securing_the_future_of_mangroves_high_res.pdf>.
[15] D.R. Richards, D.A. Friess, Rates and drivers of mangrove deforestation in Southeast Asia, 2000–2012, Proc. Natl. Acad. Sci. U.S.A. 113 (2) (2016) 344–349. Available from: http://www.pnas.org/lookup/doi/10.1073/pnas.1510272113.
[16] E.B. Barbier, Valuing ecosystem services as productive inputs, Econ. Policy 22 (49) (2007). Available from: http://www.jstor.org/stable/pdf/3601036.pdf.
[17] S. Sathirathai, E.B. Barbier, Valuing mangrove conservation in Southern Thailand, Contemp. Econ. Policy 19 (2) (2001) 109–122.
[18] B.K. van Wesenbeeck, T. Balke, P. van Eijk, F. Tonneijck, H.Y. Siry, M.E. Rudianto, et al., Aquaculture induced erosion of tropical coastlines throws coastal communities back into poverty, Ocean Coast Manage. 116 (2015) 466–469. Available from: https://doi.org/10.1016/j.ocecoaman.2015.09.004.
[19] E.B. Barbier, I. Strand, S. Sathirathai, Do open access conditions affect the valuation of an externality? Estimating the welfare effects of mangrove-fishery linkages in Thailand, Environ. Resour. Econ. 21 (4) (2002) 343–367.
[20] C. Vázquez-González, P. Moreno-Casasola, A. Juárez, N. Rivera-Guzmán, R. Monroy, I. Espejel, Trade-offs in fishery yield between wetland conservation and land conversion on the Gulf of Mexico, Ocean Coast Manage. 114 (2015) 194–203.
[21] D.M. Alongi, Present state and future of the world's mangrove forests, Environ. Conserv. 29 (3) (2002) 331–349.
[22] L. Creel, Ripple Effects: Population and Coastal Regions, Mak Link, 2003, p. 8.
[23] S.J.M. Blaber, D.P. Cyrus, J.J. Albaret, C.V. Ching, J.W. Day, M. Elliott, et al., Effects of fishing on the structure and functioning of estuarine and nearshore ecosystems, ICES J. Mar. Sci. 57 (3) (2000) 590–602.
[24] B. van Jan-Willen, E. Sullivan, T. Nakamura, The Importance of Mangroves to People: A Call to Action. United Nations Environment Programme World Conservation Monitoring Centre, 2014, pp. 33–103.

Available from: <http://newsroom.unfccc.int/es/el-papel-de-la-naturaleza/la-onu-alerta-de-la-rapida-destruccion-de-los-manglares/>.

[25] J.A. Baisre, Z. Arboleya, Going against the flow: effects of river damming in Cuban fisheries, Fish. Res. 81 (2006) 283–292 (December 2005).

[26] A. Malik, R. Fensholt, O. Mertz, Mangrove exploitation effects on biodiversity and ecosystem services, Biodivers. Conserv. 24 (14) (2015) 3543–3557.

[27] C.M. Hall, Trends in ocean and coastal tourism: the end of the last frontier? Ocean Coastal Manage. 44 (2001) 601–618.

[28] S. Islam, M. Haque, The mangrove-based coastal and nearshore fisheries of Bangladesh: ecology, exploitation and management, Rev. Fish Biol. Fish. 14 (2005) 153–180.

[29] B. Kamali, R. Hashim, Mangrove restoration without planting, Ecol. Eng. 37 (2) (2011) 387–391. Available from: https://doi.org/10.1016/j.ecoleng.2010.11.025.

[30] J.O. Bosire, F. Dahdouh-Guebas, M. Walton, B.I. Crona, R.R. Lewis, C. Field, et al., Functionality of restored mangroves: a review, Aquat. Bot. 89 (2) (2008) 251–259.

[31] E.L. Gilman, J. Ellison, V. Jungblut, H. Van Lavieren, L. Wilson, F. Areki, et al., Adapting to Pacific Island mangrove responses to sea level rise and climate change, Clim. Res. 32 (3) (2006) 161–176.

[32] E.L. Gilman, J. Ellison, N.C. Duke, C. Field, Threats to mangroves from climate change and adaptation options: a review, Aquat. Bot. 89 (2) (2008) 237–250.

[33] C.E. Lovelock, D.A. Friess, K.W. Krauss, The vulnerability of Indo-Pacific mangrove forests to sea-level rise, Nature 526 (2015) 559–563.

[34] M.D.P. Godoy, L.D.D.E. Lacerda, Mangroves response to climate change: a review of recent findings on mangrove extension and distribution, Ann. Braz. Acad. Sci. 87 (2015) 651–667.

[35] M.C. Badjeck, E.H. Allison, A.S. Halls, N.K. Dulvy, Impacts of climate variability and change on fishery-based livelihoods, Mar. Policy 34 (3) (2010) 375–383. Available from: https://doi.org/10.1016/j.marpol.2009.08.007.

[36] J.C. Ellison, Vulnerability assessment of mangroves to climate change and sea-level rise impacts, Wetl. Ecol. Manage. 23 (2) (2015) 115–137.

[37] J.L. Anderson, C.M. Anderson, J. Chu, J. Meredith, F. Asche, G. Sylvia, et al., The fishery performance indicators: a management tool for triple bottom line outcomes, PLoS One 10 (5) (2015) 1–20.

[38] R.D. Ward, D.A. Friess, R.H. Day, R.A. MacKenzie, Impacts of climate change on mangrove ecosystems: a region by region overview, Ecosyst. Heal. Sustain. 2 (4) (2016). Available from: http://doi.wiley.com/10.1002/ehs2.1211.

[39] S.D. Sasmito, D. Murdiyarso, D.A. Friess, S. Kurnianto, Can mangroves keep pace with contemporary sea level rise? A global data review, Wetl. Ecol. Manage. 24 (2) (2016) 263–278.

[40] T. Spencer, M. Schuerch, R.J. Nicholls, J. Hinkel, D. Lincke, A.T. Vafeidis, et al., Global coastal wetland change under sea-level rise and related stresses: the DIVA Wetland Change Model, Global Planet Change 139 (2016) 15–30. Available from: https://doi.org/10.1016/j.gloplacha.2015.12.018.

[41] P. Shearman, J. Bryan, J.P. Walsh, Trends in deltaic change over three decades in the Asia-Pacific Region, J. Coastal Res. 290 (2013) 1169–1183. Available from: http://www.bioone.org/doi/abs/10.2112/JCOASTRES-D-12-00120.1.

[42] A. Raha, S. Das, K. Banerjee, A. Mitra, Climate change impacts on Indian Sunderbans: a time series analysis (1924-2008), Biodivers Conserv. 21 (5) (2012) 1289–1307.

[43] O.O. Omo-Irabor, S.B. Olobaniyi, J. Akunna, V. Venus, J.M. Maina, C. Paradzayi, Mangrove vulnerability modelling in parts of Western Niger Delta, Nigeria using satellite images, GIS techniques and spatial multi-criteria analysis (SMCA), Environ. Monit. Assess. 178 (1–4) (2011) 39–51.

[44] IPCC, Climate Change 2013: The Physical Sciences Basis, University Press, Cambridge, New York, 2013.

[45] V.W.Y. Lam, W.W.L. Cheung, G. Reygondeau, U. Rashid Sumaila, Projected change in global fisheries revenues under climate change, Sci. Rep. 6 (2016) 6–13. Available from: https://doi.org/10.1038/srep32607.

[46] L. Comte, J.D. Olden, Climatic vulnerability of the world's freshwater and marine fishes, Nat. Clim. Change 7 (10) (2017) 718–722.
[47] J.E. Cinner, T.R. McClanahan, N.A.J. Graham, T.M. Daw, J. Maina, S.M. Stead, et al., Vulnerability of coastal communities to key impacts of climate change on coral reef fisheries, Global Environ. Change 22 (1) (2012) 12–20. Available from: https://doi.org/10.1016/j.gloenvcha.2011.09.018.
[48] M. Sheaves, R. Baker, I. Nagelkerken, R.M. Connolly, True value of estuarine and coastal nurseries for fish: incorporating complexity and dynamics, Estuar. Coasts 38 (2) (2014) 401–414.
[49] B.B. Walters, P. Rönnbäck, J.M. Kovacs, B. Crona, S.A. Hussain, R. Badola, et al., Ethnobiology, socio-economics and management of mangrove forests: a review, Aquat. Bot. 89 (2) (2008) 220–236.
[50] D.D. Miller, Y. Ota, U.R. Sumaila, A.M. Cisneros-Montemayor, W.W.L. Cheung, Adaptation strategies to climate change in marine systems, Global Change Biol. 24 (1) (2018) e1–e14.
[51] R. Badola, S. Barthwal, S.A. Hussain, Attitudes of local communities towards conservation of mangrove forests: a case study from the east coast of India, Estuar. Coastal Shelf Sci. 96 (1) (2012) 188–196. Available from: https://doi.org/10.1016/j.ecss.2011.11.016.

CHAPTER 28

Ocean policy on the water—incorporating fishers' perspectives and values

Julia G. Mason

Hopkins Marine Station, Stanford University, Pacific Grove, CA, United States

Chapter Outline

As the profound and complex impacts of climate change on marine and coastal ecosystems in turn affect human communities, conservation scientists predicting future oceans have been encouraged to consider social dimensions, particularly those pertaining to fishing communities. Understanding fishers' needs, concerns, and values may make research both more accurate and more actionable [1,2], and conservation interventions more effective and equitable [3,4]. For example, a top-down, non-participatory conservation intervention engendered anticonservation attitudes and even violent conflict between fishers and managers in Brazil [5], whereas a comanagement approach for a marine park in the Comoros Islands fostered stewardship, with the community initiating a second park [6]. In this chapter I will argue for the need for research that centers on fishers' perspectives and outline approaches that conservation scientists hoping to engage fishing communities in their research may find useful. While most relevant for individuals trained in natural or physical sciences, these approaches are aimed to be broad enough that they may be applicable for anyone working across disciplines or across sectors.

28.1 The value of values

Currently, most studies of people's ability to adapt to change tend toward high-level, globally-relevant characterizations of external determinants of adaptation. These may

Predicting Future Oceans.
DOI: https://doi.org/10.1016/B978-0-12-817945-1.00030-7

include qualities of the community that affect its vulnerability to threats (e.g., [7]) or access to assets and/or attributes that make them more or less resilient [8,9]. While there is much value in delineating broad adaptation strategies, challenges, and enabling conditions for framing research and conceptualizing policy options [8,10], rarely have these frameworks resulted in tangible policy outcomes [11]. Increasingly scholars in sociology and human geography entreat us to consider internal, individual determinants as well, arguing that what more immediately drives or constrains adaptation has to do with an individual's agency, worldview, and values, including the cultural factors and power dynamics that shape them [12–15]. For a given community, or individual therein, high-level characterization of capacity and adaptation pathways is likely less relevant than questions of how they perceive these changes, whether they want to respond and prioritize responding, what specific cultural or institutional contexts may promote or inhibit action, what values and goals dictate that adaptation, and how the benefits of that adaptation are distributed within the community [12–14,16]. Coulthard [13] gives the example of alternative livelihoods offered to fishers as an adaptation strategy. If said fisher doesn't perceive that livelihood and associated lifestyle as valuable, they are likely to keep fishing, likely counter to the goals and expectations of those offering said alternatives.

Both these external and internal determinants interact to shape fishers' perceptions and priorities for action in the face of change. Researchers in this group have anecdotally identified additional factors that may preoccupy fishers—and thus drive their behaviors—much more than large-scale climatic changes, including individual day-to-day concerns, such as the availability of ice and access to loans, and broader community issues, including local politics or drug trafficking. This phenomenon is applicable to general fisheries management as well: fishers are unlikely to comply with management that they deem impractical [17], illegitimate [18], unfair [19], or counter to their perceptions or knowledge of the system [2]. Inadequate consideration of the on-the-ground factors that fishers or community members prioritize and value has been identified as a key challenge driving disconnects between mitigation and adaptation frameworks and operational policies [20]. Nonetheless, management decisions and research results that do not adequately consult or include fishers' perspectives remain the norm [21], resulting in ineffective or inequitable policy decisions [2]. This is partly because values and perceptions are inherently subjective and thus often dismissed by the scientific community [20]. The attitude that fishers' knowledge is unscientific and therefore inferior or untrustworthy remains pervasive in the scientific literature [22].

How might we better incorporate fishers' perceptions and values? One way forward is more participation in research and in policy processes. Typologies of participation have been reviewed elsewhere [23,24], and the level and type of participation will depend on the context and goals. For researchers, methods for eliciting perceptions and values can be open and unstructured, allowing participants to suggest values through interviews, workshops, or focus groups [telling]; physical interactive processes like mapping and photovoice [showing] [25–27]; more directed ranking exercises of predetermined options [choosing] [28,29]; or experimental, through scenarios or games [revealing] [30,31] (for more tools, see [2,32]).

Nevertheless, as this researcher discovered, merely expanding the toolbox is insufficient—training and insight is still needed in how to use these tools and interpret their results. Given these challenges, I outline below some simple approaches that natural science trained conservation scientists hoping to engage fishers in their research may find useful, regardless of the particular method used. These approaches are not comprehensive, and share elements with other frameworks relating to stakeholder engagement and participatory research [24,33–35]. Where applicable I provide concrete examples of how these principles may be enacted, but stress that there is no one right way to do it.

28.2 Approaches to participatory research

These lessons are primarily drawn from my experiences fielding a quantitative survey with fishers, seafood industry members, and marine and coastal policymakers in the United States, in which I asked them to prioritize a series of ocean research questions, to compare their responses to a previously published survey of scientists [29]. The feedback I received from commercial fishers via comments in the survey and external emails and phone calls revealed key mistakes and oversight that a well-meaning scientist might make and spoke to deeper underlying issues with communication and values. Quotes from fishers responding to the survey will be presented below with original wording intact, followed by broader conclusions and lessons learned. Those conclusions have also been drawn from subsequent experiences engaging US fishers in informal interviews for public outreach, semistructured interviews and participatory mapping exercises with fishers in Peru, and surveys done with fishing communities in Peru, and supported with literature from other contexts whenever possible.

28.2.1 Start with trust

> "Very few scientists are honest. Trust is hard to earn in the real ocean."

Building trust is a critical first step for working with communities in any context, especially for participatory research [36] and comanagement [34,35]. Trust is key to participation in the process, quality of responses during the process, and acceptance of results and any policy outcomes at the end. Trust is often hard won and easily eroded [36], and how exactly to build and maintain it is not always apparent. As I learned with my survey, merely stating that you have no agenda or are not being funded by conservation agencies in the requisite consent language may be insufficient.

> "Is this legit or some ploy by someone we do not want to deal with."

In ideal circumstances, trust is built through long-term personal interactions. Some researchers recommend spending at least 2 weeks just "hanging out" in a community before attempting any interviews; several months to over a year may be necessary for covering sensitive issues [37]. For those researchers who may not have the time to invest in lugging crates or mending nets in order to engage other communities in their research, it is also possible to gain trust associatively through building a relationship with a trusted key informant or gatekeeper. In my work, these individuals have included fishers' associations leaders, government fisheries scientists, and fishery consultants or advocates. I connected with one such key informant by reaching out to an individual who had presented on behalf of a fishery I was studying in US fishery management council meetings, whom I found by reading his online blog. Because he vetted me and facilitated initial introductions, I was able to interview fishers who had had previous negative experiences with scientists and would have been unlikely to engage with me had I contacted them directly. In the absence of (or in addition to) such a key informant, employing local field assistants can also help build trust [38].

> "The questions are too academically worded. I won't have this coming out of my office. I just can't see sending this out [to the fleet]. I think it would do more harm than good."

Beyond personal relationships, other aspects of study design affect trust with respondents. Attention to inclusive wording, formats, and settings is key, as these aspects of study design may inadvertently reinforce power dynamics between researchers and interviewees, and among communities [24]. My online survey was a familiar task for scientists and policymakers, but not necessarily for fishers; in surveys in Peru, questions involving a Likert scale (ranking statements from strongly disagree to strongly agree), which was entirely new to my respondents, resulted in much unnecessary confusion and frustration. Behind the language and methods we use are values and worldviews that may be so familiar as to be invisible to us: although I had tried to remove jargon from my online survey where possible (I had to keep the wording of the research priority topics intact for comparability), I was surprised to be accused of promoting a conservationist agenda in what I thought were basic research questions. It had not occurred to me to examine the underlying values in my research questions and consider how they might alienate respondents. As a result a key gatekeeper, despite understanding my position and goals, ultimately refused to take the survey and discouraged his constituents from doing so as well.

Even more subtle aspects of the research design and approach can affect trust. Colleagues have suggested that Facebook can be a more effective way of engaging fishers than direct email, purchasing familiar pens and stationary in a foreign country might help make respondents more comfortable [39], and the comportment and dress of the researcher should be attuned to the particular context [37]. Running research questions and study design by key informants or a small subsample of respondents who can give feedback (depending on the analysis methods used, these beta testers may need to be excluded from the final sample) is a valuable way to ensure you and your questions will be received as intended and to catch potentially alienating language, assumptions, or other details.

> "... the survey pigeonholes you to positions ... that have not been agreed to or accepted by fishermen ... a tremendous amount [of] hubris or assumptions that preconceived positions are universally accepted by all."

28.2.2 Listen

Scientists are increasingly exhorted to become better communicators, but in engaging communities, their most important job is listening. An attitude of humility is a crucial but oft overlooked tool in the researcher's kit [36]. Emphasizing that you as the researcher are there to learn from the respondents is key to generating dialogue, and also works to build trust [37]. Allowing space for fishers to share their knowledge, perceptions, concerns, and values—as equal partners in knowledge production [34]—may fill gaps or address oversights in scientific conclusions, averting potentially misguided policy decisions [1]. In addition to painting a more realistic picture of adaptation responses and options, listening to fishers' values may foster more engagement and acceptance of results or policy suggestions. Research suggests that in controversial settings, people are able to accept and process new information only once they feel their values have been acknowledged [20,40]. Explicit attention to values may ameliorate conflict by reframing solutions to speak to shared values [41], or at least allowing for more honest and transparent discussion of the inevitable trade-offs that arise in the face of changing oceans [20].

Like trust, listening takes time. Carving out space during or ideally before directed research for more informal and unstructured interviewing techniques that allow the respondent to drive the conversation, free from the tyranny of randomized sampling and prewritten questionnaires [1], is a good place to start [37]. For researchers pressed for time or quantitative techniques, it helps to add the question at the beginning or end of an interview (I prefer end), "Do you have any questions or comments for me?" [39]. The resulting discussions are often surprising and (almost) always worthwhile.

> "I'm a fisherman all my income comes from fishing and feel that over regulation has effected [sic] my life ... the science is always two years behind what is happening at sea I think their out to put small boat fishermen out like the small farmers."

28.2.3 Integrate and iterate

Researchers must be flexible and adaptive for the inevitable changes to questions, hypotheses, and methods that may arise after all this trust building, listening, and experiencing what's happening on the ground or on the water [42]. Integrating what you learn, sharing, and seeking feedback throughout the process further improves the relevance and value of your research, while reinforcing trust. At minimum, reporting back to any participating communities at the end of your research is an act of courtesy and respect, in formats tailored to the specific audience [39]. In Peru, I used this reporting back to ask the group of fishers what they thought of my conclusions and if there was anything I had missed. Reed et al. [34] suggest sharing deliverables with stakeholders as early and often as possible, even brief summary reports or new data, to get feedback and ensure accountability.

Over the longer term, sustained dialogue and collaboration strengthen trust and allow more opportunities for perspectives and values to surface. As referenced in the quote above, science may move on a different timescale than the needs of fishers, or the needs of policymakers, so iterative work can keep researchers up to date on what fishers are experiencing out on the water and identify emerging priorities. Long-term collaboration and communication may better situate scientists and communities to take advantage of "policy windows" when they arise [43]. Finally, iterative work is necessary to continuously reassess outcomes and goals as communities adjust their values and priorities in a changing environment [14].

28.3 Conclusion

To understand how climate change affects coastal communities we invariably must grapple with people's behavior, perceptions, and values. Working with communities to elicit these perceptions and values may be time-consuming and poorly rewarded in the natural sciences, but is nonetheless critical for more realistic understanding of outcomes, goodwill and collaboration between science and society, and effective and equitable ocean policies. It may also draw out subtle underlying assumptions in scientists' research rationale—those of us engaged in research to predict future oceans would do well to think critically about who we are predicting for, and why. These approaches and lessons are a starting point that may help foster transdisciplinary and collaborative work for our future oceans.

References

[1] R.E. Johannes, M.M.R. Freeman, R.J. Hamilton, Ignore fishers' knowledge and miss the boat, Fish Fish. 1 (2000) 257–271.

[2] N.J. Bennett, Using perceptions as evidence to improve conservation and environmental management, Conserv. Biol. (2016). Available from: https://doi.org/10.1111/cobi.12681.

[3] E. Le Cornu, J.N. Kittinger, J.Z. Koehn, E.M. Finkbeiner, L.B. Crowder, Current practice and future prospects for social data in coastal and ocean planning, Conserv. Biol.: Soc. Conserv. Biol. 28 (4) (2014) 902–911. Available from: https://doi.org/10.1111/cobi.12310.

[4] N.C. Ban, M. Mills, J. Tam, C.C. Hicks, S. Klain, N. Stoeckl, et al., A Social – Ecological Approach to Conservation Planning: Embedding Social Considerations, 2013. <https://doi.org/10.1890/110205>.

[5] T. Almudi, D.C. Kalikoski, Traditional fisherfolk and no-take protected areas: the Peixe Lagoon National Park dilemma, Ocean Coastal Manage. 53 (5–6) (2010) 225–233. Available from: https://doi.org/10.1016/J.OCECOAMAN.2010.04.005.

[6] E.E. Granek, M.A. Brown, Co-management approach to marine conservation in Mohéli, Comoros Islands, Conserv. Biol. 19 (6) (2005) 1724–1732. Available from: https://doi.org/10.1111/j.1523-1739.2005.00301.x.

[7] B.L. Turner, R.E. Kasperson, P.A. Matson, J.J. McCarthy, R.W. Corell, L. Christensen, et al., A framework for vulnerability analysis in sustainability science, Proc. Natl. Acad. Sci. U.S.A. 100 (14) (2003) 8074–8079. Available from: https://doi.org/10.1073/pnas.1231335100.

[8] J.E. Cinner, W.N. Adger, E.H. Allison, M.L. Barnes, K. Brown, P.J. Cohen, et al., Building adaptive capacity to climate change in tropical coastal communities, Nat. Clim. Change 8 (2018). Available from: https://doi.org/10.1038/s41558-017-0065-x.

[9] G. Yohe, R.S.J. Tol, Indicators for social and economic coping capacity—moving toward a working definition of adaptive capacity, Global Environ. Change 12 (1) (2002) 25–40. Available from: https://doi.org/10.1016/S0959-3780(01)00026-7.

[10] D.D. Miller, Y. Ota, U.R. Sumaila, A.M. Cisneros-Montemayor, W.W.L. Cheung, Adaptation strategies to climate change in marine systems, Global Change Biol. 24 (1) (2018) e1–e14. Available from: https://doi.org/10.1111/gcb.13829.

[11] C.K. Whitney, N.J. Bennett, N.C. Ban, E.H. Allison, D. Armitage, J.L. Blythe, et al., Adaptive capacity: from assessment to action in coastal social-ecological systems, Ecol. Soc. 22 (2) (2017). Available from: https://doi.org/10.5751/ES-09325-220222.

[12] W.N. Adger, S. Dessai, M. Goulden, M. Hulme, I. Lorenzoni, D.R. Nelson, et al., Are there social limits to adaptation to climate change? Clim. Change 93 (93) (2009). Available from: https://doi.org/10.1007/s10584-008-9520-z.

[13] S. Coulthard, Can We Be Both Resilient and Well, and What Choices Do People Have? Incorporating Agency into the Resilience Debate from a Fisheries Perspective, 2012. Available from: https://doi.org/10.5751/ES-04483-170104.

[14] K.L. O'Brien, Do values subjectively define the limits to climate change adaptation? in: W.N. Adger, I. Lorenzoni, K.L. O'Brien (Eds.), Adapting to Climate Change: Thresholds, Values, Governance, 180, Cambridge University Press, Cambridge, 2009, p. 164.

[15] K. Brown, E. Westaway, Agency, capacity, and resilience to environmental change: lessons from human development, well-being, and disasters, Annu. Rev. Environ. Resour. 36 (2011) 321–342. Available from: https://doi.org/10.1146/annurev-environ-052610-092905.

[16] S. Burch, J. Robinson, A framework for explaining the links between capacity and action in response to global climate change, Clim. Policy 7 (4) (2007) 304–316. Available from: https://doi.org/10.1080/14693062.2007.9685658.

[17] M.L. Miller, J. Van Maanen, "Boats don't fish, people do": some ethnographic notes on the federal management of fisheries in Gloucester, Hum. Organ. 38 (4) (1979) 377–385.

[18] S. Jentoft, Legitimacy and disappointment in fisheries management, Mar. Policy 24 (2) (2000) 141–148. Available from: https://doi.org/10.1016/S0308-597X(99)00025-1.

[19] M. Fabinyi, S. Foale, M. Macintyre, Managing inequality or managing stocks? An ethnographic perspective on the governance of small-scale fisheries, Fish Fish. (2013) 1–15. Available from: https://doi.org/10.1111/faf.12069.

[20] K.L. O'Brien, J. Wolf, A values-based approach to vulnerability and adaptation to climate change, Ltd. WIREs Clim. Change 1 (2010) 232–242. Available from: https://doi.org/10.1002/wcc.30.

[21] E.J. Hind, A review of the past, the present, and the future of fishers' knowledge Research : a challenge to established fisheries science, ICES J. Mar. Sci. 72 (2014) 341–358. Available from: https://doi.org/10.1093/icesjms/fsu169.

[22] C.G. Soto, Socio-cultural barriers to applying fishers' knowledge in fisheries management: an evaluation of literature cases. Retrieved from <http://research.rem.sfu.ca/theses/SotoCristina_2006>, 2006.

[23] S.R. Arnstein, A ladder of citizen participation, J. Am. Inst. Plann. 35 (4) (1969) 216–224. Available from: https://doi.org/10.1080/01944366908977225.

[24] M.S. Reed, Stakeholder participation for environmental management: a literature review, Biol. Conserv. 141 (10) (2008) 2417–2431. Available from: https://doi.org/10.1016/J.BIOCON.2008.07.014.

[25] A.S. Levine, C.L. Feinholz, Participatory GIS to inform coral reef ecosystem management: Mapping human coastal and ocean uses in Hawaii, Appl. Geogr. 59 (2015) 60–69. Available from: https://doi.org/10.1016/J.APGEOG.2014.12.004.

[26] N.J. Bennett, P. Dearden, A picture of change: using photovoice to explore social and environmental change in coastal communities on the Andaman Coast of Thailand, Local Environ. 18 (9) (2013) 983–1001. Available from: https://doi.org/10.1080/13549839.2012.748733.

[27] L.C.L. Teh, L.S.L. Teh, M.J. Meitner, Preferred resource spaces and fisher flexibility: implications for spatial management of small-scale fisheries, Hum. Ecol. 40 (2012) 213–226.

[28] C.C. Hicks, N. a J. Graham, J.E. Cinner, Synergies and tradeoffs in how managers, scientists, and fishers value coral reef ecosystem services, Global Environ. Change 23 (6) (2013) 1444–1453. Available from: https://doi.org/10.1016/j.gloenvcha.2013.07.028.

[29] J.G. Mason, M.A. Rudd, L.B. Crowder, Ocean research priorities: similarities and differences among scientists, policymakers, and fishermen in the United States, BioScience 67 (5) (2017) 418–428. Available from: https://doi.org/10.1093/biosci/biw172.

[30] T.M. Daw, S. Coulthard, W.W.L. Cheung, K. Brown, C. Abunge, D. Galafassi, et al., Evaluating taboo trade-offs in ecosystems services and human well-being, Proc. Natl. Acad. Sci. U.S.A. (2) (2015) 201414900. Available from: https://doi.org/10.1073/pnas.1414900112.

[31] E.M. Finkbeiner, F. Micheli, A. Saenz-Arroyo, L. Vazquez-Vera, C.A. Perafan, J.C. Cárdenas, Local response to global uncertainty: Insights from experimental economics in small-scale fisheries, Global Environ. Change 48 (2018) 151–157. Available from: https://doi.org/10.1016/j.gloenvcha.2017.11.010 (May 2017).

[32] T. Lynam, W. de Jong, D. Sheil, T. Kusumanto, K. Evans, A review of tools for incorporating community knowledge, preferences, and values into decision making in natural resources management, Ecol. Soc. (2007). Available from: https://doi.org/5.

[33] L.A. Mease, A. Erickson, C. Hicks, Engagement takes a (fishing) village to manage a resource: principles and practice of effective stakeholder engagement, J. Environ. Manage. 212 (2018) 248–257. Available from: https://doi.org/10.1016/J.JENVMAN.2018.02.015.

[34] M.S. Reed, L.C. Stringer, I. Fazey, A.C. Evely, J.H.J. Kruijsen, Five principles for the practice of knowledge exchange in environmental management, J. Environ. Manage. 146 (2014) 337–345. Available from: https://doi.org/10.1016/j.jenvman.2014.07.021.

[35] M. Trimble, F. Berkes, Participatory research towards co-management: lessons from artisanal fisheries in coastal Uruguay, J. Environ. Manage. 128 (2013) 768–778. Available from: https://doi.org/10.1016/J.JENVMAN.2013.06.032.

[36] M. Cargo, S.L. Mercer, The value and challenges of participatory research: strengthening its practice, Annu. Rev. Public Health 29 (2008) 325–350. Available from: https://doi.org/10.1146/annurev.publhealth.29.091307.083824.

[37] H.R. Bernard, Research Methods in Anthropology: Qualitative and Quantitative Approaches, fourth ed, AltaMira Press, Oxford, 2006. Available from: https://doi.org/10.1017/CBO9781107415324.004.

[38] E.M. Finkbeiner, Survival and Sustainability in Small-Scale Mexican Fisheries: A Cross-Scale Examination of Resilience in Marine Social-Ecological Systems, Stanford University, 2014.

[39] C.B. Barrett, J.W. Cason, Overseas research II: a practical guide, in: Overseas Research II: A Practical Guide, 2010. Available from: https://doi.org/10.4324/9780203856277.

[40] D.M. Kahan, D. Braman, Cultural cognition and public policy, Yale Law Policy Rev. 24 (1) (2006) 149–172.

[41] R.A. Pielke, The Honest Broker: Making Sense of Science in Policy and Politics, 2007. Available from: https://doi.org/10.1017/CBO9780511818110.

[42] S. Hertel, M.M. Singer, D.L. Van Cott, Field research in developing countries: hitting the road running, PS – Political Science and Politics, vol. 42, Sage Publications, 2009, pp. 305–309. Available from: https://doi.org/10.1017/S1049096509090611.

[43] D.C. Rose, N. Mukherjee, B.I. Simmons, E.R. Tew, R.J. Robertson, A.B.M. Vadrot, et al., Policy windows for the environment: tips for improving the uptake of scientific knowledge, Environ. Sci. Policy (2017). Available from: https://doi.org/10.1016/J.ENVSCI.2017.07.013.

CHAPTER 29

Integration of traditional knowledge in policy for climate adaptation, displacement and migration in the Pacific

Marjo Vierros[1] **and Yoshitaka Ota**[2]

[1]Nippon Foundation Nereus Program, Institute for the Oceans and Fisheries, AERL, University of British Columbia, Vancouver, BC, Canada [2]Nippon Foundation Nereus Program, School of Marine and Environmental Affairs, University of Washington, Seattle, WA, United States

Chapter Outline

29.1 Traditional knowledge in ocean and climate change policy

This chapter addresses the role of traditional knowledge in assisting coastal communities in Pacific Island countries in responding to the impacts of climate change through the development of adaptation strategies and the maintenance of cultural integrity. The chapter considers the way in which traditional knowledge itself could adapt and keep pace with the forecasted rapid changes, including in the event of displacement and migration. Ultimately, the chapter aims to emphasize the importance of co-creation of knowledge, recognizing the value of both scientific and traditional knowledge equally, as the future policy direction for Pacific Islands. In doing so, the chapter acknowledges that the relationship between island nations and their marine environment is deeply cultural, and goes beyond viewing the ocean as simply a provider of food and other services.

Consideration of traditional knowledge in policy is not new, and is a key part of various international conventions, including in particular the Convention on Biological Diversity and its Article 8(j) on Traditional Knowledge, Innovations and Practices, as well as the

Predicting Future Oceans.
DOI: https://doi.org/10.1016/B978-0-12-817945-1.00031-9

National Biodiversity Strategies and Action Plans of most Pacific Island countries. Various other policy processes, both global and regional, such as the Arctic Council, the Pacific Regional Ocean Policy, the Small Island Developing States Accelerated Modalities of Action (Samoa Pathway), and the Sendai Framework for Disaster Risk Reduction provide for Indigenous participation and acknowledge the importance of both traditional knowledge and traditional systems of stewardship. Indigenous groups are actively participating in these policy processes, and have, in the context of Agenda 2030, proposed that traditional knowledge should be considered "on an equal footing with science and other knowledge systems for 21st century solutions to contemporary crises" [1], as well as in the implementation of Sustainable Development Goal (SDG) 14 on oceans [2].

Traditional knowledge has been considered, but in a relatively limited way, in the reports of the Intergovernmental Panel on Climate Change, resulting in calls for greater involvement of Indigenous authors [3]. In contrast, the Arctic Council also has long incorporated traditional knowledge into its work, for example, through the 2004 Arctic Climate Impact Assessment, the Arctic Shipping Assessment, and the Arctic Biodiversity Assessment. The Arctic Council could also be upheld as a model of inclusivity, where Indigenous Peoples organizations are able to participate equally with government representatives.

The following table provides examples of the ways in which traditional knowledge has been incorporated into policy processes.

Policy process	How incorporated
CBD	• Article 8(j) on Traditional Knowledge, Innovations and Practices • Working Group on implementation of Article 8(j) and related provisions • Traditional knowledge considered a cross-cutting issue in CBD implementation • -Indigenous Peoples participate in CBD meetings as observers
Agenda 2030 and SDGs	• Indigenous Peoples' needs and involvement reflected in the Agenda, including implementation, monitoring and review • Indigenous Peoples specifically mentioned in the context of SDGs 2 and 4
Sendai Framework for Disaster Risk Reduction	• Indigenous Peoples' role in development and implementation of disaster risk reduction plans acknowledged • Traditional knowledge seen as part of disaster risk reduction, together with scientific knowledge
Arctic Council	• Indigenous Peoples' organizations have Permanent Participants status, and have full consultation rights in connection to the Council's negotiations and decisions[a] • Traditional knowledge is incorporated into Arctic Council assessments, including on climate change and biodiversity
Pacific Islands Regional Ocean Policy and Framework for Pacific Oceanscape	"Throughout the region, customary association with the sea forms the basis of present day social structures, livelihoods and tenure systems and traditional systems of stewardship governing its use"

(Continued)

(Continued)

Policy process	How incorporated
Small Island Developing States Accelerated Modalities of Action (Samoa Pathway)	Recognizes: • Role of communities, customary resource owners, and customary practices • Role of traditional knowledge in better understanding the ocean • The need to protect traditional knowledge from exploitation • Vulnerabilities due to cultural dilution and loss of traditional knowledge and practices • Recognizes that indigenous and traditional knowledge and cultural expression, which underscores the deep connections among people, culture, knowledge, and the natural environment, can meaningfully advance sustainable development and social cohesion • Promotes actions to conserve, promote, protect, and preserve traditional knowledge • Reaffirms effective participation of indigenous peoples in climate action

CBD, Convention on biological diversity; *SDGs*, sustainable development goals.
[a]https://arctic-council.org/index.php/en/about-us/permanent-participants.

29.2 Traditional knowledge under a changing climate

It has been argued that traditional knowledge could assist communities in preparing to adapt to the impacts of climate change [4–6]. This argument is increasingly accepted in international policy, as described in the table above, but has not yet been fully applied in practice through climate adaptation strategies nationally and locally [7]. The following sections discuss the role of traditional knowledge in building local capacity for adaptation and mitigation; and its relevance in the event of migration. Finally, the importance of traditional knowledge to the cultural identity of a group or people, as well as policy responses, is discussed.

29.2.1 Traditional knowledge in climate change adaptation and mitigation

Observations of Indigenous Peoples and local communities about climate change and its impacts can become the basis for developing local adaptation strategies. Such adaptation strategies have long been practiced as a response to cyclones, drought, and other environmental disasters that may wipe out the food supplies of a village. In Vanuatu, for example, villagers might prepare special "famine foods" that were long lasting and were stored for use in a time of need. Methods for overcoming food shortages included storing fermented fruits and utilizing alternative foods not normally eaten [8,9].

Another strategy was to create "giant clam gardens," with fishers gathering giant clams into discrete areas on reef flats for their exclusive use in times of need. This also served to increase the reproductive success of clams by maintaining a close proximity of a breeding population dependent on external fertilization [8]. In addition, traditional building methods designed to weather extreme events are common in the Pacific Islands. In many villages in Vanuatu, family units build a sturdy "cyclone house" further inland, where they retreat in the event of a serious storm. In Samoa, the traditional *fale tele* is mounted on a high stone foundation to protect against flooding and storm surges [10,11]. Such strategies are a common way of distributing environmental risk, and they generally include traditional agriculture systems that enhance diversity and prevent erosion, scattering food production sites, and shifting target species and catch amounts in fisheries [12].

Traditional area-based marine management systems can also increase the food security of communities, and thus their ability to adapt to the impacts of climate change. Customary fisheries management practices in the Pacific have long included seasonal bans on harvesting, temporary closed (no-take) areas, and restrictions on time, places, and species or taking by certain classes of persons. Closed areas include the tabu areas of Fiji, Vanuatu, and Kiribati, the ra'ui in Cook Islands, the masalai in Papua New Guinea, and the bul in Palau [13]. Contemporary village-based management prohibitions continue to be locally monitored and enforced by village leaders [8].

The potential for scaling up such systems in a contemporary context is demonstrated by the resurgence of local marine managed areas in Pacific Island countries. In Fiji, for example, locally managed marine areas (LMMAs) now involve over 400 communities, and have provided a range of fisheries and livelihoods benefits [14], though their ability to produce such benefits depends on the effectiveness of monitoring and enforcement locally [15]. Villages have seen direct benefits in increased fisheries catches. For example, by imposing a closed, *tabu* area around a mangrove island, Sawa (Fiji) villagers found that the numbers of the mangrove lobster (*Thalassina anomala)* increased by roughly 250% annually, with a spillover effect of roughly 120% outside the *tabu* area. The increase in fishery resources not only improves nutrition but also raises household income from market sales. Marine resources, on average, make up more than 50% of the household income for these villages, raising it far above the median income level of F$4000 a year in Fiji [16].

LMMAs provide for the application of both traditional knowledge and science, and for community control of resource management, thus avoiding many of the social problems of government-led MPAs that have, in some cases, included restricting local livelihoods [17] and transferring control of local resources to a central government [18]. While government-led MPAs can deliver considerable biodiversity and fisheries benefits, their failure in some cases to take into account social consequences of limited participation and inequitable sharing of costs and benefits of conservation have led to them being criticized as "biological successes and social failures" [19]. The early successes of LMMAs have resulted in their spread throughout the Pacific Islands and into the Indian Ocean, with an estimated 30,000 km^2 of the Pacific covered by marine managed areas in 2009 [14,20,21].

LMMAs have now become part of Pacific Islands' strategy for reaching policy goals such as the Samoa Pathway and SDG 14, as demonstrated by the registering of more than 10 United Nations Ocean Conference voluntary commitments relating either directly or indirectly to LMMAs.[1]

These examples demonstrate some of the ways in which Indigenous Peoples and local communities have used their traditional knowledge to increase their resilience to environmental change, including climate variability. While climate adaptation strategies will likely utilize diverse approaches, including ones based on traditional knowledge and others based on contemporary science, solutions that are inclusive of communities and their knowledge are more likely to achieve both the desired social and ecological outcomes [22].

Pacific Islanders are no strangers to sudden and often catastrophic environmental changes, and have a long history of learning how to adapt and survive in the face of this change. However, the impacts of future climate change are expected to be unprecedented, and could rapidly alter marine and coastal ecosystems. It is not yet known whether traditional knowledge is able to evolve fast enough to keep up with the coming change. At the same time, the ontological perspective of communities would also be reshaped by external influences such as economy and politics that are under flux due to the environmental change. While traditional knowledge provides communities with a rich source of adaptation strategies, these strategies will be tested by economic, social, and political upheavals, which may cause deterioration of societal structures. Changing societies may affect both the transmission of knowledge, as well as activities, such as fishing, that are key for accumulating knowledge about the environment. While many open questions remain, the next section on migration will start exploring some of these questions.

29.2.2 Traditional knowledge in displacement and migration

Place-based adaptation is the best-case response scenario to climate change. Some estimates have indicated that between 665,000 and 1.7 million people in the Pacific alone could be displaced or forced to migrate by 2050 because of rising sea levels associated with climate change [23]. While these figures are contested (see, for example, [24]), migration related to climate impacts is already taking place, for example, in the Carteret Islands of Papua New Guinea and away from low-lying coral atolls of countries such as Kiribati. The impacts of sea level rise are made worse by additional climate-related impacts, such as ocean warming, deoxygenation, and acidification, and by other human pressures on the coastal environment. In addition, while the impacts of climate change result in changing ecological conditions, the outcomes are further affected by social and political economic circumstances on the

[1] See registry of voluntary commitments at https://oceanconference.un.org/commitments/.

ground. Thus, the impacts of climate change are difficult to separate from the multiple social, political, economic, and environmental changes confronting present-day Indigenous communities, which interact with often cumulative or cascading effects. Where environmental degradation takes place slowly, it is difficult to draw a line between what constitutes "economic migration" (presumed to be voluntary) and what constitutes "displacement" (presumed to be involuntary or forced) [23]. Opportunities and vulnerabilities are unevenly distributed in society. Those that are able to do so may leave in anticipation of coming change and in search of new opportunities, while others may wait until there are no other options. Some may not have the means to migrate or may not be able to do so, and may be left behind.

Case studies relating to government-mandated forced relocations in the United States from the 1700s to the 20th century show that these relocations limited the ability of tribes to draw on their traditional knowledge, leaving them more vulnerable to changing weather patterns and climate impacts, and in many cases unable to undertake traditional subsistence practices [25]. Where support for communities is not in place, forced relocation has intensified community impoverishment, negative economic and health impacts, and loss of place, social networks, and culture [25,26]. In South Africa, the Indigenous communities removed from protected areas experienced an erosion of local institutions and fragmentation of communities, though the relocated people retained much of their traditional knowledge about plants, animals, and ecosystems and were able to apply it in their new location [27].

As Indigenous Peoples relinquish their traditional territories during relocation, their inherited knowledge of the land as passed on from older generations may also be lost. This is particularly the case when elders die and the younger generation adapts to their new situation, resulting in social changes and weakening older cultural forms [28]. Some of this change is already evident from economic migration. For example, in Niue and the Cook Islands large-scale migration has resulted in more islanders living in New Zealand than in the islands, yet the cultures of New Zealand-based islanders have not been wholly displaced, nor have the cultures of the islands themselves. However, some aspects of the cultures have weakened and changed through changes such as the replacement of many traditional food procurement strategies by supermarkets and local shops [29]. Traditional knowledge, too, will become devalued if not actively used as part of daily lives and livelihoods [28].

Similar impacts from migration have also been observed in Vanuatu. The village of Lamen Bay in Vanuatu has seen a relatively high participation in seasonal migration to New Zealand for employment due to limited prospects locally, which while increasing cash income has also increased vulnerability in the sending community. Returning migrants were more unwilling to hold on to traditional ways of life and have lost respect for the traditional governance structure within the community. This, in turn, has impacted decision-making effectiveness, collective action capacity, and cohesion, making it more difficult to undertake community projects or manage common property resources, such as fisheries [30]. When put in the context of climate change, these changes result in increased vulnerability, as a community's capacity to adapt to climate change or to manage environmental resources is linked with their ability to act collectively. Thus migration, particularly by one component of society (in this case mainly young males), can weaken the ability of communities to undertake environmental and other management actions that might help them adapt to climate change [31,32]. At the same time, it has been argued that returning migrants bring with them new skills, for example knowledge related to agricultural production, that may positively impact climate adaptation in their home communities [33].

Much will depend on specific circumstances of migration, for example, whether communities are able to plan their own relocation with support from government and whether they are able to select their preferred site [25]. Experiences with economic migration also show that maintaining community cohesion is important for preserving traditional knowledge, practices, and governance structures, and thus planned migration of entire communities to the same destination is more likely to prevent fragmentation. Documenting traditional knowledge and practices of the community prior to migration is also likely to be helpful, as was done in the Carteret Islands of Papua New Guinea [34]. This effort, a collaboration between a researcher and Tulele Peisa, a local NGO, demonstrated the importance of documenting traditions before they are further eroded or lost due to relocation, and as an encouragement for the community to reassess the value of its own traditional knowledge, its importance for adaptation, and the importance of transmitting this knowledge to the next generation [34]. Similarly, maintaining some contact with the ancestral land in order to allow people to visit their original community provides for a cultural connection [35]. And once relocation has taken place, community and cultural restoration, as well as traditional livelihood development, will likely be necessary for the community to be once again self-sustaining [25].

While migration is generally thought to have negative impacts on traditional knowledge, it should be noted that with specific cultural and historical contexts, in which the community took an autonomous initiative of relocation, migration can be part of an adaptation strategy to sustain peoples way of life. This is the case, for example, in Tuvalu [36,37]. In such places, migration can be considered as a part of cultural adaptation, given its historical experience and social memory of the past. It is often argued that migration within the same area or region in a country is likely less detrimental for traditional knowledge and cultural integrity than migration long distances off-island or to a different country. Nonetheless, there is a need to expand our understanding of migration and its impacts, including in the cases of managed migration that is carried out according to cultural needs and priorities [25].

29.3 Traditional knowledge as cultural identity and in policy

For decades, anthropological and ethnobiological studies have recognized that the cultural relationship of island nations to the marine environment is more than its utilization for food, transport and recreation [38–42]. These studies reveal that the profound interactions between peoples and their oceans cannot be encompassed within a worldview that treats nature and culture separately. The interactions between coastal communities and the marine environment are not limited to fisheries but they are embedded in social organization and cultural identity, including kinship networks and cosmological beliefs, and are practiced and developed through traditional knowledge [43,44].

These points are important for developing climate change adaptation strategies. Traditional knowledge and tenure systems can be both flexible in accommodating even unprecedented events, but can also leave communities inherently vulnerable while responding to challenges on a global scale [45]. While it is important for adaptation plans for coastal livelihood needs to incorporate key cultural attributes [46], including for example coastal stonework and other strategies, a balance needs to be found that also considers global forces of ecological and economic change.

Co-creation between traditional knowledge and science is thus important to develop response strategies that fully incorporate both the local and global. While the sociocultural provision of oceans to coastal communities may seem intangible and not easily measured through economic or ecological metrics, there have been hybrid studies undertaken that address the use of traditional knowledge and scientific studies on an equal basis for enhancing adaptation and resilience on the coast. Examples from the Pacific include studies relating to fisheries [47–49] monitoring populations for co-management [50] and climate adaptation [4]. The study by Janif et al. [4] found that both traditional and contemporary environmental knowledge co-exist in study communities in Fiji, and that strengthening traditional knowledge for future climate change strategies is likely to improve resilience, particularly in remote villages.

While desirable, knowledge co-creation is challenging in practice given epistemological differences between knowledge systems (see, for example, [51]). Traditional knowledge cannot be separated from its context, and unlike science it also includes cultural, moral, and spiritual components [52]. Treating traditional knowledge as data may devalue both the knowledge and the culture from which it originated. The better option is to involve traditional knowledge holders themselves in the process of knowledge co-creation, understanding both the divergences and common ground between knowledge systems, and acknowledging and addressing power imbalances between practitioners [51].

One example of involving traditional knowledge holders in knowledge co-creation consisted of combining Inuit knowledge and observations of sea ice conditions with scientific observations. In this case, Inuit observations provided a long location-specific historical time series based on traditional knowledge, while scientific observations provided data on broader spatial scales and shorter timescales, with the combined observations proving helpful to better understanding climate change and its impacts [53]. Another example was the use of fishers' traditional knowledge in mapping benthic habitats and creating a marine species registry in Kerala, India, where scientists worked together with fishers at sea to identify locations of specific features, such as coral reefs, and to identify marine species, some of which were new to science [54]. In both of these cases there was active and equal participation by both traditional knowledge holders and scientists.

At the present time, climate change adaptation policies take into account traditional knowledge and cultural survival in a limited way [6]. Yet it is evident from the above studies that adaptation policies and strategies that are based on communities' own knowledge and culture, and that empower communities to make their own decisions, are more likely to provide positive societal outcomes. In the Pacific, adopting this approach to policy making could include coastal adaptation strategies that are both based on and enhance communities' traditional knowledge and management systems. These systems could include the respectful co-creation/co-learning of knowledge to provide a way for scientists and traditional knowledge holders to collaborate and produce a new form of hybrid knowledge that would assist in designing place-based adaptation strategies. Combined, these approaches have the potential to safeguard the rights and property of the community regarding the use of knowledge [55,56].

The maintenance of traditional knowledge will require policies that support community self-determination, allowing them to make their own decisions regarding the most appropriate adaptation strategies to climate change [57]. Recognizing the role of traditional knowledge as part of this broader cultural interaction will assist in maintaining community identity in a time of change, and may allow for conservation of knowledge and further development of a cultural and social narrative that fosters community-based initiatives. With the recognition of the importance of traditional knowledge, we conclude with two policy suggestions: (1) adaptation strategies ought to be undertaken through participatory processes and according to communities' needs and priorities [25,36] and guided by a human rights framework; and (2) adaptation strategies are best implemented with appropriate policy, legislative, financial, and institutional support that takes into account the preservation of cultural identity and traditional knowledge, ensuring support for sociocultural institutions, and continuity of community rights to traditional activities and resources [25,26]. These suggestions should be integrated into strategies for reaching overarching policy goals, such as the United Nations SDGs, as well as national and local policies for biodiversity conservation, climate change adaptation, and sustainable development.

References

[1] IPMG, Indigenous Peoples Major Group Policy Brief on Sustainable Development Goals and Post-2015 Development Agenda: A Working Draft. Online at <https://sustainabledevelopment.un.org/content/documents/7036IPMG%20Policy%20Brief%20Working%20Draft%202015.pdf>, 2015.

[2] IPMG, Indigenous Peoples Major Group statement on SDG 14 at the United Nations High-Level Political Forum on Sustainable Development. Online at <https://www.indigenouspeoples-sdg.org/index.php/english/all-resources/ipmg-position-papers-and-publications/ipmg-statements-and-interventions/30-statement-of-ipmg-on-goal-14/file>, 2017.

[3] J.D. Ford, L. Cameron, J. Rubis, M. Maillet, D. Nakashima, A.C. Willox, et al., Including indigenous knowledge and experience in IPCC assessment reports, Nat. Clim. Change 6 (4) (2016) 349.

[4] S.Z. Janif, P.D. Nunn, P. Geraghty, W. Aalbersberg, F.R. Thomas, M. Camailakeba, Value of traditional oral narratives in building climate-change resilience: insights from rural communities in Fiji, Ecol. Soc. 21 (2) (2016) 7.

[5] J. Barnes, et al., Contribution of anthropology to the study of climate change, Nat. Clim. Change 3.6 (2013) 541–544.

[6] E.M. Finkbeiner, K.L.L. Oleson, J.N. Kittinger, Social Resilience in the Anthropocene Ocean, Conservation for the Anthropocene Ocean, Academic Press, 2017, pp. 89–106.

[7] T. Weir, L. Dovey, D. Orcherton, Social and cultural issues raised by climate change in Pacific Island countries: an overview, Reg. Environ. Change 17 (4) (2016) 1–12.

[8] F.R. Hickey, Traditional marine resource management in Vanuatu: acknowledging, supporting and strengthening indigenous management systems, in: SPC Traditional Marine Resource Management and Knowledge Information Bulletin, No. 20, 2006, pp. 11–23. Available from: http://www.spc.int/coastfish/News/Trad/20/Trad20_11_Hickey.pdf.

[9] F.R. Hickey, Traditional marine resource management in Vanuatu: world views in transformation, in: N. Haggan, B. Neis, I.G. Baird (Eds.), Knowledge in Fisheries Science and Management. Coastal Management Sourcebooks, 4, UNESCO, Paris, 2007, pp. 147–168.

[10] Government of Samoa, SAMOA: Post-Disaster Needs Assessment Cyclone Evan 2012, 2013.

[11] UNESCO, Traditional knowledge for adapting to climate change, in: Safeguarding Intangible Cultural Heritage in the Pacific, UNESCO Office for the Pacific States, Samoa, and International Information and Networking Centre for Intangible Cultural Heritage in the Asia-Pacific Region under the auspices of UNESCO (ICHCAP), 2013.

[12] K.G. McLean, D. Nakashima, Weathering Uncertainty Traditional Knowledge for Climate Change Assessment and Adaptation, United Nations University Traditional Knowledge Initiative, 2012.

[13] M. Vierros, A. Tawake, F. Hickey, A. Tiraa, R. Noa, Traditional Marine Management Areas of the Pacific in the Context of National and International Law and Policy, United Nations University, Darwin, Australia, 2010. Traditional Knowledge Initiative.

[14] H. Govan, A. Tawake, K. Tabunakawai, A. Jenkins, A. Lasgorceix, A.M. Schwarz, et al., Status and Potential of Locally-Managed Marine Areas in the South Pacific: Meeting Nature Conservation and Sustainable Livelihood Targets Through Wide-spread Implementation of LMMAs: Study Report, 2009.

[15] S.D. Jupiter, G. Epstein, N.C. Ban, S. Mangubhai, M. Fox, M. Cox, A social–ecological systems approach to assessing conservation and fisheries outcomes in Fijian locally managed marine areas, Soc. Nat. Resour. 30 (9) (2017) 1096–1111.

[16] W.G.L. Aalbersberg, A. Tawake, T. Parras, Village by village-recovering Fiji's coastal fisheries, UNDP, UNEP, WRI World Resources 2005 – The Wealth of the Poor: Managing Ecosystems to Fight Poverty, World Resources Institute, Washington, DC, 2005, pp. 144–152.

[17] E.J. Hind, M.C. Hiponia, T.S. Gray, From community-based to centralised national management—a wrong turning for the governance of the marine protected area in Apo Island, Philippines? Mar. Policy 34 (1) (2010) 54–62.

[18] K. Brondo, W. Laura, Garifuna Land Rights and ecotourism as economic development in Honduras' Cayos Cochinos Marine Protected Area, Ecol. Environ. Anthropol. 3 (1) (2007) 2–18.
[19] P. Christie, Marine protected areas as biological successes and social failures in Southeast Asia, in: American Fisheries Society Symposium, vol. 42, pp. 155–164, 2004.
[20] S.D. Jupiter, P.J. Cohen, R. Weeks, A. Tawake, H. Govan, Locally-managed marine areas: multiple objectives and diverse strategies, Pac. Conserv. Biol. 20 (2) (2014) 165–179.
[21] J. Veitayaki, B. Aalbersberg, A. Tawake, E. Rupeni, K. Tabunakawai, Mainstreaming Resource Conservation: The Fiji Locally Managed Marine Area Network and its Influence on National Policy Development, 2003.
[22] E. Le Cornu, A.N. Doerr, E.M. Finkbeiner, D. Gourlie, L.B. Crowder, Spatial management in small-scale fisheries: A potential approach for climate change adaptation in Pacific Islands, Mar. Policy 88 (2018) 350–358.
[23] E. Ferris, M.M. Cernea, D. Petz, On the front line of climate change and displacement – Learning from and with Pacific Island Countries. The Brookings Institution, London School of Economics Project on Internal Displacement, 2011, 42 pp.
[24] G. Bettini, Climate barbarians at the gate? A critique of apocalyptic narratives on 'climate refugees', Geoforum 45 (2013) 63–72.
[25] J.K. Maldonado, C. Shearer, R. Bronen, K. Peterson, H. Lazrus, The impact of climate change on tribal communities in the US: displacement, relocation, and human rights, Clim. Change 120 (3) (2013) 601–614.
[26] R. Bronen, Climate-induced community relocations: creating an adaptive governance framework based in human rights doctrine, NYU Rev. Law Soc. Change 35 (2011) 356–406.
[27] D. Chatty, M. Colchester (Eds.), Conservation & Mobile Indigenous Peoples: Displacement, Forced Settlement and Sustainable Development, Berghahn Books, 2002.
[28] J. McLean, Conservation and the impact of relocation on the Tharus of Chitwan, Nepal, HIMALAYA 19 (2) (1999) 8.
[29] W.N. Adger, J. Barnett, F.S. Chapin Iii, H. Ellemor, This must be the place: underrepresentation of identity and meaning in climate change decision-making, Global Environ. Polit. 11 (2) (2011) 1–25.
[30] L.K. Craven, Migration-affected change and vulnerability in rural Vanuatu, Asia Pac. Viewpoint 56 (2) (2015) 223–236.
[31] W.N. Adger, Social capital, collective action, and adaptation to climate change, Econ. Geogr. 79 (4) (2003) 387–404.
[32] A. Agrawal, Local institutions and adaptation to climate change, in: R. Mearns, A. Norton (Eds.), Social Dimensions of Climate Change: Equity and Vulnerability in a Warming World, The World Bank, Washington, DC, 2009, pp. 173–198.
[33] O. Dun, N. Klocker, L. Head, Recognising knowledge transfers in 'unskilled' and 'low-skilled' international migration: insights from Pacific Island seasonal workers in rural Australia, Asia Pac. Viewp. 59 (3) (2018) 276–292.
[34] F.R. Hickey, The Carteret Islands of Papua New Guinea – An Assessment of Traditional Resource Management Systems and the use of Traditional Knowledge. Vanuatu Cultural Centre. A Report Prepared for the United Nations University, Institute of Advanced Studies, Traditional Knowledge Institute, 2013.
[35] Louisiana Workshop, Stories of change: Coastal Louisiana tribal communities' experiences of a transforming environment, Input to the National Climate Assessment. Participating tribes: Grand Bayou Village, Grand Caillou/Dulac Band of the Biloxi-Chitimacha Confederation of Muskogees, Isle de Jean Charles Band of the Biloxi-ChitimachaConfederation of Muskogees, Pointe-au-Chien Indian Tribe. Maldonado JK (Ed.) Lynn K, Daigle J, Hoffman J, 2012. <https://downloads.globalchange.gov/nca/technical_inputs/CoastalLouisianaTribalCommunities2012StoriesOfChange.pdf>.

[36] C. Farbotko, H. Lazrus, The first climate refugees? Contesting global narratives of climate change in Tuvalu, Glob. Environ. Chang 22 (2012) 382–390.
[37] H. Lazrus, Sea change: climate change and island communities, Ann. Rev. Anthropol. 41 (2012) 285–301.
[38] R.E. Johannes, Working with fishermen to improve coastal tropical fisheries and resource management, Bull. Mar. Sci. 31 (1981) 673–680.
[39] R.E. Johannes, Words of the Lagoon: Fishing and Marine Lore in the Palau District of Micronesia, Univ. of California Press, 1981.
[40] K. Ruddle, Systems of knowledge: dialogue, relationships and process, Environ. Dev. Sustain. 2 (3-4) (2000) 277–304.
[41] E. Hviding, Guardians of Marovo Lagoon: practice, place, and politics in maritime Melanesia, Vol. 14, University of Hawaii Press, 1996.
[42] S. Foale, The intersection of scientific and indigenous ecological knowledge in coastal Melanesia: implications for contemporary marine resource management, Int. Soc. Sci. J. 58 (187) (2006) 129–137.
[43] M.D. Chapman, Women's fishing in Oceania, Hum. Ecol. 15 (3) (1987) 267–288.
[44] Y. Ota, Custom and fishing-Cultural meanings and social relations of Pacific fishing, Republic of Palau, Micronesia. University of London, University College London (United Kingdom), 2006.
[45] M. Lauer, S. Albert, S. Aswani, B.S. Halpern, L. Campanella, D. La Rose, Globalization, Pacific Islands, and the paradox of resilience, Global Environ. Change 23 (1) (2013) 40–50.
[46] P.D. Nunn, K. Mulgrew, B. Scott-Parker, D.W. Hine, A.D. Marks, D. Mahar, et al., Spirituality and attitudes towards Nature in the Pacific Islands: insights for enabling climate-change adaptation, Climatic Change 136 (3-4) (2016) 477–493.
[47] S. Aswani, R.J. Hamilton, Integration indigenous ecological knowledge and customary sea tenure with marine and social science for conservation of bumphead parrotfish (*Bolbometopon muricatum*) in the Roviana Lagoon, Solomon Islands, Environ. Conserv. 31 (2004) 69–83.
[48] S. Aswani, M. Lauer, Incorporating fishermen's local knowledge and behavior into geographical information systems (GIS) for designing marine protected areas in Oceania, Hum. Organ. (2006) 81–102.
[49] M. Lauer, S. Aswani, Indigenous ecological knowledge as situated practices: understanding fishers' knowledge in the western Solomon Islands, Am. Anthropol. 111 (2009) 317–329.
[50] H. Moller, F. Berkes, P.O.B. Lyver, M. Kislalioglu, Combining science and traditional ecological knowledge: monitoring populations for co-management, Ecol. Soc. 9 (3) (2004).
[51] M. Tengö, E.S. Brondizio, T. Elmqvist, P. Malmer, M. Spierenburg, Connecting diverse knowledge systems for enhanced ecosystem governance: the multiple evidence base approach, AMBIO 43 (2014) 1–13.
[52] C. Mondragón, Te hurihuri o te Ao: cycles of change, Traditional calendars for informing climate change policies, A background paper for an international experts meeting organized by Climate Frontlines, UNESCO and New Zealand National Commission for UNESCO, 2015.
[53] G.J. Laidler, Inuit and scientific perspectives on the relationship between sea ice and climate change: the ideal complement? Clim. Change 78 (2-4) (2006) 407.
[54] R. Panipilla, A.A. Benedict, The Sea Around Us, Samudra 67, 2014. Online at <https://www.icsf.net/images/samudra/pdf/english/issue_67/243_Sam67en_ALL.pdf>.
[55] T. Williams, P. Hardison, Culture, law, risk and governance: contexts of traditional knowledge in climate change adaptation, Clim. Change 120 (3) (2013) 531–544.
[56] S. von der Porten, R.C. de Loë, D. McGregor, Incorporating indigenous knowledge systems into collaborative governance for water: challenges and opportunities, J. Can. Stud. 50 (1) (2016) 214–243.
[57] G. Hugo, Future demographic change and its interactions with migration and climate change, Global Environ. Change 21 (2011) S21–S33.

CHAPTER 30

Coastal indigenous peoples in global ocean governance

Andrés M. Cisneros-Montemayor[1] and Yoshitaka Ota[2]

[1]Nippon Foundation Nereus Program, Institute for the Oceans and Fisheries, The University of British Columbia, Vancouver, BC, Canada [2]Nippon Foundation Nereus Program, School of Marine and Environmental Affairs, University of Washington, Seattle, WA, United States

Chapter Outline

30.1 Indigenous peoples within global oceans

As the impacts of climate and economic change on marine social–ecological systems intensify and increase in scope, there has been a concurrent effort to address governance issues at international and interinstitutional forums [1–3]. This is a clearly necessary development, as global challenges must involve global responses, yet it is important that this shift to higher scale governance recognizes and integrates the perspectives and needs of coastal communities and historically marginalized groups with the strongest links to marine systems, particularly including the almost 30 million Indigenous Peoples living throughout the world's coasts [4]. Coastal Indigenous Peoples include specific communities (in both developed and developing countries), ethnic minority groups, and Small Island Developing States with majority noncolonial populations. Though these communities unfortunately continue to be often relegated in sustainability policy discussions, they form part of the people that most urgently require attention and support, as stated in the central argument of the UN 2030 Agenda for Sustainable Development, "no one must be left behind" [3].

Issues related to Indigenous Peoples and communities and, in this case, their surrounding seascapes are particularly important to address due to the quite unique relationships

Predicting Future Oceans.
DOI: https://doi.org/10.1016/B978-0-12-817945-1.00028-9

between such peoples and natural systems, relationships that include aspects of but go beyond commercial, traditional, and livelihood linkages. Traditional ecological knowledge, a common characteristic of indigeneity [5], has been highlighted for its role in underpinning historical management systems that could be highly successful in establishing sustainable (and self-defined equitable) resource use by precolonial societies. These can be strategies that are analogous to "Western" policies, such as temporary restrictions on harvest (e.g., rotating use of fishing and hunting areas by groups from Oceania to the Arctic; [6,7]), limitations on fishing methods (e.g., the Seri of Mexico allowing only hand collection by women on clam beds that are exposed at low tide, pers. obs.), or tenure rights over resources (including systems combining area types and familial groups throughout Pacific Island Indigenous peoples; [8]). Other recent examples of newly implemented management actions by Indigenous communities include habitat restoration (e.g., of mangrove forests by communities in the Saloum Delta, Senegal) or the reassertion of stewardship over local marine resources (e.g., herring stocks by the Haida and Heiltsuk Nations of British Columbia; [9]). Some of these cases have received the UN Development Programme Equator Initiative Prize, which highlights and supports sustainability efforts by local community groups [10], showing the potential close overlap between "mainstream" conservation actions and those of coastal Indigenous communities.

Such management-oriented traditions notwithstanding, for Indigenous Peoples fisheries and marine ecosystems further form a fundamental part of culture and can be used to convey knowledge and traditions beyond the use or management of resources themselves. In this context, in Indigenous communities the practice of fishing itself can be as important as fish catch. In Madagascar, for example, Vezo fishers can leave their community for jobs elsewhere, but at that point cease to be Vezo unless they return and resume fishing; in this way, fishing is obviously a fundamental part of Vezo livelihoods but also for identity itself [11].

These cultural links and practices to marine ecosystems—underpinning the perpetuation of Indigenous groups as distinct Peoples—add urgency to projected future impacts from climate change and associated economic and social changes, which are further compounded by the current sociopolitical situation of many Indigenous groups. Projected shifts in fish distributions along the Pacific coast of North America are estimated to have overall negative effects on seafood supply for coastal Indigenous communities in the region, though some species may become more locally abundant [12]. However even if net changes were positive in other regions, reductions in particular species with a high cultural value would very likely fundamentally impact practices of Indigenous communities, in addition to the effects of changes in catch for subsistence or commercial purposes [12]. Similarly, the gradual decline in Arctic sea ice thickness and extent has been discussed in the context of potential new shipping lanes [13] and the management of new fishing areas [14]. Aside from these new economic and ecological considerations (and potentially even more disastrous consequences of further human pressures; [14]), sea ice is a vital component of

Indigenous ways of life [15], which would be irrevocably altered as the land and seascape in which they developed literally melts away.

It is important to highlight that many of the challenges faced by coastal Indigenous peoples are shared with other communities with relatively high vulnerability and low adaptive capacity, as considered and addressed by climate change literature [16]. This defines vulnerability as the combination of exposure and sensitivity to projected impacts, and adaptive capacity as the ability to mitigate the potential negative effects of such impacts through their own actions. Similar issues have been more extensively documented for the world's artisanal or "small-scale fisheries," that are loosely defined as having strong links to marine ecosystems and traditions of fishing compared to industrialized fisheries [17]. Nevertheless, as noted above, the cultural importance of fishing as a fundamental aspect of self-identity is particularly salient among coastal Indigenous peoples, for whom the practice of fishing can be as important as the catching of fish.

An initial step in connecting the knowledge and perspectives of Indigenous communities to national and international ocean policy initiatives is to highlight the global-scale magnitude of these communities' practices. A first meta-analysis of seafood consumption by coastal Indigenous peoples considered around 1900 such communities in 83 countries, representing almost 30 million people [18]. Using available data on per capita seafood consumption and a step-wise value transfer approach based on the similarity between communities of the same ethnic group, and groups in close geographical proximity, it was estimated that this subsistence portion of fisheries catch by coastal Indigenous peoples around the world amounts to around 2.1 (1.5–2.8) million metric tons per year [18]. This total is equal to about 2% of global reported fisheries catch [19] and is almost entirely absent from national statistics [20]. Note here, however, that although many other forms of nonindigenous catch are unreported worldwide, including because of logistical difficulties or illicit activity [21], there can be legitimate reasons for Indigenous peoples not to report fish catch or share information regarding their resource use [22], for example, as a form of assertion of self-governance, or when it is not required for cultural or subsistence activity. Nevertheless, the global scale and breadth of Indigenous seafood consumption (again, a small subset of resource use that can also include cultural or commercial catch) means that Indigenous issues must be central to global discussions on ocean policy, particularly in the context of addressing the needs of vulnerable communities (i.e., as stated in the UN SDGs) [3].

Recognizing the global extent of Indigenous fisheries and associated communities, there are a number of international agreements and documents—and some ongoing policy negotiations—that are highly relevant for coastal Indigenous Peoples and would greatly benefit from the recognition and integration of their knowledge, perspectives, and contexts. These include the UN Declaration on the Rights of Indigenous Peoples, the Convention on

Biological Diversity's Aichi Targets, the UN Sustainable Development Goals, and the ongoing negotiations on areas (and biodiversity) beyond national jurisdiction. Importantly, Indigenous issues must be envisioned when considering potential outcomes from even the most promising agreements. For example, the WTO's negotiations on limiting fisheries subsidies for fisheries that are illegal, overfished, or at overcapacity could lead to a highly beneficial agreement in terms of global (and national) fisheries sustainability [23]. However it is not uncommon for Indigenous fishers to operate without licenses due to their marginalization from or lack of input into "formal" fisheries norms [24]; these issues must be anticipated and addressed within formal agreements.

30.2 Appropriate use of global data

Current global statistics are necessary for contextualizing the scale and urgency of Indigenous recognition in high-level ocean policy and discussions, but historical (and ongoing) dynamics are perhaps most important for recognizing and addressing socioecological challenges and adaptation strategies. Perhaps the most pertinent issue in this context is that of seafood consumption trends over time within Indigenous communities around the world that can reveal patterns of change with potentially crucial impacts on cultural practices, social cohesion, and public health. There are multiple qualitative and quantitative analyses of declining seafood consumption rates in coastal Indigenous communities due both to changing norms (e.g., with younger generations consuming higher proportions of processed foods) [25] and changing ecosystems and wild food supply [12,26,27]. These shifts contribute to social impacts, including a loss of practices and language, and also to health impacts (e.g., increased risks of diabetes and heart disease; [28–30]) partly due to less nutritious diets compared with fish (and other wild foods). At the same time, global marine pollution has led to increased concentrations of contaminants in seafood (e.g., mercury, PCBs) with potentially quite serious health effects including increased child mortality and impaired cognition [31]. Some strategies have been developed to mitigate some of these impacts, for example, by limiting consumption of some seafoods during specific stages of pregnancy, and the reduction of marine pollution itself is explicitly included in the UN SDGs (goal 14, target 6) [3]. These issues, however, highlight the incredibly complex dynamics inherent to Indigenous contexts, that nonetheless must be addressed at multiple governance scales.

A recent review of seafood consumption trends includes historical (on average, 1970–2010) data for over 50 Indigenous communities and, though data are mostly available for Arctic and Subarctic regions [32], allows for analyses of trends over that time span. No significant trends in seafood consumption are evident over the time range of available data (Fig. 30.1) but consumption rates remain much higher in these Indigenous

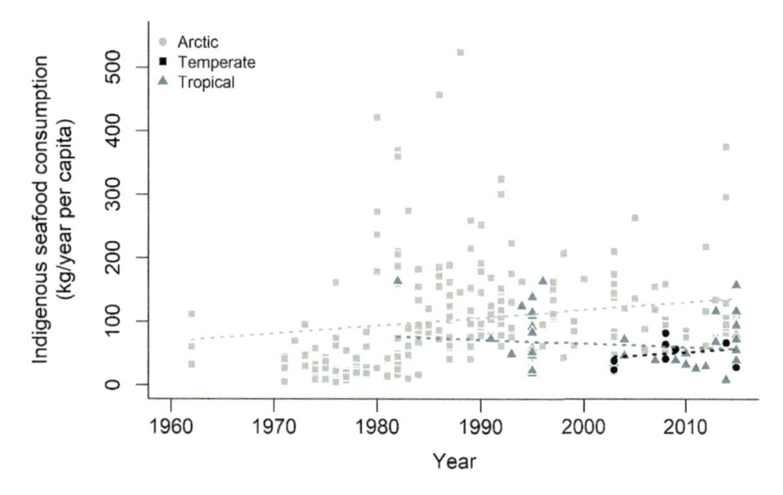

Figure 30.1

Reported yearly seafood consumption per capita for indigenous communities by region. *Data from A.M. Cisneros-Montemayor, T.-A. Kenny, Y. Ota, Dynamics in seafood consumption by coastal Indigenous peoples, in prep.*

communities than corresponding country averages [33], reflecting the enduring importance of marine living resources to such communities despite reported shifts in earlier time periods and the aforementioned continuing climate and social changes. Improved qualitative and quantitative information would clearly help in better reflecting the dynamic links between Indigenous peoples and marine systems, but are not needed to appreciate the importance of such links, and subsequently the need to incorporate these into appropriate social and environmental policies.

30.3 Including Indigenous voices in ocean policies

Moving forward, the full inclusion of Indigenous peoples into ocean policies will require different sets of specific strategies depending on regional contexts, but must involve some general actions. First, the rights of Indigenous peoples must be legally recognized and supported by corresponding nations and the international community, including access to resources and the ability to continue to carry out their cultural practices. (In this respect, there have been important advances in international agreements, including the UN Declaration on the Rights of Indigenous Peoples [34] and the International Labor Organization's Convention on Indigenous and Tribal Peoples [35], but much less so on national laws and specific agreements with local Indigenous communities [36].) Second, traditional knowledge—and the systems and practices to acquire it—must be recognized as a highly beneficial and in some cases fundamental contributor to wider management and scientific frameworks. Although many cases exist where such knowledge contributes directly to management reference points [22,37], it is more important to appreciate traditional knowledge as having emerged as a form of highly complex adaptive management [38], with principles that can be more important than specific "data points" and a knowledge system with strengths and weaknesses equivalent to—and forming synergies with—the scientific method [22].

Acknowledgments

The authors gratefully acknowledge the support from the Nippon Foundation Nereus Program, and thank the many program members whose insights and discussions helped in the preparation of this chapter.

References

[1] E.H. Allison, Big laws, small catches: global ocean governance and the fisheries crisis, J. Int. Dev. 13 (2001) 933–950. Available from: https://doi.org/10.1002/jid.834.
[2] R. Blasiak, J. Pittman, N. Yagi, H. Sugino, Negotiating the use of biodiversity in marine areas beyond national jurisdiction, Front Mar Sci 3 (2016). Available from: https://doi.org/10.3389/fmars.2016.00224.
[3] U. N. Transforming our world: the 2030 Agenda for Sustainable Development, United Nations, New York, 2015.
[4] A.M. Cisneros-Montemayor, Y. Ota, Coastal indigenous peoples fisheries database, in: I.J. Davidson-Hunt, H. Suich, S.S. Meijer, N. Olsen (Eds.), People Nat. Valuing Divers. Interrelat. People Nat. IUCN International Union for Conservation of Nature, 2016, p. 88. <https://doi.org/10.2305/IUCN.CH.2016.05.en>.
[5] J.R. Martínez Cobo, Mr. José R. Martínez Cobo Final Report (Last Part) submitted by the Special Rapporteur, United Nations, New York, 1983.
[6] F. Berkes, Shifting perspectives on resource management: resilience and the reconceptualization of 'natural resources' and 'management', Mast 9 (2010) 13–40.
[7] R.E. Johannes, Traditional marine conservation methods in Oceania and their demise, Annu. Rev. Ecol. Syst. 9 (1978) 349–364.
[8] K. Ruddle, E. Hviding, R.E. Johannes, Marine resources management in the context of customary tenure, Mar. Resour. Econ. 7 (1992) 249–273.
[9] S. von der Porten, D. Lepofsky, D. McGregor, J. Silver, Recommendations for marine herring policy change in Canada: aligning with Indigenous legal and inherent rights, Mar. Policy 74 (2016) 68–76. Available from: https://doi.org/10.1016/j.marpol.2016.09.007.
[10] UNDP, Equator Initiative, 2018. <https://www.equatorinitiative.org/>.
[11] R. Astuti, Learning to be Vezo. The Construction of the Person Among Fishing People of Western Madagascar (Ph.D. dissertation), University of London, 1991.
[12] L.V. Weatherdon, Y. Ota, M.C. Jones, D.A. Close, W.W.L. Cheung, Projected scenarios for coastal first nations' fisheries catch potential under climate change: management challenges and opportunities, PLoS One 11 (2016) e0145285. Available from: https://doi.org/10.1371/journal.pone.0145285.
[13] CNN, Maersk to send first container ship through Arctic 2018. <https://money.cnn.com/2018/08/21/news/companies/maersk-line-arctic-container/index.html>.
[14] H. Hoag, Nations agree to ban fishing in Arctic Ocean for at least 16 years, Science (2017). Available from: https://doi.org/10.1126/science.aar6437.

[15] W.N. Meier, G.K. Hovelsrud, B.E.H. van Oort, J.R. Key, K.M. Kovacs, C. Michel, et al., Arctic sea ice in transformation: a review of recent observed changes and impacts on biology and human activity, Arctic sea ice: review of recent changes, Rev. Geophys. 52 (2014) 185–217. Available from: https://doi.org/10.1002/2013RG000431.

[16] IPCC, Climate change 2014: synthesis report, Contribution of Working Groups I, II and III to the Fifth Assessment Report of the Intergovernmental Panel on Climate Change. IPCC, Geneva, 2014.

[17] FAO, Voluntary Guidelines for Securing Sustainable Small-Scale Fisheries in the Context of Food Security and Poverty Eradication, Food and Agriculture Organization of the United Nations, Rome, 2015.

[18] A.M. Cisneros-Montemayor, D. Pauly, L.V. Weatherdon, Y. Ota, A global estimate of seafood consumption by coastal indigenous peoples, PLoS One 11 (2016) e0166681. Available from: https://doi.org/10.1371/journal.pone.0166681.

[19] FAO, The state of world fisheries and aquaculture 2018, Meeting the Sustainable Development Goals, Rome, 2018.

[20] A.M. Cisneros-Montemayor, Y. Ota, Indigenous marine fisheries: a global perspective, Glob. Atlas Mar. Fish. Crit. Apprais. Catches Ecosyst. Impacts, Island Press, Washington, D.C. 2016, p. 18.

[21] D. Pauly, Major trends in small-scale marine fisheries, with emphasis on developing countries, and some implications for the social sciences, MAST 4 (2006) 7–22.

[22] H.P. Huntington, Using traditional ecological knowledge in science: methods and applications, Ecol. Appl. 10 (2000) 1270–1274. Available from: https://doi.org/10.1890/1051-0761(2000)010[1270:UTEKIS]2.0.CO;2.

[23] A.L. Mattice, The fisheries subsidies negotiations in the world trade organization: a win-win-win for trade, the environment and sustainable development, Gold Gate Univ. Law Rev.34 (2004) 573.

[24] A.M. Cisneros-Montemayor, T. Cashion, D.D. Miller, T.C. Tai, N. Talloni-Álvarez, H.W. Weiskel, et al., Achieving sustainable and equitable fisheries requires nuanced policies not silver bullets, Nat. Ecol. Evol. 2 (2018) 1334. Available from: https://doi.org/10.1038/s41559-018-0633-0.

[25] H.V. Kuhnlein, O. Receveur, Dietary change and traditional food systems of indigenous peoples, Annu. Rev. Nutr. 16 (1996) 417–442.

[26] W.W.L. Cheung, V.W.Y. Lam, D. Pauly, Modelling Present and Climate-Shifted Distribution of Marine Fishes and Invertebrates, 2008.

[27] A. Frainer, R. Primicerio, S. Kortsch, M. Aune, A.V. Dolgov, M. Fossheim, et al., Climate-driven changes in functional biogeography of Arctic marine fish communities, Proc. Natl. Acad. Sci. U.S.A. 114 (2017) 12202–12207. Available from: https://doi.org/10.1073/pnas.1706080114.

[28] B.V. Howard, E.T. Lee, L.D. Cowan, R.B. Devereux, J.M. Galloway, O.T. Go, et al., Rising tide of cardiovascular disease in american indians the strong heart study, Circulation 99 (1999) 2389–2395.

[29] Y.T. Kue, C.D. Schraer, E.V. Shubnikoff, E.J.E. Szathmary, Y.P. Nikitin, Prevalence of diagnosed diabetes in circumpolar indigenous populations, Int. J. Epidemiol. 21 (1992) 730–736.

[30] M. Naqshbandi, S.B. Harris, J.G. Esler, F. Antwi-Nsiah, Global complication rates of type 2 diabetes in Indigenous peoples: a comprehensive review, Diabetes Res. Clin. Pract. 82 (2008) 1–17. Available from: https://doi.org/10.1016/j.diabres.2008.07.017.

[31] A. Roe, Fishing for identity: mercury contamination and fish consumption among indigenous groups in the United States, Bull. Sci. Technol. Soc. 23 (2003) 368–375. Available from: https://doi.org/10.1177/0270467603259787.

[32] Alaska Department of Fish and Game, Harvest information for community, ADFG Subsist Hunt Fish Harvest Data Rep 2016. <https://www.adfg.alaska.gov/sb/CSIS/index.cfm?ADFG = harvInfo.harvestCommSelComm>.

[33] A.M. Cisneros-Montemayor, T.-A. Kenny, Y. Ota, Dynamics in Seafood Consumption by Coastal Indigenous Peoples, in prep.

[34] U. N. United Nations Declaration on the Rights of Indigenous Peoples, United Nations, New York, 2008.

[35] ILO, Convention Concerning Indigenous and Tribal Peoples in Independent Countries, International Labour Organization, Geneva, 1989.

[36] R.C.G. Capistrano, A.T. Charles, Indigenous rights and coastal fisheries: a framework of livelihoods, rights and equity, Ocean Coast Manage. 69 (2012) 200–209. Available from: https://doi.org/10.1016/j.ocecoaman.2012.08.011.

[37] K. Ruddle, The context of policy design for existing community-based fisheries management systems in the Pacific Islands, Ocean Coast Manage. 40 (1998) 105–126.

[38] F. Berkes, J. Colding, C. Folke, Rediscovery of traditional ecological knowledge as adaptive management, Ecol. Appl. 10 (2000) 1251–1262.

CHAPTER 31

The relevance of human rights to socially responsible seafood

Lydia C.L. Teh[1,2], Richard Caddell[3], Edward H. Allison[4,5], John N. Kittinger[6,7,8], Katrina Nakamura[9] and Yoshitaka Ota[4]

[1]Nippon Foundation Nereus Program, Vancouver, BC, Canada [2]Institute for the Oceans and Fisheries, University of British Columbia, Vancouver, BC, Canada [3]School of Law and Politics, Cardiff University, Cardiff, United Kingdom [4]Nippon Foundation Nereus Program, School of Marine and Environmental Affairs, University of Washington, Seattle, WA, United States [5]CGIAR Research Program on FISH, WorldFish, Bayan Lepas, Malaysia [6]Center for Oceans, Conservation International, Honolulu, HI, United States [7]Center for Biodiversity Outcomes, Life Sciences Center, Julie Ann Wrigley Global Institute of Sustainability, Arizona State University, Tempe, AZ, United States [8]Conservation International, Betty and Gordon Moore Center for Science, Arlington, VA, United States [9]The Sustainability Incubator, Honolulu, HI, United States

Chapter Outline

31.1 Introduction

Sustainable seafood is a growing market. In 2015, seafood that was certified as environmentally sustainable constituted 14% of global seafood production, compared to just 0.5% in 2005 [1]. However, it has become increasingly clear that environmental sustainability is not the only challenge facing seafood production. Recent exposures of exploitative labor practices in the seafood supply chain, including slavery and human trafficking, clearly demonstrate a systemic disregard for human well-being within some sectors of the fishing industry [2,3]. Egregious and highly publicized violations like slavery

Predicting Future Oceans.
DOI: https://doi.org/10.1016/B978-0-12-817945-1.00026-5

are, however, but one of a pervasive set of social injustices experienced by fisheries workers [4]. More latent violations include actions that perpetuate discrimination, deny fair access and sharing of benefits, and threaten food and livelihood security in fishing communities [5]. Actions that deny people their socioeconomic rights contribute to increased vulnerability and insecurity in people's lives, factors which impede marine stewardship for resource sustainability [6]. Therefore continued human rights violations undermine the socioeconomic sustainability of fisheries and ought to be addressed with the same concentrated efforts accorded to other threats to rational management, such as overfishing and ecologically unsustainable practices [6]. However, to date seafood certifications have predominantly focused on promoting environmental sustainability while largely ignoring socioeconomic considerations [7,8].

Demand is rising for the promotion of socially responsible seafood more centrally within the marketplace [9]. The social discourse has relied on the overarching framework provided by human rights laws to shape guiding principles and to provide the necessary legal mechanisms to protect and enforce human rights in seafood work. This demand is pushing the seafood sustainability movement into new territory. Thus far, seafood sustainability has largely been equated with fisheries that are managed to achieve particular ecological and environmental objectives. Social responsibility brings stakeholders accustomed to ecological issues into the less familiar realm of human rights and social development. A steep learning curve therefore lies ahead for seafood sector stakeholders to reconcile these two disparate fields, and to effectively navigate the nexus between sustainable seafood and human rights.

This chapter explores the extent to which human rights are capable of supporting the pursuit of socially responsible seafood. In particular, we evaluate the extent to which socially responsible seafood objectives are accommodated within existing human rights commitments. In this chapter we (1) review the definition of "socially responsible seafood," (2) assess the scope of social concerns in the seafood supply chain, and (3) examine the opportunities and challenges to implementing socially responsible seafood through the broad framework of human rights, considering the most pertinent legal and policy instruments. Finally, we draw upon these qualitative analyses to identify future avenues through which to incentivize the production of socially responsible seafood and to discourage exploitative and discriminatory practices.

31.2 The human rights landscape of socially responsible seafood

31.2.1 Principles of socially responsible seafood

Socially responsible seafood is defined by the Monterey Framework as comprising three core principles: (1) protect human rights and dignity, and respect access to resources; (2) ensure equality and equitable opportunities to benefit from such access; and (3) improve food and livelihood security [9]. These criteria necessarily extend across production modes (i.e., are inclusive of both aquaculture and wild-capture fisheries) and therefore apply to seafood supply chains in their entirety, from the point of capture to the point of sale to the end consumer. The Monterey Framework recognizes a spectrum of human rights from flagrant violations of established civil and political (CP) rights, such as freedom from slavery and forced labor, to more subtle infringements of economic, social, and cultural (ESC) rights, such as systemic discrimination, shortfalls in due process and inequitable outcomes in the allocation of fishing rights, access, and benefits. Next, we explore the scope for potential human rights violations in fisheries within the three principles of the Monterey Framework.

31.2.2 Social concerns in the seafood supply chain

31.2.2.1 Protect human rights and dignity, and respect access to resources

This principle calls for respecting basic human rights and dignity and protecting labor rights, as well as securing rights to resources. Failure to adhere to the first two criteria of this principle can result in violations, such as forced evictions, child labor, forced labor, detention without trial, and violence against fishing communities, all of which were evident in a review of human rights abuses in fisheries [5]. Incidences of physical abuse and unsafe working conditions for fishers, corrupt manning agents, victimization, unpaid wages, and unlawful detention were further documented by the International Transport Workers' Federation (ITF) [10]. Since 2000, a number of reports have focused on forced labor and human trafficking in the fishing sector, and are described by International Labor Organization (ILO) [11], while recent media exposure has highlighted modern-day slavery aboard Thai fishing vessels [3,12–14].

31.2.2.2 Ensure equality and equitable opportunities to benefit from resource access

This principle is concerned with social justice and fairness in the seafood supply chain. It pertains to issues of governance such as the equitable distribution of benefits to workers and inclusiveness in representation and participation in decision-making, the lack of which can negatively impact socioeconomic and ecological objectives [15–17]. Governance inadequacies can also expose fishers to potential inequities and erosion in access to fish resources, income, food security, and livelihoods, which in aggregate can drive the loss of livelihood and human security [18]. These issues are particularly relevant to small-scale fishers who often include migrants, indigenous people, or other marginalized groups that lack representation in decision-making processes [19]. De facto privatization of fisheries by the creation of marine protected areas and other "ocean grabbing" techniques also constitute violations to fishing communities' collective rights [20].

31.2.2.3 Improve food and livelihood security in fisheries

The objective of this principle is to ensure that seafood supply chains do not operate in ways that disrupt social structures or threaten people's ability to meet their sustenance and livelihood requirements, for example, through the imposition of low prices and economies of scale. Overall, it is concerned with creating conditions that improve socioeconomic stability and security in communities, including protecting tenure systems and access rights to resources and markets that can increase economic opportunities. Failure to consider this principle may include consequences like jeopardizing local food security if the targeted fishery is one that local communities have relied on for food [21], imposed changes to household nutrition and buying patterns [22], or other external influences that have uncertain impacts on the local social structure and marine resource base [23,24].

31.3 Challenges and opportunities in applying human rights to implement socially responsible seafood

Creating a definition of socially responsible seafood that encompasses the full spectrum of CP and ESC rights is a first step in drawing attention to the distinct aspects, thus different treatments, that is required to protect fisheries workers' rights and well-being [9]. There has been a relatively limited tradition of attempting to use the current tapestry of human rights instruments to promote socially responsible practices in the seafood sector. A primary concern is that the legal reach of human rights does not always extend to cover the issues and stakeholders that are involved in ensuring social responsibility along the entire seafood supply chain. The human rights instruments developed in the decades following the founding of the United Nations were primarily intended to prevent the types of atrocities against humanity that were witnessed during World War II. Guiding human rights principles that could more clearly underpin commitments toward producing socially responsible seafood have emerged more recently and, as such, have not attained the same legal traction as longstanding CP rights.

In practice, socially responsible seafood is a market-driven initiative in which seafood companies bear primary duties. Human rights treaties rarely impose legal obligations on private actors, which raises the question of how to hold seafood companies accountable for failing to secure socially responsible outcomes. In many key respects, international law remains a fundamentally consensual system and the decision to participate in treaty regimes that may be of particular value to fishers' rights remains an individual choice exercised by states based on their political and cultural sensitivities. As evidenced by the limited participation in the ILO's Work in Fishing Convention, which took 9 years to muster 10 signatories, especially by "developed states," there has been a marked reluctance to endorse treaties that directly impact upon the working conditions of fishers. Moreover, where human rights treaties have been ratified, they must be assiduously implemented by the national authorities and expressly applied to the maritime sector in order to inspire meaningful outcomes. Legally, seafarers have traditionally been treated in isolation to the terrestrial workforce as a special class of labor and a disappointingly high number of countries have failed to apply equivalent protections for vital rights such as fair and timely wages, freedom of assembly and association, and safe and healthy working conditions for ship-based work [25]. Likewise, there may be loopholes within key legal provisions that may have unintended negative consequences for seafarers. Notably, "developing countries" are not obliged to extend economic rights granted under the International Covenant on Economic, Social and Cultural Rights (ICESCR)[1] to non-nationals, thereby allowing particular signatories to legitimately fail to protect the interests of some of their most marginalized at-sea workers [26]. Similarly, the Work in Fishing Convention allows for a series of exceptions that "effectively exclude a significant number of fishing vessels from the scope" of the Convention [27], thereby curtailing its prospective impact significantly even if it were to be widely ratified.

Compounding these difficulties, while international instruments may establish particular entitlements, the enforcement of these principles is not straightforward. The right to individual petition is often an optional entitlement that many flag states have chosen not to accept [26]. Likewise, the International Tribunal for the Law of the Sea does not accept individual petitions and has routinely disregarded *amicus curiae* ("friend of the court" or neutral third party) interventions raising matters concerned specifically with human rights [28]. On a national level, however, domestic provisions may prove to be rather more far-sighted. In the United States, the Department of Homeland Security has the authority to initiate an investigation into forced labor or child labor in seafood or any goods based upon a petition submitted by any person. A list of goods made with a significant incidence of forced labor is available free of charge to the public in the Sweat & Toil smartphone application published by the US Department of Labor.

Beyond these conceptual challenges, it is also apparent that significant practical difficulties have been encountered in seeking to apply the network of human rights treaties addressing both CP and ESC rights of fishers. Among seafarers there is generally a chronic lack of awareness of the range of rights to which they are entitled [25,26]. Moreover, there is little active support or assistance for persons whose human rights have been violated to bring a legal claim in the jurisdictions in which they are most at risk. Where advocacy groups or activists are active in such jurisdictions, they may face harassment and intimidation, including through strategic lawsuits and amorphously defined legal provisions, such as defamation, civil disobedience, and the expansive concept of national insult [29].

[1] The ICESCR is a part of the United Nations International Bill of Human Rights.

Notwithstanding the challenges described above, there are opportunities for using human rights laws to achieve the promise of socially responsible seafood. Governments can draw upon international instruments to frame national laws that can then be applied to address human rights violations in fisheries. Recent examples include the United Kingdom, which in 2015 enacted the Modern Slavery Act [30], which strengthens law enforcement against instances of modern slavery, including those seen in global fisheries. At a late stage in the debate, a well-publicized report on labor violations within the national fishing fleet raised awareness of the hitherto little-considered plight of fishers [31], leading to the addition of further powers of at-sea enforcement. In New Zealand, the government responded to a series of non-legislative measures against forced labor at sea with the Fisheries (Foreign Charter Vessels and Other Matters) Amendment Act which came into force on May 1, 2016. This Act requires any foreign fishing vessel operating in New Zealand waters to reflag as a New Zealand ship, removing their right to fish in New Zealand waters until they do so, and is intended to help identify forced labor and ensure New Zealand employment law is applied uniformly at sea.

The United States has also sought to combat illegal fishing by passing the Illegal, Unreported, and Unregulated (IUU) Fishing Enforcement Act [32]. Since 2017, aspects of the legislation are supported by a risk-based traceability program, the Seafood Import Monitoring Program, which establishes reporting and recordkeeping requirements for imports of certain seafood products, to combat IUU-caught and/or misrepresented seafood from entering US commerce. There is also a US federal law that prohibits the import of goods that have been produced by forced labor (Smoot Hawley Tariff Act 1930). While these advances are to be celebrated, the violations that national laws in market countries address tend to be associated with the civil rights of individuals who often are foreign workers. While enacting relevant laws can be effective at national or subnational levels in forcing seafood companies to seriously look at accountability and transparency in their supply chain [33], to ensure that human rights are respected at every stage, there has to be cumulative buy-in and motivation across institutional levels to enforce them consistently.

Where legal regimes have been ineffective, non-binding "soft" laws can fill the gap left by "hard" binding legislation. Although nonbinding in nature, soft laws can contain elements that foster compliance and are especially suited to address issues that lack consensus or for which governments are wary of making legally binding commitments [34]. They thus represent an intriguing avenue to promote objectives that have not gained universal and binding support. The FAO's Voluntary Guidelines on the Right to Food in 2004 was a soft law approach to advance the right to food. It has had positive impacts on several fronts, including improving implementation of the right to food at the policy level, shifting prevailing perceptions around the impracticality of ESC rights, and promoting discourse between international development and human rights agendas [35]. A similar approach may likewise be appropriate for progressing to socially responsible seafood.

31.4 Conclusion

We showed that the seafood supply chain is currently operating in a landscape fraught with social abuses that impinge upon the full spectrum of the CP and ESC rights of individuals and groups. These violations have persisted in part due to the absence of a coherent vision of the desired outcome, and ambiguity or a lack of knowledge about the nature of the issues at hand and the process and tools that can be used to achieve the desired outcome. Aspirations for socially responsible seafood provide a clear objective for the seafood sector in which social abuses are to be at least reduced and preferably eliminated. This goal demands that the fundamental human rights of all individuals should be upheld; realizing human rights for fishery workers is thus an obligation and forms the premise for pursuing socially responsible seafood. Nonetheless, there are limitations in the extent to which the human rights framework can be harnessed in order to realize this objective.

There is no shortage of international human rights instruments that support the guiding principles of socially responsible seafood. However, human rights enshrined in international treaties may not always address the full scope of issues to ensure that seafood is considered socially responsible under varying local contexts, and will likely need to be complemented by improved methods of recognizing bottom-up approaches and market-based alternatives. Soft laws or voluntary standards can effectively fill in the gaps left by the weak application of human rights laws. With sufficient endorsement by stakeholders along the seafood supply chain, voluntary standards can become the industry operating norm and thus achieve further legitimacy and compliance. For fisheries workers, human rights serve to legitimize their claims and instigate corrective action against social abuses where they occur, even if the human rights framework is in and of itself imperfect. Market reforms alone are insufficient in the context of small-scale fisheries to achieve the principles of socially responsible seafood. Rather, protecting ESC rights will require broader institutional changes in local governance and social development, such as opportunities for legal pluralism like co-management, a perspective that is shared in the FAO's Small-Scale Fisheries (SSF) Guidelines. In conclusion, implementing socially responsible seafood on the ground will require actions on multiple fronts, where desired social outcomes are traced to target a broader set of rights and parties that have the capacity to influence those rights.

References

[1] J. Potts, A. Wilkings, M. Lynch, State of Sustainability Initiatives Review, International Institute for Sustainable Development, Winnipeg, 2016 (cited 7 May 2018). Available from: http://public.eblib.com/choice/publicfullrecord.aspx?p = 4532673.

[2] T. Sutton, A. Siciliano, Seafood Slavery (Internet). Center for American Progress (cited 7 May 2018). Available from: <https://www.americanprogress.org/issues/green/reports/2016/12/15/295088/seafood-slavery/>.

[3] M. Marschke, P. Vandergeest, Slavery scandals: unpacking labour challenges and policy responses within the off-shore fisheries sector, Mar. Policy 68 (2016) 39–46.

[4] A. Couper, H.D. Smith, B. Ciceri, Fishers and Plunderers (Internet), Pluto Press, 2015. Available from: http://www.jstor.org/stable/j.ctt183p451.

[5] B.D. Ratner, B. Åsgård, E.H. Allison, Fishing for justice: human rights, development, and fisheries sector reform, Global Environ. Change 27 (2014) 120–130.

[6] E.H. Allison, B.D. Ratner, B. Åsgård, R. Willmann, R. Pomeroy, J. Kurien, Rights-based fisheries governance: from fishing rights to human rights, Fish Fish. 13 (1) (2012) 14–29.

[7] S. Blackstock, Ecological Benefits of the Marine Stewardship Council Ecolabel: A Global Analysis, University of Rhode Island, 2014.

[8] Y. Stratoudakis, P. McConney, J. Duncan, A. Ghofar, N. Gitonga, K.S. Mohamed, et al., Fisheries certification in the developing world: locks and keys or square pegs in round holes? Fish. Res. 182 (2016) 39–49.

[9] J.N. Kittinger, L.C.L. Teh, E.H. Allison, N.J. Bennett, L.B. Crowder, E.M. Finkbeiner, et al., Committing to socially responsible seafood, Science 356 (6341) (2017) 912–913.

[10] ITF International Transport Workers' Federation, Out of Sight, Out of Mind: Seafarers, Fishers & Human Rights (Internet). Available from: <http://www.itfseafarers.org/files/extranet/-1/2259/humanrights.pdf>, 2006.

[11] ILO, Caught at Sea: Forced Labour and Trafficking in Fisheries, ILO, Geneva, 2013, p. 85.

[12] Human Rights Watch, From the Tiger to the Crocodile: Abuse of Migrant Workers in Thailand (Internet). New York, 2010, p. 16. Available from: <https://www.iom.int/jahia/webdav/shared/shared/mainsite/activities/countries/docs/thailand/Trafficking-of-Fishermen-Thailand.pdf>.

[13] IOM, Trafficking of Fishermen in Thailand (Internet), IOM, Thailand, 2011. Available from: https://www.iom.int/jahia/webdav/shared/shared/mainsite/activities/countries/docs/thailand/Trafficking-of-Fishermen-Thailand.pdf.

[14] I. Urbina, 'Sea Slaves': the human misery that feeds pets and livestock, The New York Times (Internet). 27 July 2015 (cited 8 May 2018). Available from: <https://www.nytimes.com/2015/07/27/world/outlaw-ocean-thailand-fishing-sea-slaves-pets.html>.

[15] C.F. Gaymer, A.V. Stadel, N.C. Ban, P.F. Cárcamo, J. Ierna, L.M. Lieberknecht, Merging top-down and bottom-up approaches in marine protected areas planning: experiences from around the globe, Aquat. Conserv.: Mar. Freshw. Ecosyst. 24 (S2) (2014) 128–144.

[16] J.A. Aburto, C.F. Gaymer, G. Cundill, Towards local governance of marine resources and ecosystems on Easter Island, Aquat. Conserv.: Mar. Freshw. Ecosyst. 27 (2) (2017) 353–371.

[17] N.J. Bennett, P. Dearden, Why local people do not support conservation: community perceptions of marine protected area livelihood impacts, governance and management in Thailand, Mar. Policy 44 (2014) 107–116.

[18] C. Sharma, Securing Economic, Social and Cultural Rights of Small-Scale and Artisanal Fisherworkers and Fishing Communities, 2011, p. 21.

[19] G. Acciaioli, H. Brunt, J. Clifton, Foreigners everywhere, nationals nowhere: exclusion, irregularity, and invisibility of stateless Bajau Laut in Eastern Sabah, Malaysia, J. Immigr. Refugee Stud. 15 (3) (2017) 232–249.

[20] N.J. Bennett, H. Govan, T. Satterfield, Ocean grabbing, Mar. Policy 57 (2015) 61–68.

[21] R.M. Garcias, J. Grobler, Spain's hake appetite threatens Namibia's most valuable fish (Internet). The Center for Public Integrity, October 2011. Available from: <http://www.publicintegrity.org/2011/10/04/6769/spain-s-hake-appetite-threatens-namibia-s-most-valuable-fish>.

[22] M. Fabinyi, W.H. Dressler, M.D. Pido, Fish, trade and food security: moving beyond 'availability' discourse in marine conservation, Hum. Ecol. 45 (2) (2017) 177–188.

[23] J.E. Cinner, T.R. McClanahan, T.M. Daw, N.A.J. Graham, J. Maina, S.K. Wilson, et al., Linking social and ecological systems to sustain coral reef fisheries, Curr. Biol. 19 (3) (2009) 206–212.

[24] J.E. Cinner, C. Huchery, M.A. MacNeil, N.A.J. Graham, T.R. McClanahan, J. Maina, et al., Bright spots among the world's coral reefs, Nature 535 (7612) (2016) 416–419.

[25] D. Fitzpatrick, M. Anderson (Eds.), Seafarers' Rights, Oxford University Press, Oxford, New York, 2005. 690 pp.

[26] U. Khaliq Jurisdiction, ships and human rights treaties, in: H. Ringbom (Ed.), Jurisdiction Over Ships: Post-UNCLOS Developments in the Law of the Sea (Internet). Nijhoff, 2015 (cited 25 July 2018). Available from: <https://brill.com/abstract/book/edcoll/9789004303508/B9789004303508_014.xml>.

[27] I. Papanicolopulu, International Law and the Protection of People at Sea, Oxford University Press, Oxford, New York, 2018. 304 pp.

[28] R. Caddell, Platforms, protestors and provisional measures: the arctic sunrise dispute and environmental activism at sea, Neth. Yearbook Int. Law, 45, 2014, pp. 358–384.

[29] Al Jazeera News, HRW condemns libel verdict against rights worker Andy Hall, 28 Mach 2018 (cited 14 December 2018). Available from: <https://www.aljazeera.com/news/2018/03/hrw-condemns-libel-verdict-rights-worker-andy-hall-180328191439164.html>.

[30] GOV.UK, Modern Slavery Act 2015 (Internet). GOV.UK (cited 23 October 2018). Available from: <https://www.gov.uk/government/collections/modern-slavery-bill>.

[31] E. McSweeney, F. Lawrence, Irish trawlers accused of "alarming" abuses of migrant workers, The Guardian (Internet). 8 February 2017 (cited 8 May 2018). Available from: <http://www.theguardian.com/global-development/2017/feb/08/irish-trawlers-abuses-migrant-workers>.

[32] ODU Magazine, United States achieves major milestone in efforts to combat illegal fishing (Internet) (cited 26 July 2018). Available from: <https://www.odumagazine.com/united-states-achieves-major-milestone-in-efforts-to-combat-illegal-fishing/>, 2015.

[33] S.T. Khadaroo, California wants to know who's harvesting your shrimp, Christian Science Monitor (Internet), 18 January 2016 (cited 26 July 2018). Available from: <https://www.csmonitor.com/USA/Justice/2016/0118/California-wants-to-know-who-s-harvesting-your-shrimp>.

[34] B. Choudhury, Balancing soft and hard law for business and human rights, Int. Comp. Law Q. 67 (4) (2018) 961–986.

[35] M. von Engelhardt, Opportunities and challenges of a soft law track to economic and social rights—the case of the voluntary guidelines on the right to food, Verfassung und Recht in Übersee/Law and Politics in Africa, Asia and Latin America 42 (4) (2009) 502–526.

CHAPTER 32

The emergence of corporate social responsibility in the global seafood industry: potentials and limitations

Wilf Swartz

Nippon Foundation Nereus Program, Institute for the Oceans and Fisheries, University of British Columbia, Vancouver, BC, Canada

Chapter Outline

32.1 Introduction

Today, the world's fisheries operate across most of our planet's productive waters [1]. In addition to the environmental consequences of overfishing, the socioeconomic impacts of declining fisheries catches to regional food security [2,3], labor, and cash income in resource-poor regions [4,5], and the inherent vulnerability of fishing people and communities dependent on such uncertain and variable food production systems are beginning to be voiced in global ocean governance and policy discussions [6]. As social concerns of marine overexploitation increase, we are seeing an emergence of innovative and diverse series of private initiatives to attaining sustainability outside of the existing regulation of fishing fleets at national, regional, and international levels [7,8].

Thus, private initiatives which can directly influence the incentives, decision-making, and structure of the fishing industry and seafood supply chain through the use of market interventions have the potential to fill the policy gaps in public governance. However, the current model of private initiatives, dominated by the use of voluntary certification

Predicting Future Oceans.
DOI: https://doi.org/10.1016/B978-0-12-817945-1.00029-0

standards (VCS) (ecolabels), may be insufficient in shaping the corporate behavior. With the seafood industry's strategies for sustainability (i.e., corporate social responsibility, CSR) driven by their commitment to subscribe to existing private, third-party certification programs, we argue that CSR as practiced in the seafood industry today is limiting the industry's ability to fully integrate sustainability throughout the supply chains, not only in terms of resource use but also in the improvement of human livelihoods.

32.2 Rise of private governance in seafood industry

Seafood is the most highly traded food commodity internationally—35% of the world's total production was exported in 2016 [9]. The growth of international seafood trade has facilitated an emergence of highly complex global supply chain networks, in many cases involving third-country processors that import unprocessed fish for the sole purpose of processing and then re-exporting it (see [10]). Faced with such a business environment, it is argued that private institutions based on generating incentives for better corporate behaviors can fill this governance gap [11,12]. There is increasing awareness that neither national nor international governance frameworks are sufficient to oversee the ever-expanding activities of global industries [13]. As large, multinational corporations gain greater market shares and cast influence over national economies, there exists an imbalance between the power of these firms and the capacity of state governments to adequately regulate and hold them accountable [14].

As concerns over depleting fish stocks grew, additional focus on market-incentives for sustainable (i.e., properly managed and low environmental impact) seafood products increased [15]. Started in the 1990s, one of the earliest seafood ecolabels was "dolphin-safe" canned tuna, which aimed to inform consumers about the bycatch impacts of some tuna fisheries [16]. At the time it was suggested that inducing consumer awareness would be paramount in the success of ecolabeling programs in seafood, and an emphasis was placed on the idea of program modifications based on region, species, and targeted buyer. Today ecolabeling is one of the key approaches used to inform consumers about the environmental impacts and management conditions of their seafood choices.

The key function of VCS is to establish a set of standards beyond the minimum requirements of government or international regulatory bodies (e.g., Regional Fisheries Management Organizations for managing the high seas and transboundary fish stocks), thus enabling the market to differentiate fisheries engaged in sustainable fishing practices. Seafood sourced from fisheries deemed to have met such standards would then bear the program's label, informing consumers that they are more ecologically sustainable than other products. In theory, ecocertifications contribute to sustainability by generating financial incentives for fisheries, through preferential market access or price premiums, and this encourages others to invest and follow suit in order to keep pace with their competitors

[15,17,18]. The success of these programs requires consumers that are willing to pay a premium to offset the additional costs associated with sustainable practices (e.g., bycatch mitigation). However, in practice empirical evidence for such price premiums remains weak, and, when present, confined to highly specialized markets (e.g., [19,20]).

Nevertheless, the success of the Marine Stewardship Council (MSC), the most ubiquitous of the private certification programs, led to the emergence of similar programs catering to various markets (both in terms of geography and consumer types, i.e., restaurants and retails), commercial species, and focused on differing yet overlapping components of sustainability. The subsequent proliferation of private ecolabels is causing confusion for stakeholders throughout the supply chains, from production to retails and consumers, as well as in public governance [21]. For the fishing industry, and particularly for small-scale fishers, the selection of appropriate programs to adhere to is now a considerable economic decision with varying compliance costs and potential market benefits. Wholesalers and retailers must identify programs that have the most credence that meet their market niches and comply with their risk management objectives (e.g., labor practices), while consumers must navigate through the specifics of these programs to satisfy their personal sustainability requirements [1]. For government, questions still remain as to whether these private initiatives support or undermine the public regulatory framework [22].

32.3 Voluntary certification standards and corporate social responsibility

Another area where private governance models have thrived is through business practices and sustainability strategies, or as CSR. CSR requires firms to make expenditures and undertake commitments associated with broadening public expectations on how they should behave. Corporations view CSR as a necessary commitment for maintaining the trust of both shareholders and stakeholders, and for retaining the *social license* to operate [23]. Under this line of argument a business case for CSR exists simply because societies are increasingly expecting businesses to behave in a socially responsible manner, and failure to do so will lead to negative responses (e.g., consumer avoidance of their products). This type of incentive is prevalent in resource-based industries where the support of local communities is pertinent to the exploitation of the resource. For the seafood industry CSR commitments must therefore extend beyond resource sustainability and to the impact of their activities to local communities, many of whom may be heavily dependent on fisheries resources not only as the basis for their economic well-being but more critically for their food security.

In analyzing the CSR policies of the global seafood industry this trend of compliance with VCS is prominent. Of the top 150 seafood firms (by revenue, [24]), two-thirds of the firms referred to some form of CSR policies, with the majority using MSC for benchmarks (or their aquaculture equivalent, Aquaculture Stewardship Council, or ASC, certification, Tables 32.1 and 32.2). This trend is particularly notable for larger corporations (i.e., annual

Table 32.1: Major seafood voluntary certification standards (VCS) and their key characteristics.

	Formation	Organization	Description	Website
General VCS				
UN Global Compact	2000	UN	The world's largest corporate sustainability initiative focused on 10 principles in the areas of human rights, labor, the environment, and anticorruption	www.unglobalcompact.org
ISO14001	2004	International Organization for Standardization (independent, nongovernmental organization)	Core set of standards used for designing and implementing an effective EMS. Does not state requirements for environmental performance	www.iso.org/iso/iso14000
Fisheries VCS				
Marine Stewardship Council	1997	Marine Stewardship Council (nonprofit organization)	Science-based certification program for fisheries focused on sustainable fish stocks, minimal environmental impact, and effective management	www.msc.org
Friend of the Sea	2008	Friend of the Sea (nonprofit organization)	Certification based on stock status (no seafood from overexploited stocks), habitat impact, bycatch considerations, and compliance	www.friendofthesea.org
Seafood Watch		Monterey Bay Aquarium	Sustainable seafood advisory program based on three tied recommendations ("Best Choice," "Good Alternative," and "Avoid"). Criteria-based stock status and vulnerability, discards, habitat and ecosystem impacts, and effective management	www.seafoodwatch.org
Aquaculture VCS				
Aquaculture Stewardship Council	2010	Aquaculture Stewardship Council (nonprofit organization)	Standards for responsible aquaculture with a set of requirements covering planning, development, and operation of aquaculture production systems	www.asc-aqua.org

(*Continued*)

Table 32.1: (Continued)

	Formation	Organization	Description	Website
Global Aquaculture Alliance Best Aquaculture Practice	2004	Global Aquaculture Alliance (nonprofit trade association)	Certification system for aquaculture facilities based on environmental impact, social responsibility, food safety, animal welfare, and traceability	www.gaaliance.org
Global GAP Aquaculture Standard	2004	Global GAP (farm assurance program)	A subgroup of agricultural standards that assess aquaculture production based on legal compliance, food safety, worker occupational health and safety, animal welfare, and environmental and ecological care	www.globalgap.org
Global Standard and Certification Programme for the Responsible Supply of Fishmeal and Fish Oil	2009	Marine Ingredients Organization (nonprofit industry organization)	Certify responsible practice for sustainable raw material sourcing and safe production of fishmeal and fish oil	www.iffo.net

EMS, Environmental Management System; *GAP*, Good Agriculture Practice.

Table 32.2: Use of voluntary certification standards (VCS) and other external guidelines by seafood companies in their corporate social responsibility commitments.

			General VCS		Fisheries VCS			Aquaculture VCS			
Revenue (in million US$)	**No. firms**	**CSR**	**UNGP**	**ISO14001**	**MSC**	**FoS**	**SW**	**ASC**	**GAA BAP**	**Global GAP**	**IFFO**
≥1000	30	22	6	7	11	5	1	3	8	1	3
500–999	37	29	1	6	18	3	0	2	7	3	1
350–499	28	20	2	5	17	3	0	3	4	3	1
250–349	25	15	2	3	5	2	0	2	3	5	1
100–249	30	25	1	7	8	3	1	3	4	8	4
Total	150	111	12	28	59	16	2	13	26	20	10

Revenue as of 2013, from Intrafish 150 Report as descriptions of the VCSs are provided in Table 32.1. Global GAP = Global GAP Aquaculture Standard; IFFO = Global Standard and Certification Programme for the Responsible Supply of Fishmeal and Fish Oil. See appendix for descriptions of the programs. *ASC*, Aquaculture Stewardship Council; *FoS*, Friends of the Sea; *GAA BAP*, Global Aquaculture Alliance Best Aquaculture Practices; *GAP*, Good Agriculture Practice; *MSC*, Marine Stewardship Council; *SW*, Seafood Watch; *UNGP*, United Nations Global Compact.

revenue over US$350 million). Other fisheries standards include Friends of the Sea and the Monterey Bay Aquarium Seafood Watch guidelines, and non-fisheries standards such as the UN Global Compact and ISO14001 for environmental management are also commonly used.

As noted in the previous section, MSC and other programs are fisheries-focused programs with environmental sustainability as their core focus. As such they lack mechanisms to monitor social and ethical standards or environmental standards throughout the supply chains. Although MSC requires certifications of participation through the supply chain, this is primarily to ensure the transparency and chain of custody for the specific certified seafood product, and not to certify the company's business practices. As consumer-facing ecolabel programs designed to differentiate between fisheries, their assessment criteria apply to a given fishery collectively, not for individual firms.

Compared to the fishing industry, aquaculture firms have a much broader scope in VCS, both in terms of their range of commitments and programs. The ASC, for example, includes criteria that explicitly address the labor standards of fish farms, as well as community relations and interactions ("Principle 7: Be a good neighbor and conscientious coastal citizen," ASC 2012). In addition to the ASC certification, aquaculture standards include the Global Aquaculture Alliance Best Aquaculture Practices, Global Good Agriculture Practice Module for Aquaculture, Federation of European Aquaculture Producers Code of Conduct, and a Global Standard and Certification Programme for the Responsible Supply of Fishmeal and Fish Oil (IFFO RS). Members of the Global Salmon Initiative (globalsalmoninitiative.org), for example, have committed to having all their salmon aquaculture production be ASC certified by 2020. The discrepancy in the level of CSR commitment between fishing and aquaculture firms may be largely due to the nature of these two operations. Unlike fisheries, aquaculture operations tend to operate at a much smaller scale and often in closer proximity to communities. Furthermore, since it is a fairly new industry, aquaculture firms' need for social license may be stronger than for fisheries.

32.4 Discussion

CSR in the seafood industry suffers from a lack of holistic and strategic approaches. For many CSR is reduced to resource sustainability, and implemented through a series of parallel yet fragmented programs that fail to consider the connections between environmental—that is, ecological sustainability—and social systems—that is, human rights, food security, and community impacts. Moreover, their reliance on VCS exposes them to criticisms common to these programs, such as the use of inadequate or deceptive measurements and the erosion or outsourcing of responsibility [25], or that the incentives to comply with existing standards can stifle innovation [26]. Such criticisms may be particularly pertinent in the seafood industry where the most widely used CSR VCS—the MSC Standard—is designed as a consumer-facing ecolabel for product differentiation based on sustainable practices, rather than as a mechanism for corporate self-regulation.

Moreover, some scholars question the effectiveness of industry self-regulation via CSR, arguing that this approach may lead to companies being selective with their responsibilities [27], whereby they focus their CSR effort on projects deemed profitable rather than those that are most needed by society [28]. Often CSR is reduced to a list of specific benchmarks and targets, with accomplishment limited to those that "tick the list." CSR programs based on compliance or adherence to voluntary standards can potentially overemphasize the trading relationships between buyers and suppliers (i.e., market-based mechanisms), leading to an imbalance in the sharing of costs and benefits in favor of large international buyers. Fisheries in particular, with a large number of suppliers and relatively small number of buyers, may result in disproportionate distribution of compliance costs to suppliers [36]. Fishers, with no explicit guarantee from buyers, are effectively bearing all the risk of compliance. Overreliance on VCS for corporate policies may also contribute to outsourcing of supply chain risks, with third-party certification bodies held responsible for compliance failure. In other words, large international buyers may be using VCS as accountability buffers.

Undoubtedly, private sustainability initiatives can compensate for some of the shortcomings of public governance. All international fisheries-focused VCS have integrated the concept of ecosystem-based management in their assessments, including considerations for vulnerable species and fishery impacts via trophic interactions, and fisheries have responded to such requirements [29] and have contributed to expanded practices in assurance and enforcement of standards [30]. Yet they are not a substitute for the more effective implementation of state-led management regimes both at the national and international levels. The long-term impact of private sustainability initiatives will ultimately depend on the extent to which the new business environment and corporate behaviors that are advanced under these initiatives are integrated with and reinforced by public policies [31].

Fisheries in the high seas are areas in which industry-led CSR commitments can supplement and reenforce the existing management frameworks. With no explicit mandates, current international governing regimes such as the International Maritime Organization (IMO) and RFMOs are finding themselves with limited capacity to enforce their management measures. The vessel registration and automatic identification system (AIS) vessel tracking programs by IMO, for example, do not cover a large proportion of the world's fishing vessels. Commitments by the seafood industry to voluntarily adhere to such programs (i.e., full AIS implementation aboard their fishing fleet) can greatly enhance RFMOs' capacity to monitor vessel activities and combat illegal, unreported, and unregulated fisheries in the high seas. Similarly with the adoption of the FAO Agreement on Port State Measures, the industry should take a proactive role in ensuring transparency for its activities. In this sense, CSR should not be viewed as a "to-do" list but as a continuously evolving strategy, reflecting the governance requirements of the time.

32.5 Conclusion

Given the magnitude, diversity, and nature of the challenges facing the world's oceans, private initiatives that depend on market-driven incentives through consumer awareness and preference or industry self-regulation via CSR may not be sufficient to ensure fisheries sustainability given existing governance gaps. Issues such as climate change impacts on ocean ecosystems [32], global and regional food security, and public health, including equitable access to fisheries resources for artisanal fisheries and coastal communities [33], as well as the protection of fundamental human rights throughout seafood supply chains [34] require systematic and structural transformations in our global seafood systems that cannot be brought about solely via advancements in VCS or CSR.

Lund-Thomsen and Lindgreen [35] note the need to shift from a "compliance" paradigm in CSR to a "cooperative" paradigm that emphasizes long-term relationships between corporations and local partners and investments in capacity building throughout the supply chains. While such cooperative approaches may not alter the power relationship in global value chain, stronger public commitments from buyers, as well as explicit and formalized partnerships with local suppliers can enable the greater sharing of benefits and accountability for buyers. Such a transformation may be especially beneficial in the seafood industry since it operates across a wide spectrum of environmental and socioeconomic conditions.

References

[1] W. Swartz, E. Sala, S. Tracey, R. Watson, D. Pauly, The spatial expansion and ecological footprint of fisheries (1950 to present), PLoS One 5 (12) (2010) e15143–e15146.

[2] D. Pauly, R. Watson, J. Alder, Global trends in world fisheries: impacts on marine ecosystems and food security, Philos. Trans. R. Soc. B: Biol. Sci. 360 (1453) (2005) 5–12.

[3] U.T. Srinivasan, W.W.L. Cheung, R. Watson, U.R. Sumaila, Food security implications of global marine catch losses due to overfishing, J. Bioecon. 12 (3) (2010) 183–200.

[4] C. Bene, B. Hersoug, E.H. Allison, Not by rent alone: analysing the pro-poor functions of small-scale fisheries in developing countries, Dev. Policy Rev. 28 (3) (2010) 325–358.

[5] C. Bene, G. Macfadyen, E.H. Allison, Increasing the Contribution of Small-Scale Fisheries to Poverty Alleviation and Food Security, Food and Agriculture Organization of the United Nations, Rome, 2007.

[6] E.H. Allison, B.D. Ratner, B. Åsgård, W. Willmann, R. Pomeroy, J. Kurien, Rights-based fisheries governance: from fishing rights to human rights, Fish Fish. 13 (1) (2011) 14–29.

[7] E.H. Allison, Big laws, small catches: global ocean governance and the fisheries crisis, J. Int. Dev. 13 (7) (2001) 933–950.

[8] A. Kalfagianni, P. Pattberg, Fishing in muddy waters: exploring the conditions for effective governance of fisheries and aquaculture, Mar. Policy 38 (C) (2013) 124–132.

[9] FAO, The State of World Fisheries and Aquaculture 2018: Opportunities and Challenges, FAO, Rome, 2018, p. 243.

[10] C. Stringer, G. Simmons, E. Rees, Shifting post production patterns: Exploring changes in New Zealand's seafood processing industry, N. Z. Geogr. 67 (3) (2011) 161–173.

[11] L.H. Gulbrandsen, Overlapping public and private governance: can forest certification fill the gaps in the global forest regime? Global Environ. Polit. 4 (2) (2004) 75–99.

[12] D. O'Rouke, Outsourcing regulation: analyzing nongovernmental systems of labor standards and monitoring, Policy Stud. J. 31 (1) (2003) 1–29.

[13] D. Vogel, The private regulation of global corporate conduct: achievements and limitations, Bus. Soc. 49 (1) (2010) 68–87.

[14] H. Österblom, J.B. Jouffray, C. Folke, B. Crona, M. Troell, A. Merrier, et al., Transnational corporations as 'keystone actors' in marine ecosystems, PLoS One 10 (5) (2015) e0127533.

[15] K. Sainbury, *Review of ecolabelling schemes for fish and fishery products from capture fisheries* (No. 533), FAO Fisheries and Aquaculture Technical Paper, FAO, Rome, 2010, p. 93.

[16] J.L. Jacquet, D. Pauly, The rise of seafood awareness campaigns in an era of collapsing fisheries, Mar. Policy 31 (3) (2007) 308–313.

[17] M. Thrane, F. Ziegler, U. Sonesson, Eco-labelling of wild-caught seafood products, J. Clean. Prod. 17 (3) (2009) 416–423.

[18] M.F. Tlusty, Environmental improvement of seafood through certification and ecolabelling: theory and analysis, Fish Fish. 13 (1) (2011) 1–13.

[19] C.A. Roheim, F. Asche, J.I. Santos, The elusive price premium for ecolabelled products: evidence from seafood in the UK market, J. Agric. Econ. 62 (3) (2011) 655–668.

[20] H. Uchida, Y. Onozaka, T. Morita, S. Managi, Demand for ecolabeled seafood in the Japanese market: a conjoint analysis of the impact of information and interaction with other labels, Food Policy 44 (C) (2014) 68–76.

[21] S. Washington, L. Ababouch, Private standards and certification in fisheries and aquaculture: current practice and emerging issues, FAO Fisheries and Aquaculture Technical Paper, vol. 553, FAO, Rome, 2010, p. 203.

[22] J. Konefal, Environmental movements, market-based approaches, and neoliberalization: a case study of the sustainable seafood movement, Organ. Environ. 26 (3) (2013) 336–352.

[23] N. Gunningham, R.A. Kagan, D. Thornton, Social license and environmental protection: why businesses go beyond compliance, Law Soc. Inq. 29 (2) (2004) 307–341.

[24] Intra Fish Media, IntraFish 150 Report: Industry Report 2013, Intra Fish, 2014.

[25] S. de Colle, A. Henriques, S. Sarasvathy, The paradox of corporate social responsibility standards, J. Bus. Ethics 125 (2) (2013) 177–191.

[26] N. Dew, S.D. Sarasvathy, Innovations, stakeholders & entrepreneurship, J. Bus. Ethics 74 (3) (2007) 267–283.

[27] K. Bondy, K. Starkey, The dilemmas of internationalization: corporate social responsibility in the multinational corporation, Br. J. Manage. 25 (1) (2012) 4–22.

[28] C. Parkes, J. Scully, S. Anson, CSR and the "undeserving": a role for the state, civil society and business? Int. J. Sociol. Soc. Policy 30 (11/12) (2010) 697–708.

[29] L.M. Bellchambers, B.F. Phillips, M. Pérez-Ramírez, From certification to recertification the benefits and challenges of the Marine Stewardship Council (MSC): a case study using lobsters, Fish. Res. 182 (2016) 88–97.

[30] P. Castka, C.J. Corbett, Governance of eco-labels: expert opinion and media coverage, J. Bus. Ethics 135 (2) (2016) 309–326.

[31] M. Bailey, H. Packer, L. Schiller, M. Tlusty, W. Swartz, The role of corporate social responsibility in creating a Seussian world of seafood sustainability, Fish Fish. 19 (5) (2018) 782–790.

[32] W.W.L. Cheung, V.W.Y. Lam, J.L. Sarmiento, K. Kearney, R. Watson, D. Zeller, et al., Large-scale redistribution of maximum fisheries catch potential in the global ocean under climate change, Global Change Biol. 16 (1) (2010) 24–35.

[33] J.D. Bell, V. Allain, E.H. Allison, S. Andréfouët, N.L. Andrew, M.J. Batty, et al., Diversifying the use of tuna to improve food security and public health in Pacific Island countries and territories, Mar. Policy 51 (C) (2015) 584–591.

[34] A. Couper, H.D. Smith, B. Ciceri, Fishers and Plunderers: Theft, Slavery and Violence at Sea, Pluto Press, 2015.

[35] P. Lund-Thomsen, A. Lindgreen, Corporate social responsibility in global value chains: where are we now and where are we going? J. Bus. Ethics 123 (1) (2014) 11–22.

[36] W. Swartz, L. Schiller, U.R. Sumaila, Y. Ota, Market-based sustainability pathway: challenges and opportunities for seafood certification programs in Japan, Mar. Policy 76 (2017) 185–191.

SECTION 6

Governance and Well-Being in Changing Oceans

CHAPTER 33

The opportunities of changing ocean governance for sustainability

Henrik Österblom

Stockholm Resilience Centre, Stockholm University, Stockholm, Sweden

Chapter Outline

33.1 Introduction

The ocean is under threat. This is an increasingly frequent statement from scientists, civil society organizations, government representatives, from the private sector, and from people in general. Personally, I have heard this statement many times, almost to the point that I find it boring. Yes it is true, and it is critical to jointly identify problems and reach consensus that there is a problem—but how long do we have to talk about problems before we start describing, and also engaging in solutions? How many individuals do scientists need to convince that there is a problem before "they" start acting, whether or not "they" are corporations, governments, communities, or any other stakeholder with a stake in the ocean? What is our own role as scientists in stimulating change? Everyone, including scientists, has a stake in the ocean. Every second breath taken by oxygen-breathing inhabitants on earth is produced by the ocean. But how have we governed that stake thus far, is it changing, and are there any prospects for scaling up and increasing the necessary process of change?

Predicting Future Oceans.
DOI: https://doi.org/10.1016/B978-0-12-817945-1.00033-2

33.2 Ocean governance—what does it look like?

The preceding chapters illustrate the many ways in which the ocean is governed, through international cooperation by governments in regional fisheries organizations (see Chapter 36: New actors, new possibilities), and by civil society (see Chapter 42: Legitimacy has risks and benefits for effective international marine management). Presently an ongoing process of negotiation is likely to have substantial implications for how large parts of the ocean are governed (see Chapter 34: Climate change vulnerability and ocean governance, and Chapter 41: BBNJ and the open ocean). Other parts of this book have explored the role of market mechanisms, in the form of seafood trade (see Chapter 23: Future global seafood markets), consumption (see Chapter 17: Fisheries and seafood security under changing oceans, and Chapter 24: Climate change, contaminants, and country food), and corporate social responsibility (see Chapter 32: Corporate social responsibility in the global seafood industry).

The complexity of ocean challenges has been addressed in numerous recent publications (e.g., [1–3]). Ways in which such complex challenges can be effectively governed have also recently been documented [4]. The potential futures of governance have been explored in this book, for example, by looking at development and the risks associated to climate change [5], or conflict [6]. A number of possible, but not completely likely scenarios have also been explored—using science fiction methodology (see Chapter 48: Beyond prediction—radical ocean futures). These radical forms of scenarios can perhaps help us think differently about the future, and represent a fun and creative way to communicate science.

In the early phases of the Nippon Foundation Nereus Program, we started to explore the many aspects of the human dimension of the ocean—a mapping exercise to better understand where to focus our attention when thinking about the future of ocean governance [7]. The components we identified of perceived relevance for study were, among many things, the *implementing organizations*, the *political and institutional setting*, and the *societal norms*. This also identified the importance of understanding the *actors* involved—the resource users themselves, the *economy* (including *markets* and *technology*), and the *knowledge systems*. This was a starting point for our studies of global ocean governance—a framework within which we have spent the last few years developing and studying relevant case studies for an improved understanding of the future of the ocean. These human dimension parts are not stand-alone and independent from the ocean ecosystem, but rather are influenced by, and influence, the ocean.

Historical analysis of governance within the Nereus Program and beyond [8–10] illustrates how resource use commonly expands much faster than institutions can adapt to changing practices of resource users. The expansion of high seas fishing in the deep sea, for instance, resulted in the collapse of fish stocks, years before any institutions were in place [9]. Reactive governance consistently tried to catch up to expanding fishing activities, and will likely continue to struggle to keep up with changing resource use in the high seas, whether it is deep sea mining or any of the other rapidly developing "blue ocean economy" sectors [8,11]. The *implementing organizations* thus have appeared to struggle with some generic challenges, which are not unique for the ocean.

The real question now is whether or not an increased awareness of the problem will reduce this gap, and enable governance that is precautionary, anticipatory, and proactive, rather than reactive, to problems already in place, and with effects already influencing ecosystems and in societies.

History tells us that problems at scale are eventually dealt with, although in many different ways, and not always with the desired outcome if the goal is a more sustainable and equal world. In a case study aimed to investigate the *political and institutional setting*, we focused on the dynamics and geopolitics of the Soviet Union. Soviet overfishing in national coastal waters, for instance, resulted in a rapid expansion towards the coastal waters of neighboring countries, and beyond—first expanding to Norway and North America, and then all the way down to the end of the world [10]. This pattern of "roving bandits" [12], has been described for multiple fisheries, and for multiple nations (e.g., Eriksson et al. [13]). The current expansion of Chinese fishing activities toward a globally extensive distant water fleet [14] is following a similar pattern—stocks depleted, move on—generating substantial conflict with countries close to and far away from China [6]

Scientists are increasingly starting to clarify that there are biophysical limits to the pressure that ecosystems can take, not only as a consequence of overfishing [15], but also more generally [16,17]. Human pressure has been empirically shown to have large-scale and detrimental effects on ecosystems, over long time periods [18]. It is clearly not a sustainable strategy to continue with the current pressure, or moving the problem around, on a finite planet.

Institutions develop, albeit slowly—but is that development accelerating, and if not, are there ways in which they can accelerate? The establishment of national economic zones (Exclusive Economic Zones, EEZs) during the 1970s and 1980s represented a first foundation, where governments were better able to manage their own resources, or simply excluded the activities of other nations. These took several years to develop. To create *societal norms* and rules for who is part of your community, and clear rules for excluding those who are not, represented a foundation for common pool resource (CPR) management, as explained by Nobel Laureate Elinor Ostrom in her Design Principles for CPRs [19]. Within such EEZs governments have an extensive ability to influence how costal resources are governed—regulations and policies in the United States and Europe are now extensive. This does not mean that all stocks and all natural resources in these regions are in good shape, but it does mean that there are mechanisms in place that can substantially improve the coastal ocean in this region, provided that governments are interested in using their political power and will to ensure such development.

The UN Sustainable Development Goals (SDGs) are an explicit expression of the political will to ensure a safe, just, and sustainable planet for all. It is a new societal expression of intent for our relationship with the planet. This historically unique framework is expressing that the planet is fundamental for human well-being—and that all people on this planet have the same fundamental rights to development. How then, can we get there?

33.3 Is it fair?

The vast majority of the coastal regions in the world have much less developed *implementing organizations* for ocean governance than those in North America and Europe. The corresponding capacity to monitor and enforce compliance in, for example, sub-Saharan Africa or among small island development states, is in places terrifyingly limited. Participating in a conference in 2012, I attended a seminar where a charismatic government official demonstrated the capacity to monitor and enforce noncompliance of fisheries legislation in the coastal waters of his nation. He had four vessels at his disposal for the entire coastline of his country.

First, there was the 5 hp, outboard boat that probably could travel at a speed of 5 kn or so. Then there was the dugout canoe, which needed a few colleagues to power. Then there was the flat bottomed aluminum boat which had previously been used to transport cattle. I don't think it could top the 5 kn of the first boat. Finally, he had a slightly more modern boat, also in aluminum, and with a 50 hp engine. This was the pride of his fleet. A European NGO had bought that for him, and he was very happy about it. It was fast, and could probably reach 20 kn on a good day, but since it was less than 20 ft long, there is no way it could move out toward the larger waves, where the distant nation vessels were catching his fleet. Needless to say: no, this was not fair, and he did not stand a chance against the distant water fleets depleting the resources within his jurisdiction.

The global capacity to extract resources, monitor such extraction, and enforce laws (to the extent there are laws) is thus very differently distributed. Australia, for instance, spends hundreds of millions of dollars annually to monitor and enforce their Southern Ocean fisheries with high seas vessels, satellite monitoring, flyovers, and other advanced military technology [20]. A dramatically higher capacity than that of many countries, but for only one of its fisheries. The world is clearly not equal.

Neither are impacts on coastal communities equal from climate change. Analyses from Allison et al. [21] and Blasiak et al. [22] have illustrated the disproportionate impacts of climate change on vulnerable coastal communities in the Global South. These regions and their issues must then clearly be prioritized in international development aid programs, one would think? Instead, recent analysis of fisheries related overseas development aid illustrate that financial resources for the ocean have been declining substantially during the last few years [23]. As the problem is increasing, the support to deal with it is declining. Not the right direction to take if we are to meet the SDGs according to the agreed deadline. Especially if highly mobile resource users of the oceans consistently move toward less well governed areas to reduce the burdens from regulations, and as these regions are simultaneously becoming more vulnerable to the impacts to climate change.

33.4 Are there shortcuts?

Luckily, there are some shortcuts to this unequal monitoring and management capacity. And there may also be alternative and complementary ways to govern the ocean. There has been an increased attention to the *economy*, looking at *technology* and *markets* (e.g., [24]). An increased access to, and reduced cost of, advanced technological solutions is creating new prospects for ocean governance (see Chapter 35: (Re)constructing an ocean future). The emerging approaches include open access satellite data, monitoring of distant areas with drones, collecting information throughout complex and global supply chains using blockchain technology and also small, handheld devices (digital cameras, cell phones) to stimulate real-time monitoring of multiple resource users, each photographing suspected fishing vessels and sending their individual information to a coordinating authority who can then receive information movement on vessels who shut off their Automatic Identifier System in order not to be detected when fishing illegally in coastal waters of developing nations. Substantial and international education and training of enforcement personnel, and joint investigation of complex international fisheries crime cases across national borders [25] has been instrumental for creating incentives for illegal vessel operators to operate within the existing legal framework. Collaboration in the International Monitoring Control and Surveillance Network and increased cooperation within Interpol [26] are all signs that governments are actively exploring new ways of collaboration for improved fisheries governance.

The ocean is not only fisheries, but a range of other sectors to be managed. Most of these activities take place within national economic boundaries and are thus subject to national and regional (e.g., in the EU) law. During the Nereus Program, there has not only been an increasingly expressed concern for the future of the ocean, there have also been many voices raised for the importance of a blue economy (see Chapter 38: A Blue Economy). In 2018, the Norwegian Prime Minister Erna Solberg launched a High Level Group on the Ocean as a way to stimulate increased international cooperation for the ocean. In September of 2018, 12 heads of state announced their joint ambition during the United Nations General Assembly meeting in New York. This group will work together to present a report in 2020, aiming to launch a vision for future ocean governance and ways to combine conservation with use, in line with the SDGs. Similarly Deputy Prime Minister Isabella Lövin launched a complementary cooperative initiative—Friends of the Ocean Action Network—aimed at stimulating increased action across a wide range of organizations from all over the world for a sustainable ocean. This initiative is a logical follow-up of the first UN Ocean conference in 2017, which Sweden hosted together with Peter Thompson from Fiji, Special UN Envoy for the Ocean. These are promising, global responses to the ocean challenge in relation to the SDGs, but it remains to be seen how ambitious these government-led initiatives, in partnership with many actors, will be and what they will be able to deliver.

The United Nations Action Platform for Sustainable Ocean Business is a private sector initiative for global ocean governance, also launched in 2018. Industries from the seafood, shipping, oil and gas, minerals, insurance, and finance sectors are joining forces with science, as a means to improve ocean governance through voluntary commitments by companies. This will be good for business, they think. The process will follow the same time table as the above—aiming for a main output by 2020, although a first report from the group has already been published—mapping the global ocean governance landscape in relation to these sectors [27]. The group will build on numerous existing platforms for collaboration within the respective industries, and on the principles for the UN Global Compact. The work has only begun and it will be interesting to follow its development. It will be particularly interesting to see how these global companies, primarily headquartered in the North, will be able to address the global challenges of inequality, and the flow of benefits from the ocean.

Another and also relatively new approach to investigating and interaction with global *actors* in ocean governance is an ongoing collaboration between science and business for global ocean stewardship, with a focus on seafood. This initiative, called the Seafood Business for Ocean Stewardship (SeaBOS) builds on multiple existing voluntary agreements from aquaculture, wild capture fisheries, and feeds, and aims to stimulate global cooperation and learning between companies—such that they can all take on a leadership role in sustainability. This collaborative platform is the result of, first, the scientific identification of (what we refer to as) the *Keystone actors* who had the largest ability to stimulate systemic change [28], and then, a strategic and long-term engagement with them through dialogue and collaborative learning [29]. After three global dialogues among the CEOs of these largest companies of the world—from Japan, Korea, Norway, Thailand, the Netherlands, and the United States (between 2016 and 2018)—this platform for collaboration, learning, and action is starting to accelerate. This is not only an initiative that challenges the normal ways in which scientists interact with business, it also challenge us as scientists and helps us to learn many new things. It also requires careful attention to the possible negative side effects of working with such powerful actors, and how to avoid them.

33.5 The role of science in ocean governance

Given the scale and scope of the ocean challenge, a collaborative effort is clearly needed to reach the SDGs. This is where the last piece of the puzzle, the *knowledge systems* come in. Scientists have traditionally focused on producing and publishing their results in the most respectable scientific journals, and ensured that they learn how to effectively communicate their results. This is no longer enough, and the expectations for scientists are changing. Members of the scientific community have engaged in many different ways for an improved ocean governance. This is something we can all learn from and further develop. A critically important research question for sustainability is this: how can scientists actively engage in positive change, while still maintaining integrity and independence? Ocean science, is part of the *knowledge systems*, and should be regarded as part of ocean governance. We need to continue talking about the existing problems and identify new ones. But we also need to speak about the solutions. To neighbors, politicians, companies, and schoolchildren. We need to be part of shaping the solutions. Not only for the well-being of coastal communities, but also for the well-being of ourselves and the community we are part of.

The Nereus Program has been one of science, capacity building, and of communication. Its scientists have not been content with only writing papers. The program has illustrated the diverse roles that scientists can take and how we can also exercise our own agency to stimulate change. This engagement includes active participation in regional fisheries management organizations, and otherwise acting as advisors in national delegations to international negotiations. Nereus Program scientists engage in complex modeling and time-consuming processes in the International Panel for Climate Change—the global assessments that synthesize knowledge about the state of the ocean [1]. It has also involved practical engagement in on the ground fisheries management, as in the case with development of "move on rules" in the implementing organization for the New England fisheries [30]. Members of the group have been engaged in global biodiversity assessments and the identification of Ecological and Biological Significant Areas, an important preparatory work for the negotiations associated with Biodiversity Beyond National Jurisdiction.

We have engaged to advance the science of the past, present, and future of the ocean, and these results have been communicated to political leaders and CEOs. The Nereus Program community has also engaged with coastal and indigenous communities, schoolchildren, and scientific colleagues from diverse disciplines. The future of the ocean is also about the future of its people. Targeted scientific publications to schoolchildren—the leaders of the future—has thus been prioritized as part of the effort to stimulate change [31,32]. There are multiple options for changing ocean governance for sustainability. A rapidly developing field of science is exploring how collaborative research between scientists and practitioners can stimulate positive change. The key words are co-design and co-production of knowledge for action [33,34]. Our own creativity should be the only limitation.

References

[1] J.-P. Gattuso, A. Magnan, R. Billé, W.W.L. Cheung, E.L. Howes, F. Joos, et al., Contrasting futures for ocean and society from different anthropogenic CO_2 emissions scenarios, Science 349 (2015) aac4722.

[2] B.I. Crona, T.M. Daw, W. Swartz, A.V. Norström, M. Nyström, M. Thyresson, et al., Masked, diluted and drowned out: how global seafood trade weakens signals from marine ecosystems, Fish Fish. 17 (2016) 1175–1182.

[3] H. Österblom, B. Crona, C. Folke, M. Nyström, M. Troell, Marine ecosystem science on an intertwined planet, Ecosystems 20 (2017) 54–61. Available from: https://doi.org/10.1007/s10021-016-9998-6.
[4] L. Schultz, C. Folke, H. Österblom, P. Olsson, Adaptive governance, ecosystem management, and natural capital, Proc. Natl. Acad. Sci. U.S.A. 112 (2015) 7369–7374.
[5] R. Blasiak, J.-B. Jouffray, C.C.C. Wabnitz, E. Sundström, H. Österblom, Corporate control and global governance of marine genetic resources, Sci. Adv. 4 (6) (2018) eaar5237.
[6] J. Spijkers, T.H. Morrison, R. Blasiak, G.S. Cumming, M. Osborne, J. Watson, et al., Marine fisheries and future ocean conflict, Fish Fish. 19 (5) (2018) 798–806.
[7] H. Österblom, M. Merrie, M. Metian, W.J. Boonstra, T. Blenckner, J.R. Watson, et al., Modeling social-ecological scenarios in marine systems, BioScience 9 (2013) 735–744.
[8] A. Merrie, C. Daniel, D.C. Dunn, M. Metian, A.M. Boustany, Y. Takei, et al., Human use trends, potential surprise and governance challenges in the global marine commons, Global Environ. Change 27 (2014) 19–31.
[9] E.A. Norse, S. Brooke, W.W.L. Cheung, M.R. Clark, I. Ekeland, R. Froese, et al., Sustainability of deep-sea fisheries, Mar. Policy 36 (2012) 307–320.
[10] H. Österblom, C. Folke, Globalization, marine regime shifts and the Soviet Union, Philos. Trans. R. Soc. B 370 (2015) 1–8.
[11] J.S. Golden, J. Virdin, D. Nowacek, P. Halpin, L. Bennear, P.G. Patil, Making sure the blue economy is green, Nat. Ecol. Evol. 1 (2017) 0017.
[12] F. Berkes, T.P. Hughes, R.S. Steneck, J.A. Wilson, D.R. Bellwood, B. Crona, et al., Globalization, roving bandits, and marine resources, Science 311 (2006) 1557–1558.
[13] H. Eriksson, H. Österblom, B. Crona, M. Troell, N. Andrew, J. Wilen, et al., Contagious exploitation of marine resources, Front. Ecol. Environ. 13 (2015) 435–440. Available from: https://doi.org/10.1890/140312.
[14] D. Pauly, D. Belhabib, R. Blomeyer, W.W.W.L. Cheung, A. Cisneros-Montemayor, D. Copeland, et al., China's distant water fisheries in the 21st century, Fish Fish. 15 (2014) 474–488.
[15] S.J. Lade, S. Niiranen, J. Hentati-Sundberg, T. Blenckner, W.J. Boonstra, K. Orach, et al., An empirical model of the Baltic Sea reveals the importance of social dynamics for ecological regime shifts, Proc. Natl. Acad. Sci. U.S.A. 112 (2015) 11120–11125.
[16] J. Rockström, W. Steffen, K. Noone, Å. Persson, F.S. Chapin III, E.F. Lambin, et al., A safe operating space for humanity, Nature 461 (2009) 472–475.
[17] W. Steffen, J. Rockström, K. Richardson, T.M. Lenton, C. Folke, D. Liverman, et al., Trajectories of the Earth System in the Anthropocene, Proc. Natl. Acad. Sci. U.S.A. 115 (2018) 8252–8259.
[18] J.B. Jackson, M.X. Kirby, W.H. Berger, K.A. Bjorndal, L.W. Botsford, B.J. Bourque, et al., Historical overfishing and the recent collapse of coastal ecosystems, Science 293 (2001) 629–637.
[19] E. Ostrom, T. Dietz, N. Dolšak, P.C. Stern, S. Stonich, E.U. Weber, et al., The Drama of the Commons, The National Academies Press, Washington, DC, 2002.
[20] H. Österblom, C. Folke, Emergence of global adaptive governance for stewardship of regional marine resources, Ecol. Soc. 18 (2) (2013) 4.
[21] E.H. Allison, A.L. Perry, M. Badjeck, N.W. Adger, K. Brown, D. Conway, et al., Vulnerability of national economies to the impacts of climate change on fisheries, Fish Fish. 10 (2009) 173–196.
[22] R. Blasiak, J. Spijkers, K. Tokunaga, J. Pittman, N. Yagi, H. Österblom, Climate change and marine fisheries: least developed countries top global index of vulnerability, PLoS One 12 (2017) e0179632.
[23] R. Blasiak, C.C.C. Wabnitz, Aligning fisheries aid with international development targets and goals, Mar. Policy 88 (2018) 86–92.
[24] D. McCauley, P. Woods, B. Sullivan, B. Bergman, C. Jablonicky, A. Roan, et al., Ending hide and seek at sea, Science 351 (6278) (2016) 1148–1150.
[25] V. Galaz, H. Österblom, Ö. Bodin, B.I. Crona, Global networks and global change induced tipping points, Int. Environ. Agreements 1 (2015) 33.
[26] H. Österblom, Catching up on fisheries crime, Conserv. Biol. 28 (2014) 877–879.

[27] B. Pretlove, R. Blasiak, Mapping Ocean Governance and Regulation, UN Global Compact, Sustainable Ocean Business, 2018.
[28] H. Österblom, J.B. Jouffray, C. Folke, B. Crona, A. Merrie, M. Troell, et al., Transnational corporations as 'Keystone Actors' in marine ecosystems, PLoS One 10 (5) (2015) e0127533.
[29] H. Österblom, J.-B. Jouffray, C. Folke, J. Rockström, Emergence of a global science-business initiative for ocean stewardship, Proc. Natl. Acad. Sci. U.S.A. 114 (2017) 9038–9043.
[30] D.C. Dunne, A.M. Boustany, J.J. Roberts, E. Brazer, M. Sanderson, B. Gardner, et al., Empirical move-on rules to inform fishing strategis: a New England case study, Fish Fish. 15 (2013) 359–375.
[31] H. Österblom, J.-B. Jouffray, C. Folke, J. Rockström, Can science and business work together to save the ocean? Sci. J. Teens (in press).
[32] D. Pauly, W. Cheung, Is climate change shrinking our fish, Environ. Sci. J. Teens (2017). <http://www.sciencejournalforkids.org/uploads/5/4/2/8/54289603/gills_article.pdf>.
[33] M. Tengö, E.S. Brondizio, T. Elmqvist, P. Malmer, M. Spierenburg, Connecting diverse knowledge systems for enhanced ecosystem governance: the multiple evidence base approach, AMBIO 43 (2014) 579–591.
[34] M. Tengö, R. Hill, P. Malmer, C.M. Raymond, M. Spierenburg, F. Danielsen, et al., Weaving knowledge systems in IPBES, CBD and beyond: lessons learned for sustainability, Curr. Opin. Environ. Sustain. 26-27 (2017) 17–25.

CHAPTER 34

Climate change vulnerability and ocean governance

Robert Blasiak

Stockholm Resilience Centre, Stockholm University, Stockholm, Sweden

Chapter Outline

34.1 A growing storm

Human well-being is inextricably linked with the ocean. It shapes local weather patterns and influences the global climate. It provides the largest habitat for life on Earth, fostering a diversity of organisms whose genetic material contains the keys to a rapidly expanding biotechnology sector. It has shaped rich cultural traditions and practices in coastal communities around the world. It is also the last major source of foraged food on the planet, providing 12% of the world's population with livelihoods, while supplying 17% of its animal protein [1]. From prehistoric times its currents and winds have been harnessed by human ingenuity to allow the movement of people and goods across much of the Earth's surface. Over millennia, it has captured the human imagination as a place of riches and mysteries, or as filmmaker Werner Herzog once put it: "What would an ocean be without a monster lurking in the depths? It would be like sleep without dreams."

The expansion of human activities into the most remote parts of the ocean suggests that the ocean today is itself a vast social–ecological system [2]. One of the forces rapidly shaping the ecosystems of the ocean is climate change driven by anthropogenic greenhouse gas (GHG) emissions. The myriad impacts of climate change are fundamentally changing the ocean's biogeochemistry—leading among other things to changes in primary production,

Predicting Future Oceans.
DOI: https://doi.org/10.1016/B978-0-12-817945-1.00039-3

salinity, acidity levels, and sea surface temperature. These factors, in turn, result in shifts in species distribution, size, migratory paths, and abundance levels [3,4]. Coastal communities dependent on local marine resources are subsequently affected [5], and the impacts eventually ripple across supply chains and global markets connecting all the nations of the world [6].

This chapter introduces recent work to understand how the ocean and the communities dependent upon it are changing, and what is being done to adapt to this evolving context. The focus starts at the global level, where climate change signals are perhaps the most extensively studied, and then shifts to a local lens to understand what these global trends mean at a community level. Finally, new research is introduced on how people perceive the value of the oceans, and what this could mean for the future of ocean governance.

34.2 Hotspots of vulnerability in an ocean under climate change

The public availability of climate model output data of ever increasing quality has provided researchers with the means to anticipate future ocean conditions. Significant initial steps have been taken [7,8] to couple climate model output data across exclusive economic zones with national level socioeconomic variables to understand which countries will be most vulnerable to climate change impacts on fisheries. These efforts build on several assumptions, most prominently, that climate change will alter the distribution and abundance of fish stocks in the future, and that vulnerability is a function of exposure, sensitivity, and adaptive capacity (in line with the framework of the Fourth Report of the Intergovernmental Panel on Climate Change [9]). In Blasiak et al. [6] this methodology was expanded and updated, resulting in a vulnerability index for 147 countries indicating the vulnerability of national economies to climate change impacts on fisheries (Fig. 34.1).

Some novel steps were taken in the construction of this vulnerability index, in addition to updating the data used in previous analyses [7,8], often by over a decade. In particular, steps were taken to gain greater insight into the impact of the exposure variable on vulnerability. Multimodel ensemble (MME) means were constructed for 14 different climate models for sea surface temperature anomalies against a historical reference time frame. These data were collected from Phase 5 of the Coupled Model Intercomparison Project (CMIP5), which defines a common framework and shared protocols to support the development and comparison of climate models. These MME means were calculated for three different possible GHG concentration trajectories (so called representative concentration pathways (RCPs)) ranging from optimistic (RCP 2.6, where GHG emissions peak between 2010 and 2020) to pessimistic (RCP 8.5, with emissions rising throughout the 21st century). In addition, near- and distant-future scenarios were considered by calculating MME means across two time periods: 2016–50 and 2066–2100.

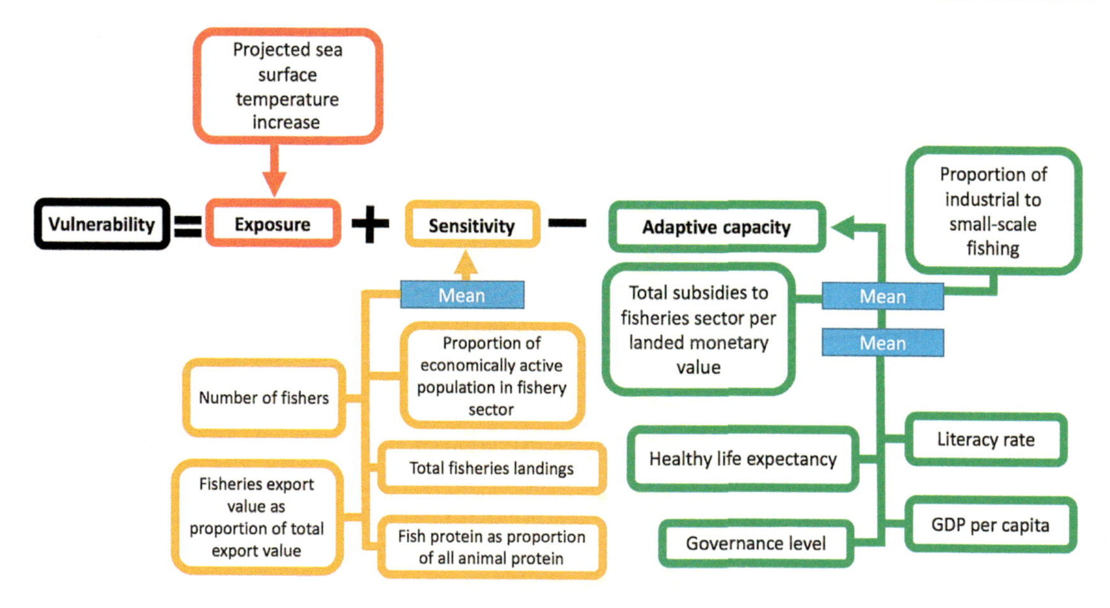

Figure 34.1
Variables used to construct index of vulnerability to climate change impacts on fisheries. *Reproduced from R. Blasiak, J. Spijkers, K. Tokunaga, J. Pittman, N. Yagi, H. Österblom, Climate change and marine fisheries: least developed countries top global index of vulnerability, PLoS One 12 (6) (2017) e0179632.*

Blasiak et al.'s analysis suggests that climate change impacts on fisheries pose the greatest threat to countries at the lower end of the development spectrum, many of which are highly dependent on fish as a source of nutrition and livelihoods. The study furthermore supports previous findings that the countries most responsible for driving climate change may also be among the least vulnerable to resulting impacts: a negative correlation was identified between per capita carbon emissions and scores on the vulnerability index (Spearman's $\rho = -0.60$, $R^2 = 0.300$, $P < .0001$) [6]. Moreover, the inherent unpredictability of extreme climatological events, such as hurricanes or heatwaves, exacerbated by climate change raises the specter of fishery conflicts [10,11] and instability in cooperative management bodies and agreements aimed at sustainable management of shared, straddling, and highly migratory fish stocks [12–14].

34.3 Resilience and vulnerability at the local level

While climate change is a global phenomenon and these studies help anticipate and plan for potential future conditions, they come with a substantial list of caveats and limitations. Most prominently, they use future projections of biogeochemical ocean conditions, but couple these with static socioeconomic variables. Yet factors such as employment in the fisheries sector, reliance on fish as a source of protein, and value of fisheries exports will not remain static in a dynamic world. A comparative study [15] of people's perceptions in Matsushima Bay (in northeastern Japan) and the Salish Sea (straddling the west coast of Canada and the United States) underscores these factors. Both regions have seen rapid changes over the past decades—one senior fisheries manager emphasized that "Our social systems are in as much change, if not more, as our ecological systems" [15]. Respondents' justifications for deciding to enter, remain in, or leave the fisheries sector were deeply personal and various, ranging from pragmatic questions of entry costs and economic insecurity to emotional connections to nature and the sea [15,16]. In several cases the globalized nature of seafood value chains, while providing opportunities for increased benefits from new markets, had also created a new source of vulnerability, as communities were increasingly reliant on export markets—corresponding price fluctuations for these products and for fuel represented factors beyond the influence or control of local communities [15,17]. As with other social–ecological systems, resilience is therefore inextricably linked to both the changes in ocean conditions and the communities themselves [18], allowing them to "absorb disturbance and reorganize while undergoing change so as to still retain essentially the same function, structure, identity and feedbacks" [19].

People's perceptions of the value of the ocean are also changing, and vary widely between different nations and regions of the world. A survey of 1434 residents of the United States, for instance, showed strong correlations between age and engagement with various forms of sustainable ocean management, but no correlation between proximity to the ocean and perceptions of the indispensability of marine ecosystem services. Such services—described by some as "nature's contribution to people"—are typically grouped into three categories: (1) Provisioning services providing tangible products, such as fish and timber; (2) regulating services, such as coastal protection or carbon storage; and (3) cultural services (e.g., recreation). In the above survey, provisioning services such as seafood and energy were considered the most indispensable benefit provided by the ocean [12]. Conversely, with regard to willingness to act to conserve and sustainably manage marine ecosystems, similar mass surveys in Japan found a stronger correlation with cultural ecosystem services than provisioning ecosystem services [20,21]. Such perceptions take on an additional layer of importance when trying to understand the added value consumers see in seafood products that have received sustainability or stewardship certifications. Home to several of the world's top producers and consumers of fish, East Asia is of crucial importance for achieving ocean stewardship, yet international certification mechanisms have struggled in these markets [22]. The creation of a large number of international, regional, and domestic schemes for the ecolabeling and certification of fishery and aquaculture products has the potential to confuse consumers and retailers alike, but may be a first step toward allowing a diversity of mechanisms tailored to local and regional norms—a corresponding benchmarking of these mechanisms to ensure that they align with underlying internationally agreed policy documents like the FAO Code of Conduct for Responsible Fisheries could then help to distinguish reliable schemes [23].

34.4 Potential mechanisms to promote sustainable ocean and coastal management

The countries most vulnerable to climate change impacts on fisheries also face particularly challenging development contexts—many are remote small island states with few alternative industries that are simultaneously facing the risk of severe natural disasters like cyclones and hurricanes, while seeing the damage caused by these natural disasters aggravated by rising sea levels. ODA has been one of the primary mechanisms for donor governments to reach development targets [24]. Encouragingly ocean issues seem to be rising up the international development agenda, as reflected in a rapidly expanding list of international commitments to the conservation and sustainable use of marine resources [25], but a recent survey of world leaders found SDG 14 on Life Below Water to be considered the lowest priority among the SDGs [26]. The urgency of mobilizing support and the prominence of ocean policy issues do not seem to be mirrored in decisions on how to allocate ODA related to sustainable fisheries—from 2010 to 2015, such funding decreased by 30%, with states in Oceania seeing a disproportionate drop (−43%) [25]. Allocations of ODA for fisheries research decreased by 77% over the same time frame. Such decreases would be less surprising if overall ODA across all sectors were also dropping, but it increased by over 13% to set a new record in 2015 [25]. A network analysis of donor–recipient linkages suggests that allocation decisions have less to do with a systematic assessment of need, vulnerability, or development potential, and are more likely directed by short-term agendas, geopolitical circumstances, and a lack of coordination [27]. Meanwhile, philanthropic support for ocean issues has grown rapidly in recent years to now exceed ODA, although the geographic overlap in the two forms of support is minimal [28]. The swift pace of change with regard to ocean finance suggests opportunities for more positive development and conservation outcomes, but depends on commitments to transparency and coherence as well as a renewed focus on establishing the impacts of supported projects and initiatives [29,30].

A parallel challenge is the management of marine areas beyond national jurisdiction that comprise nearly half of the Earth's surface (ABNJ), where a fragmented institutional framework has left a substantial set of gaps with regard to the conservation and sustainable use of biodiversity found there (BBNJ). In 2004 the UN General Assembly (UNGA) created a BBNJ working group, and over a decade of negotiations led up to the December 2017 UNGA decision to start brokering an international treaty on BBNJ. The treaty is likely to address the establishment of marine protected areas in ABNJ, environmental impact assessments, access and benefit sharing related to marine genetic resources, and capacity building and transfer of marine technology [31]. As with the UN Fish Stocks Agreement, this new treaty is a response to rapidly expanding exploitation [32]. In the case of marine genetic resources, for instance, patents associated with gene sequences from over 800 marine species have already been filed, many from hydrothermal vent systems found in ABNJ [33]. Yet the filing of patents associated with MGRs is limited to entities in just 30 countries, while over 160 countries are inactive in this form of ocean use [33]. While the negotiation of a new treaty to conserve and sustainably use BBNJ is a positive and encouraging step, significant imbalances in state representation across the 2005–17 BBNJ negotiations [31,34] suggest that the outcome may be inequitable unless these imbalances are addressed in future negotiations. Scientists are engaged in a range of both commercial and non-commercial activities associated with MGRs and, along with industry actors, should bemore explicitly incorporated into negotiation processes [35,36].

Simply allocating more resources or strengthening international agreements for sustainable ocean and coastal management, however, does not guarantee positive outcomes. Regionally and locally appropriate approaches are needed to effectively assess and promote healthy ecosystems and resilient communities. The Ocean Health Index, for instance, began as a global assessment of social and ecological factors relevant to ocean health, but has sought to tailor its methodology to smaller-scale regional and local assessments [23]. The international effort to identify ecologically or biologically significant areas has also provided governments with a better understanding of marine areas requiring tailored management approaches, including restrictions or bans on activities such as fishing or mining. Both efforts have faced criticism for insufficiently addressing social issues in their methodologies. Yet a suite of recent initiatives and frameworks, including the principles included in the FAO Guidelines on Small-scale Fisheries, provide insight into how these efforts can be bolstered in the future.

Acknowledgments

I would like to express thanks to my colleagues within the Nereus Program for many enriching discussions and collaborations, and to the Nippon Foundation for supporting ocean science and making all of this possible. Special thanks are due to Andrés Cisneros-Montemayor, William W.L. Cheung, and Colette C.C. Wabnitz for their constructive comments on drafts of this chapter.

References

[1] F.A.O. Fisheries, The State of World Fisheries and Aquaculture. Food and Agriculture Organization of the United Nations, Rome, 2016.

[2] H. Österblom, B.I. Crona, C. Folke, M. Nyström, M. Troell, Marine ecosystem science on an intertwined planet, Ecosystems 20 (1) (2017) 54–61.

[3] W.W. Cheung, V.W. Lam, J.L. Sarmiento, K. Kearney, R.E.G. Watson, D. Zeller, et al., Large-scale redistribution of maximum fisheries catch potential in the global ocean under climate change, Global Change Biol. 16 (1) (2010) 24–35.

[4] H.O. Pörtner, M.A. Peck, Climate change effects on fishes and fisheries: towards a cause-and-effect understanding, J. Fish Biol. 77 (2010) 1745–1779.
[5] M.C. Badjeck, E.H. Allison, A.S. Halls, N.K. Dulvy, Impacts of climate variability and change on fishery-based livelihoods, Mar. Policy 34 (3) (2010) 375–383.
[6] R. Blasiak, J. Spijkers, K. Tokunaga, J. Pittman, N. Yagi, H. Österblom, Climate change and marine fisheries: least developed countries top global index of vulnerability, PLoS One 12 (6) (2017) e0179632.
[7] E.H. Allison, A.L. Perry, M.C. Badjeck, W. Neil Adger, K. Brown, D. Conway, et al., Vulnerability of national economies to the impacts of climate change on fisheries, Fish Fish. 10 (2) (2009) 173–196.
[8] M. Barange, G. Merino, J.L. Blanchard, J. Scholtens, J. Harle, E.H. Allison, et al., Impacts of climate change on marine ecosystem production in societies dependent on fisheries, Nat. Clim. Change 4 (3) (2014) 211.
[9] J.J. McCarthy (Ed.), Climate Change 2001: Impacts, Adaptation, and Vulnerability: Contribution of Working Group II to the Third Assessment Report of the Intergovernmental Panel on Climate Change., Cambridge University Press, 2001.
[10] J. Spijkers, T.H. Morrison, R. Blasiak, G.S. Cumming, M. Osborne, J. Watson, et al., Marine fisheries and future ocean conflict, Fish Fish. 19 (5) (2018) 798–806.
[11] J. Spijkers, G. Singh, R. Blasiak, T.H. Morrison, P. Le Billon, H. Österblom, Global patterns of fisheries conflict: Forty years of data, Global Environmental Change 57 (2019). Available from: https://doi.org/10.1016/j.gloenvcha.2019.05.005.
[12] R. Blasiak, N. Yagi, H. Kurokura, K. Ichikawa, K. Wakita, A. Mori, Marine ecosystem services: perceptions of indispensability and pathways to their sustainable use, Mar. Policy 61 (2015) 155–163.
[13] R. Blasiak, Balloon effects reshaping global fisheries, Mar. Policy 57 (2015) 18–20.
[14] M.L. Pinsky, G. Reygondeau, R. Caddell, J. Palacios-Abrantes, J. Spijkers, W.W.L. Cheung, Preparing ocean governance for species on the move, Science 360 (2018) 1189–1191.
[15] A. Minohara, C. Cooling, R. Blasiak, Coastal communities and livelihoods in a changing world: a comparison of the fisheries and aquaculture sector in Matsushima Bay and the Salish Sea, Satoyama Initiative Thematic Review, vol. 3, UNU Press, Tokyo, Japan, 2018.
[16] A. Minohara, R. Blasiak, Socio-ecological linkages in Japan's Urato Islands, Satoyama Initiat. Thematic Rev. 1 (2015) 29–36.
[17] Y.H. Lu, N. Yagi, R. Blasiak, Factors contributing to effective management in the Sakuraebi (*Sergia lucens*) fishery of Donggang, Taiwan, Mar. Policy 86 (2017) 72–81.
[18] N. Bergamini, R. Blasiak, P. Eyzaguirre, K. Ichikawa, D. Mijatovic, F. Nakao, et al., Indicators of resilience in socio-ecological production landscapes (SEPLs), United Nations University Policy Report, 2013, 44 pp.
[19] B. Walker, C.S. Holling, S.R. Carpenter, A. Kinzig, Resilience, adaptability and transformability in social–ecological systems, Ecol. Soc. 9 (2) (2004).
[20] Z. Shen, K. Wakita, T. Oishi, N. Yagi, H. Kurokura, R. Blasiak, et al., Willingness to pay for ecosystem services of open oceans by choice-based conjoint analysis: a case study of Japanese residents, Ocean Coastal Manage. 103 (2015) 1–8.
[21] K. Wakita, Z. Shen, T. Oishi, N. Yagi, H. Kurokura, K. Furuya, Human utility of marine ecosystem services and behavioural intentions for marine conservation in Japan, Mar. Policy 46 (2014) 53–60.
[22] R. Blasiak, J.H.W. Huang, H. Ishihara, I. Kelling, S. Lieng, H. Lindoff, et al., Promoting diversity and inclusiveness in seafood certification and ecolabelling: prospects for Asia, Mar. Policy 85 (2017) 42–47.
[23] R. Blasiak, E.J. Pacheco, K. Furuya, C. Golden, A.R. Jauharee, Y. Natori, et al., Local and regional experiences with assessing and fostering ocean health, Mar. Policy 71 (2016) 54–59.
[24] W. Hynes, S. Scott, The Evolution of Official Development Assistance., OECD, Paris, France, 2013, p. 28.
[25] R. Blasiak, C.C.C. Wabnitz, Aligning fisheries aid with international development targets and goals, Mar. Policy 88 (2018) 86–92.

[26] S. Custer, M. DiLorenzo, T. Masaki, T. Sethi, A. Harutyunyan, Listening to Leaders 2018: Is Development Cooperation Tuned-In or Tone-Deaf? AidData at the College of William & Mary, Williamsburg, VA, 2018.

[27] J. Pittman, C.C.C. Wabnitz, R. Blasiak, A global assessment of structural change in development funding for fisheries, Mar. Policy (in review).

[28] M. Berger, V. Caruso, E. Peterson, New perspectives on a historical underestimation of marine conservation funding, Mar. Policy (in review).

[29] R. Blasiak, C.C. Wabnitz, T. Daw, M. Berger, A. Blandon, G. Carneiro, B. Crona, M.F. Davidson, S. Guggisberg, J. Hills, F. Mallin, Towards greater transparency and coherence in funding for sustainable marine fisheries and healthy oceans, Marine Policy (2019). Available from: https://doi.org/10.1016/j.marpol.2019.04.012.

[30] C. Wabnitz, R. Blasiak, The rapidly changing world of ocean finance, Marine Policy (2019). Available from: https://doi.org/10.1016/j.marpol.2019.103526.

[31] R. Blasiak, J. Pittman, N. Yagi, H. Sugino, Negotiating the use of biodiversity in marine areas beyond national jurisdiction, Front. Mar. Sci. (2016). Available from: https://doi.org/10.3389/fmars.2016.00224.

[32] R. Blasiak, N. Yagi, Shaping an international agreement on marine biodiversity beyond areas of national jurisdiction: lessons from high seas fisheries, Mar. Policy 71 (2016) 210–216.

[33] R. Blasiak, J.B. Jouffray, C.C. Wabnitz, E. Sundström, H. Österblom, Corporate control and global governance of marine genetic resources, Sci. Adv. 4 (6) (2018) eaar5237.

[34] R. Blasiak, C. Durussel, J. Pittman, C.A. Senit, M. Petersson, N. Yagi, The role of NGOs in negotiating the use of biodiversity in marine areas beyond national jurisdiction, Mar. Policy 81 (2017) 1–8.

[35] R. Blasiak, International regulatory changes poised to reshape access to marine genes, Nature Biotechnology 37 (2019) 357–358.

[36] R. Blasiak, J.B. Jouffray, C.C.C. Wabnitz, H. Österblom, Scientists should disclose origin in marine gene patents, Trends in Ecology and Evolution 34 (5) (2019) 392–395.

CHAPTER 35

The last commons: (re)constructing an ocean future

Katherine Seto and Brooke Campbell

Australia National Centre for Ocean Resources and Security (ANCORS), University of Wollongong, Wollongong, NSW, Australia

Chapter Outline

35.1 Introduction

Recent research on the global ocean has emphasized diverse, dramatic, and largely ominous shifts in the ways that oceans function as a socionatural system. Frequently cited studies indicate that global fish stocks are generally in decline, with 90% of assessed commercial stocks fully or overfished [1]. Other research suggests that due to this systematic and long-term overharvesting, as well as pollution, habitat loss, and other anthropogenic stressors, marine biodiversity in general is in decline [2,3]. Still more research indicates that global environmental drivers like climate change will lead to ocean warming, sea level rise, ocean acidification, increased storm intensity, ecosystem simplification and degradation, and overall unparalleled disruptions to the biophysical properties of most major oceans [4–7]. Attending these alarming findings in marine systems are further projections of their effects on human and coupled socionatural systems. Recent literature has tied degradation of marine systems to such diverse social outcomes as food security and food sovereignty [8–11], loss of livelihoods [12,13], conflict [14,15], and human rights abuses [16,17].

Predicting Future Oceans.
DOI: https://doi.org/10.1016/B978-0-12-817945-1.00037-X

While all of this research provides a critical body of evidence for understanding what "future ocean" we will experience, the future of nature–society relations is ultimately about shaping *human* systems and behaviors [18,19]. One of the fundamental ways in which these relations are shaped and mediated is through the notion of property; as Mansfield states, "property has become the central mode of regulating multiple forms of nature" [20]. Property, understood not as an object itself, but as a social relation, is often defined as "a claim that will be enforced by society or the state, by custom or convention or law... Property is a political relation between persons" [21]. As such, it is clear that neither space (e.g., the sea) nor resources (e.g., fish) are themselves intrinsically "property," but become such through their articulation in the nature–society relationship through the conferring of rights.

The last several decades have seen notable changes in the ways that we think about ocean spaces and resources. New property regimes have come into play, accompanied by novel institutions and modes of governance, with considerable consequences for individuals, communities, and political and economic systems. Despite the early dominance of diverse oceanic property arrangements (e.g., open access, communal, individual, state, etc.), a growing body of evidence suggests that property configurations at sea are increasingly resembling the established neoliberal political economic relations seen on land, including the enclosure, privatization, commodification, and marketization of previously untargeted forms of "natural capital" [22–25]. Well-developed literatures in agrarian political economy, political ecology, critical physical geography, and environmental governance have demonstrated that these neoliberal configurations have in many cases resulted in increasing resource disparities, concentrations of wealth, and the dispossession of poor and marginalized individuals and groups [26].

In this chapter we identify historical trends in the enclosure of ocean spaces and resources, how these enclosures emerged, and the discourses that have facilitated their expansion. We contrast the original goals and expectations of actors in creating these enclosures with the reality of their consequences in practice, highlighting some of the outcomes for developing countries and marginalized actors. We then briefly explore two trends in contemporary marine enclosure, focusing on the role of privatization in both the nature of enclosure and its social consequences. Finally, we conclude by contrasting two possible "future oceans": one brought forth from continued support for neoliberal approaches to managing our relationship with the sea, and an alternate future, in which we purposefully reshape nature–society relations with the sea around goals of equity and empowerment.

35.2 "The largest single enclosure in history": a changing ocean regime

As "projects associated with territory making and unmaking," efforts to define property in seaspace have occurred for millennia [27–29]. However, most of these efforts remained localized and relatively small in scale until the 16th and early 17th centuries, when expanding maritime claims from multiple European powers prompted the juridical debates now considered the foundation of modern property relations at sea [27,28,30]. This debate emerged at a time of conflicting state practice wherein some states (e.g., England, Spain, Portugal) sought to promote their interests through increased appropriation of resources and control of seaspaces, whereas others (e.g., Netherlands) supported open access to enhance freedom of navigation and capitalist trade [27,30]. As contemporary publicists sought to engender support for these divergent causes, the debate became characterized by two opposing views. *Mare liberum*, or "freedom of the seas," championed by Hugo Grotius, asserted that oceans cannot be appropriated because of their abundance and because they cannot be occupied, as "all property has arisen from occupation" [31]. In contrast, *mare clausum*, or "closed seas," attributed to John Selden, claimed that state practices of maritime enclosure and control were not only legitimate, but long-standing [32]. While the Grotian view of ocean property came to dominance, this perspective prevailed not simply on its theoretical superiority, but because of the specific limitations of these claims at the time. As a social relation, property is comprised of both the assertion of property by the rights holder (this is mine) and the concession as property from those without rights (I agree to act as though that is yours) [20,33,34]. Friedheim and others suggest that these state claims to ocean enclosure in the 17th century failed precisely because of (1) the limited capacity of states to actually control activity within the claimed spaces, and (2) the rejection of the claims on the part of other state actors [30,34]. The ocean, commonly perceived at the time as unoccupied, unbounded, limitless, and abundant, neither necessitated nor lent itself to the notion of enclosure [34,35].

This *mare liberum* perspective prevailed until the mid-20th century, when several changes shifted these dominant conditions and discourses. First, rapid changes in vessel and surveillance technology fundamentally changed states' perceptions of their ability to appropriate and control ocean spaces [34,36]. Accompanying this change was a similar shift in the overall acceptability of maritime enclosure [30,37]. Finally, in addition to the newly perceived *ability* of states to control ocean territories, global environmental and economic discourse began to suggest the *necessity* of this enclosure. Legal theorists had for centuries debated the seas as either *res nullius* (property of no one, e.g., "open access"), or *res communis* (common possession, e.g., "common heritage") [34], understood as the difference between "open access" resources under no property arrangement versus commonly managed resources under communal property arrangements [18,38]. In the 20th century this debate shifted almost entirely to discourses of ocean resources as "open access," not only susceptible to overexploitation in terms of Hardin's Tragedy of the Commons [39], but also at risk of losing their economic potential, just as states came to perceive them as capable of exclusive exploitation [36]. This framing of oceans and their wealth as "open access"—inevitably culminating in overexploitation and depletion—produced a necessity for closed access, or the introduction of property rights to secure potential economic investment and ensure sustainability [18,20,25,40].

These emerging discourses presupposed a need for property rights that remains strong today. This necessity was gradually realized through a series of international and national policy shifts that culminated in the UN Convention on the Law of the Sea (UNCLOS). Under UNCLOS, sovereign rights to the resources of the continental shelf and exclusive economic zone (EEZ) extending 200 nm were granted to the coastal state, in an act which has been called "the largest single enclosure in history" [27]. This shift toward granting sovereign rights to states, where they had never previously existed, resulted in both expected and unanticipated consequences, explored in the following section.

35.3 Sustainable efficiency or inequitable dispossession: tracing the legacies of neoliberal enclosure

The material consequences of the property regime shift from the "property of no one" to state property were substantial. With the establishment of UNCLOS, in a single act, 36% of the worlds ocean surface, more than a third of its seabed, and 90% of its fisheries resources were enclosed as state property [36,41,42]. However, these material consequences also engendered significant discursive and normative consequences, and Alcock states that "few seem to fully appreciate the ongoing evolution in property rights systems that has been triggered by this change" [36]. The shift in ocean wealth to individual state property created, for the first time, the potential for states to extract ground rent, or the right to charge fees for access to the space and resources within its EEZ [27]. Prevailing fisheries economics and property theories at the time suggested that this conferral of rights and responsibilities would result in increased investments in sustainability and maximization of resource benefits [43,44]. However, as diverse scholars have demonstrated, instead of increasing stewardship and improving the welfare of citizens, the shift to state control largely engendered a series of neoliberal incentives of maximizing profits. Indeed, many have argued that this shift to state control was only a necessary precondition for a more complete shift toward the "perfect right," a private right [45]. Friedheim states that "from the earliest discussions of what analysts hoped UNCLOS III might accomplish, the idea was to enclose as much space and resources as possible to designated stakeholders with exclusive rights of access... The hope of some of them was to take it a step further and use state power to privatize ownership or access rights" [41,46]. What followed were substantial projects of industrialization, capitalization, and subsidization that, rather than protect local resources and resource users, frequently jeopardized traditional institutions and practices, increased overall resource exploitation, and introduced new industrial actors into previously isolated spaces [36].

Many of the social costs of enclosure and privatization observed for hundreds of years in terrestrial systems (e.g., agricultural, pastoral, forest) have recently been noted in marine systems. These studies illustrate that while private property rights may increase investment and security for those with rights, they are often accompanied by processes of violence and dispossession for those for whom rights are withheld [20]. Instead of the platitude that "a high tide raises all boats," a substantial body of research suggests that the creation of private property rights often increases disparities and concentrates wealth [47,48]. Within marine systems, the introduction of private property has been shown in multiple cases to increasingly marginalize traditional and small-scale fishers [23,49], women [50], developing countries and small island developing states, and indigenous peoples [51]. Two cases in marine fisheries and "blue growth" industries below illustrate the pervasiveness of contemporary marine enclosure and the growing awareness of their social impacts. Here we focus briefly on the increasing role of privatization in reconfiguring the power and access dynamics behind current marine property distribution, resource use, and governance decision-making. Each case highlights similarities in socioeconomic outcomes for less powerful and more resource-dependent actors, and paints a concerning picture for the future of equitable resource use in the world's oceans.

35.3.1 The rise of individual transferable quotas and privatization of a public good

In the mid-20th century, following the rise of EEZs and rights-based fisheries discourses, many states set about privatizing fishing rights within their waters by creating and sustaining market-based mechanisms that commodified, limited, and controlled resource access to this previously public good [25,52]. A popular privatizing mechanism for "securing" fishing rights was the individual transferable quota (ITQ). ITQs are a type of catch share whereby the regulator (i.e., government) sets a total allowable catch over a period of time and allocates shares of this quota to private entities who may then purchase, sell, or lease them depending on the set conditions. ITQs, and variants thereof, were widely promoted and implemented in the global north from the 1970s into the 1990s in countries like Canada, the United States, Iceland, and New Zealand [52]. The presumed ability of ITQs to usher in a new era of responsible ocean stewardship was twofold. First, individuals or groups allocated the right to fish would, as rational and economically-driven entities, be predisposed to conserve their exclusive property right in order to fish for profit in perpetuity. Second, the efficiency of markets would address worldwide overfishing by tackling excess capacity. Such markets could "weed out" less efficient fishers and incentivize them to sell their rights to more efficient fishers before voluntarily exiting the fishery.

The ultimate effect of ITQ policies on fisheries sustainability around the world is the subject of debate. Many economists assert that such schemes did reduce overcapitalization, increase operational efficiency, and facilitate long-term planning (e.g., [36,45,53,54]). However, others find no compelling evidence that property rights schemes like ITQs create resource conservation incentives [55], or are even primarily oriented towards resource protection [56]. What are increasingly clear are the well-documented negative social effects that ITQ schemes have facilitated through the consolidation of resource use rights.

By their very nature ITQ programs prioritize economic values, subjugating other possible social and cultural values into "trade-offs" to achieve economic efficiency. However, this operational model fails to consider the more sociocultural ethic of many small-scale fishery (SSF) actors and interests [55]. As such, the negative distributive impacts of ITQ schemes has often fallen most heavily onto "inefficient" and "irrational" indigenous and small-scale resource users whose operational priorities may incorporate more of a balance of values. These groups are also typically highly reliant on fisheries for their livelihoods and often unable to compete for fishery benefits with the wealthier and capital-rich entities operating within ITQ regimes [25,52,55].

A brief review of the history of ITQ fisheries in Canada, the United States, and New Zealand illustrates the common occurrence of private corporate interests becoming majority owners of ITQ fisheries through state-supported consolidation of quota ownership and vertical integration [25,47,49,51,57]. This power was subsequently exercised to dictate new terms for allocating fishing rights in state-owned waters: in some cases consolidated catch shares were leased by private entities back to less powerful SSF actors at wildly inflated prices; in others local actors were excluded entirely as the right to fish was sold to foreign interests [47,48]. This pattern of corporate consolidation and exclusion, and the negative impacts on indigenous and SSF communities, was sufficiently consistent and acute that some states eventually intervened in ITQ markets by placing limits on transferability and consolidation or by implementing ITQ moratoriums [25,47,57]. While many of these ITQ fisheries still exist in some form today, in many cases this has come at great sociocultural cost. In British Columbia, Canada, for example, the historic coastal salmon canning industry permanently collapsed as the majority corporate owner sent fish to be processed more cheaply abroad [55, p. 4]. Beyond the impacts of power consolidation on the ability of smaller actors to participate in their traditional fisheries, is the disruption and even illicitization of traditional cultural values and practices in some fisheries following the implementation of ITQ schemes [49,51]. For example, in Alaska, the high financial cost of staying in ITQ fisheries eroded crew and kin-based traditions around operational and capital succession and exacerbated existing class divisions between those with and without external financial backing [49]. BC halibut and salmon fisheries witnessed a new focus on wealth accumulation by boat owners at the expense of crew wages and other benefits [52]. Many indigenous Māori fishermen today find themselves structurally excluded from fishing, processing, or selling despite owning a significant portion of fishing quota in New Zealand [51].

ITQ's operationalize the privatization and marketization of the right to fish. That this act is promoted by the state as being "for the public good" is paradoxical given that this championing of exclusivity must necessarily emphasize the rights of some at the expense of others. As Mansfield [25, p. 323] notes: "all of the forms [of privatization] entail reducing the options of those who once relied on public fisheries, while giving to those who qualify a form of wealth that can then be used for further gain." The continued focus on economic efficiency as the primary determinant of "rationality," and of a fishery's inherent or even sole value, ensures that even the most thoughtfully designed ITQs will continue to favor profit-driven actors and interests and deliver unclear outcomes for all other values inherent within a fishery.

35.3.2 Marine protected areas: the social cost of enclosing the ocean to save it

The lead up to and adoption of the intergovernmental Convention of Biological Diversity in 1992 ushered in a new era of state-led commons enclosure in the name of biodiversity protection and conservation. While considerable attention has been paid to the subsequent patterns of privatization, dispossession, appropriation, and power capture that have accompanied the designation of terrestrial parks around the world (e.g., [57,58,59,60]), literature documenting similar trends in marine spaces is only more recently emerging (e.g., [61,62,63]). This growing focus on the similarities in sociocultural impacts of conservation-oriented marine and coastal enclosure comes at a time of renewed state interest in meeting global biodiversity conservation policy commitments.

With the approval of the Aichi biodiversity targets in 2010, nearly 200 states agreed to collectively "set aside" 10% of coastal and marine areas for protection worldwide [64]. Some scholars argue that these targets have provided a rationale and justification for states to enclose common property into a growing number of restricted marine protected areas (MPA) in "local" spaces, sometimes without local consultation [65,66]. Examples from Honduras, Kenya, Madagascar, Malaysia, Tanzania, and the Indian Ocean illustrate that the "designation" of marine conservation spaces, regardless of the intent, is inherently a form of "primitive accumulation" that converts public property into private property, dispossesses previous custodians of the right to use ocean spaces and freely access the benefits therein, and redirects the balance of power and capital accumulation away from local communities and marginalized groups (e.g., women, fishers) toward more powerful actors [61,62,63,66]. As Benjamin and Bryceson note, the majority of these actors include "rent-seeking state officials, transnational conservation organizations, tourism operators, and the state" [61]. The similarities noted between recent marine conservation enclosure and capital accumulation trends, and land-based "green grabs," has given rise to the term "blue" or "ocean" grabbing [67].

Small-scale coastal fishing communities in particular have borne much of the social brunt of marine biodiversity conservation initiatives, with women often the most affected. Echoing similar arguments against "irrational" "open access" fishing used to justify enclosure in state-owned waters with ITQs, narratives of "overfishing" and resource degradation have been used to justify the implementation of marine protected areas in countries such as Malaysia, Tanzania, and Madagascar [61,62,66]. MPAs in these (and other) countries have effectively provided the framework to restrict or entirely exclude traditional SSF through the use of mechanisms such as no-take zones, gear restrictions, and fenced-off areas that are endorsed not just by the state, but sometimes by coastal communities themselves. Often this dispossession is accompanied by promises of alternative opportunities with profit-oriented and privately-controlled marine "ecotourism" that may or may not be realized [61,63,66]. In essence, MPAs exclude SSF from the ability to derive benefits from the same state-owned natural resources that contribute to capital accumulation by tourist operators and indirectly the state itself (xx).

By demonstrating that conservation priorities have often outweighed sociocultural considerations, we do not suggest that conservation is not normatively "good" or that conservation cannot also bring social, economic, or environmental benefits [66,68]. Nor do we suggest that the historic practices of coastal communities are ideal, benign, or inclusive. What we, and other emerging scholarly works suggest is that marine conservation often has a very real human impact, and that more needs to be done by policy makers to acknowledge and understand the implications of implementing fundamentally exclusionary enclosures within marine areas.

35.4 Choosing a future ocean

In this chapter, we have presented a series of recent trends in the enclosure and privatization of ocean spaces and resources. We have discussed the material, discursive, and normative effects that these ocean property regimes shifts have had, and highlighted the asymmetrical consequences for particular individuals and communities. The final task is to is to consider a "future ocean" based on a critical analysis of this recent history.

In a business as usual scenario, our future ocean would look much like our terrestrial past; based on colonial histories, it would align with currently observed neoliberal principles of commodification, privatization, and marketization of ocean spaces and resources. Reflecting familiar trends in terrestrial resources from mining, agriculture, and forestry, this neoliberal ocean future predicts the concentration of ocean resource wealth into the hands of more privileged actors with the power, influence, and capital to maximize "efficient" economic use. Consequently, this *neoliberal ocean future* presages growing resource disparities, whereby those with privileged or exclusive access are able to derive increasing wealth from marine systems, while those lacking rights will see access to diverse benefits (e.g., nutrition, employment, etc.) diminished. This growing emphasis on efficiency, privatization, and exclusive rights will further weaken traditional claims to ocean spaces and uses, pushing noneconomic values and practices to the margins or condemning them as illicit activities.

In contrast, we suggest an *alternate ocean future* whereby scholars, practitioners, policy makers, and user groups purposefully shape nature–society relations with the sea around goals of equity and empowerment. As stated above there is no property relation nor management system that is objective and ideal, without emphasizing certain values and actors over others. The neoliberal ocean future stresses privatization and profit maximization, accounting for other noneconomic values by way of trade-offs and sacrifices in economic efficiency. However, just because this approach is dominant does not mean it is inevitable or unchangeable. Other diverse principles besides economic efficiency have shaped natural resource rights and responsibilities, and many are based on principles of equity and distributional justice (e.g., allocating rights based on resource dependence, development status, or cultural significance). These principles occur not just in small coastal villages and towns, but also within global resource regimes that seek to acknowledge the consequences of neoliberal policies and reclaim the diverse benefits obtained from the sea. Many of these natural resource regimes actively seek to recover the benefits of communal management, essentially reasserting these resources as *res communis*. There is no way to exploit finite marine resources without making judgments about when, and by whom, resources should be used. However, it is possible to make those judgments explicit and reflective of chosen values around resource equity, rather than implicit and hidden under the guise of "optimal" or "efficient" use. To do so may lead to more than just improved social outcomes, but better overall socionatural outcomes that recognize the inextricable linkages between human well-being and our future ocean.

References

[1] FAO, The State of World Fisheries and Aquaculture 2016, 2016, pp. 1–204.

[2] J.B. Jackson, M.X. Kirby, W.H. Berger, K.A. Bjorndal, L.W. Botsford, B.J. Bourque, et al., Historical overfishing and the recent collapse of coastal ecosystems, Science 293 (2001) 629–637. Available from: https://doi.org/10.1126/science.1059199.

[3] D.J. McCauley, M.L. Pinsky, S.R. Palumbi, J.A. Estes, F.H. Joyce, R.R. Warner, Marine defaunation: animal loss in the global ocean, Science 347 (2015) 1255641. Available from: https://doi.org/10.1126/science.1255641.

[4] P.U. Clark, J.D. Shakun, S.A. Marcott, A.C. Mix, M. Eby, S. Kulp, et al., Consequences of twenty-first century policy for multi-millennial climate and sea-level change, Nat. Clim. Change 6 (2016) 360–369. Available from: https://doi.org/10.1038/nclimate2923.

[5] S. Koenigstein, F.C. Mark, S. Gößling-Reisemann, H. Reuter, H.-O. Poertner, Modelling climate change impacts on marine fish populations: process-based integration of ocean warming, acidification and other environmental drivers, Fish and Fisheries 17 (2016) 972–1004. Available from: https://doi.org/10.1111/faf.12155.

[6] E.H. Allison, H.R. Bassett, Climate change in the oceans: human impacts and responses, Science 350 (2015) 778–782. Available from: https://doi.org/10.1126/science.aac8721.

[7] S.C. Doney, M. Ruckelshaus, J. E. Duffy, J.P. Barry, F. Chan, C.A. English, et al., Climate change impacts on marine ecosystems, Annu. Rev. Mar. Sci. 4 (2012) 11–37. Available from: https://doi:10.1146/annurev-marine-041911-111611.

[8] A. Bennett, P. Patil, D. Rader, J. Virdin, X. Basurto, Contribution of Fisheries to Food and Nutrition Security, Duke University, Durham, NC, 2018. Available from: http://nicholasinstute.duke.edu/publicaon.

[9] C.D. Golden, E.H. Allison, W.W.L. Cheung, M.M. Dey, B.S. Halpern, D.J. McCauley, et al., Nutrition: fall in fish catch threatens human health, Nature 534 (2016) 317–320. Available from: https://doi.org/10.1038/534317a.

[10] C.D. Golden, K.L. Seto, M.M. Dey, O.L. Chen, J.A. Gephart, S.S. Myers, et al., Does aquaculture support the needs of nutritionally vulnerable nations? Front. Mar. Sci. 4 (2017) 151–157. Available from: https://doi.org/10.3389/fmars.2017.00159.

[11] K.E. Charlton, J. Russell, E. Gorman, Q. Hanich, A. Delisle, B. Campbell, et al., Fish, food security and health in Pacific Island countries and territories: a systematic literature review, BMC Public Health 16 (2016) 1–26. Available from: https://doi.org/10.1186/s12889-016-2953-9.

[12] R.J. Stanford, B. Wiryawan, D.G. Bengen, R. Febriamansyah, J. Haluan, The fisheries livelihoods resilience check (FLIRES check): a tool for evaluating resilience in fisher communities, Fish Fish. 18 (2017) 1011–1025. Available from: https://doi.org/10.1111/faf.12220.

[13] M.C. Badjeck, E.H. Allison, A.S. Halls, N.K. Dulvy, Impacts of climate variability and change on fishery-based livelihoods, Mar. Policy 34 (2015) 375–383. Available from: https://doi.org/10.1016/j.marpol.2009.08.007.

[14] J. Spijkers, T.H. Morrison, R. Blasiak, G.S. Cumming, M. Osborne, J. Watson, et al., Marine fisheries and future ocean conflict, Fish Fish. 10 (2018) 173–179. Available from: https://doi.org/10.1111/faf.12291.

[15] M.G. Collins, Preparing Ocean Governance for Species on the Move, 2018, pp. 1–4.

[16] D. Tickler, J.J. Meeuwig, K. Bryant, F. David, J.A.H. Forrest, E. Gordon, et al., Modern slavery and the race to fish, Nat. Commun. (2018) 1–9. Available from: https://doi.org/10.1038/s41467-018-07118-9.

[17] J.S. Brashares, B. Abrahms, K.J. Fiorella, C.D. Golden, C.E. Hojnowski, R.A. Marsh, et al., Wildlife decline and social conflict, Science 345 (2014) 376–378. Available from: https://doi.org/10.1126/science.1256734.

[18] K. Ruddle, E. Hviding, R.E. Johannes, Marine resources management in the context of customary tenure, Mare 7 (1992) 249–273. Available from: https://doi.org/10.1086/mre.7.4.42629038.

[19] R. Hilborn, Managing fisheries is managing people: what has been learned? Fish Fish. 8 (2007) 285–296. Available from: https://doi.org/10.1111/j.1467-2979.2007.00263_2.x.

[20] B. Mansfield, Privatization: property and the remaking of nature-society relations, Antipode 39 (2007) 393–405. Available from: https://doi.org/10.1111/j.1467-8330.2007.00532.x.

[21] C.B. Macpherson, Property, Mainstream and Critical Positions, University of Toronto Press, 1978.

[22] S.J. Langdon, Foregone harvests and neoliberal policies: creating opportunities for rural, small-scale, community-based fisheries in southern Alaskan coastal villages, Mar. Policy 61 (2015) 347–355. Available from: https://doi.org/10.1016/j.marpol.2015.03.007.

[23] B. Tolley, M. Hall-Arber, Tipping the scale away from privatization and toward community-based fisheries: policy and market alternatives in New England, Mar. Policy 61 (2015) 401–409. Available from: https://doi.org/10.1016/j.marpol.2014.11.010.

[24] S.J. Breslow, Accounting for neoliberalism: "Social drivers" in environmental management, Mar. Policy 61 (2015) 420–429. Available from: https://doi.org/10.1016/j.marpol.2014.11.018.

[25] B. Mansfield, Neoliberalism in the oceans: "rationalization," property rights, and the commons question, Geoforum 35 (2004) 313–326. Available from: https://doi.org/10.1016/j.geoforum.2003.05.002.

[26] M. Fairbairn, J. Fox, S.R. Isakson, M. Levien, N. Peluso, S. Razavi, et al., Introduction: new directions in agrarian political economy, J. Peasant Stud. 0 (2014) 1–14. Available from: https://doi.org/10.1080/03066150.2014.953490.

[27] L. Campling, E. Havice, The problem of property in industrial fisheries, J. Peasant Stud. 41 (2014) 707–727. Available from: https://doi.org/10.1080/03066150.2014.894909.

[28] T.W. Fulton, The Sovereignty of the Sea, Blackwood, Edinburgh, London, 1911.
[29] J.C. Cordell, A sea of small boats, Cultural Survival (1989) 1–418.
[30] D. Rothwell, T. Stephens, The International Law of the Sea, Hart Publishing, Oxford and Portland, Oregon, 2010.
[31] H. Grotius, The Freedom of the Seas or the Right which Belongs to the Dutch to Take Part in the East Indian Trade, 1633rd ed., Oxford University Press, New York, 2006.
[32] J. Selden, Mare Clausum seu De Dominio Maris, W. Du-Gard, London, 1652.
[33] C.M. Rose, Property and Persuasion: Essays on the History, Theory, and Rhetoric of Ownership, Westview Press, Boulder, CO, 1994.
[34] R.L. Friedheim, Managing the second phase of enclosure, Ocean Coastal Manage. 17 (1992) 217–236. Available from: https://doi.org/10.1016/0964-5691(92)90011-9.
[35] L. Campling, A. Colás, Capitalism and the sea: sovereignty, territory and appropriation in the global ocean, Environ. Plan. D (2017). Available from: https://doi.org/10.1177/0263775817737319. 026377581773731–19.
[36] F. Alcock, UNCLOS, property rights, and effective fisheries management, in: S. Oberthür, O.S. Stokke (Eds.), Managing Institutional Complexity Regime Interplay and Global Environmental Change, The MIT Press, 2011, pp. 255–284. Available from: https://doi.org/10.7551/mitpress/9780262015912.003.0010.
[37] W.S. Ball, The old grey mare national enclosure of the oceans, Ocean Dev. Int. Law 27 (1996) 97–124. Available from: https://doi.org/10.1080/00908329609546077.
[38] E. Ostrom, Governing the Commons, Cambridge University Press, 1990.
[39] G. Hardin, The tragedy of the commons, Science 162 (1968) 1243–1248. Available from: https://doi.org/10.1126/science.159.3818.920-a.
[40] E.M. Finkbeiner, N.J. Bennett, T.H. Frawley, J.G. Mason, D.K. Briscoe, C.M. Brooks, et al., Reconstructing overfishing: moving beyond Malthus for effective and equitable solutions, Fish Fish. 13 (2017) 1180–1191. Available from: https://doi.org/10.1111/faf.12245.
[41] R.L. Friedheim, Ocean governance at the millennium: where we have been—where we should go, Ocean Coastal Manage. 42 (1999) 747–765. Available from: https://doi.org/10.1016/s0964-5691(99)00047-2.
[42] United Nations, United Nations Convention on the Law of the Sea (UNCLOS), Montego Bay, 1982.
[43] H.S. Gordon, The economic theory of a common-property resource: the fishery, J. Polit. Econ. 62 (1954) 124. Available from: https://doi.org/10.1086/257497.
[44] P.A. Neher, R. Árnason, N. Mollett (Eds.), Rights Based Fishing, Kluwer Academic Publishers, Dordrecht, Boston, London, 1989.
[45] R. Árnason, Property Rights as a Means of Economic Organization, Rome, 2000.
[46] R.D. Eckert, The Enclosure of Ocean Resources, Hoover Institution Press, Stanford, CA, 1979.
[47] B. Mansfield, Property, Markets, and Dispossession: the Western Alaska Community Development Quota as Neoliberalism, Social Justice, Both, and Neither, Antipode 39 (2007) 479–499.
[48] B.L. Endemaño Walker, Engendering Ghana's Seascape: Fanti Fishtraders and Marine Property in Colonial History, Soc. Natur. Resour. 15 (2002) 389–407. Available from: https://doi.org/10.1080/08941920252866765.
[49] C. Carothers, Fisheries privatization, social transitions, and well-being in Kodiak, Alaska, Mar. Policy 61 (2015) 313–322. Available from: https://doi.org/10.1016/j.marpol.2014.11.019.
[50] Q.R. Grafton, D. Squires, J.E. Kirkley, Private Property Rights and Crises in World Fisheries: Turning the Tide? Contemp. Econ. Policy 14 (1996) 90–99.
[51] H. Bodwitch, Challenges for New Zealand's individual transferable quota system: processor consolidation, fisher exclusion, & Māori quota rights, Mar. Policy 80 (2017) 88–95. Available from: https://doi.org/10.1016/j.marpol.2016.11.030.
[52] E. Pinkerton, R. Davis, Neoliberalism and the politics of enclosure in North American small-scale fisheries, Mar. Policy 61 (2015) 303–312. Available from: https://doi.org/10.1016/j.marpol.2015.03.025.
[53] M. Barbesgaard, Blue growth: savior or ocean grabbing? J. Peasant Stud. 45 (2017) 130–149. Available from: https://doi.org/10.1080/03066150.2017.1377186.

[54] C. Chambers, C. Carothers, Thirty Years After Privatization: a Survey of Icelandic Small-Boat Fishermen, Mar. Policy 80 (2017) 69–80. Available from: https://doi.org/10.1016/j.marpol.2016.02.026.

[55] E. Pinkerton, Hegemony and resistance: disturbing patterns and hopeful signs in the impact of neoliberal policies on small-scale fisheries around the world, Mar. Policy 80 (2017) 1–9. Available from: https://doi.org/10.1016/j.marpol.2016.11.012.

[56] H. Saevaldsson, S.B. Gunnlaugsson, The Icelandic Pelagic Sector and Its Development Under an ITQ Management System, Mar. Policy 61 (2015) 207–215. Available from: https://doi.org/10.1016/j.marpol.2015.08.016.

[57] N.L. Peluso, "Coercing Conservation?: the Politics of State Resource Control" 3, no. 2 (January 1, 1993): 199–217. Available from: https://doi:10.1016/0959-3780(93)90006-7.

[58] C. Corson, K.I. MacDonald, Enclosing the Global Commons: the Convention on Biological Diversity and Green Grabbing, J. Peasant Stud. 39 (2) (April 2012) 263–283. Available from: https://doi.org/10.1080/03066150.2012.664138.

[59] J. Fairhead, M. Leach, I. Scoones, Green Grabbing: a New Appropriation of Nature? J. Peasant Stud. 39 (2) (April 2012) 237–261. Available from: https://doi.org/10.1080/03066150.2012.671770.

[60] B. Büscher, Letters of Gold: Enabling Primitive Accumulation Through Neoliberal Conservation, Human Geography 2 (3) (January 1, 2009) 91–93.

[61] T.A. Benjaminsen, I. Bryceson, Conservation, Green/Blue Grabbing and Accumulation by Dispossession in Tanzania, J. Peasant Stud. 39 (2) (April 2012) 335–355. Available from: https://doi.org/10.1080/03066150.2012.667405.

[62] M. Baker-Médard, Gendering Marine Conservation: the Politics of Marine Protected Areas and Fisheries Access, Soc. Natur. Resour. 30 (6) (October 31, 2016) 723–737. Available from: https://doi.org/10.1080/08941920.2016.1257078.

[63] C.A. Loperena, Conservation by Racialized Dispossession: the Making of an Eco-Destination on Honduras's North Coast, Geoforum 69 (C) (February 1, 2016) 184–193. Available from: https://doi.org/10.1016/j.geoforum.2015.07.004.

[64] Conference of the Parties to the Convention on Biological Diversity. Conference of the Parties (COP) 10 Decision X/2, 2010.

[65] E.M. De Santo, P.J.S. Jones, A.M.M. Miller, Fortress Conservation at Sea a Commentary on the Chagos Marine Protected Area, Mar. Policy 35 (2) (March 1, 2011) 258–260. Available from: https://doi.org/10.1016/j.marpol.2010.09.004.

[66] A. Hill, Blue grabbing: Reviewing marine conservation in Redang Island Marine Park, Malaysia, Geoforum 79 (2017) 97–100.

[67] N.J. Bennett, H. Govan, T. Satterfield, Ocean Grabbing, Mar. Policy 57 (C) (July 1, 2015) 61–68. Available from: https://doi.org/10.1016/j.marpol.2015.03.026.

[68] D. Hall, P. Hirsch, L.T. Murray, Powers of Exclusion: Land Dilemmas in Southeast Asia, National University of Singapore, Singapore, 2011.

CHAPTER 36

New actors, new possibilities, new challenges—nonstate actor participation in global fisheries governance

Matilda Tove Petersson
Stockholm Resilience Centre, Stockholm University, Stockholm, Sweden

Chapter Outline

36.1 Introduction

Transboundary policy problems in the world's oceans, such as overfishing and unsustainable exploitation patterns of shared fish stocks [1,2], represent significant governance challenges. Importantly, states engage in international cooperation to manage transboundary fish stocks, notably by establishing and delegating authority to regional fisheries management organizations (RFMOs). These RFMOs have the mandate to adopt legally binding conservation and management measures for highly migratory and straddling fish stocks moving between the high seas and national jurisdiction under the United Nations (UN) Convention on the Law of the Sea (1982) and Fish Stocks Agreement (1995). However, in spite of these international efforts, RFMOs have displayed a mixed track record in their ability to manage highly migratory and straddling fish stocks sustainably [3], implement an ecosystem approach to fisheries management [4], and respond to resource fluctuations from climate change [5]. Today, there is increasing demand for global fisheries governance that can enable sustainable and effective management of these transboundary fish resources.

Predicting Future Oceans.
DOI: https://doi.org/10.1016/B978-0-12-817945-1.00038-1

This chapter argues that scholars should pay more attention to the participation of nonstate actors and the implications of such participation for effectiveness, i.e., the problem-solving capacity (e.g., [6–8]) of international organizations to manage transboundary natural resources. The term nonstate actors refers both to nonprofit organizations, like nongovernmental organizations (NGOs), private research-based organizations, and business associations (who themselves are nonprofit, but that represent for-profit organizations), and for-profit organizations, like multinational corporations and consultancies [9]. In the past three decades we have witnessed a participatory turn in global governance, whereby international organizations across policy areas have opened up access to nonstate actors [10,11]. Subsequently these actors increasingly participate in international organizations [12,13], with implications for the outcomes achieved by these organizations [14,15]. This book chapter follows previous international relations literature and assumes that nonstate actors can shape governance outcomes, even though state actors remain central actors in the global political system [14–17]. Despite significant scholarly interest in the role and influence of nonstate actors in global environmental [9,18–21] and fisheries governance [22–24], relatively few studies look at nonstate actor participation from a comparative perspective (however, see [25,26]). As a consequence, questions such as "what are the patterns of nonstate actor participation?" and "what conditions shape nonstate actor participation?" in global fisheries governance remain largely unanswered. This book chapter addresses these questions, with the aim of contributing to existing debates about the role of nonstate actors in global environmental governance and sustainability science literature.

The chapter starts with a review of relevant literature on the participation of nonstate actors in global environmental governance and the links to the effectiveness of international organizations. Thereafter it summarizes the findings from two recent studies focusing on the participation of nonstate actors across RFMOs and discusses the implications for the effectiveness of international organizations.

36.2 Literature review

Previous literature on nonstate actor participation is dominated by case studies, focusing either on individuals or coalitions of actors [9,18,19], in specific policy processes or international organizations [23,24]. There are some recent comparative studies of the participatory patterns of nonstate actor population over time in international organizations addressing trade [26], biodiversity, and climate policy [25]. These comparative works study representational diversity in terms of nonstate actor type (e.g., NGOs or business actors) and country of origin. In terms of nonstate actor type, these studies have found skewed patterns of participation, whereby business interests dominate nonstate actor participation in international organizations, demonstrating similar findings as the studies of interest group representation at both national [27] and EU level [28]. The patterns of participation by different types of nonstate actors may however differ between different policy areas. For example, a recent case study of the role of nonstate actors in the UN General Assembly working group on the conservation and sustainable use of biodiversity in marine areas beyond national jurisdiction (BBNJ Working Group) shows that NGOs (and not business actors) currently dominate nonstate actor participation [29]. In terms of country of origin a recent study found that the majority of nonstate actors participating in international organizations concerning biodiversity and climate policies come from high-income countries in North America and Western and Northern Europe [25]. This suggests that nonstate actors from low-income countries may lack the resources necessary to be able to mobilize at the global level [30]. Taken together these findings raise a number of questions about the relationship between patterns of representation diversity and the ability of different types of nonstate actors from different countries to influence global policy processes across policy areas. They also suggest that there is a need for additional comparative studies of the populations of nonstate actors across international organizations and in previously understudied policy areas, such as global fisheries policy.

Previous literature on nonstate actors has often focused either on the role and implications of NGOs or business actors. Earlier literature on nonstate actors often associates NGO participation with positive environmental outcomes, assuming these actors can enhance the effectiveness of global environmental governance, as they are nonprofit, motivated on normative grounds, and pursue public policy goals [9,31]. By contrast, industry or business actors are usually assumed to limit effectiveness, as they are for-profit actors and may want to protect their own private interests at the expense of public interests [20,32]. However, some authors instead view the participation by a diversity of nonstate actors as something that can contribute to effectiveness, since different types of nonstate actors can contribute with different information, expertise, and resources that can enhance governance capacity to manage transboundary resources [24,33]. Conversely, other authors view increasing participation and diversity of nonstate actors in international organizations as something that can limit effectiveness, as competing views may make consensus decision-making more difficult [17,34].

The literature on international regimes have extended our knowledge about how and in what ways institutional factors and conditions can shape regime effectiveness [e.g., 6-8]. Although this literature generally acknowledges that nonstate actors increasingly participate and can impact outcomes of international regimes (see [6] for an overview) the role of these actors and its implications for effectiveness has so far received relatively limited attention. International regimes are implicit or explicit collections of rules, norms, and decision-making procedures which prescribe and proscribe certain state behavior in a particular area [35]. This book chapter considers international regimes that are formalized in international organizations with an attached bureaucracy, by discussing the implication of nonstate actor participation on the effectiveness of RFMOs. Bodin and Österblom [24] have studied nonstate actor participation and its implications for the effectiveness of the Commission for the Conservation of Antarctic Marine Living Resources to reduce illegal, unreported, and unregulated fishing in the Southern Ocean. They find that nonstate actors can play important roles when they are active in monitoring compliance, and when they engage both in public awareness campaigns and in direct and extensive dialogue with policy makers. This aligns well with previous findings in international relations that nonstate actors often combine inside and outside strategies and pursue multiple goals in global governance, such as advocacy, visibility, networking, and information gathering and exchange [36,37]. Inside strategies refer to activities whereby nonstate actors directly engage with policy makers, by offering policy expertise and information, and try to shape policy making in line with the interests and needs of their members or constituencies. Outside strategies instead refer to activities like campaigning and naming and shaming, whereby nonstate actors indirectly try to shape policy making by influencing public opinion. Inside strategies, enabled by access to international organization or policy makers, are generally considered to be more influential, compared to outside strategies where access is lacking [36,37].

Taken together, previous literature on nonstate actor participation and its implications for international regime effectiveness has contributed to our common understanding of the characteristics, goals, and strategies of individual nonstate actors, as well as their impacts on particular international organizations, policy areas, and policy processes. This literature has also generated theoretical expectations about nonstate actor participation. However, to further our understanding about whether and in what ways nonstate actor participation can shape effectiveness, there is a need for additional comparative studies that can test theoretical expectations concerning patterns and conditions for the participation of nonstate actors in the first place. In the following section this chapter attempts to contribute to this gap by summarizing two recent comparative studies on nonstate actor participation in global fisheries governance.

36.3 Nonstate actor participation in global fisheries governance: patterns, conditions, and implications

This section presents and summarizes the findings of two recent studies of the participation of nonstate actors within RFMOs, and discusses the implications of that participation in relation to broader debates on representation, influence, and effectiveness.

The first study by Petersson et al. [38] is a comparative study of the patterns of nonstate actor participation across the five tuna RFMOs (2004–11). In this study, the authors find that there are two distinct patterns of nonstate actor participation across the five tuna RFMOs that are particularly relevant to consider in future studies considering implications for effectiveness: nonstate actor type and country of origin.

In relation to nonstate actor type, the authors find that industry representatives (companies and industry associations) dominate participation, generally participate as part of member state delegations, and participate with continuity in tuna RFMO meetings. By contrast, NGOs are much fewer in number, generally participate as observers, and with less continuity. According to the authors these trends indicate that industry representatives are better positioned when compared to NGOs to influence RFMO policy making, being in close proximity to RFMO policy makers and thus able to build interpersonal relationships as a means of lobbying policy makers through repeated participation. Prominent observers have also criticized RFMO member states for protecting the interests of national fishing fleets, thereby limiting the effectiveness of RFMOs in attempting to achieve sustainable fisheries [3,39]. This could be linked to the dominant participation by fishing industry representatives. The dominant participation of industry representatives can also be understood against the background of the large incentives these actors have to lobby policy makers concerning measures that directly impact their business operations, such as access to stock allocations, stock allocation quotas, fishing effort restrictions, the imposition of closed areas, and restrictions on gear types. Industry representatives are also likely to possess larger financial and organizational resources compared to NGOs, enabling them to participate in RFMO meetings to a larger extent (cf. [25,26]). NGOs may also prefer to use their resources for other activities, such as campaigning and naming and shaming, rather than participating in tuna RFMO meetings. Indeed RFMO policy processes are often perceived as bureaucratic and lengthy, and participation is often associated with both travel costs and participation fees [40].

In terms of country of origin of nonstate actors, the authors find that nonstate actors from high-income countries participate more often than nonstate actors from low- and lower-middle-income countries, even though several tuna RFMOs have developing coastal states as members. This finding corroborates a recent comparative study of nonstate actors participating in international organizations addressing biodiversity and climate policy (see [25]). According to the authors it also suggests that opening up access to RFMOs for nonstate actors does not necessarily improve the representation of nonstate actors from developing countries, which may lead to policy decisions at the expense of the interests of these countries. In the context of the tuna fisheries the interests of developing countries are critically important since some of the world's richest tuna waters lie within the exclusive economic zones of developing countries (e.g., within the Pacific Ocean region), and since several developing countries sell access rights to distant water fishing fleets [41] and/or are increasing their national tuna catches (e.g., Indonesia, the Philippines, and Ecuador) [42].

The second study by Dellmuth et al. [43] explores the conditions for NGO participation across seven RFMOs, through a comparative approach considering both institutional and ecological factors potentially driving or hindering NGO participation. Four results stand out. First, the authors find that NGO participation is greater in RFMOs with larger financial resources, suggesting that policy makers themselves shape NGO participation, not only by allowing access to nonstate actors, but also by supplying financial resources to RFMOs enabling NGO participation. Second, they find that NGO participation is greater in RFMOs with a relatively similar member state composition in terms of domestic democratic norms, corroborating previous studies arguing that similarity in national political systems makes compromises concerning NGO participation easier, as compared to more heterogeneous member state compositions [17,30]. Third, the study shows that NGO participation is greater in RFMOs with fewer experts from research-based organizations participating, indicating trade-offs between NGOs and experts, and that these actors play similar functions in providing policy makers with information and expertise. Fourth, the study does not find that ecological factors related to target fish stock health stock (indicated by measures of fishing pressure and biomass status)[1] drive NGO participation in RFMOs. This means that NGOs do not appear to increase their participation in RFMOs when the status of target fish stocks worsen. Instead the authors suggest that NGOs may participate in RFMOs for other reasons, for example, over concerns of the biological status of nontarget species, i.e., related to bycatch [44] and ecosystem impacts, such as habitat destruction, resulting from fishing [45]. NGOs may also choose to use their scarce resources in other ways than to participate on the inside of RFMOs and by directly engaging with policy makers, in order to achieve goals related to the high-value target fish stocks managed by RFMOs. They may prefer to use outside strategies to achieve such goals and try to influence policy making indirectly by shaping public opinion through campaigning, and naming and shaming strategies. Indeed, previous studies find that NGOs pursue multiple goals and combine "inside" and "outside" strategies in global governance, such as visibility, networking, and information gathering and exchange [36,37].

36.4 Conclusions

This chapter summarizes two recent studies on the patterns of and conditions for nonstate actor participation. Taken together, these studies show that nonstate actors extensively participate in international fisheries organizations, but that actor type and country of origin shape patterns of participation, even though formal access requirements are similar for all nonstate actors. They also show that NGO participation is conditioned by RFMO financial resources, member state similarity, and by the participation of experts from research-based organizations, but not by target fish stock health.

[1] The authors use two indicators of target fish stock health, *fishing pressure* (F/F_{MSY})*biomass status* (B/B_{MSY}). Fishing pressure refers to the ratio of fishing mortality (F) to the level that would allow for eventual maximum sustainable yield (F_{MSY}). Biomass status is measured by comparing the biomass (B) or spawning stock biomass (SSB) to the respective levels that would produce maximum sustained yield (B_{MSY} or SSB_{MSY}).

While the question as to whether and in that case under what conditions nonstate actor participation shapes effectiveness in global fisheries governance remains, this chapter demonstrates the value of comparative studies for better understanding the patterns of participation and representational diversity, and its implications for effectiveness. Moreover, the chapter draws attention to how factors in the institutional environment shape nonstate actor participation. This is particularly important given that the global institutional landscape is increasingly described as complex in nature, for example, in the literature on fragmented [46,47] and polycentric [48,49] governance. Importantly it has been suggested that an increasing complexity of the global institutional landscape can increase the opportunities for nonstate actors to pursue advocacy, since there are more political venues, and more opportunities for collaboration between involved actors that share similar interests [41,50–52]. This indicates that the global environmental governance and sustainability science literature focusing on the linkages between nonstate actors and effectiveness would benefit from additional comparative studies of nonstate actor participation operating at national, regional, and global level in relation to transboundary environmental resources managed by multiple international organizations. This chapter also encourages future studies to systematically test the potential relationship between the variation in the status of nontarget species and ecosystem impacts from fishing as an ecosystem signal with importance for NGO participation in global fisheries governance.

References

[1] D. Pauly, V. Christensen, S. Guénette, T.J. Pitcher, U.R. Sumaila, C.J. Walters, et al., Toward sustainability in world fisheries, Nature 418 (2002) 689–695.

[2] T.J. Pitcher, W.W.L. Cheung, Fisheries: hope or despair? Mar. Pollut. Bull. 74 (2) (2013) 506–516.

[3] S. Cullis-Suzuki, D. Pauly, Failing the high seas: a global evaluation of regional fisheries management organizations, Mar. Policy 34 (5) (2010) 1036–1042.

[4] M.J. Juan-Jordá, H. Murua, H. Arrizabalaga, N.K. Dulvy, V. Restrepo, Report card on ecosystem-based fisheries management in tuna regional fisheries management organizations, Fish Fish. (2017) 1–19.

[5] B. Pentz, N. Klenk, S. Ogle, J.A.D. Fisher, Can regional fisheries management organizations (RFMOs) manage resources effectively during climate change? Mar. Policy (2018). Available from: https://doi.org/10.1016/j.marpol.2018.01.011.

[6] O.R. Young, Effectiveness of international environmental regimes: existing knowledge, cutting-edge themes, and research strategies, Proc. Natl. Acad. Sci. U.S.A. 108 (50) (2011) 19853–19860.

[7] H. Breitmeier, A. Underdal, O.R. Young, The effectiveness of international environmental regimes: comparing and contrasting findings from quantitative research, Int. Stud. Rev. 13 (4) (2011) 579–605.

[8] A. Underdal, Methodological challenges in the study of regime effectiveness, in: A. Underdal, O.R. Young (Eds.), Regime Consequences: Methodological Challenges and Research Strategies., Springer, Dordrecht, 2004, pp. 27–48.

[9] M.M. Betsill, E. Corell, NGO Diplomacy: The Influence of Nongovernmental Organizations in International Environmental Negotiations., MIT Press, 2008.

[10] K.D. Reimann, A view from the top: international politics, norms and the worldwide growth of NGOs, Int. Stud. Q. 50 (1) (2006) 45–68.

[11] J. Tallberg, T. Sommerer, T. Squatrito, C. Jönsson, The opening up of international organizations: transnational access in global governance, Cambridge Univ. Press, Cambridge, 2013.

[12] O. Willets, From Stockholm to Rio and beyond: the impact of the environmental movement on the United Nations consultative arrangements for NGOs, Rev. Int. Stud. 22 (1996) 57–80.
[13] R.O. Keohane, J.S. Nye, Introduction, in: J.S. Nye, J.D. Donahue (Eds.), Governance in a Globalizing World, Visions of Governance for the 21st Century, Cambridge, MA, 2000, pp. 1–41.
[14] F. Biermann, P. Pattberg, Global environmental governance: taking stock, moving forward, Annu. Rev. Environ. Resour. 33 (1) (2008) 277–294.
[15] T. Risse, Transnational actors and world politics, in: W. Carlsnaes, T. Risse, B.A. Simmons (Eds.), Handbook of International Relations, second ed., Sage Publications, Thousand Oaks, CA, 2012, pp. 426–452.
[16] R.O. Keohane, D.G. Victor, The regime complex for climate change, Perspect. Polit. 9 (1) (2011) 7–23.
[17] K. Raustiala, States, NGOs, and international environmental institutions, Int. Stud. Q. 41 (4) (1997) 719–740.
[18] K. Rietig, The power of strategy: environmental NGO influence in international climate negotiations, Global Gov. 22 (2) (2016) 269–288.
[19] T. Princen, M. Finger, M. Finger, Environmental NGOs in World Politics, Routledge, London, 1994.
[20] D. Levy, P. Newell, Business strategy and international environmental governance: toward a neo-Gramscian synthesis, Global Environ. Polit. 2 (4) (2002) 84–101.
[21] J. Clapp, Global environmental governance for corporate responsibility and accountability, Global Environ. Polit. 5 (3) (2005) 23–34.
[22] T. Skodvin, S. Andresen, Nonstate influence in the international whaling commission, 1970-1990, in: M. M. Betsill, E. Corell (Eds.), NGO Diplomacy: The Influence of Nongovernmental Organizations in International Environmental Negotiations., MIT Press, 2008, pp. 61–86.
[23] K. Orach, M. Schlüter, H. Österblom, Tracing a pathway to success: how competing interest groups influenced the 2013 EU Common Fisheries Policy reform, Environ. Sci. Policy 76 (2017) 90–102.
[24] Ö. Bodin, H. Österblom, International fisheries regime effectiveness—activities and resources of key actors in the Southern Ocean, Global Environ. Change 23 (5) (2013) 948–956.
[25] A.N. Uhre, Exploring the diversity of transnational actors in global environmental governance, Interest Groups Advocacy 3 (1) (2014) 59–78.
[26] M. Hanegraaff, J. Beyers, C. Braun, Open the door to more of the same? The development of interest group representation at the WTO, World Trade Rev. 10 (04) (2011) 447–472.
[27] F.R. Baumgartner, B.L. Leech, Interest niches and policy banwagons: patters of interest group involvement in national politics, J. Polit. 634 (2001) 1191–1213.
[28] H. Klüver, Europeanization of lobbying activities: when national interest groups spill over to the European level, J. Eur. Integr. 322 (2010) 175–191.
[29] R. Blasiak, C. Durussel, J. Pittman, C.A. Sénit, M. Petersson, N. Yagi, The role of NGOs in negotiating the use of biodiversity in marine areas beyond national jurisdiction, Mar. Policy 81 (2017) 1–8.
[30] J. Tallberg, L.M. Dellmuth, H. Agné, A. Duit, NGO influence in international organizations: information, access and exchange, Br. J. Polit. Sci. 48 (1) (2018) 213–238.
[31] M.E. Keck, K. Sikkink, Activists Beyond Borders: Advocacy Networks in International Politics., Cornell University Press, Ithaca, NY, 1998.
[32] B. Arts, Business and NGOs – new styles of self-regulation., Corp. Soc. Responsib. Environ. Manage. 9 (2002) 26–36.
[33] C. Folke, T. Hahn, P. Olsson, J. Norberg, Adaptive governance of social-ecological systems, Annu. Rev. Environ. Resour. 30 (1) (2005) 441–473.
[34] J.S. Barkin, E.R. DeSombre, Do we need a global fisheries management organization? J. Environ. Stud. Sci. 3 (2) (2013) 232–242.
[35] R.O. Keohane, After Hegemony: Cooperation and Discord in the World Political Economy, Princeton Univ. Press, Princeton, NJ, 1984.
[36] M. Hanegraaff, J.A.N. Beyers, I.D.E. Bruycker, Balancing inside and outside lobbying: the political strategies of lobbyists at global diplomatic conferences, Eur. J. Polit. Res. 55 (2016) 568–588.

[37] L.M. Dellmuth, J. Tallberg, Advocacy strategies in global governance: inside versus outside lobbying, Pol. Stud. (2017). Available from: https://doi.org/10.1177/0032321716684356.

[38] M.T. Petersson, L.M. Dellmuth, A. Merrie, H. Österblom, Patterns and trends in non-state actor participation in regional fisheries management organizations, Mar Policy 104 (2019) 146–156.

[39] G.D. Hurry, M. Hayashi, J.-J. Maguire, Report of the Independent Review International Commission for the Conservation of Atlantic Tunas (ICCAT). [cited 2018 November 1]. Available from: <https://www.iccat.int/Documents/Other/PERFORM_%20REV_TRI_LINGUAL.pdf>.

[40] K. Barclay, I. Cartwright, Governance of tuna industries: the key to economic viability and sustainability in the Western and Central Pacific Ocean, Mar. Policy 31 (3) (2007) 348–358.

[41] E. Havice, L. Campling, Shifting tides in the Western and Central Pacific Ocean tuna fishery: the political economy of regulation and industry responses, Global Environ. Polit. 10 (1) (2010) 89114.

[42] M.P. Miyake, P. Guillotreau, H.-C. Sun, G. Ishimura, Recent developments in the tuna industry: stocks, fisheries, management, processing, trade and markets, FAO Fisheries and Aquaculture Technical Paper 543, Rome, 2010.

[43] L.M. Dellmuth, M.T. Petersson, D.C. Dunn, A. Boustany, P.N. Halpin, Institutional Environments Drive Non-Governmental Organization Participation in Regional Fisheries Management Organizations, 2018 (unpublished paper).

[44] J.L. Jacquet, D. Pauly, The rise of seafood awareness campaigns in an era of collapsing fisheries, Mar. Policy 31 (3) (2007) 308–313.

[45] R. Hilborn, Defining success in fisheries and conflicts in objectives, Mar. Policy 31 (2) (2007) 153–158.

[46] F. Biermann, P. Pattberg, H. van Asselt, F. Zelli, The fragmentation of global governance architectures: a framework for analysis, Global Environ. Polit. 9 (4) (2009) 14–40.

[47] F. Zelli, The fragmentation of the global climate governance architecture, Wiley Interdiscip. Rev. Clim. Change 2 (2) (2011) 255–270.

[48] E. Ostrom, Polycentric systems for coping with collective action and global environmental change, Global Environ. Change 20 (4) (2010) 550–557.

[49] K.W. Abbott, Strengthening the transnational regime complex for climate change, Transnatl. Environ. Law 3 (1) (2014) 57–88.

[50] J. Hadden, Networks in Contention: The Divisive Politics of Global Climate Change., Cambridge University Press, Cambridge, 2015.

[51] J. Allan, J. Hadden, Exploring the framing power of NGOs in global climate politics, Environ. Polit. 26 (4) (2017) 600–620.

[52] M.T. Petersson, Transnational partnerships' strategies in global fisheries governance, Interest Groups & Advocacy, (2019), https://doi.org/10.1057/s41309-019-00056-x.

CHAPTER 37

Exploring the knowns and unknowns of international fishery conflicts

Jessica Spijkers[1,2]

[1]Stockholm Resilience Centre, Stockholm University, Stockholm, Sweden [2]ARC for Coral Reef Studies, James Cook University, Townsville, QLD, Australia

Chapter Outline

Threats occurring in the maritime domain are raising concerns among policy makers for the future of national security in the ocean space. Maritime security threats can take many forms, from illegal fishing to maritime terrorism and smuggling [1], some of which directly impact human security through, for example, compromised safety of employment at sea. Resource conflict centered around fish is one such security threat as it can endanger not only national security but also compromise different dimensions of human security due to the vital role fisheries play in providing employment and nutrition. Conflict over fishery resources can make it unsafe for fishermen to venture out to sea, for example, or displace fishing pressure to adjacent areas, threatening the sustainability of neighboring fisheries [2]. Ensuring the cooperative use of fisheries resources is not only vital for future maritime security, it is also a necessity to avoid environmentally undesirable consequences as fishery conflicts raise concerns for the sustainability of fishery resources [3]. Although there are cases where conflict is met with proactive agreements between opposing parties to improve the overall sustainability of fishery resources (such as the Pacific Salmon Interception Treaty in 1985 for the Pacific Northwest, revised in 1999 [4]), conflicts can also allow for increased or uncontrolled fishing efforts by third parties [5], as exemplified by the increased overfishing of the northeast Atlantic mackerel for an extent of the conflict [3].

Predicting Future Oceans.
DOI: https://doi.org/10.1016/B978-0-12-817945-1.00035-6

This chapter explores what is known about the historical occurrence and likely current drivers of international fishery conflict (Section 37.1), then moves into what is not known about fishery conflict and how one can start exploring those unknowns (Section 37.2), and ends with a discussion on the vitality of understanding conflict to understand other maritime threats as those can interlink and produce wider regional instability (Section 37.3).

37.1 The anatomy of international fishery conflict

37.1.1 Historical instances of international fishery conflicts

Disagreements over fishing quotas and maritime boundaries have not only driven many militarized post-World War II conflicts in the past (such as the infamous Cod Wars between Great Britain and Iceland of the 1950s and 1970s) [6], we have also witnessed many fishery disputes more recently. Diplomatic tensions between China and its neighboring countries, for example, have become prominent due to incursions from the Chinese fishing fleet into foreign or disputed waters [7]. Particularly in the South China Sea, which harbors rich fishing grounds that supply the livelihoods of people across the region, spats over fishery resources (occurring in the backdrop of competing claims over territories by China, the Philippines, Taiwan, Vietnam, Brunei, and Malaysia) have not only become numerous but, in a few instances, have led to deaths at sea [8]. Fishery disputes are not limited to the developing world however, as can be learnt from the dispute over northeast Atlantic mackerel (*Scomber scombrus*) between the European Union, Norway, the Faroe Islands, and Iceland. The northeast Atlantic mackerel stocks began spawning further towards the northwest of the Nordic Seas and their surrounding waters around 2007, sparking a dispute that (at the time of writing) has still not been fully resolved [3].

In certain areas governance mechanisms are in place to avoid or de-escalate disputes (such as long-term management plans where quota allocations and access are set for long stretches of time, the availability of international courts and tribunals to resolve disputes, or side payment schemes between states to compensate the other party for shifts in available fish resources), but even if they do exist, they are likely insufficient to contain conflict in the face of climate change. New approaches to contain conflict have been suggested, from increased cooperation between Regional Fisheries Management Organizations to fisheries permits tradable across boundaries [9], and such new governance strategies will be necessary to ensure maritime, national, and human security in the future.

37.1.2 Accelerating drivers of conflict

Likely environmental drivers of fishery conflict such as climate change and increased fishery resource scarcity are ramping up. Climate change is driving unprecedented geographic shifts in marine animals by altering water temperatures, ocean currents, and coastal upwelling patterns, with fish already shifting into new territory at an average of 70 km per decade [10] and shifts only expected to accelerate [9]. Shifting fish species have already posed serious governance challenges resulting in international conflict, as exemplified by the shift in migration and spawning area of the northeast Atlantic mackerel mentioned earlier that sparked an ongoing inter-state conflict [3]. Such international disputes, triggered by shifting species, are likely to increase in the future [9]. Many of the world's Exclusive Economic Zones (EEZs) are likely to receive one to five new climate-driven transboundary stocks by the end of the century, and the number of EEZs receiving the stocks increases with global temperature [9].

At the same time catches from wild capture fisheries have declined [11] which can pose additional security challenges. Declining fish catch and deteriorating coastal environments can incentivize an illegal race to fish [12] and illegal, unreported, and unregulated (IUU) fishing has become increasingly problematic in Asia, with especially the Chinese fleet being pinpointed by Northeast Asian neighbors Japan, South Korea, and North Korea for transboundary poaching [8]. As actual incidences of international fishery conflict continuously occur in both the developed and lesser developed areas of the world [3,8,13], and suspected drivers of fishery conflict are ramping up, policy makers are warned to anticipate an increase in clashes over fishery resources [9].

37.2 Learning from the environmental security literature

Despite growing concerns about future fishery conflict, there is still limited knowledge on the occurrence and nature of international fishery conflicts (more specifically, the frequency of conflict occurrences over time, between whom they have occurred and what types of conflict have occurred), nor is there a consensus on fundamental causes or mechanisms connecting fishery resources to conflict, with existing causal explanations deemed "too simplistic" [14,15]. This is despite the progress made in the environmental security literature, a scholarship where both renewable as well as nonrenewable resources have been studied for their possible linkages with conflicts, which conflict scholars can draw from to better understand fishery conflicts. International conflict over freshwater resources (such as river basins) has especially received a great deal of scholarly attention, which can be emulated by the fishery conflict literature [16–19].

From the environmental security literature, one can learn that assuming a *direct* link between environmental change (such as climate change) and conflict is often incorrect, as the discipline has seen scholars move beyond simple cause-and-effect explanations for resource-based conflict [19] into research acknowledging that *causal pathways are complex and contingent on a host of additional factors* [20, pp. 316]. Heavily critiqued, for example, is the existence of a direct link between environmental scarcity and conflict [21], with scholars claiming any direct linkage is impossible to draw as environmental degradation is a mere side effect caused by political, economic, and institutional malfunctioning [22–25]. Seter et al. [26] analyzed what caused conflicts between resource-user groups in arid and semiarid areas in sub-Saharan Africa and found that environmental factors such as resource scarcity (induced by drought) were mere contributing factors, never the important cause, in the 11 case studies they explored; nor did the environmental factors explain conflict intensity.

To understand more about the occurrence and nature of international fishery conflict as well as the risk potential of commonly cited drivers such as scarcity, then, there are a few fundamental gaps that still need addressing, which the environmental security literature can be consulted for [14]. Firstly, the lack of precise and shared definitions for fishery conflict that distinguish among different degrees of fishery conflict intensity (such as trade or import bans, vessel seizures, or violent actions), for example, have inhibited comparability in the identification and characterization of fishery conflict [14]. Based on examples from the environmental security literature on freshwater resources, such as the BAR-scale used by Yoffe et al. [17], Spijkers et al. [14] have developed a five-point intensity scale, with each intensity linked to different observable actions and behaviors, which can be even further expanded on for other scales of fishery conflict (see Table 37.1).

Secondly, the fishery conflict literature generally lacks theoretical framings that explicitly recognize nonlinear and dynamic feedbacks, multiple causes, effects and intervening variables [14], which has often led to a linear representation of cause and effect. This is despite the evolution witnessed within the environmental security literature: the move from direct causation hypotheses to a more nuanced, detailed proposition of causality recognizing the myriad of variables (socioeconomic, ecological, political) influencing the pathway from natural resources and environmental change to conflict [27]. Some of the "mediators" of conflict addressed in the environmental security literature include GDP per capita and regime type [28]; vulnerable livelihoods, poverty, weak states, and migration [29]; development, state strength, and dysfunctional institutions [25]; and institutional design [30]. The literature on fishery conflict should follow suit and expand on the advances made in other literatures, and explicitly address causal complexity by comprehensively assessing multiple potential conflict drivers and feedback mechanisms to avoid drawing simplistic conclusions about the causality that underlies conflict. Linked to that is the currently sparse use of higher order systems terminology to describe the complexity of the marine system and its link to conflict (including, for example, terms such as "sensitivity," "feedbacks," "tipping points," and "thresholds"), which again reflects the often oversimplified view scholars have of fishery conflict [14].

Lastly, methodological gaps also exist in the literature, as most research on fishery conflict has used single, qualitative case studies on fishery conflict to draw conclusions from; and no comprehensive quantitative studies have been performed to assess the occurrence, nature, and causes behind fishery conflict [14]. The literature on freshwater conflicts, in

Table 37.1: Intensity of observed behavior/action.

Intensity	Description
5	Military acts causing death
	Attack of foreign vessels, crew members, or Coast Guards, with resulting deaths
4	Military acts
	Attack of foreign vessels, crew members, or Coast Guards, no death toll
3	Political—military hostile acts
	Sending out police vessels/warships
	Seize vessel and/or crew
	Gear destruction
	Reinforcing borders
2	Diplomatic—economic hostile acts
	Breaking or not adhering to existing agreement
	Lawsuit
	Trial in court
	Seeking international arbitration
	Trade ban
	Fishing ban
	Landing ban
	Monetary penalties
	Close ports
1	Verbal expressions displaying discord or hostility in interaction
	Failing to reach an agreement
	Making threatening demands and accusations
	Threatening sanctions
	Condemning specific actions, behaviors, or policies
	Requesting change in policy
	Civilian protests
0	Nonsignificant acts

Modified from: J. Spijkers, T.H. Morrison, R. Blasiak, G.S. Cumming, M. Osborne, J. Watson, et al., Marine fisheries and future ocean conflict, Fish Fish. 19 (5) (2018) 798–806.

contrast, has seen meta-analytical, quantitative studies on conflict, and they have shed light on the linkages and dynamics across multiple case studies over time (by using, for example, the Transboundary Freshwater Dispute Database) [14]. Implementing similar innovative methodological approaches to understand fishery conflict, such as the development and use of a global database of fishery conflicts, provides the large-scale comparative data suitable for quantitative analysis needed to complement existing research and better understand the occurrence, nature, and causality behind fishery conflict.

37.3 Maritime threats and regional instability

Maritime threats such as fishery conflict, illegal fishing, arms and drug trafficking, or labor abuses at sea are increasingly recognized as interconnected, and attempts are being made to better understand the linkages between such threats to maritime security. From the rapid expansion of literature on such threats to maritime security [31], scholars have learned more about the role of fishery conflict in the wider context of maritime security through, for example, the work by Pomeroy et al. [13], where the authors attempted to piece together certain maritime threats, drivers, and conditions to describe the so-called "fish wars cycle". From such work, one learns that in regions such as Southeast Asia, IUU fishing can not only exacerbate diplomatic, and sometimes even military tensions between nations leading to conflict at sea, it can also perpetuate maritime threats such as slave labor and drugs and arms smuggling, destabilizing marine ecosystems as well as vulnerable communities [13,32]. Maritime threats feeding into each other to spark wider regional instability and insecurity is particularly problematic in areas where people are vulnerable (due to, for example, a high proportion of the population being economically active in the fisheries sector), and governance is weak (and unable to govern their EEZ) [2,33]. Getting a grasp on the driving variables (and the linkages between them) underlying such regional instability and volatility is difficult. For example, does competition over fishery resources (or even competing claims over other natural resources such as oil or minerals) spark diplomatic and military actions between countries, or are such resource disputes "mere" results and proxies for a more deeply rooted battle over naval superiority, regional power, and ownership of territory?

More in-depth understanding is necessary to truly understand how maritime threats can become interlinked to create wider regional instability. Providing large-scale evidence for such linkages could help confirm or nuance certain a priori assumptions and create a more in-depth understanding that can shape maritime policies to ensure ocean security. Large-scale evidence has been recognized as an important part of national maritime strategies in certain countries such as the United Kingdom, for example, which seeks to gather intelligence, analyze data, and identify concerning maritime threats in a holistic manner to maximize governance impact in the maritime space. Academics working in the arena of maritime security are well-positioned to make meaningful contributions to policy development by considering novel ways of collecting large-scale, comparative data on maritime threats and developing more realistic models that help policy makers understand the drivers of regional maritime instability.

Acknowledgments

I would like to express my gratitude to the Nereus Program for their continued support of my projects and ideas, and to the Nippon Foundation for contributing funding to my research. I also wish to thank the Australian Centre of Excellence for Coral Reef Studies and the Stockholm Resilience Centre for their support. Special thanks to Andrés Cisneros-Montemayor for the constructive comments he provided on this chapter.

References

[1] C. Bueger, What is maritime security? Mar. Policy [Internet]. 53 (2015) 159–164. Available from: https://doi.org/10.1016/j.marpol.2014.12.005.

[2] T. Mcclanahan, E.H. Allison, J.E. Cinner, Managing fisheries for human and food security, Fish Fish. 16 (1) (2015) 78–103.

[3] J. Spijkers, W.J. Boonstra, Environmental change and social conflict: the northeast Atlantic mackerel dispute, Reg. Environ. Change [Internet] (2017). Available from: <http://link.springer.com/10.1007/s10113-017-1150-4>.

[4] R.A. Rogers, C. Stewart, Prisoners of their histories: Canada-U.S. conflicts in the Pacific salmon fishery, Am. Rev. Can. Stud. 27 (2) (1997) 253–269.

[5] C.S. Hendrix, S.M. Glaser, Civil conflict and world fisheries, 1952-2004, J. Peace Res. [Internet]. 48 (4) (2011) 481–495. Available from: <http://jpr.sagepub.com/content/48/4/481.abstract>.

[6] S.M. Mitchell, B.C. Prins, Beyond territorial contiguity: issues at stake in democratic militarized interstate disputes, Int. Stud. Q. 43 (1) (1999) 169–183.

[7] H. Zhang, Fisheries cooperation in the South China Sea: evaluating the options, Mar. Policy [Internet] 89 (2018) 67–76. Available from: https://doi.org/10.1016/j.marpol.2017.12.014 (December 2017).

[8] A. Dupont, C.G. Baker, East Asia's maritime disputes: fishing in troubled waters, Wash Q. [Internet] 37 (1) (2014) 79–98. Available from: <http://www.tandfonline.com/doi/abs/10.1080/0163660X.2014.893174>.

[9] B.M.L. Pinsky, G. Reygondeau, R. Caddell, J. Palacios, J. Spijkers, W. William, Preparing ocean governance for species on the move, Science 360 (6394) (2018) 1189–1192 (80-).

[10] L.M. Robinson, H.P. Possingham, A.J. Richardson, Trailing edges projected to move faster than leading edges for large pelagic fish habitats under climate change, Deep. Res. Part II Top. Stud. Oceanogr. 113 (2015) 225–234. Available from: https://doi.org/10.1016/j.dsr2.2014.04.007.

[11] D. Pauly, D. Zeller, Catch reconstructions reveal that global marine fisheries catches are higher than reported and declining, Nat. Commun. [Internet] 7 (2016) 1–9. Available from: https://doi.org/10.1038/ncomms10244.

[12] W.J. Boonstra, H. Österblom, A chain of fools: or, why it is so hard to stop overfishing, Marit. Stud. [Internet] 13 (1) (2014) 15. Available from: <http://www.maritimestudiesjournal.com/content/13/1/15>.

[13] R. Pomeroy, J. Parks, K.L. Mrakovcich, C. LaMonica, Drivers and impacts of fisheries scarcity, competition, and conflict on maritime security, Mar. Policy [Internet] 67 (2016) 94–104. Available from: <http://linkinghub.elsevier.com/retrieve/pii/S0308597X16000105>.

[14] J. Spijkers, T.H. Morrison, R. Blasiak, G.S. Cumming, M. Osborne, J. Watson, et al., Marine fisheries and future ocean conflict, Fish Fish. 19 (5) (2018) 798–806.

[15] R. Penney, G. Wilson, L. Rodwell, Managing sino-ghanaian fishery relations: a political ecology approach, Mar. Policy 79 (2017) 46–53.

[16] P.R. Hensel, M. Brochmann, Peaceful management of international river claims, Int. Negot. [Internet]. 14 (2) (2009) 393–418. Available from: <http://booksandjournals.brillonline.com/content/journals/10.1163/157180609x432879>.

[17] S. Yoffe, A.T. Wolf, M. Giordano, Conflict and cooperation over international freshwater resources: indicators of basins at risk, J. Am. Water Resour. Assoc. [Internet]. 39 (2003) 1109–1126. Available from: <http://onlinelibrary.wiley.com.proxy.library.oregonstate.edu/doi/10.1111/j.1752-1688.2003.tb03696.x/abstract>.

[18] L. De Stefano, P. Edwards, L. De Silva, A.T. Wolf, Tracking cooperation and conflict in international basins: historic and recent trends, Water Policy 12 (6) (2010) 871–884.

[19] C. Devlin, C.S. Hendrix, Trends and triggers redux: climate change, rainfall, and interstate conflict, Polit. Geogr. [Internet] 43 (2014) 27–39. Available from: https://doi.org/10.1016/j.polgeo.2014.07.001.

[20] I. Salehyan, From climate change to conflict? No consensus yet, J. Peace Res. 45 (3) (2008) 315–326.

[21] T.F. Homer-Dixon, On the threshold: environmental changes as causes of acute conflict, Int. Secur. 16 (2) (1991) 76–116.

[22] J. Selby, C. Hoffmann, Beyond scarcity: rethinking water, climate change and conflict in the Sudans, Global Environ. Change [Internet] 29 (2014) 360–370. Available from: https://doi.org/10.1016/j.gloenvcha.2014.01.008.

[23] A. Ciccone, Economic shocks and civil conflict: a comment, Am. Econ. J. Appl. Econ. 3 (4) (2011) 215–227.

[24] N.L. Peluso, M. Watts, Violent Environments, Cornell University Press, Ithaca, NY, 2001.

[25] O.M. Theisen, Blood and soil? Resource scarcity and internal armed conflict revisited, J. Peace Res. 45 (6) (2008) 801–818.
[26] H. Seter, O.M. Theisen, J. Schilling, All about water and land? Resource-related conflicts in East and West Africa revisited, GeoJournal 83 (1) (2016) 169–187.
[27] S. Yoffe, G. Fiske, M. Giordano, M. Giordano, K. Larson, K. Stahl, et al., Geography of international water conflict and cooperation: data sets and applications, Water Resour. Res. 40 (5) (2004) 1–12.
[28] W. Hauge, T. Ellingsen, Beyond environmental scarcity: causal pathways to conflict, J. Peace Res. 35 (3) (1998) 299–317.
[29] J. Barnett, W.N. Adger, Climate change, human security and violent conflict, Polit. Geogr. 26 (6) (2007) 639–655.
[30] S. Dinar, Environmental security, in: G. Kütting (Ed.), Global Environmental Politics: Concepts, Theories and Case Studies., Routledge, London, New York, 2011, pp. 56–69.
[31] B. Germond, The geopolitical dimension of maritime security, Mar. Policy [Internet] 54 (2015) 137–142. Available from: https://doi.org/10.1016/j.marpol.2014.12.013.
[32] World Wildlife Fund, Illegal Fishing: Which Fish Species are at Highest Risk from Illegal and Unreported Fishing? 2015.
[33] R. Blasiak, J. Spijkers, K. Tokunaga, J. Pittman, N. Yagi, O. Henrik, Climate change and marine fisheries: least developed countries top global index of vulnerability, PLoS One 12 (6) (2017) 1–15.

CHAPTER 38

A Blue Economy: equitable, sustainable, and viable development in the world's oceans

Andrés M. Cisneros-Montemayor

Nippon Foundation Nereus Program, Institute for the Oceans and Fisheries, The University of British Columbia, Vancouver, BC, Canada

Chapter Outline

38.1 Background

The "Blue Economy" has become a popular term in governmental, intergovernmental, and academic literature, but remains poorly defined. Depending on the case, it can represent an end goal of development or a development pathway [1]; primarily involve economic accounting [2] or marine spatial planning [3]; or focus on integrating environmental sustainability or social equity goals into economic development [1]. (Equity is used here to refer to the recognition of all stakeholders impacted by developments, and their meaningful inclusion in planning and implementation, leading to a fair distribution of costs and benefits.) This chapter briefly reviews key issues surrounding the Blue Economy term and its potential implications, while also providing examples of how it might be applied. The aim is not to settle what are quite current and valid debates surrounding its definition, included sectors, or implementation. However, I do argue for the use of "Blue Economy" as a development pathway for marine sectors that strictly and explicitly incorporates social

Predicting Future Oceans.
DOI: https://doi.org/10.1016/B978-0-12-817945-1.00034-4

equity and environmental sustainability (which should already be part of any modern development plan) as well as economic viability.

The concept of a Blue Economy was first prominently expressed at the United Nations Rio +20 Conference and has since evolved into several (sometimes perhaps competing) development frameworks [4]. The main theme of the Rio conference was the formal development and proposal of the Green Economy framework, which emphasizes ecologically sustainable and socially equitable economic development [5]. This involves significant investments in development and implementation of technology and agricultural practices that reduce greenhouse gas emissions and directly link to mitigating global climate change, habitat impacts, and biodiversity loss, with a view to ensuring that benefits and costs of new development are distributed in an equitable manner (though this aspect often receives less focus [6]). Within this context, during the meeting a group of Small Island Developing States highlighted the limitations of the Green Economy framework to such nations given their limited available terrestrial area. They noted, however, that their comparatively very large marine areas could instead be a basis for sustainable development. Although it was debated if a seemingly separate (though linked) plan was necessary to incorporate ocean resources into Green Economy thinking, this discussion led to the adoption of a parallel terminology, the "Blue Economy" [1].

As will be discussed below, there are several interpretations of the Blue Economy concept, but in this chapter the term is used to refer to ocean (and coastal) resource-based development that is explicitly planned and implemented in a way that considers social equity, ecological sustainability, and economic viability. A related but quite different term that is commonly used in marine development discussions is the "Ocean Economy," that is, the economic benefits (e.g., revenue, employment) provided by any industry related to marine environments [7], regardless of how they are carried out. This is generally used quite broadly to include (to name only a few sectors) fisheries and tourism, oil and gas drilling and exploration, mining operations, engineering, technology and software companies serving marine applications and environments, shipping, marine restoration, etc. [7]. In that context Blue Growth refers to environmentally sustainable economic growth of Ocean Economy sectors [2]. These and other concepts are further discussed in the section below.

38.2 Definitions and discourses of a Blue Economy

A key argument throughout this chapter is that under a Blue Economy industries must be developed with local people so that their needs and perspectives are fully integrated, rather than literally or figuratively displacing locals to make way for development. However, this is certainly not the only view and since the idea's conception the Blue Economy concept has diversified and diverged, been linked with other ocean development concepts, and gradually lost its explicit focus on social equity issues in favor of more traditional industrial development [4]. For example, a recent special issue on the Blue Economy published by

The Economist omits mention of social considerations entirely, stating instead that a "sustainable ocean economy emerges when economic activity is in balance with the long-term capacity of ocean ecosystems to support this activity and remain resilient and healthy" [8]. Recent reports by the World Bank have also defined the Blue Economy as "sustainable development of the ocean economy" or one "comprising the range of economic sectors that together determine whether the use of oceanic resources is sustainable" [2,9]. Some Small Island Developing States themselves have somewhat shifted their proposals, focusing on receiving a share of benefits from industrial development rather than having such development be self-led [4,10]. It must be noted that environmental sustainability indeed figures quite prominently in many of these marine development plans, and yet this should hardly be a revolutionary or even optional aspect of planning given the current state of marine environments and the extensively-documented impacts of industry on them [11,12].

The various, sometimes competing, discourses surrounding the Blue Economy (and Ocean Economy, for this section) have deep implications for the concept as a development strategy and for its implementation across global regions. First, there is a fundamental tension between Blue (or Ocean) Economy interpretations that do or do not explicitly consider social equity as a core aspect of new development frameworks, essentially mirroring debates on trickle-down versus bottom-up development. In the trickle-down view ocean industries that become profitable will eventually provide increased benefits across society, even if a majority might accrue to a limited part of the population or private enterprises. Environmental impacts may well occur but would eventually be reduced through new technologies. The opposing argument, which would seem central to the original proposal for a Blue Economy, is that uneven economic development has led to historical marginalization, the extraction of benefits from local resources away from local communities, and lasting environmental impacts as industries used up resources and left. A more socially equitable, and culturally-appropriate, development of new industries and the guidelines to share in their benefits (procedural justice) is thus vital to address past inequities and enable sustained and equitable benefits from natural resources.

38.3 Sectors of a Blue Economy

A related, and deceptively complex, issue related to discourses on the Blue (versus Ocean) Economy is the definition of specific industrial sectors to be included (or not) in planning. The definition used in this chapter, for example, focuses on the inclusion of social equity concerns but also takes a strict view on the Blue Economy as requiring environmental sustainability. In that sense, activities that are by definition unsustainable, such as seabed mining or oil and gas extraction, would not be included, under the rationale that including unsustainable industries in a plan meant to be a novel strategy for sustainable development would be contradictory and ultimately self-defeating for the framework itself. However, this does in no way imply that they should not be pursued, and they would certainly be included in Ocean Economy accounting. Indeed there are many ways in which existing and future operations and technologies may reduce their negative impacts on the environment and human health, irrespective of the particular development framework under which they are carried out (if any).

Under Blue Economy conceptualizations that place focus on economic growth and innovation (e.g., "Blue Growth"), sectors of most interest are those most likely to yield high returns on investments (Fig. 38.1). Thus, offshore wind energy, seafood trade, and infrastructure projects are often a focus of intergovernmental and national planning. Fisheries are a good example of the practical implications of competing discourses on implementation of Blue Economy plans. As the most important marine sector in terms of employment (Fig. 38.1A) and food provision [13], fisheries would of course be a vital component for any Blue Economy plan centered on social equity and human well-being. However, current mismanagement has resulted in many fisheries being unprofitable (Fig. 38.1B; [14,15]) and having limited potential for employment gains (Fig. 38.1A) and therefore of less interest to plans focused solely on economic growth or in further increasing employment [16]. Though this could be seen as an example of different priorities stemming from different overarching goals, it also highlights a misleading perceived trade-off between social and economic concerns, as there is much evidence that sustainable economies themselves depend on implementing socially equitable development [17–20]. This is perhaps best linked to the parallel debate on the best way to achieve human well-being in the context of the cross-scale, interdisciplinary, and highly interrelated UN Sustainable Development Goals (SDGs), where human well-being is the ultimate goal and is bolstered by but also facilitates economic growth and environmental sustainability [22].

Under the Blue Economy framework used here (again, explicitly incorporating social equity concerns as well as environmental and financial ones), basic needs, alternative livelihoods, and income for coastal regions would be provided by marine-based industries that are relatively new or have not yet been widely adopted, such as offshore wind or tidal energy, integrated aquaculture, or ecotourism (in the strict sense, involving both local benefits and support for conservation [23]), or existing industries (e.g., fisheries) operating under the new guidelines. This is particularly important for coastal communities that are often geographically isolated or otherwise marginalized and yet comprise large populations and economic sectors, such as coastal Indigenous Peoples [24] and artisanal fishers [25,26]. Economic development for these regions and peoples is particularly important when benefits from fisheries—historically the most important provider of food and employment along the world's coasts—are generally declining [27]. At local scales, for example, offshore wind turbines could provide power in places where costs may otherwise be prohibitive (i.e., "last-mile electrification"); an added value of these developments is that

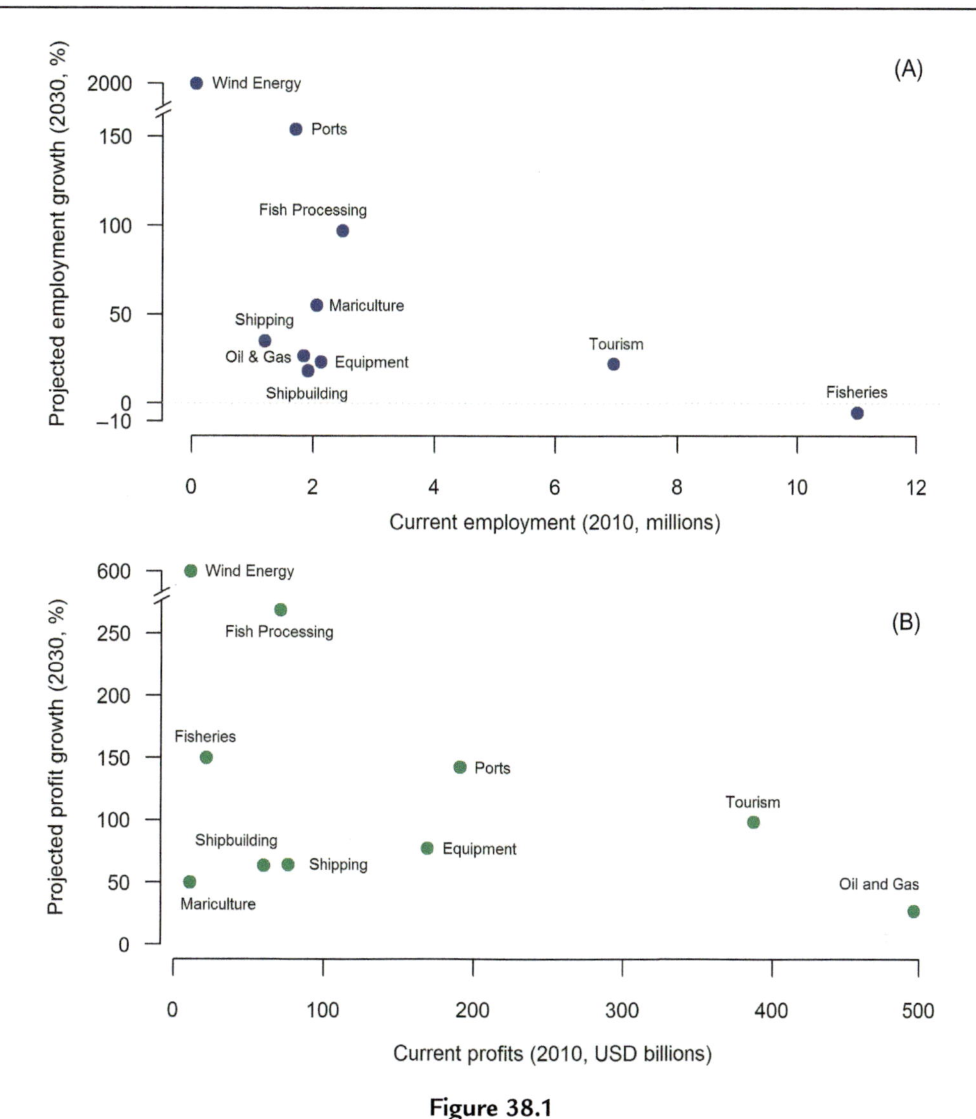

Figure 38.1

Current estimates and projections of (A) employment and (B) profits from ocean economy sectors. *Data from OECD, The Ocean Economy in 2030, OECD Publishing, 2016. <https://doi.org/10.1787/9789264251724-en> [21].*

they could represent employment opportunities in such communities [28]. At the national and international scale there are already extensive negotiations underway to catch up to technological developments and lay down the legal frameworks for sharing benefits from marine-based chemical (including pharmaceutical) compounds that can only be realistically identified and developed by a few private firms but often represent common resources [29].

38.4 Implementing a Blue Economy

It is important to remember that, though a Blue Economy approach could imply important changes in how development is carried out, many of the sectors that would potentially be included are already very large and distributed across the world (Fig. 38.2). In that sense, the Blue Economy does not only imply that new developments incorporate social equity and environmental sustainability concerns, but that existing and potentially quite significant industries be transformed. Perhaps the biggest barrier to this type of change will likely be existing and entrenched interests by individuals, firms, or groups that profit from current inequitable (and/or unsustainable) activities.

In operational terms, considering and ensuring that marine industries are developed and carried out in an equitable and sustainable manner need not be an onerous task, and there are a range of existing guidelines and best practices that already exist for various specific sectors and at a broader international level. These include the FAO Small Scale Fisheries

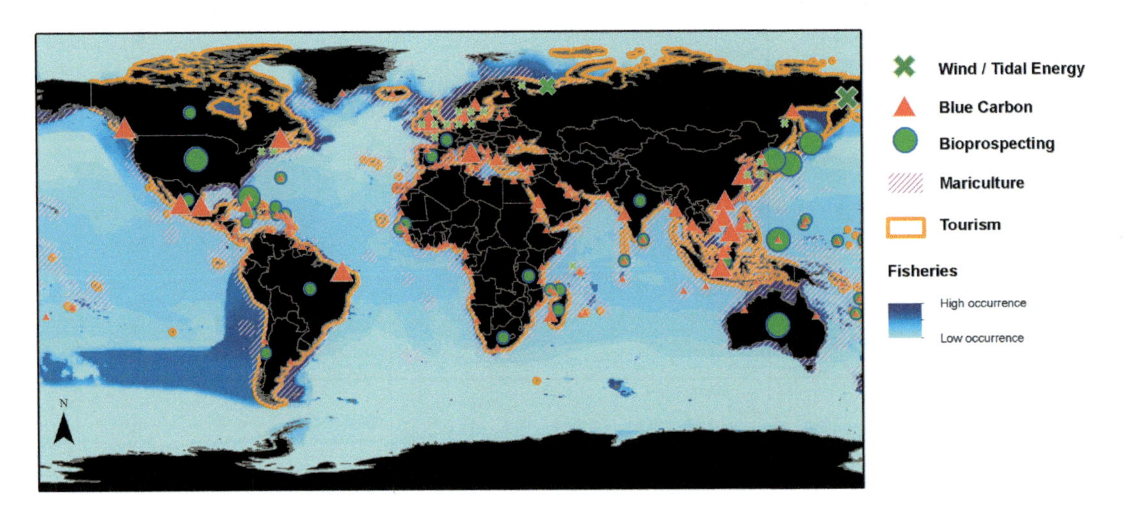

Figure 38.2

Current sites and distribution of some marine industrial sectors potentially considered under a Blue Economy. *Data from W.W.L. Cheung, V.W.Y. Lam, J.L. Sarmiento, K. Kearney, R. Watson, D. Zeller, et al., Large-scale redistribution of maximum fisheries catch potential in the global ocean under climate change, Global Change Biol. 16 (2010) 24–35, doi:10.1111/j.1365-2486.2009.01995.x; A.M. Cisneros-Montemayor, U.R. Sumaila, A global estimate of benefits from ecosystem-based marine recreation: potential impacts and implications for management, J. Bioeconomics 12 (2010) 245–268; S. Thomas, Blue carbon: knowledge gaps, critical issues, and novel approaches, Ecol. Econ. 107 (2014) 22–38, doi:10.1016/j.ecolecon.2014.07.028; B. Hunt, A.C.J. Vincent, Scale and sustainability of marine bioprospecting for pharmaceuticals, AMBIO J. Hum. Environ. 35 (2006) 57–64, doi:10.1579/0044-7447(2006)35[57:SASOMB]2.0.CO;2; M.A. Oyinlola, G. Reygondeau, C.C. Wabnitz, M. Troell, W.W. Cheung, Global estimation of areas with suitable environmental conditions for mariculture species, PLoS One 13 (2018) e0191086 [32–34].*

Guidelines [26], specific provisions within the Convention on Biological Diversity's Aichi Targets [35], UN Declaration on the Rights of Indigenous Peoples [30,31], FAO Ecosystem Approach to Fisheries [36], and Code of Conduct for Responsible Fisheries [37], and, as a much broader but highly useful cross-scale framework, the UN SDGs [38].

The examples above provide some potential strategies for incorporating social equity and environmental sustainability into economic development plans, but this does not imply that it is always an easy or straightforward process. A recent case is quite illustrative of the many challenges that can be faced in that regard, and is not intended as a criticism but an example of the operational difficulties of fundamental change. The Canadian government has for some years been a vocal supporter of long-term sustainability and renewable industries, and an early and strong supporter of the Paris Agreement. However, faced with a challenging economic and political environment, it invested heavily in oil production, including new pipelines to link production areas with marine transport routes. While oil extraction is an important part of many national economies, it is by definition a nonsustainable industry. There is ample potential for renewable energy in Canada [39], and continued oil extraction will negatively impact global climate change and ensure that national emissions goals will not be met [40]. Concerns about equity and sustainability spurred legal actions, and the Federal Court of Appeal of Canada suspended the project and required planners to meaningfully consult with Indigenous communities that would be impacted and consider risks to marine ecosystems including endangered killer whale populations [41]. The argument here isn't that this example is worse than others around the world but, rather, that even highly developed nations that recognize the importance of social and environmental issues may find it challenging to act on these concerns within a globalized economy without international support. As the world's largest commons, ocean environments make this all the more difficult.

The example noted above also reflects a final important distinction to be made when considering alternative views of a Blue Economy: whether it represents an end goal for development (e.g., "a Blue Economy has been achieved") or an approach to development (e.g., "this sector was developed following Blue Economy principles"). If the former, it could be argued that unsustainable industries, namely oil and gas production or seabed mining, can be an important part of a transition towards sustainable activities and thus could be further expanded in the short term. Ultimately such industries would be phased out as the new Blue Economy is fully implemented. A key issue with this approach is that the global environment is in a critical state due to emissions from oil extraction and use, so further expansion may be catastrophic for any future development [40]. Understanding the Blue Economy as an approach to development would require that industries are undertaken in both an equitable and sustainable manner and thus would likely exclude some sectors to the benefit of others, such as renewable energy, that could be much more profitable (Fig. 38.1B) but require investments in innovation and currently could not compete with existing fossil fuel industries operating without accounting for the full costs of their activities. As noted earlier, however, any industry can and should transform its operations to promote more equitable outcomes and to reduce its environmental impacts, regardless of whether it depends on renewable or nonrenewable resources or whether it is "classified" as part of the Blue Economy.

38.5 Conclusion

This chapter argues for the Blue Economy as an approach to ocean and coastal development, that strictly and explicitly incorporates social equity in addition to environmental sustainability and economic viability. It is vitally important, however, that discussions on definitions, implementation strategies, and included sectors, not lose sight of critical priorities for ocean-based development. Historical and current inequities have contributed to conflict and difficulties in managing ocean resources [17,42] and must be addressed if human well-being is to be achieved across the world. Marine ecosystems have been negatively impacted on many fronts, including through local industries and global climate change, and require urgent actions to minimize ongoing risks and restore and protect ecosystems and species [43–46]. Adapting to these new realities is critical for future ecosystem functions and human well-being [47–49], and many changes to industrial development, policy planning, and attitudes regarding social and environmental issues must be included. In that context, however, the Blue Economy approach should not be seen as a slight improvement on business-as-usual but as an opportunity for a fundamental shift towards ambitious social and environmental goals.

Acknowledgments

The author gratefully acknowledges the support from the Nippon Foundation Nereus Program and thanks the many insights of collaborators in this program, and Marcia Moreno-Baez for her help in preparing Fig. 38.2.

References

[1] J.J. Silver, N.J. Gray, L.M. Campbell, L.W. Fairbanks, R.L. Gruby, Blue Economy and competing discourses in international oceans governance, J. Environ. Dev. 24 (2015) 135–160. Available from: https://doi.org/10.1177/1070496515580797.
[2] P.G. Patil, J. Virdin, C.S. Colgan, M.G. Hussain, P. Failler, T. Vegh, Toward a Blue Economy: A Pathway for Bangladesh's Sustainable Growth, World Bank, Washington, DC, 2018.
[3] Government of Grenada, Blue Grenada, A Commitment to Blue Growth, Sustainability, and Innovation, 2017.
[4] M. Voyer, G. Quirk, A. McIlgorm, K. Azmi, Shades of blue: what do competing interpretations of the Blue Economy mean for oceans governance? J. Environ. Policy Plan. (2018) 1–22. Available from: https://doi.org/10.1080/1523908X.2018.1473153.
[5] United Nations Environment Programme, Towards a Green Economy: Pathways to Sustainable Development and Poverty Eradication, UNEP, Nairobi, Kenya, 2011.
[6] C. Corson, K.I. MacDonald, B. Neimark, Grabbing "green": markets, environmental governance and the materialization of natural capital, Hum. Geogr. 6 (1) 1–15.

[7] D.K.S. Park, D.J.T. Kildow, Rebuilding the classification system of the ocean economy, J. Ocean Coast Econ. 2014 (2015). Available from: https://doi.org/10.15351/2373-8456.1001.
[8] Economist Intelligence Unit, The blue economy. Growth, opportunity and a sustainable ocean economy, The Economist, 2015.
[9] World Bank, UN-DESA. The Potential of the Blue Economy. Increasing Long-term Benefits of the Sustainable Use of Marine Resources for Small Island Developing States and Coastal Least Developed Countries, World Bank, Washington, D.C, 2017.
[10] M.R. Keen, A.-M. Schwarz, L. Wini-Simeon, Towards defining the Blue Economy: practical lessons from pacific ocean governance, Mar. Policy (2017). Available from: https://doi.org/10.1016/j.marpol.2017.03.002.
[11] A. Stock, L.B. Crowder, B.S. Halpern, F. Micheli, Uncertainty analysis and robust areas of high and low modeled human impact on the global oceans: human-impact areas, Conserv. Biol. 32 (2018) 1368–1379. Available from: https://doi.org/10.1111/cobi.13141.
[12] B.S. Halpern, C. Longo, D. Hardy, K.L. McLeod, J.F. Samhouri, S.K. Katona, et al., An index to assess the health and benefits of the global ocean, Nature 488 (2012) 615–620. Available from: https://doi.org/10.1038/nature11397.
[13] FAO, The state of world fisheries and aquaculture 2018, in: Meeting the Sustainable Development Goals, Rome, 2018.
[14] U.R. Sumaila, W. Cheung, A. Dyck, K. Gueye, L. Huang, V. Lam, et al., Benefits of rebuilding global marine fisheries outweigh costs, PLoS One 7 (2012) e40542. Available from: https://doi.org/10.1371/journal.pone.0040542.
[15] World Bank, The Sunken Billions Revisited: Progress and Challenges in Global Marine Fisheries., The World Bank, 2017. Available from: https://doi.org/10.1596/978-1-4648-0919-4.
[16] ECORYS, Deltares, Oceanic Dévelopement, Blue Growth, Scenarios and drivers for sustainable growth from the oceans, Seas and coasts, 2012.
[17] E.M. Finkbeiner, N.J. Bennett, T.H. Frawley, J.G. Mason, D.K. Briscoe, C.M. Brooks, et al., Reconstructing overfishing: moving beyond Malthus for effective and equitable solutions, Fish Fish. (2017). Available from: https://doi.org/10.1111/faf.12245.
[18] N.J. Bennett, Navigating a just and inclusive path towards sustainable oceans, Mar. Policy 97 (2018) 139–146. Available from: https://doi.org/10.1016/j.marpol.2018.06.001.
[19] P. Christie, N.J. Bennett, N.J. Gray, T. Aulani Wilhelm, N. Lewis, J. Parks, et al., Why people matter in ocean governance: incorporating human dimensions into large-scale marine protected areas, Mar. Policy 84 (2017) 273–284. Available from: https://doi.org/10.1016/j.marpol.2017.08.002.
[20] U. Pascual, J. Phelps, E. Garmendia, K. Brown, E. Corbera, A. Martin, et al., Social equity matters in payments for ecosystem services, BioScience 64 (2014) 1027–1036. Available from: https://doi.org/10.1093/biosci/biu146.
[21] OECD, The Ocean Economy in 2030, OECD Publishing, 2016. Available from: https://doi.org/10.1787/9789264251724-en.
[22] G.G. Singh, A.M. Cisneros-Montemayor, W. Swartz, W. Cheung, J.A. Guy, T.-A. Kenny, et al., A rapid assessment of co-benefits and trade-offs among Sustainable Development Goals, Mar. Policy (2017). Available from: https://doi.org/10.1016/j.marpol.2017.05.030.
[23] H.M. Donohoe, R.D. Needham, Ecotourism: the evolving contemporary definition, J. Ecotourism. 5 (2006) 192–210. Available from: https://doi.org/10.2167/joe152.0.
[24] A.M. Cisneros-Montemayor, D. Pauly, L.V. Weatherdon, Y. Ota, A global estimate of seafood consumption by coastal Indigenous Peoples, PLoS One 11 (2016) e0166681. Available from: https://doi.org/10.1371/journal.pone.0166681.
[25] L.C.L. Teh, U.R. Sumaila, Contribution of marine fisheries to worldwide employment: global marine fisheries employment, Fish Fish. 14 (2013) 77–88. Available from: https://doi.org/10.1111/j.1467-2979.2011.00450.x.
[26] FAO, Voluntary Guidelines for Securing Sustainable Small-Scale Fisheries in the Context of Food Security and Poverty Eradication, Food and Agriculture Organization of the United Nations, Rome, 2015.

[27] B. Worm, Averting a global fisheries disaster, Proc. Natl. Acad. Sci. U.S.A. 113 (2016) 4895–4897. Available from: https://doi.org/10.1073/pnas.1604008113.
[28] G. Hiriart Le Bert, Potencial energético del Alto Golfo de California, Bol. Soc. Geológica Mex. 61 (2009) 143–146. Available from: https://doi.org/10.18268/BSGM2009v61n1a13.
[29] R. Blasiak, J.-B. Jouffray, C.C. Wabnitz, E. Sundström, H. Österblom, Corporate control and global governance of marine genetic resources, Sci. Adv. 4 (2018) eaar5237.
[30] W.W.L. Cheung, V.W.Y. Lam, J.L. Sarmiento, K. Kearney, R. Watson, D. Zeller, et al., Large-scale redistribution of maximum fisheries catch potential in the global ocean under climate change, Global Change Biol. 16 (2010) 24–35. Available from: https://doi.org/10.1111/j.1365-2486.2009.01995.x.
[31] A.M. Cisneros-Montemayor, U.R. Sumaila, A global estimate of benefits from ecosystem-based marine recreation: potential impacts and implications for management, J. Bioeconomics 12 (2010) 245–268.
[32] S. Thomas, Blue carbon: knowledge gaps, critical issues, and novel approaches, Ecol. Econ. 107 (2014) 22–38. Available from: https://doi.org/10.1016/j.ecolecon.2014.07.028.
[33] B. Hunt, A.C.J. Vincent, Scale and sustainability of marine bioprospecting for pharmaceuticals, AMBIO J. Hum. Environ. 35 (2006) 57–64. Available from: https://doi.org/10.1579/0044-7447(2006)35[57:SASOMB]2.0.CO;2.
[34] M.A. Oyinlola, G. Reygondeau, C.C. Wabnitz, M. Troell, W.W. Cheung, Global estimation of areas with suitable environmental conditions for mariculture species, PLoS One 13 (2018) e0191086.
[35] UNEP, The strategic plan for biodiversity 2011–2020 and the Aichi Biodiversity Targets, Nagoya, Japan, 2010.
[36] S. Garcia, K. Cochrane, Ecosystem approach to fisheries: a review of implementation guidelines, ICES J. Mar. Sci. 62 (2005) 311–318. Available from: https://doi.org/10.1016/j.icesjms.2004.12.003.
[37] FAO, FAO Code of Conduct for Responsible Fisheries, FAO Fisheries and Aquaculture Department, 1995.
[38] UN, Transforming our world: the 2030 Agenda for Sustainable Development, United Nations, New York, 2015.
[39] Natural Resources Canada,. Renewable Energy Facts, 2018. <https://www.nrcan.gc.ca/energy/facts/renewable-energy/20069>.
[40] S.I. Seneviratne, M.G. Donat, A.J. Pitman, R. Knutti, R.L. Wilby, Allowable CO_2 emissions based on regional and impact-related climate targets, Nature 529 (2016) 477–483. Available from: https://doi.org/10.1038/nature16542.
[41] Federal Court of Appeal, Tsleil-Waututh Nation v. Canada (Attorney General), 2018, FCA 153.
[42] T.F. Homer-Dixon, Environmental scarcities and violent conflict: evidence from cases, Int. Secur. 19 (1994) 5. Available from: https://doi.org/10.2307/2539147.
[43] D. Pauly, V. Christensen, S. Guénette, T.J. Pitcher, U.R. Sumaila, C.J. Walters, et al., Towards sustainability in world fisheries, Nature 418 (2002) 689–695.
[44] E.L. Gilman, J. Ellison, N.C. Duke, C. Field, Threats to mangroves from climate change and adaptation options: a review, Aquat. Bot. 89 (2008) 237–250. Available from: https://doi.org/10.1016/j.aquabot.2007.12.009.
[45] J.M. Pandolfi, Global trajectories of the long-term decline of coral reef ecosystems, Science 301 (2003) 955–958. Available from: https://doi.org/10.1126/science.1085706.
[46] E.H. Allison, H.R. Bassett, Climate change in the oceans: human impacts and responses, Science 350 (2015) 778–782.
[47] W.N. Adger, J. Barnett, K. Brown, N. Marshall, K. O'Brien, Cultural dimensions of climate change impacts and adaptation, Nat. Clim. Change 3 (2012) 112–117. Available from: https://doi.org/10.1038/nclimate1666.
[48] W.N. Adger, N.W. Arnell, E.L. Tompkins, Successful adaptation to climate change across scales, Global Environ. Change 15 (2005) 77–86.
[49] D.D. Miller, Y. Ota, U.R. Sumaila, A.M. Cisneros-Montemayor, W.W.L. Cheung, Adaptation strategies to climate change in marine systems, Global Change Biol. (2017). Available from: https://doi.org/10.1111/gcb.13829.

CHAPTER 39

Can aspirations lead us to the oceans we want?

Gerald G. Singh

Nippon Foundation Nereus Program, Institute for the Oceans and Fisheries, University of British Columbia, Vancouver, BC, Canada

Chapter Outline

39.1 The future we want... but we do not always get what we want

In 2015 there was a great optimism in the international community as the sustainable development goals (SDGs) were established. Surpassing the previous Millennium Development Goals, the SDGs set ambitious policy goals at an international scale, aiming to address a comprehensive set of topics related to human prosperity and its long-term maintenance [1]. Simply agreeing on goals and targets in itself signaled an achievement of international cooperation and indicated that there is a converging international view of how the world should look. However, optimism was tempered by a sobering realism (sometimes conflated with pessimism): almost no one is confident that they will be reached.

The SDGs encompass various environmental, economic, social, and governance domains that potentially interrelate to support each other and are in themselves important normative goals to achieve (Table 39.1). The importance of the SDGs as "the future we want" is readily apparent, but there are many pitfalls and obstacles to their achievement, including the ways in which we envisage how to achieve the SDGs. This chapter will outline three major obstacles on the way to achieving the SDGs, with a particular focus on SDG 14: Life Below Water, and discuss a way forward.

Predicting Future Oceans.
DOI: https://doi.org/10.1016/B978-0-12-817945-1.00032-0

Table 39.1: The sustainable development goals.

Sustainable development goal	Description
1: No poverty	End poverty in all its forms everywhere
2: Zero hunger	End hunger, achieve food security and improved nutrition, and promote sustainable agriculture
3: Good health and well-being	Ensure healthy lives and promote well-being for all at all ages
4: Quality education	Ensure inclusive and equitable quality education and promote lifelong learning opportunities for all
5: Gender equality	Achieve gender equality and empower all women and girls
6: Clean water and sanitation	Ensure availability and sustainable management of water and sanitation for all
7: Affordable and clean energy	Ensure access to affordable, reliable, sustainable and modern energy for all
8: Decent work and economic growth	Promote sustained, inclusive, and sustainable economic growth, full and productive employment and decent work for all
9: Industry, innovation, and infrastructure	Build resilient infrastructure, promote inclusive and sustainable industrialization, and foster innovation
10: Reduced inequalities	Reduce inequality within and among countries
11: Sustainable cities and communities	Make cities and human settlements inclusive, safe, resilient, and sustainable
12: Responsible consumption and production	Ensure sustainable consumption and production patterns
13: Climate action	Take urgent action to combat climate change and its impacts
14: Life below water	Conserve and sustainably use the oceans, seas, and marine resources for sustainable development
15: Life on land	Protect, restore, and promote sustainable use of terrestrial ecosystems, sustainably manage forests, combat desertification, and halt and reverse land degradation and halt biodiversity loss
16: Peace, justice, and strong institutions	Promote peaceful and inclusive societies for sustainable development, provide access to justice for all and build effective, accountable, and inclusive institutions at all levels
17: Partnerships for the goals	Strengthen the means of implementation and revitalize the global partnerships for sustainable development

The ambition of the SDGs lies, in part, in their short timescales. While established in 2015, the first targets are due to be achieved by 2020, and most are due by 2030. The second major aspect of the ambition lies in its focus on transformation away from a deeply entrenched world system of economic inequalities, environmental stress, and social tension. The second aspect was always known to be difficult, but it was the main purpose of the goals. The international community determined that the current state of the world is no longer acceptable, and proposed a new version of the world to achieve. However, the radical scale of transformation of our economy, institutions, and politics needed to achieve the SDGs is often not acknowledged or appreciated. The first aspect (the short timescales) is often pointed to as a problem of political will [2,3]. For example, one of the goal targets for ocean sustainability is 10% ocean coverage in marine protected areas. So far only 3.6%

has been protected, with 2 years now to go to achieve the rest [4]. Additionally, the areas selected for protection are often problematic, preserving areas with no conservation value or even having no enforcement, and only 2% of the ocean is in fully protected waters [4,5]. While the lack of political will is undoubtedly a contributor to our lack of achievement, this fails to explain everything and indeed is often used dismissingly as a way to explain a situation we have not adequately analyzed and cannot fully account for. In this case I argue that the current framing of our sustainability goals may also prevent progress.

A focus on achievement may itself be too strict a consideration around sustainable development, particularly because what achievements mean in concrete terms is vague. Indeed some may say that the journey is itself the destination—that just striving to progress toward the goals is enough. If any action taken to address the goals takes the world closer to the goal's realization, then we will be closer to a sustainable planet so long as we strive—progress is progress.

However, not every action to plan for and address the goals will necessarily contribute toward their progress. Some efforts may produce no results, and some may have negative consequences. Many goals—especially ones associated with human rights abuses—are absolute in their standards. Take, for example, the goal addressing modern slavery: the target is to eradicate slavery in all its forms. This is undeniably a noble goal, and policy makers understandably had to give no indication that *any* amount of slavery is acceptable (setting a goal to limit slavery to some total or proportion of the population could establish an idea that having some slaves is part of an ideal and sustainable future). Current studies estimate that 45.8 million people are subject to some form of modern slavery across 167 countries [6], also underpinning the importance of this target. But the rise in slavery also points to how difficult it would be to achieve. Having slaves is profitable in a world where consumers desire the cheapest products and are unaware of (or dissociated from) the conditions of production, and slaveholders have and will continue to elude efforts to stop them. The same issues will be faced by the target to eliminate illegal fishing efforts (SDG 14.4), which sometimes uses forced labor.

While strongly worded aspirational targets in the SDGs may drive action—while weak, incremental targets may fail to inspire—failing to pair the targets with strategic implementation can hamper progress. There will be easier and harder slavery rackets to bring down and easier and harder contexts to prevent people from ending up trapped as slaves. Strategically assessing the opportunities for addressing the target of ending slavery can ensure that real progress is made, but no such strategy is embedded in the SDGs. In fact the absolutism that frames the targets may encourage high risk and potentially wasteful actions [7]; if all slavery must be ended by 2030 then surely resources must be spent to end the most difficult slavery rings, but these actions may not generate results, taking resources away from actions that have a greater potential to succeed, such as targeting slavery rings easier to identify and prosecute. If all slavery must end, then cases where people are at most risk and where ending slavery is most difficult may act as a vacuum for the antislavery resources, with little to no results. The confluence of absolute goals and the absence of strategy to achieve these goals means that aspirations may not be enough, and may even be counterproductive [7].

The third major reason that the SDGs may only ever be aspirational is that we do not know how these SDGs interact [8]. It may simply be too much to ask for continued GDP growth along with an elimination of inequalities. It might be too much to ask for a restoration of natural systems and a halt of species loss while simultaneously granting equity in resource allocation and decision-making. If marine protected areas were set up to their 10% target, would all other targets be more or less achievable? The history of MPA development is riddled with disadvantageous consequences for local coastal peoples as access is stripped in the name of conservation [9,10]. While conservationists may praise the SDGs for promoting conservation to the extent that it does, truly engaging with the SDGs will require hard choices where likely tradeoffs exist [8,11]. Conversely, many of the goals may be complementary, and while some work has been done to understand their trade-offs and complementarity in the SDGs (e.g., [8,11]), this work is far from comprehensive at appropriate national, subnational, and international scales relevant for policy, and as a result we do not know which SDGs are most important for coenhancement in specific contexts. Regardless of the structure of these interrelationships, we do not know what they are, and that makes the achievement of the goals less likely.

39.2 Policies of wishful thinking

Setting sustainability policy goals must navigate two competing fallacies—the naturalistic fallacy (the appeal to nature) and the moralistic fallacy (the appeal to morality). The naturalistic fallacy is the misplaced belief that understanding the way things are prescribes the way things should be [12,13]. The moralistic fallacy, conversely, is the notion that selecting a favorable way forward carries the mechanisms of the way things are [14].

39.2.1 The naturalistic fallacy

Regarding sustainable development, perhaps the most well-known example of the naturalistic fallacy is the planetary boundaries framework. Originally put forward in 2009 as a way to frame a "safe operating space" for human development, the planetary boundaries framework proposes that human development is bounded by planetary constraints, and that the planetary constraints on living a good life for people are defined by a select set of environmental processes (and now including biological characteristics) that define the Holocene [15,16]. The Holocene was the geological epoch that preceded the now-touted

"Anthropocene" (a geological epoch characterized by human influence over the world), and the Holocene was the epoch that humans became a distinct species in. That early human evolution occurred during the Holocene, the argument suggests, means that the Holocene conditions represent the conditions that humans are best suited to flourish in. There are currently nine environmental processes included in the framework (climate change, biodiversity, land-system change, freshwater use, phosphorus and nitrogen flows, ocean acidification, atmospheric aerosol loading, ozone depletion, and chemical pollution), with biodiversity, phosphorus and nitrogen flows already considered to be past their boundary [16].

There's no doubt that the planetary boundaries framework has had an effect on the formulation of the SDGs. The language of a "safe operating space" is used in much of the language of the preamble for the SDGs, and the notion that human well-being is bounded by specific environmental conditions is reflected in the fact that the SDGs with an environmental focus almost all have the nearest completion dates (2020), while this is not the case for the social and economic goals—suggesting that getting the environment right is a precondition for ensuring a sustainable society and economy [17]. With the achievement dates of environmental targets set before economic, social, and governance reforms, the structure of the SDGs may not be structured in a way to facilitate their achievement, since many economic and governance reforms may be necessary to achieve environmental goals (discussed below).

My intention is not to revisit the technical critiques of the planetary boundaries framework (for example, see [18,19]) because I think the problems with the framework are built into its foundations. Suggesting that humans can only thrive under the conditions that they evolved under throws the current human experience—in every corner of the globe—into existential crisis. There is absolutely no evolutionary rule that states that the conditions under which something evolves represents the best conditions under which they will thrive. It is a science-based twist of a genetic fallacy to claim that development policy should reflect the conditions of our fledgling species rather than the conditions that we find ourselves in now. To state that the historic conditions of the natural world should be continued into the future (the appeal to nature) neglects the fact that humanity had to circumvent and adapt around the limitations the natural environment put on it in order to evolve physically and culturally and spread around the world. According to the planetary boundaries framework, the nitrogen cycle of the earth has passed its Holocene boundary [15], but it is only because we have passed this boundary that we can feed the 7.5 billion people that currently live—in other words, our continued existence (or, our existence without mass starvation) is predicated on planetary boundaries being crossed [19,20]. Further, taking the premise of planetary boundaries as true must throw out leading theories into the invasibility of species—that some species become invasive in new environments because they become released from conditions that regulated them in the environments they evolved under [21]. No species evolves to optimize flourishing under their environment; they evolve to cope with their environment long enough to procreate.

Beyond the foundation of the framework, the consequences of the framework on development policy, humans, and the environment, is potentially problematic. Though the framework may be developed through an appeal toward science in an effort to protect nature, it is natural systems that could ironically degrade. In one example biodiversity plans to increase resilience to climate change in Europe, such as forest practices of thinning to promote growth and carbon storage, endangered old growth and biodiversity rich forests [22]. As a framework to environmentally bound the limits of development, the main concern of the framework is a safe operating space for humanity, not environmental quality; environmental quality is simply a function to promote human development.

A recent anthropological study revisits the story of Easter Island, the famous story of ecological collapse leading to societal collapse. Recent research highlights that despite systematic environmental collapse on the island, societal collapse and routine warfare among its inhabitants (the Rapa Nui people) did not follow [23,24]. Rather, this new evidence suggests that the Rapa Nui people were a functioning, tightly woven, well-fed society despite a degraded ecosystem. In fact, evidence suggests that there was less malnutrition in the Rapa Nui people than there was in larger and lusher islands such as Tonga, and also less malnutrition than found in Europe at the time [24,25]. Instead the collapse may have come from contact with European and South American explorers, who decimated the population through the introduction of disease and forced them into slavery [24,26]. The point here is that if people can prosper under incredibly denuded environments, then the relationship between environmental condition and human well-being is less tightly coupled and less linear than many environmentalists think [27]. And this is potentially problematic if you care about the environment: if human thriving can be defined as having peaceful societies with intact cultures that can meet their needs, then we can potentially achieve this with incredibly denuded environments, such as Easter Island. Of course, you can argue that this is an incomplete definition of human thriving, but the planetary boundaries makes no effort to define it other than suggesting we live within a "safe operating space" which is tautologically defined by its own dimensions.

I am not advocating that the issue of environmental catastrophe (what the planetary boundaries cautions against) is a trivial issue to worry about. However, if society indicates that it is willing to take the chance at risking a less stable climate and more volatile environment for a chance to increase material wealth, and all that matters is human material wealth, why should an appeal to a past "more stable" environment matter? The authors of the planetary boundaries framework routinely make the case that the planetary boundaries are a scientific exercise [28], yet at the same time make claims about the limits of human prosperity and (implicitly) the direction human development should go.

Fortunately, on this issue the SDGs diverged from the planetary boundaries concept by setting environmental issues as goals in themselves, so they are endpoints and not just means to some end. If the environment is to matter for policy, it must matter subjectively and we must honestly treat it so [29]. Trying to force environmental concerns as objective necessities for human prosperity can be overturned by evidence that it matters less than we thought.

39.2.2 The moralistic fallacy

The opposite problem of setting policy according to appeals to nature is setting policy according to appeals to morals. In suggesting that the 17 SDGs are interrelated there is a risk of implicitly assuming that they are positively related. People have a tendency to view favorable characteristics as complementary [30], and proponents of the SDGs may see the goals as mutually beneficial when they may not be [8,11]. The major issue in viewing sustainability policy through a lens of the moralistic fallacy is failing to recognize tradeoffs, and instead promoting policy action assuming that any action on any goal is mutually beneficial across goals [11].

In the realm of environmental management and policy, there is considerable research indicating the potential for trade-offs with protected area creation. If all SDG targets were truly all positively interrelated, then achieving the SDG target of protecting 10% of all marine ecosystems should positively influence our ability to achieve all other SDGs. Yet pursuing MPA implementation without proper consideration of social context can promote environmental protection at the expense of social progress [5,9,10]. In case studies from around the world, the implementations of MPAs have upended social cohesion and economic and political power between groups, leading to distrust and disenfranchisement of certain groups [9,10]. In some cases the negative social outcomes have led to a perceived delegitimization of MPAs, with corresponding increases in poaching and a concomitant lack in MPA enforcement [9]. Thus the negative social effects led to negative environmental effects. In fact a study of the relationship between the SDG ocean targets and other SDG goals found that pursuing the target of 10% protection of marine ecosystems had the largest number of trade-offs, primarily with goals related with equal access to resources, representation in decision-making, and reducing conflict [11].

An evaluation of policy implementation can easily reveal the difficulties of neglecting social considerations of implementing environmental policy. Environmental policy limiting access to resources can restore natural systems at the expense of poverty alleviation and political inclusion of people [11]. Often the people left behind are already marginalized [10]. Research from the field shows that these dynamics are not guaranteed, and some of these trade-offs can be mitigated by early identification of vulnerable populations and their inclusion into decision-making. The results may not protect areas that are as desirable from a purely nature conservation standpoint, but in the end there might be a more desirable outcome across goals.

Avoiding the simplistic, linear models created by the naturalistic and moralistic fallacies means navigating sustainable development recognizing and responding to deep uncertainties. Doing this requires analysis among the multiple dimensions of the SDGs and not following assumptions.

39.3 Aspirations with strategic plans

If we are to make as much progress as we can toward the SDGs, and more specifically sustainable oceans according to SDG 14: Life Below Water, the problems raised in this chapter: that the SDGs timescales are impossible, that the goals are too absolute, and that the implicit structure behind the goals are based on fallacies, will need to be addressed. Planning and action will need to step beyond aspiration and toward strategy.

Timelines to achieve goals are important, but unrealistic timelines can breed complacency. Short timelines will likely need less ambitious outcomes, but in any case governments need to make significant commitments to the SDGs in order to make progress.

Strategy, whereby progress, cost, and feasibility are explicitly explored, will need to support the lack of strategy and absolute standards of the SDGs. Governments, decisions groups, and others acting toward the SDGs will need to think strategically about where resources should be placed to achieve progress. One way to ensure progress is to couch modest progress benchmarks on the way to the grand outcomes [31].

Determining the structure of the SDGs is the area where research can best provide support to achieve the SDGs. If policy cannot rely on simple rules for implementation, such as assuming that desirable goals are synergistic or that social and economic considerations are dependent on environmental conditions, then research can help determine which SDG targets are cobeneficial and which ones lead to trade-offs, and in what contexts [11]. However, as a starting point for planning research into the SDGs, it is worth considering what conceptual model shapes the research.

Currently there are two prominent models for sustainable development. First, there is the classic "three pillars" model, whereby sustainable development is at the epicenter of environmental, social, and economic concerns [3]. This model emphasizes the "target" of sustainable development as comprising three domains including social, environmental, and economic (Fig. 39.1A). It has some support in the literature as best promoting sustainable development [11] but has been challenged for not emphasizing the importance of having intact functioning ecosystems and failing to forestall environmental degradation [17]. Indeed the competing model, with economic and social issues as nested within the

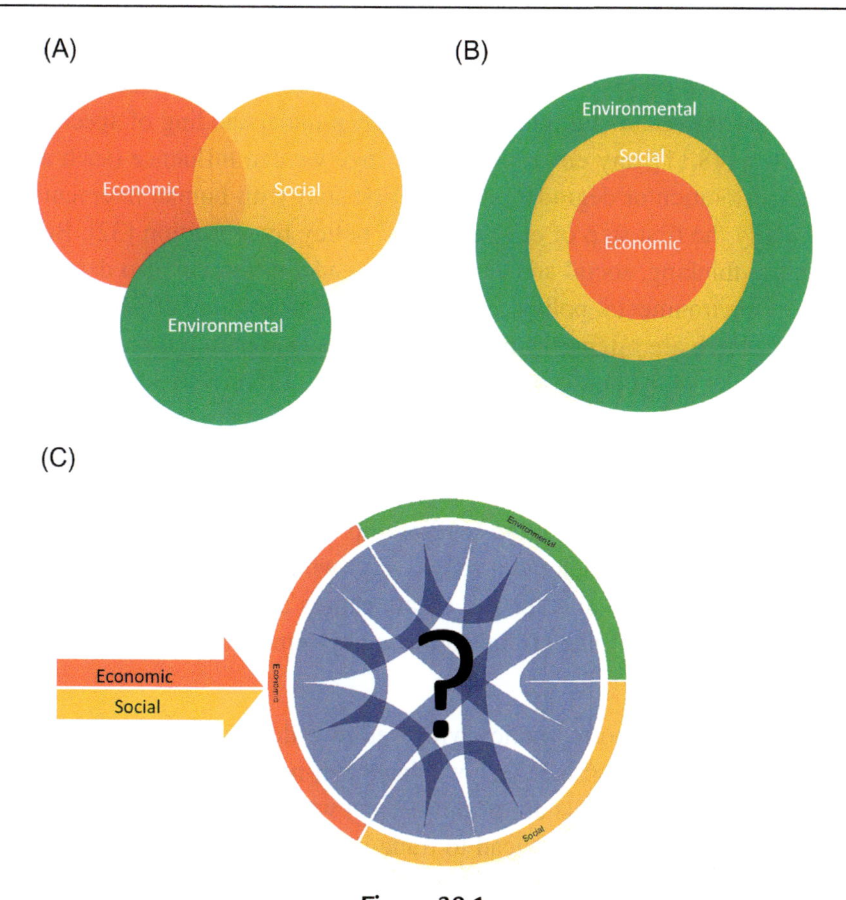

Figure 39.1
Two competing models of sustainable development (A and B), and a third proposed model (C).

environment [17], showcases that any human activity is bounded by what the natural environment can sustain (Fig. 39.1B). While this nested structure may have heuristic appeal, in reality the amount of intact environment required for continued economic development and social cohesion may be very little. Taking a purely functionalist, scientifically dispassionate approach to the environment in its role in sustainable development can ironically lead to increased environmental degradation if a degraded environment can still support and sustain development (as described in the Easter Island case above). The explicit strong subjectivity in the SDGs as a "desired future" will hopefully ensure that environmental conditions are treated as important in themselves and not simply in their degree to prevent social and economic collapse. Regardless of their strengths and drawbacks, both models (Fig. 39.1A and B) provide little to no guidance on how to frame development policy.

In contrast to these approaches, I propose that a sustainable development model be based on research determined to explore the interlinkages of SDG targets, but recognize that any action on any of the SDGs must be taken first through economic or social actions (Fig. 39.1C). While the relationships between environmental, social, and economic factors in general are very uncertain and often require a deep understanding of specific context and systematic analysis [8,11], any environmental change we can influence (such as environmental restoration or degradation) must first start with human intervention, since this is sequentially the first step of environmental policy by definition [32]. Beyond this simple sequential thinking, social and economic considerations can also determine the effectiveness of environmental policy. Even the establishment and effectiveness of marine protected areas, which are often intended to simply leave the natural environment alone, are strongly regulated by social and economic prerequisites. Effective restoration of fish biomass through marine protected areas is regulated strongly through staff and budget capacity [33], further highlighting the importance of social and economic foundations to environmental action. Beyond this first step, sequentially linking social, economic, and environmental dimensions is difficult to predict in principle, as the three dimensions may be better understood to relate to each other reflexively (where there is no linear sequence of events from one dimension to another and all are causes and effects on one another). This extreme uncertainty will need to be addressed systematically and analytically instead of theoretically.

Of course, any research supported policy will need to follow strategic priorities. Strategic planning in support of SDG 14: Life Below Water will need to address the following questions: (1) which ocean targets are priorities? (2) which SDG subject areas act to support those priorities and which ones may lead to trade-offs? (3) what policy pathways are preferential to achieve these priorities? (4) which institutions have the capacity to contribute to these pathways? and (5) which policy actions can be taken by these institutions to progress down strategic paths and avoid trade-offs? Following this strategic approach would likely need policy makers to ignore the achievement dates of the SDGs—where some SDG targets are supposed to be achieved before others—and instead determine SDG priorities based on systematic analysis based on these questions. Following the sustainable development model proposed here, the economic, social, and governance focused SDGs may need to be achieved before the environmental SDGs.

Strategic approaches do not guarantee progress on sustainability, but they can nudge probability in favor of progressing sustainability. New knowledge can change our best understanding of achieving our priorities. Indeed, despite the current international agreement on the composition of the SDGs, the world tomorrow (or next year or next decade) could disagree on what a "desired future" looks like. In spite of this uncertainty, adaptive and iterative planning processes to determine desired futures, priorities, and pathways forward are necessary [34]. While the issues of political resistance, institutional inertia, and persistent unknowns and surprises may always elude sustainable development's full achievement, there is still a potential for progress to be made—the journey can be the destination so long as the journey is planned out. Progress needs to be directed by strategic approaches; setting ambitious goals that seem inviolate and essentially *good* is not enough.

References

[1] UN, Transforming Our World: The 2030 Agenda for Sustainable Development, UN General Assembly, 2015.

[2] UN, Resilient People, Resilient Planet: A Future Worth Choosing, New York, 2012.

[3] R.W. Kates, T.M. Parris, A.A. Leiserowitz, What is Sustainable Development? Goals, indicators, values, and practice, Environ.: Sci. Policy Sustain. Dev. 47 (3) (2005) 8–21.

[4] E. Sala, J. Lubchenco, K. Grorud-Colvert, C. Novelli, C. Roberts, U.R. Sumaila, Assessing real progress towards effective ocean protection, Mar. Policy 91 (1) (2018) 11–13.

[5] T. Agardy, G.N. Di Sciara, P. Christie, Mind the gap: addressing the shortcomings of marine protected areas through large scale marine spatial planning, Mar. Policy 35 (2) (2011) 226–232.

[6] A. Marx, J. Wouters, Combating slavery, forced labour and human trafficking. Are current international, European and national instruments working? Global Policy 8 (4) (2017) 495–497.

[7] M. Hebblewhite, Billion dollar boreal woodland caribou and the biodiversity impacts of the global oil and gas industry, Biol. Conserv. 206 (2017) 102–111.

[8] M. Nilsson, D. Griggs, M. Visbeck, Policy: map the interactions between sustainable development goals, Nat. News 534 (7607) (2016) 320.

[9] P. Christie, Marine protected areas as biological successes and social failures in Southeast Asia, in: American Fisheries Society Symposium, Citeseer, 2004, pp. 155–164.

[10] S. Singleton, Native people and planning for marine protected areas: how "stakeholder" processes fail to address conflicts in complex, real-world environments, Coastal Manage. 37 (5) (2009) 421–440.

[11] G.G. Singh, A.M. Cisneros-Montemayor, W. Swartz, W. Cheung, J.A. Guy, T.-A. Kenny, et al., A rapid assessment of co-benefits and trade-offs among sustainable development goals, Mar. Policy 93 (2018) 223–231.

[12] J. Buchdahl, D. Raper, Environmental ethics and sustainable development, Sustain. Dev. 6 (2) (1998) 92–98.

[13] M. Christen, S. Schmidt, A formal framework for conceptions of sustainability—a theoretical contribution to the discourse in sustainable development, Sustain. Dev. 20 (6) (2012) 400–410.

[14] E.C. Moore, The moralistic fallacy, J. Philos. 54 (2) (1957) 29–42.

[15] J. Rockström, W. Steffen, K. Noone, Å. Persson, F.S. Chapin III, E.F. Lambin, et al., A safe operating space for humanity, Nature 461 (7263) (2009) 472.

[16] W. Steffen, K. Richardson, J. Rockström, S.E. Cornell, I. Fetzer, E.M. Bennett, et al., Planetary boundaries: guiding human development on a changing planet, Science 347 (6223) (2015) 1259855.

[17] D. Griggs, M. Stafford-Smith, O. Gaffney, J. Rockström, M.C. Öhman, P. Shyamsundar, et al., Policy: sustainable development goals for people and planet, Nature 495 (7441) (2013) 305.

[18] J.M. Montoya, I. Donohue, S.L. Pimm, Planetary boundaries for biodiversity: implausible science, pernicious policies, Trends Ecol. Evol. 33 (2) (2018) 71–73.

[19] T. Nordhaus, M. Shellenberger, L. Blomqvist, The Planetary Boundaries Hypothesis, A Review of the Evidence Breakthrough Institute, Oakland, CA, 2012.

[20] P.M. Vitousek, J.D. Aber, R.W. Howarth, G.E. Likens, P.A. Matson, D.W. Schindler, et al., Human alteration of the global nitrogen cycle: sources and consequences, Ecol. Appl. 7 (3) (1997) 737–750.

[21] D. Blumenthal, C.E. Mitchell, P. Pyšek, V. Jarošík, Synergy between pathogen release and resource availability in plant invasion, Proc. Natl. Acad. Sci. U.S.A. 106 (19) (2009) 7899–7904.

[22] A.C. Newton, Biodiversity risks of adopting resilience as a policy goal, Conserv. Lett. 9 (5) (2016) 369–376.

[23] D.F. Simpson Jr, J.A. Van Tilburg, L. Dussubieux, Geochemical and radiometric analyses of archaeological remains from Easter Island's moai (statue) quarry reveal prehistoric timing, provenance, and use of fine–grain basaltic resources, J. Pac. Archaeol. 9 (2) (2018) 12–34.
[24] T. Hunt, C. Lipo, The Statues that Walked: Unraveling the Mystery of Easter Island, Simon and Schuster, 2011.
[25] J.B. MacKinnon, The once and future world: nature as it was, as it is, as it could be. Houghton Mifflin Harcourt, 2013.
[26] L. Fehren-Schmitz, C.L. Jarman, K.M. Harkins, M. Kayser, B.N. Popp, P. Skoglund, Genetic ancestry of Rapanui before and after European contact, Curr. Biol. 27 (20) (2017) 3209–3215.e6.
[27] C. Raudsepp-Hearne, G.D. Peterson, M. Tengö, E.M. Bennett, T. Holland, K. Benessaiah, et al., Untangling the environmentalist's paradox: why is human well-being increasing as ecosystem services degrade? BioScience 60 (8) (2010) 576–589.
[28] J. Rockström, K. Richardson, Planetary boundaries: separating fact from fiction. A response to Montoya et al, Trends Ecol. Evol. 33 (4) (2018) 232–233.
[29] J. Robinson, Squaring the circle? Some thoughts on the idea of sustainable development, Ecol. Econ. 48 (4) (2004) 369–384.
[30] P. Slovic, M.L. Finucane, E. Peters, D.G. MacGregor, The affect heuristic, Eur. J. Oper. Res. 177 (3) (2007) 1333–1352.
[31] P.M. Gollwitzer, P. Sheeran, Implementation intentions and goal achievement: a meta-analysis of effects and processes, Adv. Exp. Soc. Psychol. 38 (2006) 69–119.
[32] M. Rounsevell, T. Dawson, P. Harrison, A conceptual framework to assess the effects of environmental change on ecosystem services, Biodivers. Conserv. 19 (10) (2010) 2823–2842.
[33] D.A. Gill, M.B. Mascia, G.N. Ahmadia, L. Glew, S.E. Lester, M. Barnes, et al., Capacity shortfalls hinder the performance of marine protected areas globally, Nature 543 (7647) (2017) 665.
[34] G.I. Broman, K.-H. Robèrt, A framework for strategic sustainable development, J. Clean. Prod. 140 (2017) 17–31.

SECTION 7

Ocean Governance Beyond Boundaries

CHAPTER 40

Ocean governance beyond boundaries: origins, trends, and current challenges

Erik J. Molenaar

Netherlands Institute for the Law of the Sea, Utrecht University, Utrecht, The Netherlands;
UiT The Arctic University of Norway, Tromsø, Norway

Chapter Outline

The term "ocean(s) governance" stems from the early 2000s and has gradually become increasingly *en vogue* and mainstream since then. It reflects the need and desire to pursue a holistic, integrated, and/or cross-sectoral approach to the management of the oceans, its resources, and the human activities occurring within it or affecting it. Oceans governance could be seen as the most recent phase in the evolution of the international law of the sea, and is among the key drivers and features in the currently ongoing negotiation process on an "International legally binding instrument under the United Nations Convention on the Law of the Sea (UNCLOS) on the conservation and sustainable use of marine biological diversity of areas beyond national jurisdiction (ABNJ)" (ILBI process), established by United Nations General Assembly (UNGA) Resolution 72/249 of December 24, 2017. The ILBI process, oceans governance, and the related concept of ecosystem-based fisheries management are discussed in Dunn et al. (see Chapter 41: BBNJ and the open ocean) and by Ortuño Crespo (see Chapter 46: An ecosystem-based approach to high seas fisheries management) in this section.

The international law of the sea is a domain (or "rule-complex") of public international law, which (mainly) consists of rights and obligations between states. The origins of the international law of the sea can be traced back at least as far as the early 17th century when the Dutch scholar Grotius published *Mare Liberum*. In this seminal work, Grotius postulated legal arguments in support of the interests of the Dutch East Indian Company, which was confronted by Portugal's maritime control in India and Southeast Asia. His view that the sea cannot be appropriated, belongs to all (*res communis*), and is governed by the

Predicting Future Oceans.
DOI: https://doi.org/10.1016/B978-0-12-817945-1.00040-X

"freedom of the seas" was subsequently embraced by the international community, and became the prevailing regime for the oceans until shortly after World War II. Various technological developments relating to the exploitation of living and non-living resources (e.g., sonar, onboard freezing capacity, and offshore drilling for gas and oil) then set in motion the phenomenon of "creeping coastal state jurisdiction," which involved coastal states claiming exclusive access and jurisdiction over living and nonliving resources in increasingly broader maritime zones adjacent to their coasts [1,2].

These claims by coastal states were eventually codified by the UNCLOS, which was opened for signature in 1982 and entered into force in 1994. This "Constitution for the Oceans" recognizes the sovereignty, sovereign rights, and jurisdiction of coastal states in their broader and new maritime zones (e.g., archipelagic waters, a 12 nautical mile (nm) territorial sea, and a 200 nm exclusive economic zone) and was very successful in bringing an end to unilateral coastal state claims to new maritime zones [3]. There have nevertheless been various occasions where coastal states contemplated extending their jurisdiction further. Calls to that effect arose for instance within the United States and the Soviet Union in response to the overexploitation of Pollock in the high seas of the Central Bering Sea toward the end of the 1980s [4]. Another example was the calls within Canada in the mid-2000s to claim so-called custodial management over fisheries resources on the nose and tail of the Grand Banks of Newfoundland and the Flemish Cap [3].

In situations where coastal states regard unilateral extensions of coastal state jurisdiction—whether individually or acting in concert—as undesirable, they can still exert significant control over fishing in ABNJ (the high seas and the deep-seabed beyond the continental shelves of coastal states) that are adjacent to their maritime zones. They are for example able to make use of their port and coastal state jurisdiction under existing international law. Access to their ports and fishing access to their maritime zones can be made conditional on not fishing in adjacent areas of high seas. Most regional fisheries management organizations pursue some or all of these approaches, and the membership of some of these consist entirely or mainly of coastal states [5].

Creeping coastal state jurisdiction continues to be relevant today. Rather than unilateral action by a single state or several states acting in concert, however, it occurs predominantly at the multilateral level through intergovernmental organizations pursuant to their mandate as "competent international organizations" under the UNCLOS. One example in this regard is the so-called cooperative legislative competence between the International Maritime Organization and its Members acting in their capacity as coastal states modeled on Articles 41(4) and 53(9) of the UNCLOS. Other examples are the coastal state jurisdiction beyond the outer limit of the territorial sea relating to underwater cultural heritage created by the 2001 Convention on the Protection of the Underwater Cultural Heritage, and relating to the removal of wrecks created by the 2007 Nairobi International Convention on the Removal of Wrecks [6].

Concerns by some states that this so-called multilateral creeping coastal state jurisdiction would also occur during the negotiations on the 2018 Agreement to Prevent Unregulated High Seas Fisheries in the Central Arctic Ocean were a key factor in the package deal that brought the negotiations to a successful result [6]. Similar concerns continue to exist for these and other states in the currently ongoing ILBI process. These concerns are above all based on coastal state assertions of their special roles, interests or rights in ABNJ adjacent to their maritime zones in relation to all of the four components of the package deal of the ILBI process, but in particular in relation to the identification and designation of area-based management tools—including marine protected areas—and environmental impact assessments. Such assertions are reflected in the references to "adjacency" and "adjacent (coastal/small island developing) States" in the "President's aid to negotiations" dated December 3, 2018 ([7]; see also Ref. [8]).

That coastal states can have special roles, interests, or rights in ABNJ adjacent to their maritime zones is not new per se. Articles 63(2), 64–67 and 116 of the UNCLOS as well as Article 3(1) and many other provisions of its second implementation agreement—the 1995 Fish Stocks Agreement—recognize the roles, rights, and interests of coastal states over transboundary fish stocks whose range of distribution partially overlaps with their maritime zones and partially with ABNJ. Somewhat similar are Articles 77(1) and 78(2) of the UNCLOS. These recognize that the sovereign rights of coastal states over the natural resources of their continental shelves entitle them to impose restrictions on high seas bottom fisheries that have significant adverse impacts on sedentary species (e.g., sea cucumbers or snow crab) or other benthic organisms (e.g., sponge beds or cold-water coral reefs). Reference can in this context also be made to the notions of oceanographic and migratory connectivity discussed in Dunn et al. (see Chapter 41: BBNJ and the open ocean). While these notions are not explicitly advanced or operationalized in the context of adjacency in the ILBI process, it is not inconceivable that participants in this process could do so, in particular in relation to area-based management tools.

As the negotiations on the UNCLOS were finished in 1982, and the negotiations on its two implementation agreements—the 1994 Deep-Seabed Mining Agreement and the 1995 Fish Stocks Agreement—finished more than two decades ago, the ILBI process is understandably an exciting event in oceans governance. Unfortunately, however, one should not be too optimistic on what it can achieve. For one thing, several critically important states are far from enthusiastic about the ILBI process. Moreover, from the perspective of international fisheries law—which can be seen as a branch or part of the international law of the sea—hopes should be tempered even further due to the fact that UNGA Resolution

72/249 "Recognizes that [the ILBI] process and its result should not undermine existing relevant legal instruments and frameworks and relevant global, regional and sectoral bodies." It seems that delegations primarily had the domain of international fisheries law in mind when they negotiated this constraint.

In the face of this reality, all relevant actors should continue their efforts toward incremental change in the domain of international fisheries law proper. One crucial aspect in this regard is enhancing compliance by states of their obligations under international fisheries law. As explained by Boustany and Guggisberg in this section (see Chapter 45: The trouble with tunas, and Chapter 43: Verifying and improving states' compliance), this is a challenging task due to the consensual nature of international law, which means that states cannot be bound against their will. Both authors point out that it is fundamentally flawed to think that only flag states are "free riders," as states can also pursue such behavior in their capacities as coastal, port, or market states. Moreover, Dellmuth warns that legitimacy does not only contribute to compliance, but can at some stage also lead to situations where scrutiny declines and compliance levels erode (see Chapter 42: Legitimacy has risks and benefits for marine management).

Guggisberg (see Chapter 43: Verifying and improving states' compliance) also expresses her concerns that the Voluntary Guidelines for Flag State Performance adopted by the United Nations Food and Agriculture Organization (FAO) in 2013 only apply to flag states and not also to coastal states. Unfortunately, this is merely one example of a persistent unwillingness of coastal states to consider constraints or reforms relating to their maritime zones. Another example are the efforts of the UNGA and FAO relating to the impacts of bottom fisheries on vulnerable benthic habitats and deep-sea species, which were in the end confined to the high seas even though similar or more serious impacts were occurring in coastal State maritime zones. The fact that the ILBI process is confined to ABNJ can be seen as another example.

While the compliance deficit is one of the principal "internal" challenges currently faced by international fisheries law, one "external" challenge is arguably far more troubling: climate change. The climate change-induced impacts that are most directly relevant for marine capture fisheries are increases in temperature, deoxygenation, and stratification, changing currents, and reductions in salinity (due to increased freshwater inflow) and sea ice cover [9]. In view of the negative implications that climate change is likely to have for the level or extent of certainty, reliability, and adequacy of scientific information, it would seem logical for states to pursue more conservative management practices. Potential climate change scenarios and technological innovations in fishing could perhaps also be used as variables or "wildcards" for the innovative approach of science fiction prototyping canvassed by Merrie (see Chapter 48: Beyond prediction—radical ocean futures). In view of the considerable internal and external challenges with which international fisheries law as well as oceans governance in general are confronted now and in the future, it is clear that healthy fish stocks and healthy oceans require the broadest possible legal, institutional, and political support.

References

[1] R.Y. Jennings, A changing international law of the sea, Cambridge Law J. (1972) 34–36, 32–49.
[2] B.H. Oxman, The territorial temptation: a siren song at sea, Am. J. Int. Law (2006) 830–851.
[3] E.J. Molenaar, New maritime zones and the law of the sea, in: H. Ringbom (Ed.), Jurisdiction Over Ships—Post-UNCLOS Developments in the Law of the Sea, Brill/Nijhoff, 2015, pp. 249–277, 267–271.
[4] D. Balton, The Bering sea doughnut hole convention: regional solution, global implications, in: O.S. Stokke (Ed.), Governing High Seas Fisheries: The Interplay of Global and Regional Regimes, Oxford University Press, 2001, pp. 143–177, 150–151.
[5] E.J. Molenaar, Participation in regional fisheries management organizations (forthcoming) in: R. Caddell, E.J. Molenaar (Eds.), Strengthening International Fisheries Law in an Era of Changing Oceans, Hart, 2019, pp. 103–129.
[6] E.J. Molenaar, Participation in the central arctic ocean fisheries agreement (forthcoming) in: A. Shibata, L. Zou, N. Sellheim, M. Scopelliti (Eds.), Emerging Legal Orders in the Arctic: The Role of Non-Arctic Actors, Routledge, 2019, pp. 132–170.
[7] Doc. A/CONF.232/2019/1 of 3 December 2018.
[8] A.G. Oude Elferink, Coastal states and MPAs in ABNJ: ensuring consistency with the LOSC, Int. J. Mar. Coastal Law (2018) 437–466.
[9] W.W.L. Cheung, V.W.Y. Lam, Y. Ota, W. Swartz, Modelling future oceans: the present and emerging future of fish stocks and fisheries (forthcoming) in: R. Caddell, E.J. Molenaar (Eds.), Strengthening International Fisheries Law in an Era of Changing Oceans, Hart, 2019, pp. 13–23, pp. 15–17.

CHAPTER 41

Incorporating the dynamic and connected nature of the open ocean into governance of marine biodiversity beyond national jurisdiction

Daniel C. Dunn[1,2], Guillermo Ortuño Crespo[1] and Patrick N. Halpin[1]

[1]Nicholas School of the Environment, Duke University, Beaufort, NC, United States [2]Centre for Biodiversity and Conservation Science, School of Earth and Environmental Sciences, University of Queensland, Brisbane, Australia

Chapter Outline

41.1 This planet is open ocean

The open ocean beyond the continental shelf (Fig. 41.1) accounts for 64% of the planet's surface. This pelagic realm has more than twice the surface area of all terrestrial biomes *combined* and 168 times the habitable volume. To put this in perspective, if terrestrial habitats were an ant, open ocean habitats would be the size of a person. The open ocean provides more than US$10 billion in fisheries landings [1], and represents the longest "highways" on the planet, connecting the globe and providing for the transportation of ~90% of international trade.

Predicting Future Oceans.
DOI: https://doi.org/10.1016/B978-0-12-817945-1.00041-1

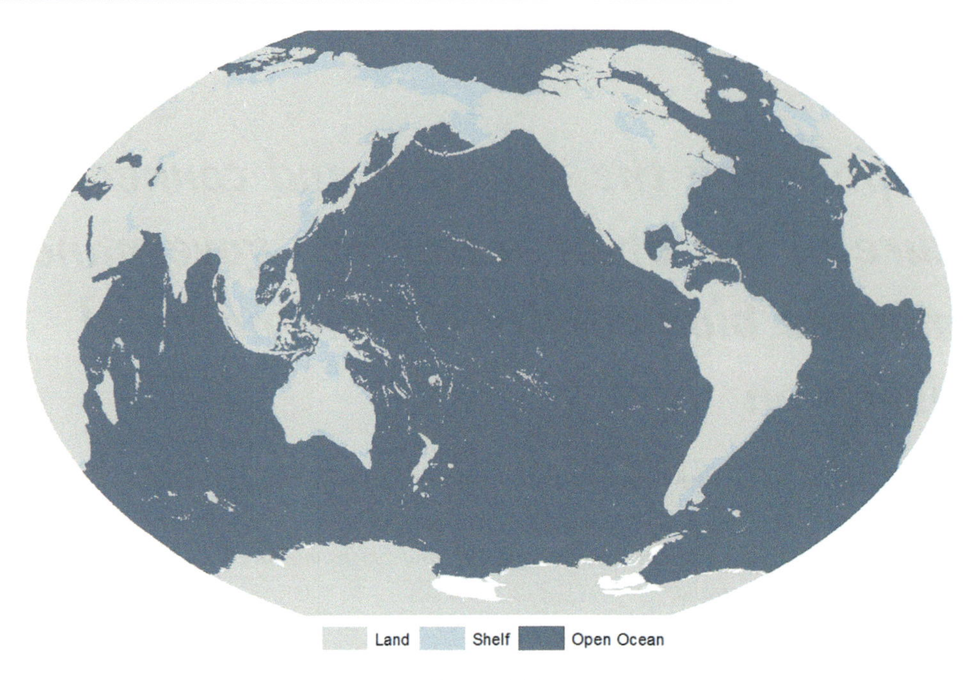

Figure 41.1
This planet is open ocean. Seventy one percent of the planet is ocean and nearly half of the planet lies beyond national jurisdictions.

Furthermore, the ocean is critical in moderating Earth's climate. It provides more than half the oxygen we breathe and mitigates impacts from anthropogenic carbon dioxide (CO_2) by absorbing 93% of the heat generated by CO_2 emissions and 26% of the CO_2 gas [2,3]. This climate mitigation service is of enormous value, but the impacts of assimilating that heat and CO_2 are strongly altering open/deep ocean environments and ecosystems: accelerating ocean warming, deoxygenation, and acidification, which affect marine life throughout the areas beyond national jurisdiction (ABNJ), by changing species' distributions, migration routes, ecosystem structures, and functions [4,5].

Climate change–induced impacts act synergistically with other impacts from human uses of the ocean, in particular, fisheries. Between 1950 and 1989, industrial marine fisheries catch in ABNJ increased by a factor of more than 40 [1,6]. This growth was an order of magnitude more than the increase in catch within exclusive economic zones (EEZs) during the same time period. Since 1990 high seas marine fisheries catches have remained relatively stagnant [7], but fishing effort, and all concomitant impacts from increasing the amount of fishing gear in the water, more than doubled between 1990 and 2006 [8]. In spatial terms, the greatest expansion of fishing effort during the second half of the 20th century has taken place primarily beyond the limits of the continental shelf and in ABNJ [9]. Long thought to be too big and distant to harm, there is now growing scientific evidence of the impacts of fisheries not just on open ocean species, but open ocean communities and ecosystems [10]. The combination of these impacts with the dynamics of a boundaryless, fluid, and changing ocean are in urgent need of attention from the international community.

From the safety of shore, the increasingly strong cumulative impacts of climate change and excessive resource extraction on the open ocean might be easily ignored by the world's human population, except that the delineation of EEZs is ecologically meaningless. Strong connectivity between the high seas and coastal states' jurisdictions results in far flung impacts frequently washing up on our shores. The very nature of open ocean ecosystems is that they are defined and constantly influenced by powerful winds and oceanic currents. These systems have no fixed physical boundaries—they move through time and horizontal and vertical space, connecting distant areas with physical flows. Many iconic migratory and pelagic species use these ecosystems as habitat for spawning, breeding, migrating, and feeding; and their movements also connect distant ecosystems and ecological processes.

41.2 Ecological connectivity

Ecological connectivity can be broadly categorized into two types: passive and active forms of movement. Oceanographic connectivity is the main form of passive connectivity and relies on ocean currents that drive larval or planktonic dispersal, and which can also transport anthropogenic impacts, such as pollutants, into and out of coastal state waters. Active dispersal, on the other hand, arises from directed movement by, inter alia, seabirds, sea turtles, marine mammals, and fish. This form of dispersal can lead to different types of transboundary movements, from transoceanic migrations through multiple EEZs and the high seas [11], to smaller-scale straddling behavior between an EEZ and the high seas. Many of the animals that engage in this type of movement rely on distinct parts of the ocean to fulfill different annual cycle or life-history stages (e.g., from nesting to foraging). Understanding how populations utilize different spaces across time is essential for the conservation and sustainable use of marine migratory species. Below we describe these two types of connectivity in more detail.

41.2.1 Oceanographic connectivity

For many marine species population connectivity is largely determined by ocean currents transporting larvae and juveniles between distant patches of suitable habitat. These long-distance connections often contribute to the genetic stability of populations and metapopulations by periodically providing new recruits from distant sources. The strength of the connections between sites may change intra- or interannually according to major climate cycles such as El Niño or La Niña oscillations [12]. Regional analyses have been conducted in the South Pacific [13] as well as the Caribbean [14] and have helped to better define the long-distance interdependence of marine ecosystems within these regions. These analyses demonstrate the importance of direct adjacency between near-shore (EEZ) areas and offshore (ABNJ) areas as well as more complicated multipath connections that may span multiple sites and jurisdictions.

In addition to planktonic larvae, ocean currents and oceanographic features also transport and redistribute nutrients, heat, and pollutants including marine debris. Marine debris, such as plastics, can impact biological diversity due to entanglement or ingestion. Plastics can also act as vectors for the transport of harmful chemicals, which can have ecological impacts in regions as isolated as the Arctic [15]. Furthermore, other pollutants that make their way into the marine environment, such as oil, can be transported across wide regions by surface currents [16].

41.2.2 Migratory connectivity

Animal migration has been broadly defined as persistent, large spatial scale movements to connect discrete home ranges that help fulfill a species' life-history objectives [17]. Migration is fundamental to marine ecosystem structure given the strong ecological imperative for animal movements to evade predation, to access spatially distributed and seasonal resources, or to access suitable habitats for different life-history purposes. Migratory connectivity emerges from persistent movement between habitat patches and frequently straddles jurisdictional boundaries. Understanding and accounting for the transboundary connectivity of migratory species is essential for their conservation and management.

Migration is common among marine species in the open ocean. Lascelles et al. [18] identified a total of 829 migratory marine species of fish, seabirds, marine mammals, and sea turtles occurring and frequently straddling ocean basins, much less jurisdictional boundaries. Satellite technologies have revealed, inter alia, the straddling behaviors of important target species such as bigeye tuna [19] or yellowfin tuna [20], as well as the transoceanic movements of nontarget species such as basking sharks [21], white sharks [22], leatherback sea turtles [23], or wandering albatross [24]. These large-scale movements result in migratory marine species moving across political boundaries on a regular basis: 18 species of marine predator in the Pacific Ocean were found to visit 94% of the EEZs in the Pacific Ocean and spent 14%–33% of their annual cycle in these waters and 53%–76% of the time in the high seas [11]. However, our understanding of marine migratory movements is still poor across taxonomic groups and geographic regions. A review of shark satellite tagging studies in the primary literature revealed that only 15 species of migratory sharks have been studied using this technology, with most of the studies having been conducted in the Pacific Ocean (50%) [25].

This illustrates that the remaining ~80 species (or 84%) of migratory sharks lack specific information on their migratory/straddling movement patterns [26].

41.3 Governance of highly connected, dynamic, open ocean ecosystems

The management and conservation of open ocean ecosystems, and the straddling and highly migratory species that use them, is a serious challenge given the large spatiotemporal distributions of the species, the cost of sampling in distant and or deep locations, and the complexities of coordination among multiple parties across jurisdictional boundaries. In 2011 the Food and Agriculture Organization (FAO) estimated that straddling stocks were overfished or experiencing overfishing at a rate twice that of stocks within national jurisdictions (64.0% vs 28.8%). Similarly, an assessment of the 48 migratory fish stocks managed by the world's tuna Regional Fisheries Management Organizations (RFMOs) concluded that 67% of these were either overfished or depleted. Chondricthyans, the most threatened vertebrate group [27], show a similar pattern. Only 14% of nonmigratory sharks were threatened (Vulnerable, Endangered, or Critically Endangered under IUCN Red List) whereas 46% of the 95 migratory sharks are threatened, with a further 21% assessed as Near Threatened [26]. To conserve and sustainably use the open ocean of ABNJ, governance measures need to account for both the dynamic nature of the system and the connectivity that it generates.

Significant gaps and deficiencies plague the current network of governance structures in charge of conserving and managing biotic resources in the open ocean. While existing tuna RFMOs cover almost the entirety of ABNJ, the spatial coverage of nontuna RFMOs is still alarmingly patchy, which results in the unmonitored and unmanaged exploitation of large areas of the open ocean [28]. Similarly, even more serious taxonomic gaps exist in international fisheries management [29]. Management of shipping lacks any spatial management in ABNJ, while cables have no governance structure under the United Nations at all. Deep-sea mining has been governed in a more proactive manner, including the development of regional environmental management plans including setting aside 30% of a region for conservation purposes, but the leasing of the seabed for exploration contracts before such plans are in place has led to suboptimal results [30]. Finally, only three Regional Seas Organizations have a mandate to address ABNJ, leaving sector-based management with no partner governing biodiversity in most parts of the open ocean [28].

Recognizing many of the potential limitations of the collage of ocean management bodies for the conservation and sustainable use of biodiversity in the high seas, the international community began a process in 2005 to assess whether the existing regional and sectoral governance scheme for marine biodiversity beyond national jurisdiction (BBNJ) was sufficient. After a decade of discussions, the UN General Assembly (UNGA) passed Resolution 69/292, stressing the need for the establishment of a global regime to better manage and conserve BBNJ (UNGA 69/292; [31]). After another 2 years of developing substantive recommendations on the elements of a draft text of an international legally binding instrument (ILBI) through a Preparatory Committee (PrepCom), the UNGA passed Resolution 72/249 calling for the opening of an Intergovernmental Conference to negotiate the new ILBI. Since 2011 there has been consensus to address a "package" of four elements in the ILBI which must be addressed "together and as a whole": (1) marine genetic resources, including questions regarding the sharing of benefits; (2) area-based management tools (ABMTs), including marine protected areas (MPAs); (3) environmental impact assessments; and (4) capacity building and the transfer of marine technology. Below we outline how consideration of well-connected and dynamic pelagic systems can be incorporated into three elements of the BBNJ package:

41.3.1 Area-based management tools

Potential ABMTs include marine spatial planning, individual and networks of MPAs, as well as sectoral measures such as areas closed to some or all fishing, mining, navigation, discharge activities, or increased reporting requirements. As stated in the PrepCom 3 Chair's overview, there is a need to define ABMTs and their objectives as they relate to the conservation and sustainable management of static and dynamic biodiversity in ABNJ. As described earlier, the high seas provide critical habitat for migratory species, which make use of open ocean ecosystems to fulfill different annual or life cycle stages. There is widespread evidence that many target and nontarget oceanic species track dynamic oceanographic features such as frontal zones or eddies, which are becoming increasingly easier to track and predict [32,33]. The wide-ranging and dynamic distribution not just of those species, but of the ecosystems they utilize means that ABMTs for their conservation may need to be sufficiently "fluid" to track their changing distributions. Two approaches to dealing with this fluidity have been put forward: large MPAs ($>10,000$ km^2; [34]), and dynamic ocean management [35,36] which allows for real-time shifting of the ABMT boundary based on environmental or socioeconomic conditions.

ABMTs, and MPAs in particular, are frequently cited as being part of a precautionary approach to management. The role of MPAs within a precautionary approach is not as a measure to be enacted in reaction to a past event with ecological impact, but as proactive insurance against unknowns in the system and errors in governance. To play this role, they should be in place before evidence of harm is found. In addition to their role in providing proactive protection in advance of harm, ABMTs can be used to build resilience and to mitigate the cumulative and synergistic impacts of human uses and climate change. For ABMTs to be effective as a precautionary measure, it is critical that monitoring programs are in place that can adequately measure environmental changes. The scale and variability of open ocean ecosystems require that the monitoring mechanisms be put in place at regional or global scales and be sustained over longer time periods than may be necessary in static systems. While challenging, this is the only way to differentiate local or short-term variability from true impacts to the ecosystem.

One major challenge to meeting this requirement for monitoring of open ocean ecosystems to be long term and over large scales comes from the diverse mandates for ecosystem monitoring in ABNJ. While RFMOs have a duty to monitor ecosystem components "associated with" target species, that does not implicate them in the monitoring of all BBNJ. Furthermore, even if RFMOs chose to tackle monitoring of all BBNJ, strong coordination is needed not just among tuna RFMOs (as represented by the advent of the Kobe process), but among all organizations with competency for managing open ocean ecosystems. A common concern across all organization with competency in ABNJ is that comprehensive monitoring of biodiversity cannot occur without vast increases in current budgets [10]. Coordination of large-scale biodiversity monitoring through programs like the Global Ocean Observing System is critical to provide ecosystem-level observing more efficiently.

An equally critical element to support effective monitoring is technology transfer and capacity building to developing states. We address these issues below to support monitoring through technology transfer and capacity building. Only by increasing cooperation and collaboration among competent organizations, industry, and academia, along with other civil society, will appropriate monitoring of the open ocean ecosystems be available to underpin effective development and management of ABMTs in open ocean ecosystems.

41.3.2 Environmental impact assessments

Ecological impacts on the deep seabed (e.g., changes in species abundance; destruction of benthic habitat) are relatively static. Conversely, ecological impacts on pelagic species, communities, or ecosystems move across the ocean as their distributions in ABNJ and EEZs change. A 2006 FAO of the United Nations report on the state of migratory straddling and high seas stocks identified up to 226 highly mobile open ocean species (Chondrichthyes and Osteichthyes), while a later Convention on the Conservation of Migratory Species of Wild Animals (CMS) and United Nations Environment Program report identified 153 migratory or potentially migratory chondrichthyan fishes [26,37]. Furthermore, a 2014 study identified 319 seabird species and 102 marine mammal species which are migratory, highly migratory, or very highly migratory [18]. These highly mobile species contribute to the ecological, social, and economic stability of socioecological systems both within and beyond national jurisdictions. Therefore any changes to the diversity, abundance, or range of these highly mobile species, and the subsequent impacts of these changes, should be tracked and assessed.

If species which migrate between coastal and oceanic ecosystems are severely depleted during their residency in the open ocean, such changes will later affect ecological relationships in coastal ecosystems. Aware of the dynamic and even transboundary nature of many open ocean species and ecosystems, and the consequent mobile nature of negative ecological impacts, various delegations throughout the second and third PrepComs expressed interest in ensuring that EIAs account for the mobility of impacts by developing transboundary environmental impact assessments (TEIAs). In the second PrepCom the African Group took a further step and opined that the ILBI should also cover activities within EEZs with impacts in ABNJ and vice versa. Other coastal states, such as the Pacific Small Island Developing States, advocated for TEIAs as a way to monitor impacts of high seas activities on adjacent coastal nations. TEIAs are particularly relevant for regions such as the Costa Rica Thermal Dome or the Sargasso Sea, among others, which move, expand, and contract across jurisdictional boundaries. In these scenarios, conservation and management measures in ABNJ will have direct implications for the resilience and health of biodiversity and ecosystems within EEZs, and vice versa.

41.3.3 Technology transfer and capacity building

The importance of capacity building and the transfer of technology is clearly a priority for numerous states, as reflected in the Chair's overview of the second session of the Preparatory Committee (http://www.un.org/depts/los/biodiversity/prepcom_files/Prep_Com_II_Chair_overview_to_MS.pdf). To quote G77 and China, the scope of capacity building and technology transfer in a new instrument should include "establishment or strengthening the capacity of relevant organizations/institutions in developing countries to deal with conservation of marine biological diversity in ABNJ; access and acquisition of necessary knowledge and materials, information, data in order to inform decision making of the developing countries." The CARICOM countries apply this more directly to monitoring, stating that the scope should include "[c]apacity building for development, implementation and monitoring of ABMTs including MPAs." Given differences in capacity for monitoring between regions and states, capacity building and technology transfer to support monitoring, as well as minimum monitoring standards across RFMOs and other international organizations could be an important component of the new ILBI.

The distant, deep, and dynamic nature of open ocean ecosystems requires ambitious commitments to monitoring as laid out above. Since open ocean systems make up the vast majority of areas to be governed under any new ILBI, the success of the ILBI may be highly dependent on strong commitments regarding technology transfer and capacity development in support of monitoring open ocean ecosystems, particularly to developing states. While much discussion of frameworks, modes and types of capacity building, and technology transfer have been discussed at the PrepCom meetings, various stakeholders, including civil society and academia, could play increasingly important roles in the implementation of any capacity building and technology transfer commitments [29]. This support can come from, for example, civil partnerships providing technical expertise by working directly with individual governmental or intergovernmental organizations, creating a task force of several countries that share information with each other, or by simply making the fishing data of ABNJ transboundary species freely available. Such partnerships will complement more traditional multilateral and bilateral technology transfer and capacity building approaches.

41.4 Conclusion

Conservation and sustainable use of marine biodiversity in ABNJ is dependent on governance that can account for the nature of the systems therein. Pelagic open ocean ecosystems in ABNJ are characterized by their dynamism and the connectivity it generates. With increasing impacts being felt from both resource extraction and climate change, the need for comprehensive (i.e., geographically and taxonomically) and nimble governance structures in ABNJ has reached a critical juncture. The intergovernmental conference to negotiate a new treaty for BBNJ is thus very timely. To fully meet the challenge and provide a more holistic governance structure for BBNJ, connectivity and the fluidity of the pelagic environment should be addressed within each of the package elements. The scale of the open ocean is orders of magnitude different than terrestrial ecosystems and, together with its dynamic nature, requires longer-term and larger-scale monitoring to understand the changes in the system. Those requirements underpin an enhanced need for increased and innovative capacity building and technology transfer. Connectivity generated by physical flows and migratory behaviors results in impacts being felt far from their source. Finally, the assessment of potential impacts from activities in ABNJ should therefore include any teleconnections within their scope including transboundary connections.

References

[1] D. Pauly, D. Zeller (Eds.), Sea Around Us Concepts, Design and Data, University of British Columbia, Vancouver, BC, 2015.

[2] L.A. Levin, N. Le Bris, The deep ocean under climate change, Science 350 (2015) 766–768 (80-.).

[3] M. Rhein, S.R. Rintoul, S. Aoki, E. Campos, D. Chambers, et al., in: T. Stocker, D. Qin, G.-K. Plattner, M. Tignor, S. Allen, et al. (Eds.), Clim. Change, 2013 Phys. Sci. Basis. Contrib. Work. Gr. I to Fifth Assess. Rep. Intergov. PanelClim. Change, 2013.

[4] W.W.L. Cheung, V.W.Y. Lam, J.L. Sarmiento, K. Kearney, R. Watson, D. Zeller, et al., Large-scale redistribution of maximum fisheries catch potential in the global ocean under climate change, Global Change Biol. 16 (2010) 24–35.

[5] E.L. Hazen, S. Jorgensen, R.R. Rykaczewski, S.J. Bograd, D.G. Foley, I.D. Jonsen, et al., Predicted habitat shifts of Pacific top predators in a changing climate, Nat. Clim. Change 3 (2012) 234–238.

[6] D. Pauly, D. Zeller, Catch reconstructions reveal that global marine fisheries catches are higher than reported and declining, Nat. Commun. 7 (2016) 10244.

[7] FAO, The State of World Fisheries and Aquaculture 2016, Contributing to Food Security and Nutrition for All, FAO, Rome, 2016.

[8] A. Merrie, D.C. Dunn, M. Metian, A.M. Boustany, Y. Takei, A.O. Elferink, et al., An ocean of surprises – Trends in human use, unexpected dynamics and governance challenges in areas beyond national jurisdiction, Global Environ. Change 27 (2014) 19–31.
[9] W. Swartz, E. Sala, S. Tracey, R. Watson, D. Pauly, The spatial expansion and ecological footprint of fisheries (1950 to present), PLoS One 5 (2010) e15143.
[10] G.O. Crespo, D.C. Dunn, A review of the impacts of fisheries on open-ocean ecosystems, ICES J. Mar. Sci. 74 (2017) 2283–2297.
[11] A.-L. Harrison, D.P. Costa, A.J. Winship, S.R. Benson, S.J. Bograd, M. Antolos, et al., The political biogeography of migratory marine predators, Nat. Ecol. Evol. 2 (2018) 1571.
[12] E.A. Treml, P.N. Halpin, D.L. Urban, L.F. Pratson, Modeling population connectivity by ocean currents, a graph-theoretic approach for marine conservation, Landsc. Ecol. 23 (2008) 19–36.
[13] E.A. Treml, J.J. Roberts, Y. Chao, P.N. Halpin, H.P. Possingham, C. Riginos, Reproductive output and duration of the pelagic larval stage determine seascape-wide connectivity of marine populations, Integr. Comp. Biol. 52 (2012) 525–537.
[14] S.R. Schill, G.T. Raber, J.J. Roberts, E.A. Treml, J. Brenner, P.N. Halpin, No reef is an island: integrating coral reef connectivity data into the design of regional-scale marine protected area networks, PLoS One 10 (2015) e0144199.
[15] C. Zarfl, M. Matthies, Are marine plastic particles transport vectors for organic pollutants to the Arctic? Mar. Pollut. Bull. 60 (2010) 1810–1814.
[16] T. Özgökmen, E. Chassignet, C. Dawson, D. Dukhovskoy, G. Jacobs, J. Ledwell, et al., Over what area did the oil and gas spread during the 2010 Deepwater Horizon oil spill? Oceanography 29 (2016) 96–107.
[17] E.J. Milner-Gulland, J. Fryxell, A. Sinclair (Eds.), Animal Migration: A Synthesis, Oxford University Press, Oxford, 2011.
[18] B. Lascelles, G. Notarbartolo di Sciara, T. Agardy, A. Cuttelod, S. Eckert, L. Glowka, et al., Migratory marine species: Their status, threats and conservation management needs, Aquat. Conserv. Mar. Freshw. Ecosyst. 24 (2014) 111–127.
[19] K.M. Schaefer, D.W. Fuller, Vertical movements, behavior, and habitat of bigeye tuna (*Thunnus obesus*) in the equatorial eastern Pacific Ocean, ascertained from archival tag data, Mar. Biol. 157 (2010) 2625–2642.
[20] K.M. Schaefer, D.W. Fuller, Vertical movement patterns of skipjack tuna (*Katsuwonus pelomis*) in the eastern equatorial Pacific Ocean, as revealed with archival tags, Fish. Bull. 105 (2007) 379–389.
[21] G.B. Skomal, S.I. Zeeman, J.H. Chisholm, E.L. Summers, H.J. Walsh, K.W. McMahon, et al., Transequatorial Migrations by Basking Sharks in the Western Atlantic Ocean, Curr. Biol. 19 (2009) 1019–1022.
[22] R. Bonfil, M. Meÿer, M.C. Scholl, R. Johnson, S. O'Brien, H. Oosthuizen, et al., Transoceanic migration, spatial dynamics, and population linkages of white sharks, Science 310 (2005) 100–103 (80-.).
[23] S.R. Benson, T. Eguchi, D.G. Foley, K.A. Forney, H. Bailey, C. Hitipeuw, et al., Large-scale movements and high-use areas of western Pacific leatherback turtles, *Dermochelys coriacea*, Ecosphere 2 (2011) art84.
[24] H. Weimerskirch, Y. Cherel, K. Delord, A. Jaeger, S.C. Patrick, L. Riotte-Lambert, Lifetime foraging patterns of the wandering albatross: Life on the move! J. Exp. Mar. Biol. Ecol. 450 (2014) 68–78.
[25] N. Hammerschlag, A.J. Gallagher, D.M. Lazarre, A review of shark satellite tagging studies, J. Exp. Mar. Biol. Ecol. 398 (2011) 1–8.
[26] S. Fowler, The Conservation Status of Migratory Sharks, UNEP/CMS Secretariat, Bonn, 2014.
[27] N.K. Dulvy, S.L. Fowler, J.A. Musick, R.D. Cavanagh, P.M. Kyne, L.R. Harrison, et al., Extinction risk and conservation of the world's sharks and rays, Elife 3 (2014) 1–35.
[28] N.C. Ban, N.J. Bax, K.M. Gjerde, R. Devillers, D.C. Dunn, P.K. Dunstan, et al., Systematic conservation planning: a better recipe for managing the high seas for biodiversity conservation and sustainable use, Conserv. Lett. 7 (2014) 41–54.

[29] D.C. Dunn, C. Jablonicky, G.O. Crespo, D.J. Mccauley, D.A. Kroodsma, K. Boerder, et al., Empowering high seas governance with satellite vessel tracking data, Fish Fish. 19 (2018) 729–739.
[30] D.C. Dunn, C.L. Van Dover, R.J. Etter, C.R. Smith, L.A. Levin, T. Morato, et al., SEMPIA workshop participants A strategy for the conservation of biodiversity on mid-ocean ridges from deep-sea mining, Sci. Adv. 4 (2018) eaar4313.
[31] G. Wright, J. Rochette, E. Druel, K. Gjerde, The long and winding road continues: towards a new agreement on high seas governance, in: Study N°01/16, Paris, France, 2016.
[32] B.A. Block, I.D. Jonsen, S.J. Jorgensen, A.J. Winship, S.A. Shaffer, S.J. Bograd, et al., Tracking apex marine predator movements in a dynamic ocean, Nature 475 (2011) 86–90.
[33] K.L. Scales, P.I. Miller, L.A. Hawkes, S.N. Ingram, D.W. Sims, S.C. Votier, On the front line: Frontal zones as priority at-sea conservation areas for mobile marine vertebrates, J. Appl. Ecol. 51 (2014) 1575–1583.
[34] N.C. Ban, T.E. Davies, S.E. Aguilera, C. Brooks, M. Cox, G. Epstein, et al., Social and ecological effectiveness of large marine protected areas, Global Environ. Change 43 (2017) 82–91.
[35] S.M. Maxwell, E.L. Hazen, R.L. Lewison, D.C. Dunn, H. Bailey, S.J. Bograd, et al., Dynamic ocean management: Defining and conceptualizing real-time management of the ocean, Mar. Policy 58 (2015) 42–50.
[36] D.C. Dunn, S.M. Maxwell, A.M. Boustany, P.N. Halpin, Dynamic ocean management increases the efficiency and efficacy of fisheries management, Proc. Natl. Acad. Sci. U.S.A. 113 (2016) 668–673.
[37] J.-J. Maguire, M. Sissenwine, J. Csirke, R. Grainger, The state of the world highly migratory, straddling and other high seas fish stocks, and associated species, U.N. Food and Agriculture Organization, 2006.

CHAPTER 42

Legitimacy has risks and benefits for effective international marine management

Lisa Maria Dellmuth

Department of Economic History and International Relations, Stockholm University, Stockholm, Sweden

Chapter Outline

42.1 Introduction

One of the most nagging criticisms of international marine institutions has been their mixed track record in effectively protecting biodiversity [1] and managing highly migratory and straddling fish stocks [2]. This variation in effectiveness has fueled social science scholarship on the sources of effective marine management. Both in sustainability science [3,4] and political science [5,6], informational, financial, and organizational factors are consistently highlighted as central drivers of effectiveness.

This chapter focuses on a factor that much social science scholarship connects with compliance and thus more effective institutions: legitimacy [7–10]. A legitimate institution

Predicting Future Oceans.
DOI: https://doi.org/10.1016/B978-0-12-817945-1.00042-3

enjoys public trust, which may increase compliance with its rules and decrease investment of scarce resources in coercion and enforcement through, for example, fines or sanctions [11].

Most sustainability research is exclusively concerned with the drivers of effectiveness in international marine institutions in terms of the potentially positive effects of "collaborative management" [12]. Participation in management by nongovernmental organizations (NGOs) and other stakeholders such as end users of resources has been shown to improve the processes [13,14], reform capacity [15,16], and problem-solving capacity [3,17] of international marine institutions. In addition, comanagement has been argued to foster trust and cooperative behavior in international institutions [18]. Despite this progress, we still know little about how legitimacy could be useful—or detrimental—in the quest for improved marine management.

To contribute to this endeavor, this chapter argues that legitimacy may be a resource for more effective marine management, but that it comes with risks. Legitimacy is a great social good. Building on early contributions in sociology [19], political scientists have suggested that legitimacy fosters human happiness as people prefer to comply with institutions on moral grounds and not for reasons of self-interest [20]. A number of previous studies suggest that international institutions should attend to their legitimacy in order to increase their viability and effectiveness [11,21,22]. However, while legitimacy certainly contributes to compliance, a large literature in political science alludes to the risks of declining public scrutiny and greater influence of special interest groups.

The remainder of this chapter is as follows. First, it defines legitimacy, drawing from political science literature. Second, it illustrates legitimacy patterns in 19 international marine institutions by drawing on original data from an expert survey among more than 300 natural and social scientists and practitioners. Third, it discusses which potential benefits and risks legitimacy may engender in international marine management. By way of conclusion, the chapter outlines a research agenda for sustainability scholars on legitimacy and effectiveness.

42.2 Defining legitimacy as beliefs

Legitimacy refers to the degree of individual beliefs among the subjects of a political institution that this institution's authority is appropriately exercised [19,22]. Legitimacy in this sense refers to beliefs in sociological terms and not on, whether, in normative terms, global institutions conform to philosophical standards such as fairness or justice. While most literature in International Relations (IR) deals with normative legitimacy [21,23,24], sociological legitimacy has increasingly attracted scholarly attention in recent years [22,25,26].

Legitimacy is a matter of degrees. Actors such as individuals, NGOs, or states may hold legitimacy beliefs to varying extents because they believe in the moral rightfulness of a political institution [26,27]. Political scientists have debated this definition in terms of whether legitimacy beliefs are exclusively grounded in moral beliefs or whether they also reflect personal self-interest. Indeed in today's more global world, international organizations often visibly affect individual lives, making it likely that individuals hold international institutions accountable both on grounds of moral conviction and self-interest [28].

42.3 The legitimacy of international marine institutions

A recent study examines legitimacy perceptions of institutions addressing fisheries and marine issues that are central hubs in their respective issue areas.[1] Existing political science literature derives legitimacy measures through a variety of methods, such as content analysis of debates in international institutions [29] or newswires [30], and public opinion polls [26]. In contrast to these methods, expert surveys are cost-efficient data collection efforts that yield comparable measures of legitimacy perceptions of different stakeholders. Because expert surveys do not require access to voting records or documents, this method can be applied to any institution in the world.

42.3.1 Research design

This study focuses on the extent to which states, NGOs, citizens, and experts trust a number of international institutions whose actions are consequential for domestic and transnational marine management: United Nations (UN) agencies, partnerships, regional organizations, and development banks (see Table 42.1). These institutions are central hubs in their respective issue areas and all deal with marine management. Data were gathered through an online survey fielded between January and April 2018 among natural and social scientists in the academic and practitioner community.

Experts were selected in two steps. First, the starting point for identifying experts was Google Scholar, programs of central academic IR and political science conferences, and the environmental social science networks "Earth System Governance" and "Mistra Geopolitics." Second, we selected experts that come from a variety of cultural backgrounds (48 countries) and that have firsthand knowledge of the marine institutions based on research or collaboration. We made this assessment based on publication records, work descriptions, and in the case of limited information from these sources by asking the experts. Consequently, legitimacy assessments are made by experts from different cultural backgrounds and largely reflect respondents' personal opinion and not exclusively

[1] This data collection was designed and conducted with Dr. Maria-Therese Gustafsson at Stockholm University.

Table 42.1: List of international institutions included in the survey.

International organization	Acronym	Questionnaire	Founding year	Number of member states
Arctic Council	AC	Single	1996	8
Asian Development Bank	ADB	Single	1966	48
African Development Bank	AFDB	Single	1964	81
Association of Southeast Asian Nations	ASEAN	Asian organizations	1967	10
European Bank for Reconstruction and Development	EBRD	Single	1991	66 member states, including the EU and the European Investment Bank
European Union	EU	Single	1951	27
Food and Agriculture Organization	FAO	Single	1945	191
Global Environment Facility	GEF	Environment	1992	183
Intergovernmental Panel on Climate Change	IPCC	Environment	1988	195
Organization for Economic Co-operation and Development	OECD	Development	1961	34
Pacific Islands Forum	PIF	Asian organizations	1971	16
South Asian Association for Regional Cooperation	SAARC	Asian organizations	1985	8
South Asia Co-operative Environment Program	SACEP	Asian organizations	1983	8
United Nations Development Programme	UNDP	Development organizations	1965	193 (UN)
United Nations Environment	UN Environment	Single	1972	193 (UN)
United Nations Framework Convention on Climate Change	UNFCCC	Environment	1992	165
Union of South American Nations	UNASUR	South American organizations	2004	12
World Bank	World Bank	World Bank	1946	185
World Health Organization	WHO	Single	1946	193

Source: *Organizations' websites. Several natural and social scientists in our survey had expertise on more than one international institution. In these cases, we bundled institutions in one questionnaires and block-randomized the order in which the questions on a specific institution occur.*

information retrieved from the media. The average completion rate was 37% (see Appendix A).

The online survey took about 5 minutes. For each institution, about 25 experts with knowledge on that specific institution were asked to rate legitimacy beliefs among different stakeholders toward that institution ("In your opinion, how much confidence do different stakeholders have in the Arctic Council?" see Appendix B). The study includes researchers affiliated with universities, research institutions, and think tanks. Think tank researchers only make up 10% of the sample, as they may be more partial compared to other types of experts. However, researchers at think tanks often work closely with international institutions and have firsthand information of meetings and internal documents, making them valuable knowledge sources.

Measures for legitimacy beliefs are four-point confidence scales inspired by the World Values Survey and range from 0 (no confidence at all), 1 (not very much confidence), 2 (quite a lot of confidence), to 3 (a great deal confidence). Where survey responses for member state confidence were split between member states in the global south and the global north, an average was produced to create one response (see Appendix B). Confidence measures are common in public opinion research on legitimacy. Importantly, "confidence" is a better measure than "support," as the latter can direct attention away from the underlying confidence in a political system and toward satisfaction with short-term distributional consequences [31]. The confidence measure was tested for scale reliability, and these checks vouch for the quality of the expert ratings (see Appendix C).

42.3.2 Key findings

Analyzing expert judgments of stakeholder confidence yields two key findings. First, Fig. 42.1 presents the variation in legitimacy perceptions across stakeholders. While the median citizen does not have very much confidence in institutions, the median elite (member states, NGO, and experts) has quite a lot of confidence. This suggests that there is a gap between how citizens and elites perceive international marine institutions. This finding corroborates previous insights on gaps between mass and elite opinion about politics more generally [32].

Second, the extent to which there is an elite–citizen gap in legitimacy beliefs differs between marine institutions (Fig. 42.2). The results suggest that member states on average have quite a lot or a great deal of confidence in the United Nations Framework Convention on Climate Change (UNFCCC) and Intergovernmental Panel on Climate Change (IPCC), most development banks, Arctic Council, Association of Southeast Asian Nations, European Union (EU), and Pacific Islands Forum. By contrast, NGOs and citizens have much lower confidence levels, with NGOs only having a mean confidence of quite a lot or a great deal in IPCC, and citizens only having a mean confidence of quite a lot or a great deal in WHO. Finally, experts themselves indicate an average confidence of quite a lot or a great deal in the UNFCCC and IPCC, some development banks, Arctic Council, and the EU. While citizen confidence on average is weaker than elite confidence, it is particularly low for South Asia Co-operative Environment Program, European Bank for Reconstruction and Development, and South Asian Association for Regional Cooperation (SAARC).

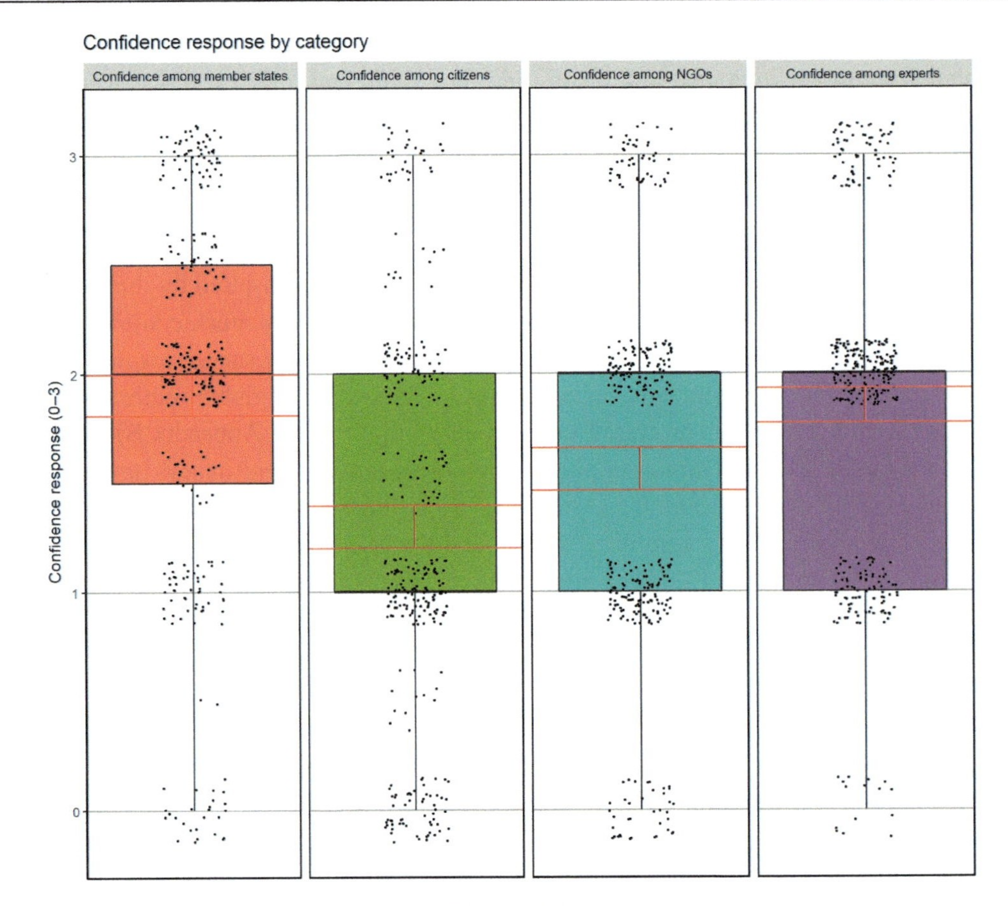

Figure 42.1

Confidence in international institutions across stakeholders. 0 (no confidence at all), 1 (not very much confidence), 2 (quite a lot of confidence), 3 (a great deal confidence). The red lines represent the 95% confidence interval, and the boxes show the interquartile range of observations. The horizontal thick black line shows the median. The point clusters depict individual expert ratings of the different audiences' legitimacy beliefs across the confidence scale $N = 478$.

These patterns indicate different legitimacy challenges for the different institutions. To begin with, elites have more confidence in *development and health* institutions than citizens with the exception of the WHO. Among institutions with a mandate in *environmental affairs*, the elite–citizen gap is less visible, which is in line with recent evidence from public opinion research. Over the past decade, exposure to several publicly salient UN climate summits may have rendered citizens on average more aware of UN activities on climate change [33] and therefore also more critical toward climate institutions than, for example, security and finance institutions [34]. Interestingly, experts tend to be critical toward all rated environmental institutions, which could be due to ongoing debates about the ability of global institutions to limit global warming to safe levels in a "post-Paris

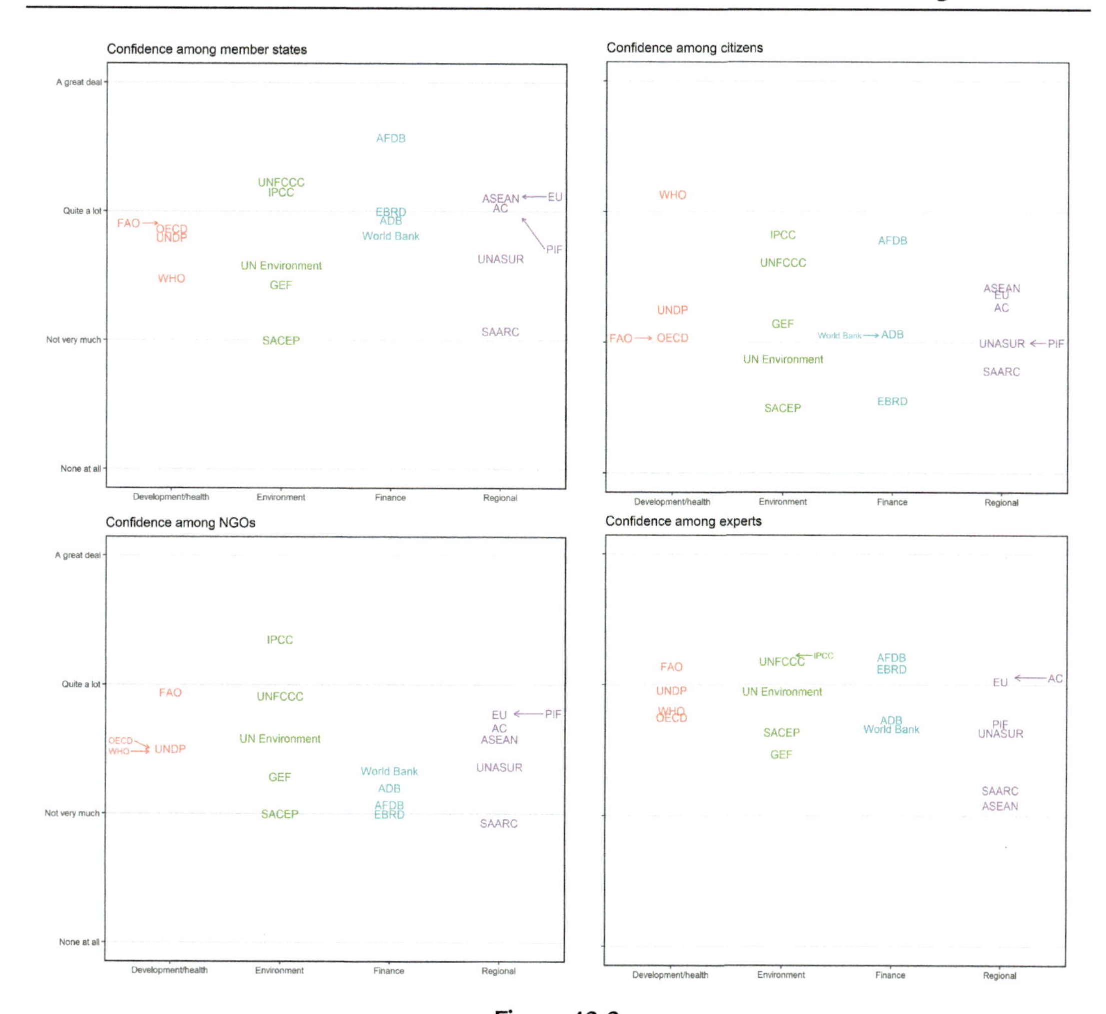

Figure 42.2

Mean confidence across international institutions and stakeholders. 0 (no confidence at all), 1 (not very much confidence), 2 (quite a lot of confidence), 3 (a great deal confidence).

world" [35,36]. In *financial affairs*, citizens and NGOs are most critical and member states least critical, with some exceptions. The African Development Bank enjoys quite a lot of confidence among citizens and member states but is viewed rather critically by NGOs and experts. The Asian Development Bank and World Bank stand out since experts have more confidence than other audiences. Finally, *regional organizations* are least popular among citizens compared to all other elite groups, which ties in with recent public research diagnosing a "crisis of trust" in regional institutions [37]. However, there is some institution-specific variation, with SAARC standing out as having relatively low confidence across all audience types.

42.4 The benefits and risks of legitimacy: a research agenda for sustainability scholars

To manage the legitimacy challenges shown above, policy makers need to consider the specific benefits—but also risks—implied in different institutional contexts and policy areas.

Two central *benefits* of legitimacy are reduction of transaction costs and willingness to internalize proenvironmental norms. In terms of reduced transaction costs, legitimacy is a form of social capital that reduces transaction costs, which is efficiency-enhancing [38,39]. In this respect, greater trust in institutions has been shown to enhance cooperative behavior and outcomes in local [10] and international institutions [18]. In terms of norms internalization, legitimacy has been shown to make people more willing to obey and defer to the government. More generally, people who trust institutions and policies are more likely to change their beliefs and norms in ways that facilitate proenvironmental behavior change [40–42].

Thus legitimate institutions have to rely less often on coercion or secrecy to enforce compliance, which is less effective and costly [43]. Legitimacy is therefore vital for international institutions that generally lack the tools to enforce compliance [11]. In terms of outcomes, international institutions trusted by state and nonstate actors may be able to raise more funds from their environment [30] and be more likely to receive necessary information on developments on the ground from stakeholders [44].

Conversely, there are two central *risks*. The first is that legitimacy may make voters less prone to acquiring information [45,46]. Institutions that voters trust to generate benefits or do the right thing even if left unattended enjoy greater legitimacy [47], but for this very reason, legitimacy may also be a challenge for marine management. In the absence of public scrutiny, collective action problems may become severe. International decision makers may misuse their leeway to make decisions that benefit them personally, with suboptimal collective outcomes. Second, as Levi et al. [48] notes, "[l]egitimacy does not signify that power will be used to promote the good of the nation or of humanity." Indeed citizens that are relatively uninformed tend to be more susceptible to elite framing of international institutions [49], implying that populist rhetoric may in turn be more likely to undermine legitimacy beliefs.

This brief overview of benefits and risks associated with increased legitimacy points to a research agenda on legitimacy and effectiveness in marine management. Below I list four areas in which sustainability scholars could contribute through their research to enabling better management and foresight of legitimacy challenges in international marine institutions.

1. *Legitimacy management.* Institutions can actively try to boost their own legitimacy, but NGOs and domestic governments are typically more effective in swaying citizen opinion [49]. Top-down legitimation through actions, speeches, and policies may not be as effective as bottom-up legitimation, which occurs when, for example, state or nonstate actors become involved in an international institution, thereby accepting the institution's exercise of power [22]. However, institutional factors such as transparency,

fairness, and effectiveness can increase the legitimacy of international institutions in the eyes of citizens [34]. Sustainability scientists could usefully examine how legitimacy could be promoted, and when this has beneficial consequences for compliance and effectiveness.

2. *Funding strategies*. Well-funded international marine institutions are able to include a greater number of NGOs in their management [50]. This has important implications for how policy makers should manage and seek funding. For example, seeking greater legitimacy among scientists and member states may increase the prospects of funding [30,39]. Yet this assumption may rarely be tenable and the extent to which this is the case is an empirical question that warrants future research.
3. *Risks for special interest influence*. We still know little about how trust operates at the international level between experts and policy makers [39] as well as between citizens and policy makers [49]. A convincing literature in economics building by Downs [45] has shown that "capture" by interest groups, such as business associations, is a risk in governance, especially when public scrutiny is low [51]. Exploring when and how legitimacy may entail risks for capture in international institutions would be an important contribution to the literature.
4. *Legitimacy beliefs*. There is a burgeoning literature on legitimacy beliefs among user groups [7], stakeholders [52], and private actors [9] in fisheries and marine institutions. Future research could usefully compare legitimacy beliefs among different stakeholders to fully understand legitimacy challenges and elite–citizen gaps in legitimacy that different institutions struggle with, with implications for compliance and effective marine management.

In sum, sustainability scholars have a role to play in enhancing our knowledge about legitimacy in international marine institutions. This chapter has shown that legitimacy entails both benefits and risks for effective marine management. Using novel expert survey evidence on the legitimacy of 19 international institutions dealing with marine issues, the key finding is twofold. First, there is a striking gap between legitimacy beliefs among elites and citizens toward international marine institutions. Second, there is variation in this elite–citizen gap across institutions, which begs questions of how institution-specific legitimacy challenges should be managed to address compliance problems. Overall, legitimacy research in sustainability science yields the prospects of a better understanding of the legitimacy–effectiveness link in marine management.

Acknowledgment

I would like to thank Hugo Faber, Alice Fasakin, Ana-Sofia Valderas, and Ognjen Zugic for their excellent research assistance.

References

[1] Food and Agriculture Organization of the United Nations, The state of world fisheries and aquaculture, Contributing to Food Security and Nutrition for All, FAO, Rome, 2016, pp. 1–200.

[2] S. Cullis-Suzuki, D. Pauly, Failing the high seas: a global evaluation of regional fisheries management organizations, Mar. Policy 34 (2010) 1036–1042. Available from: https://doi.org/10.1016/j.marpol.2010.03.002.

[3] Ö. Bodin, H. Österblom, International fisheries regime effectiveness—activities and resources of key actors in the Southern Ocean, Global Environ. Change 23 (2013) 948–956. Available from: https://doi.org/10.1016/j.gloenvcha.2013.07.014.
[4] M. Pons, M.C. Melnychuk, R. Hilborn, Management effectiveness of large pelagic fisheries in the high seas, Fish Fish. 19 (2018) 260–270. Available from: https://doi.org/10.1111/faf.12253.
[5] K.W. Abbott, J.F. Green, R.O. Keohane, Organizational ecology and institutional change in global governance, Int. Organ. 70 (2016) 247–277. Available from: https://doi.org/10.1017/S0020818315000338.
[6] J. Tallberg, L.M. Dellmuth, H. Agné, A. Duit, NGO influence in international organizations: information, access and exchange, Brit. J. Polit. Sci. 48 (2018) 213–238. Available from: https://doi.org/10.1017/S000712341500037X.
[7] S. Jentoft, Legitimacy and disappointment in fisheries management, Mar. Policy. 24 (2000) 141–148. Available from: https://doi.org/10.1016/S0308-597X(99)00025-1.
[8] E. Pinkerton, L. John, Creating local management legitimacy, Mar. Policy 32 (2008) 680–691. Available from: https://doi.org/10.1016/j.marpol.2007.12.005.
[9] A. Kalfagianni, P. Pattberg, Exploring the output legitimacy of transnational fisheries governance, Globalizations 11 (2014) 385–400. Available from: https://doi.org/10.1080/14747731.2014.888305.
[10] J. Piwowarczyk, B. Wróbel, Determinants of Legitimate Governance of Marine Natura 2000 Sites in a Post-Transition European Union Country: A Case Study of Puck Bay, Poland, 2016. <https://doi.org/10.1016/j.marpol.2016.01.019>.
[11] I. Hurd, Legitimacy and authority in international politics, Int. Org. 53 (1999) 379–408.
[12] Ö. Bodin, Collaborative environmental governance: achieving collective action in social-ecological systems, Science 357 (2017) eaan1114. Available from: https://doi.org/10.1126/science.aan1114.
[13] R. Parmentier, Role and impact of international NGOs in global ocean governance, Ocean Yearbook Online 26 (2012) 209–229.
[14] I. Sakaguchi, The roles of activist NGOs in the development and transformation of IWC regime: the interaction of norms and power, J. Environ. Stud. Sci. 3 (2013) 194–208. Available from: https://doi.org/10.1007/s13412-013-0114-3.
[15] T. Skodvin, S. Andresen, Nonstate influence in the International Whaling Commission, 1970–1990, in: M.M. Betsill, E. Corell (Eds.), NGO Diplomacy: The Influence of Nongovernmental Organizations in International Environmental Negotiations, MIT Press, Cambridge, 2008, pp. 119–148.
[16] K. Orach, M. Schlüter, H. Österblom, Tracing a pathway to success: how competing interest groups influenced the 2013 EU Common Fisheries Policy reform, Environ. Sci. Policy 76 (2017) 90–102. Available from: https://doi.org/10.1016/j.envsci.2017.06.010.
[17] C. Coffey, What role for public participation in fisheries governance? in: T.S. Gray (Ed.), Participation in Fisheries Governance, Springer, Dordrecht, 2005, pp. 27–54.
[18] R.O. Keohane, D.G. Victor, Cooperation and discord in global climate policy, Nat. Clim. Change 6 (2016) 570–575. Available from: https://doi.org/10.1038/nclimate2937.
[19] M. Weber, Economy and Society, University of California Press, Berkeley, CA, 1922 [1978].
[20] B. Gilley, The Right to Rule: How States Win and Lose Legitimacy, Columbia University Press, New York, 2009.
[21] J. Steffek, Why IR needs legitimacy: a rejoinder, Eur. J. Int. Rel. 10 (2004) 485–490. Available from: https://doi.org/10.1177/1354066104045545.
[22] J. Tallberg, K. Bäckstrand, J.A. Scholte (Eds.), Legitimacy in Global Governance: Sources, Processes, and Consequences, Oxford University Press, Oxford, 2018.
[23] D. Beetham, The Legitimation of Power, Macmillan, London, 1991.
[24] A. Buchanan, R.O. Keohane, The legitimacy of global governance institutions, Ethics Int. Aff. 20 (2006) 405–437. Available from: https://doi.org/10.1111/j.1747-7093.2006.00043.x.
[25] I. Hurd, After Anarchy. Legitimacy and Power in the United Nations Security Council, Princeton University Press, Princeton, NJ, 2007.
[26] L.M. Dellmuth, J. Tallberg, The social legitimacy of international organisations: interest representation, institutional performance, and confidence extrapolation in the United Nations, Rev. Int. Stud. 41 (2015) 451–475. Available from: https://doi.org/10.1017/S0260210514000230.

[27] M.C. Suchman, Managing legitimacy: strategic and institutional approaches, Acad. Manage. Rev. 20 (1995) 571–610. Available from: https://doi.org/10.2307/258788.

[28] T. Macdonald, Political legitimacy in international border governance institutions, Eur. J. Polit. Theor. 14 (2015) 409–428. Available from: https://doi.org/10.1177/1474885115589875.

[29] M. Binder, M. Heupel, The legitimacy of the UN Security Council: evidence from recent general assembly debates, Int. Stud. Q. 59 (2015) 238–250. Available from: https://doi.org/10.1111/isqu.12134.

[30] T. Sommerer, H. Agné, Consequences of legitimacy in global governance, in: J. Tallberg, K. Bäckstrand, J.A. Scholte (Eds.), Legitimacy in Global Governance: Sources, Processes, and Consequences, Oxford University Press, Oxford, 2018.

[31] L.M. Dellmuth, Individual sources of legitimacy beliefs: theory and data, in: J. Tallberg, K. Bäckstrand, J.A. Scholte (Eds.), Legitimacy in Global Governance: Sources, Processes, and Consequences, Oxford University Press, Oxford, 2018.

[32] M.K. Jennings, Ideological thinking among mass publics and elites, Public Opin. Q. 56 (1992) 419–441.

[33] Z. Bakaki, T. Bernauer, Do global climate summits influence public awareness and policy preferences concerning climate change? Environ. Polit. 26 (2017) 1–26.

[34] L.M. Dellmuth, J.A. Scholte, J. Tallberg, Institutional sources of legitimacy for international organizations: beyond procedure versus performance, Rev. Int. Stud. (2019) 1–20. https://doi.org/10.1017/S026021051900007X.

[35] L. Kemp, A systems critique of the 2015 Paris agreement on climate, in: M. Hossain, R. Hales, T. Sarker (Eds.), Pathways to a Sustainable Economy, Springer, Cham.

[36] D. Bodansky, The Paris climate change agreement: a new hope? Am. J. Int. Law 110 (2016) 288–319. Available from: https://doi.org/10.5305/amerjintelaw.110.2.0288.

[37] C. Foster, J. Frieden, Crisis of trust: socio-economic determinants of Europeans' confidence in government, European Union Politics, Online First 2017. <https://doi.org/10.1177/1465116517723499>.

[38] E. Weede, Legitimacy, democracy and comparative economic growth reconsidered, Eur. Sociol. Rev. 12 (1996) 217–225. Available from: https://doi.org/10.1093/oxfordjournals.esr.a018189.

[39] J. Lacey, M. Howden, C. Cvitanovic, R.M. Colvin, Understanding and managing trust at the climate science-policy interface, Nat. Clim. Change 8 (2018) 22–28. Available from: https://doi.org/10.1038/s41558-017-0010-z.

[40] J.A. Joireman, P.A.M. van Lange, M. van Vugt, A. Wood, T.V. Leest, C. Lambert, Structural solutions to social dilemmas: a field study on commuters' willingness to fund improvements in public transit, J. Appl. Soc. Psychol. 31 (2001) 504–526. Available from: https://doi.org/10.1111/j.1559-1816.2001.tb02053.x.

[41] S. Matti, Sticks, carrots and legitimate policies: effectiveness and acceptance in Swedish environmental public policy, in: P. Soderholm (Ed.), Environmental Policy and Household Behaviour: Sustainability and Everyday Life, Earthscan, James & James, London, 2010, pp. 69–98.

[42] S. Linde, Political communication and public support for climate mitigation policies: a country-comparative perspective, Clim. Policy 18 (2017) 543–555. Available from: https://doi.org/10.1080/14693062.2017.1327840.

[43] T. Franck, The Power of Legitimacy Among Nations, Oxford University Press, Oxford, 1990.

[44] K.P. Coleman, International Organisations and Peace Enforcement: The Politics of International Legitimacy, Cambridge University Press, Cambridge, 2007.

[45] A. Downs, An Economic Theory of Democracy, Cambridge University Press, Cambridge, 1957.

[46] B. Caplan, The Myth of the Rational Voter, Princeton University Press, Princeton, NJ, 2007.

[47] D. Easton, A re-assessment of the concept of political support, Br. J. Polit. Sci. 5 (1975) 435–457. Available from: https://doi.org/10.1017/S0007123400008309.

[48] M. Levi, A. Sacks, T.R. Tyler, Conceptualizing legitimacy, measuring legitimating beliefs, Am. Behav. Sci. 53 (2009) 354–375. Available from: https://doi.org/10.1177/0002764209338797.

[49] L.M. Dellmuth, J. Tallberg, Elite communication and popular legitimacy in global governance. Available at SSRN: <https://ssrn.com/abstract = 2757650> or <https://doi.org/10.2139/ssrn.2757650>, 2018.

[50] L.M. Dellmuth, M.T. Petersson, D. Dunn, A. Boustany, P. Halpin, Drivers of Non-Governmental Organization Participation in Regional Fisheries Management are Institutional, 2018 (Unpublished paper).

[51] S. Singleton, Co-operation or capture? The paradox of co-management and community participation in natural resource management and environmental policy-making, Environ. Policy 9 (2000) 1–21.

[52] C. Parés, J. Dresdner, H. Salgado, Who should set the total allowable catch? Social preferences and legitimacy in fisheries management institutions, Mar. Policy 54 (2015) 36–43. Available from: https://doi.org/10.1016/j.marpol.2014.12.011.
[53] M.R. Steenbergen, G. Marks, Evaluating expert judgments, Eur. J. Polit. Res. 46 (2007) 347–366.

Appendix

Appendix A: Completion rates

Questionnaire	International organization and acronym	Contacted	Completed	Completion rate (%)
Arctic Council	Arctic Council (AC)	66	36	54.55
Asian Development Bank	Asian Development Bank (ADB)	55	15	27.27
African Development Bank	African Development Bank (AFDB)	51	16	31.37
Environment	Global Environment Facility (GEF), Intergovernmental Panel on Climate Change (IPCC), United Nations Framework Convention on Climate Change (UNFCCC), United Nations Environment (UN Environment)	68	37	54.41
European Bank for Reconstruction and Development	European Bank for Reconstruction and Development (EBRD)	36	9	25.00
Food Governance	Food and Agriculture Organization (FAO)	54	14	29.92
Regional America	Union of South American Nations (UNASUR)	51	11	21.57
Regional Asia	Association of Southeast Asian Nations (ASEAN), South Asia Co-operative Environment Program (SACEP), South Asian Association for Regional Cooperation (SAARC), Pacific Islands Forum (PIF)	92	66	71.74
European Union	European Union (EU)	71	35	49.30
Development and humanitarian aid	United Nations Development Programme (UNDP)	60	23	38.33
OECD	Organization for Economic Co-operation and Development (OECD)	46	16	34.78
World Health Organization	World Health Organization (WHO)	79	20	25.32
World Bank	World Bank	74	27	36.49
Total contacted/completed, and average completion rate		803	325	36.52

Notes: Author's own data from expert survey. Completed questionnaires are defined in terms of substantive answers to at least 75% of the questions. Where natural and social scientists had expertise on more than one international institution, we bundled institutions in one questionnaire and block-randomized the order in which the questions on a specific institution occur.

Appendix B: Question wording presented to experts in international marine governance institutions in the order they appear in the questionnaire

The next questions focus on [*international organization*].

1. Next
2. I don't have any expertise with regard to this intergovernmental organization.

[*IF 1, then next questions, otherwise exit questionnaire on this organization.*]

In your opinion, how much confidence do different stakeholders have in the Arctic Council?

	A great deal	Quite a lot	Not very much	None at all	Don't know
Arctic Council member states	3	2	1	0	99
Citizens in Arctic Council member states	3	2	1	0	99
Nongovernmental organizations	3	2	1	0	99

We'll finish with some questions about your personal opinions about the Arctic Council and climate change.

How much confidence do you personally have in the Arctic Council?

A great deal	Quite a lot	Not very much	None at all	Don't know
3	2	1	0	99

How important do you personally think it is that the Arctic Council deal with climate risks?

Very important	Fairly important	Not very important	Not at all important
1	2	3	4

Taking all aspects of the Arctic Council's views on climate risks into account, how close is the Arctic Council to your own personal views on climate risks?

Very close	Fairly close	Not very close	Not at all close
1	2	3	4

To finish, we would like to ask you a few background questions.

Which type of organization are you employed at?

- Think tank
- Research institute

- University
- Intergovernmental organization
- If other, please specify: __________ *(open text box)*

How many years of experience with working on intergovernmental organizations or climate change do you have? Please write the years in numbers in the box below.

Appendix C: Scale reliability checks

Expert surveys are designed to generate aggregated ratings. As experts filter their ratings through their personal worldviews and preexisting knowledge, this appendix assesses the reliability with which they use the answer scales statistically. To test scale reliability, that is, the degree of uncertainty in the expert perceptions of organizational preferences, I decompose the variance across perceptions of preferences for specific institutions and experts using regression analysis. The regression model can be written as follows:

$$Y_{ik} = \mu + \delta_k + \varepsilon_{ik} \quad \text{(C.1)}$$

where μ is the grand mean of perceptions of preferences, and δ_k and δ_{ik} capture institutional and expert effects. Treating μ as a fixed effect and δ_k and δ_{ik} as random effects, the variance component models reported in Table C1 take the following generic form:

$$V[y_{ij}] = \sigma^2_\delta + \sigma^2_{\varepsilon i} \quad \text{(C.2)}$$

Table C1: Variance components analysis of expert ratings of stakeholder confidence.

Estimated parameter	Confidence among member states	Confidence among citizens	Confidence among NGOs	Confidence among experts
Fixed effects				
Grand mean μ	1.869***	1.236***	1.518***	1.828***
	(0.081)	(0.051)	(0.084)	(0.072)
Variance components				
Institutions (σ^2_δ)	0.268**	0.334***	0.279***	0.251***
	(0.082)	(0.080)	(0.079)	(0.059)
Experts ($\sigma^2_{\varepsilon i}$)	0.767	0.777	0.793	0.688***
	(0.032)	(0.032)	(0.339)	(0.031)
N (Institutions)	19	19	19	19
N (Experts)	323	325	325	359
Log likelihood	−402.774	−413.876	−417.407	−410.257
Reliability	0.860	0.880	0.857	0.826

Notes: Multilevel models of confidence items (see Appendix B) estimated with maximum likelihood. Coefficients with robust standard errors clustered at the level of institutions. I pooled the data so that the cases are institution-specific expert perceptions, implying that the number of experts is slightly higher than in Table 42.1. A constant is included in all models but not reported for the sake of brevity. *$P<.05$, **$P<.01$, ***$P<.001$.

where σ^2_{δ} is the cross-institutional variation in perceptions and $\sigma^2_{\varepsilon i}$ is the cross-expert variance.

The key result of this variance component analysis for the different confidence items is that the size of the expert-specific variation is quite limited (see Table C1). The estimated expert variance is only statistically significant with regard to one of the four items. Moreover, the size of this variation is relatively small: the estimated standard deviation is around 0.7 on the 4-point confidence scale. These results indicate that the larger share of experts renders quite similar judgments about organizational preferences. Moreover, as expected, cross-institutional variation is substantial given that the cross-institutional variance is statistically significant for all confidence items.

On the basis of the regression estimates in the variance component analysis in Table C1, I calculate the reliability of expert ratings via the Spearman–Brown formula using the average number of experts [53]. The reliability coefficients are above 0.8, indicating a very high reliability.[2]

[2] The Spearman–Brown formula is computed as $nr[1 + (n - 1)r]$, where r is the interexpert correlation from the variance components model and n is the average number of experts [53, p. 363].

CHAPTER 43

Verifying and improving states' compliance with their international fisheries law obligations

Solène A. Guggisberg[1,2]

[1]*Netherlands Institute for the Law of the Sea (NILOS), Utrecht University, Utrecht, The Netherlands*
[2]*Nippon Foundation Nereus Program, University of British Columbia, Vancouver, BC, Canada*

Chapter Outline

43.1 Introduction

The status of marine living resources is extremely concerning in terms of environmental conservation, with most fish stocks being fully exploited or suffering from overexploitation [1]. While the development of quality science and the adoption of adequate laws and conservation and management measures (CMMs) are necessary for the long-term sustainability of such marine resources, they will remain insufficient if states, as the main actors on the international scene, do not respect their obligations. The traditional regime regulating fisheries is highly fragmented and decentralized and has had difficulty in ensuring high levels of state compliance with the existing rules. It is consequently crucial to examine the manner in which states' compliance with their obligations is verified and encouraged at the international level.

This chapter presents and analyzes some recent developments in compliance procedures with regard to the different obligations by which states are bound. After introducing each

Predicting Future Oceans.
DOI: https://doi.org/10.1016/B978-0-12-817945-1.00043-5

procedure and its current state of implementation, the chapter focuses on three important constitutive elements for a compliance mechanism to verify, impartially and fairly, whether states respect their fisheries obligations and to address instances of noncompliance [2]. First, it examines who the reviewer is; this factor may have an influence on the independence of the process and the centralization of compliance assessments. Second, it looks at whether the criteria used to verify a state's behavior with regard to fishing activities are the same for all relevant states; the existence or absence of a similar benchmark may have implications for the consistency of assessments of different states. Third, the chapter discusses issues related to accountability, in particular whether states are obliged to undertake a review and what follow-up mechanisms exist to ensure that states rectify their behavior if needed. The chapter concludes by proposing options which could improve the current regime. Due to the focus on global obligations, regional compliance procedures are not examined, although it is acknowledged that some regional bodies have set up compliance committees to review their members and cooperating states' compliance [3].

States play different roles in relation to fishing activities and have, accordingly, different duties. Central to the regulatory regime is that the flag state must control its vessels and ensure that they respect the applicable CMMs. The obligation is provided for in the United Nations Convention on the Law of the Sea (UNCLOS) in Article 94 [4] and further elaborated, in relation to the fisheries sector, in other treaties [5,6]. This obligation of due diligence does not imply that a flag state will be held responsible for each violation of applicable rules by one of its vessels, but that it must "take all necessary measures to ensure compliance and to prevent IUU (illegal, unreported and unregulated) fishing by fishing vessels flying its flag" [7]. Coastal states must ensure that the resources under their jurisdictions are not overexploited [4] as well as, at least in theory, grant access to other states to any surplus in their exclusive economic zones (EEZs) [4,8]. Both coastal states—when dealing with fish stocks not exclusively located in their waters—and the flag states of vessels active in other states' EEZs or on the high seas are under an obligation to cooperate in the conservation of marine living resources [4]. UNCLOS does not state which form such cooperation should take, but states have, in practice, generally entered into agreements setting up regional fishery bodies. Some of them, the regional fisheries management organizations (RFMOs), adopt binding CMMs. The United Nations Fish Stocks Agreement (UNFSA) recognizes RFMOs as the vehicle for cooperative management of straddling and highly migratory stocks [6]. As to port states, they are not under any general obligation under UNCLOS, but parties to the Agreement on Port State Measures to Prevent, Deter and Eliminate Illegal, Unreported and Unregulated Fishing (PSMA) must deny entry into port to vessels involved in IUU fishing, or inspect vessels entering their ports and, if they determine that relevant CMMs were violated, deny the use of port services to such vessels [9].

In addition to issues caused by the unclear content of some of these obligations found in UNCLOS and by the limited ratification status of more detailed treaties [10], no global international organization or standing institutional mechanism has been created to comprehensively overview states' compliance with their international obligations in the fisheries field. Several global international mechanisms provide forums for regular discussion of fisheries issues, in particular the Committee on Fisheries (COFI) of the Food and Agriculture Organization (FAO) [11], a United Nations specialized agency; the United Nations General Assembly with its annual Sustainable Fisheries Resolution [12]; and the UNFSA (Resumed) Review Conference [13]. However, these are not compliance mechanisms in that they do not review the compliance status of each state with its international obligations and do not provide for measures to address incidences of noncompliance. Therefore it is necessary to examine the existing procedures and mechanisms meant to assess compliance, separately, with each type of obligation binding states in the field of fisheries.

43.2 Flag state: voluntary guidelines for performance review

One of the major issues impairing the achievement of sustainable fisheries and enabling IUU fishing is the unwillingness or inability of some flag states to control their vessels. Assessing and hopefully improving the compliance of flag states with their obligations in the fisheries field is the focus of a 2014 soft-law instrument, the FAO Voluntary Guidelines for Flag State Performance (flag state Guidelines) [14]. It was endorsed at a COFI meeting [15], following an inclusive procedure involving experts and states [16].

The flag state Guidelines bring together, in one general document, both substantive and procedural provisions. With regard to substantive provisions, this instrument lists, in the form of performance assessment criteria, the most accepted obligations of flag states. In terms of procedural provisions, the flag state Guidelines suggest procedures for carrying out assessments and measures to promote compliance, as well as describe the role of the FAO. As stated in paragraph 56, states should inform the FAO of their assessments and the results thereof as part of their biennial reporting on the 1995 Code of Conduct for Responsible Fisheries.

According to the results of the 2018 questionnaire on the Code of Conduct for Responsible Fisheries, some countries have already undertaken assessments of their performance as flag states within the framework of the flag state Guidelines [17]. On average, 28% of the 83 responding states have undertaken a flag state assessment and nearly 80% of those which have not yet done so expressed their intension to engage in a review process. The proportion of states having already undertaken an assessment differs from region to region, ranging from 9.5% for African countries to 60% for the South West Pacific (see Table 43.1). No further information is publicly available at present on

Table 43.1: Number of states having undertaken a flag state performance review or intending to do so.[a]

	FAO Members per region	Responded to question on assessment according to flag state Guidelines		Have already undertaken an assessment according to the flag state Guidelines		Expressed intention to do so (in remaining group)	
	Number	% of FAO Members	Number of FAO Members	% of respondents	Number of respondents	% of respondents	Number of respondents
Africa	50	42.00	21	9.52	2	94.74	18
Asia	23	56.52	13	23.08	3	80.00	8
Europe	50	14.00	7	57.14	4	100.00	3
Latin America and the Caribbean	33	72.73	24	26.09	6	76.47	14
Near East	21	28.57	6	16.67	1	40.00	2
Northern America	2	100.00	2	50.00	1	100.00	1
South West Pacific	18	55.56	10	60.00	6	50.00	2
Total/Average	197	42.13	83	28.05	23	79.66	48

FAO, Food and Agriculture Organization.
[a]Data from Regional Statistical Analysis of Responses by FAO Members to the 2018 Questionnaire on the Implementation of the Code of Conduct for Responsible Fisheries and Related Instruments, COFI/2018/SBD.1 <http://www.fao.org/3/CA0465EN/ca0465en.pdf>, 2018 (accessed 18.12.18), Tables 1 and 67.

the identity of the states in question, the procedures followed for the assessments, or their conclusions. While this lack of transparency may be detrimental to identifying which states are actively attempting to be responsible flag states, it may not be a long-term problem. Indeed, information can be made available through other avenues: Norway, which had already undertaken an assessment by 2016, reported to the United Nations Secretary General within the framework of the UNFSA (Resumed) Review Conference and provided more information on the results of its assessment [18].

There is no legally binding institutional structure under UNCLOS or another global treaty to which flag states must report on their actions and which would assess their compliance. According to the voluntary flag state Guidelines, states can choose how to assess their performance; the procedure for carrying out assessments envisages both options of self-review and external assessment. The FAO soft-law instrument also does not indicate how reviewers may be chosen. The criteria against which one's conduct is to be assessed should logically be the ones listed in the flag state Guidelines. However, this is only made explicit for external assessments, which seems to imply that states can freely use different criteria in their own self-assessments.

The flag state Guidelines are voluntary in nature, which means that states are under no obligation to undertake a review of any sort. Whether there are immediate consequences—and the nature of such consequences—to an assessment finding that a flag state is poorly performing is not fully clarified in the flag state Guidelines. This instrument provides a number of facilitative and enforcement-like measures which may be adopted [19]. However, there is no procedural certainty as to the order in which such measures can be applied or as to the level of underperformance which would trigger them. This ambiguity could be an issue if it discourages states to undergo the process. Nevertheless, the results of the FAO questionnaire are promising since 23 countries appear to have already undertaken an assessment and 48 more have expressed their intention to do so. In any case, whatever their ultimate level of implementation, the flag state Guidelines may have some constructive impacts on the current global situation, if only by contributing to identify which states are attempting to respect their international obligations. Also other procedures and mechanisms could build upon the conclusions of the performance assessments. Finally, that states have to inform the FAO of their performance reviews is positive since such a centralizing mechanism may provide an overview of the assessments and has the potential, in the mid-term, to promote the development of best procedural practices.

43.3 Coastal state: discretion of states

As provided for in Article 56(1) of UNCLOS, coastal states have sovereign rights over their EEZs for the purpose of conserving and managing living resources. These areas of the oceans were, until the adoption of UNCLOS, considered to be part of the high seas and subject to the freedom of fishing. The large enclosure of the high seas agreed to in 1982 was, at least in part, meant to improve the status of fish stocks. The rationale was that states would take greater care of resources if they were theirs [20]. While there seems to be some truth in that [21,22], the overall status of fish stocks points to some remaining issues, such as the lack of sufficient financial and technical resources necessary for efficient management and enforcement by certain states [23].

As pointed out in the introduction, coastal states are bound by international rules and their compliance could be assessed on the basis thereof. There is, however, no mechanism in place to review whether they respect their obligations in terms of sustainability or of granting access to surplus to other states. Such de facto absolute discretion left to coastal states is regrettable since the establishment of EEZs was intended to improve sustainability and came at the price of global common areas.

43.4 Cooperation through regional fisheries management organizations: widespread performance reviews

When targeting stocks that do not occur exclusively in the area under the jurisdiction of a single state, states must cooperate in the conservation of marine living resources. States generally do so through RFMOs. In light of these institutions' central role in fisheries management, their performance is crucial. However, RFMOs have been criticized for their lack of success in sustainably managing fisheries and marine ecosystems [24–27]. A major issue relates to the actions of some states that choose not to join the RFMOs and refuse to cooperate with them. This problem falls beyond the realm of the present chapter as it concerns the question of the applicability of CMMs to nonparties to RFMOs, which in turn raises the issues of the content of the obligation to cooperate in customary international law and the nature of the UNFSA [28]. However, there are other factors contributing to RFMOs' underperformance, such as outdated legal frameworks, the adoption of inadequate CMMs, and noncompliance by members with their obligations.

Recognizing the need for reforms to address these internal issues, a voluntary process known as RFMO performance review has been encouraged by the international community [29–31] and largely embraced by RFMOs. This procedure consists of a review of RFMOs' performance in fulfilling their functions, conducted by a panel of experts, who compare the situation within a specific RFMO to a set of benchmark criteria. Since the mid-2000s, performance reviews have been undertaken by most RFMOs and some have already finalized a second assessment [1,18,32]. The reports of these panels are publicly available.

RFMOs are independent organizations and do not have to file reports with any supervising body that would assess their compliance in a centralized manner. They are free to choose the members of the review panel as they wish and practice differs as between institutions. Some have selected a mix of representatives from the RFMO or their member states and external experts, while others have preferred a purely external group of people [33,34]. Overall, all panels have included at least one outside expert. The criteria used as a benchmark to assess RFMOs' performance are similar across institutions. Although the precise criteria ultimately depend on the terms of reference developed by each RFMO—and there was a debate as to the adequacy of using a similar set of benchmarks for different organizations [35]—all the performance reviews appear, in practice, to be based on the minimum standards agreed upon in 2007 within the Kobe process [1], an ongoing collaboration of the tuna RFMOs on issues of common interest.

The decision to undergo a performance review is, in all but one case, a voluntary one. Only the South Pacific RFMO (SPRFMO) provides at Article 30 of its founding document that a review procedure must be undertaken at least every 5 years [36]. Nevertheless, if one considers the widespread application of the procedure and its endorsement by the international community, it seems that performance reviews are now voluntary in name only. In terms of follow-up, most RFMOs have set up a framework to examine or implement the recommendations made by the review panel. However, no external institution is tasked with overseeing the extent to which recommendations have been taken into account and at which pace progress is made. The norm is for second performance reviews to start by assessing whether and how the recommendations of the first review have been implemented [37–39]. This ensures that some pressure is kept on RFMOs to act upon recommendations, but it does not, as a centralized mechanism would, contribute to the consistency of evaluations across RFMOs. The international community has recognized the need to "establish mechanisms for follow-up actions in response to performance reviews" [13], even if the nature of such mechanisms remains to be determined.

43.5 Port state: an integrated mechanism under development

Port state measures were originally developed with respect to vessel safety and the prevention of pollution. They have gradually extended to the field of fisheries, as a means of supporting the conservation and management of marine fisheries, in particular by fighting IUU fishing [40]. Following the trend of regional schemes set up by RFMOs, such measures recently became the focus of a dedicated treaty, the PSMA, which entered into force in 2016.

As seen above, the PSMA requires respectively at its Articles 9 and 11 that port states refuse entry into port to vessels involved in IUU activities, or undertake inspections, which may lead to a denial of landing or transshipment (i.e., the transfer of catch from one fishing vessel to another vessel) of fish, or of other port services. In addition to its substantive content, the PSMA institutionally sets up a procedure to verify states' compliance with their obligations under the treaty. This treaty indeed demands, "within the framework of FAO and its relevant bodies, the regular and systematic monitoring and review of the implementation of this Agreement." The parties have agreed that a specific web-based questionnaire, which will be filled by each state reporting on its own implementation of the PSMA, should be developed for these purposes, and that review should take place every other year [41].

The relevant procedures are not yet in place; the FAO Secretariat and an open-ended technical working group have been tasked with their development [41]. It is hence currently impossible to assess the nature of a future monitoring mechanism, if any, or the procedures that may be established to address noncompliance. However, it seems likely that there will be a centralized body to which state parties to the PSMA have to report. The criteria against which the compliance with the treaty will be reviewed, as set in the questionnaire, will be based on the treaty and will be the same for all parties. All state parties have to comply with the procedures to monitor their implementation of the PSMA since the treaty explicitly provides for such review. By setting up a regular assessment procedure, even if it is only by requiring states to share information with each other, the mechanism will create a certain level of ongoing international accountability.

43.6 Conclusion: ways forward

The compliance procedures that exist in the fisheries field are in the early stages of their development (for an overview, see Table 43.2). Even if one leaves aside the problematic absence of any procedure to assess coastal states' compliance with their obligations, the existing mechanisms still suffer from some shortcomings. In particular, the assessments of flag states and RFMOs do not have to be undertaken by external experts, do not have to use the same review criteria as applied to other similar entities, and above all are not compulsory. While the voluntary nature of these two procedures does not seem to have notably weakened their implementation, having a binding procedure integrated in the relevant treaty, as envisioned under the PSMA, is nevertheless preferable in terms of accountability.

Voluntary guidelines and schemes have, in the past, opened the path toward binding instruments. Of particular relevance is the shipping sector, where a framework for flag state performance has been developed under the ambit of the International Maritime Organization (IMO). It started in 2001 with self-assessment, followed in 2003 by the Voluntary Member State Audit Scheme, and finally became the mandatory IMO Member State Audit Scheme, which entered into force at the beginning of 2016 and which reviews states' performance as flag, port, and coastal states [42,43]. One can hope that, similarly, a centralized and multilateral review of all states obligations in the field of fisheries is in the making.

In the meanwhile, incremental improvement is probably the way forward. For example, the centralization of flag state reviews by the FAO may be used to contribute to the promotion of best procedural practices. Likewise, for RFMO performance reviews, a centralized mechanism may be established to verify the timely implementation of recommendations. Until then, the international community should encourage the setting up of second or third performance reviews, as these include an element of follow-up over the previous assessments.

Currently, there is one mechanism where all the relevant obligations of a state in the fisheries field are assessed and where noncompliance comes at the price of trade measures. The European Union (EU), in its Regulation to prevent, deter, and eliminate IUU fishing, provides for the possibility to sanction non-EU states that do not fulfill their obligations as flag, coastal, port, or market states under international law [44]. The EU first undertakes a thorough procedure and dialog with the preidentified state, supporting it to make necessary changes. Then, if that is not sufficient, the third state may be identified as noncooperating in the fight against IUU fishing. This means that it will be unable to export seafood products to the EU, as provided for in Articles 31–33 and 38 of the EU's Regulation. Up to December 2018, the EU has formally pre-identified 25 states as potentially noncooperating,

Table 43.2: Overview of compliance procedures.

	Relevant treaties and ratification status					Accountability		
	Treaty rules	Number of state parties (December 2018)	Mechanism or procedure for review	Identity of reviewer	Criteria	Voluntary or compulsory process	Centralization of reports	Follow-up on recommendations/ consequences for underperformance
Flag state	UNCLOS Art. 94(1) Compliance Agreement UNFSA	168 42 89	• FAO Voluntary Guidelines for Flag State Performance • FAO soft-law instrument adopted through comprehensive procedure	• Self-assessment or assessment by external experts • To be decided by the flag state itself	• Substantive criteria listed in the Guidelines • Explicitly recommended to use them for external assessment	• Voluntary	• FAO to be informed of assessments and results	• Some consequences for underperformance listed in the Guidelines • Lack of clarity as to triggers for consequences or order of implementation thereof
Coastal state	UNCLOS Art. 61-62	168						
Cooperation through RFMOs	UNCLOS Art. 61, 63-64,66-67, 118 UNFSA	168 89	• RFMO performance reviews • Mechanism developed ad hoc but quite established	• Panel the composition of which is decided by the RFMO itself • Generally includes at least one independent expert	• Kobe list of performance criteria is the minimum list used by RFMOs • Ad hoc development • Exact terms of reference may differ	• Generally voluntary • Binding for one RFMO, SPRFMO	• Reports are made public but not centralized	• Internal follow-up at RFMO level • Second or third performance reviews verify implementation of previous recommendations • No formal consequences for underperformance
Port state	PSMA	57	• Regular and systematic monitoring and review of the implementation of the PSMA • Mechanism provided for in the treaty • Detailed functioning under development	• Self-reporting by the port state through a questionnaire • Additional level of review to be determined, mechanism under development	• PSMA obligations	• Binding	• To be determined, mechanism under development	• To be determined, mechanism under development

FAO, Food and Agriculture Organization; *PSMA*, Agreement on Port State Measures; *RFMO*, regional fisheries management organization; *SPRFMO*, South Pacific RFMO; *UNCLOS*, United Nations Convention on the Law of the Sea; *UNFSA*, United Nations Fish Stocks Agreement.

of which 6 have moved to the next phase and have been subject to trade sanctions [45]. This trade sanction mechanism, applied by a state with a huge market for fishery products, seems to have proven effective, triggering improvements in fisheries governance in several states [46]. However, in the absence of a mechanism to review or contest the EU's decisions—short of attempting to bring a case to the World Trade Organization's Dispute Settlement Body for breach of trade law [45]—the unilateral nature of this mechanism also raises issues, in particular of (perceived) arbitrariness or of unequal treatment between states [47].

Aside from the limitations still in the way of efficiently verifying and encouraging states' compliance with their obligations in the fisheries field, it is promising that recent international initiatives focus on developing means to improve implementation and enforcement of existing obligations. This suggests a change of paradigm, where noncompliance is increasingly considered unacceptable and where states recognize the need to monitor each other's behavior to protect the living resources of the seas.

References

[1] Food and Agriculture Organization (FAO), *The State of the World Fisheries and Aquaculture*, Rome, 2018.

[2] Compliance mechanisms, which have become the norm in multilateral environmental agreements (MEAs), are far more structured than what is available in the fisheries field, at least at the global level. Hence, only selected aspects of an efficient compliance mechanism are discussed here. For more on compliance mechanisms in MEAs, see, for example, U. Beyerlin, P.-T. Stoll and R. Wolfrum (Eds.), *Ensuring Compliance With Multilateral Environmental Agreements—A Dialogue Between Practitioners and Academia*, Martinus Nijhoff Publishers, Leiden/Boston, 2006; UNEP, *Manual on Compliance With and Enforcement of MEAs*, 2006.

[3] On this, see for example M.C. Engler Palma, Non-Compliance Procedure: Can Regional Fisheries Management Organizations Learn from the Experience of Multilateral Environmental Agreements? *Ocean Yearbook* 24 (2010) 185–237.

[4] United Nations Convention on the Law of the Sea (UNCLOS), Montego Bay, 10 December 1982, in force 16 November 1994, 1833 *UNTS* 396.

[5] Agreement to Promote Compliance With International Conservation and Management Measures by Fishing Vessels on the High Seas (Compliance Agreement), Rome, 24 November 1993, in force 24 April 2003, 2221 *UNTS* 120.

[6] Agreement for the Implementation of the Provisions of the United Nations Convention on the Law of the Sea of 10 December 1982 relating to the Conservation and Management of Straddling Fish Stocks and Highly Migratory Fish Stocks (UNFSA), New York, 4 August 1995, in force 11 December 2001, 2167 *UNTS* 88.

[7] International Tribunal for the Law of the Sea (ITLOS), *Request for an Advisory Opinion Submitted by the Sub-Regional Fisheries Commission (SRFC), Advisory Opinion, 2 April 2015, ITLOS Reports 2015.*

[8] In practice, coastal states have a large leeway in determining the total allowable catch and hence can decide whether to create a surplus in relation to their fishing capacity. Moreover, there are no judicial avenues resulting in a binding decision open to third states to review such determination or to demand access.

[9] Agreement on Port State Measures to Prevent, Deter and Eliminate Illegal, Unreported and Unregulated Fishing (PSMA), Rome, 22 November 2009, in force 5 June 2016.

[10] In particular, as of December 2018, 89 states are parties to the UNFSA; 42 to the Compliance Agreement; and 57 to the PSMA.

[11] At the time of writing, the most recent of these biennial meetings had taken place in July 2018; see COFI33 Documents for an overview of the issues discussed <http://www.fao.org/about/meetings/cofi/documents-cofi33/en/>, 2018 (accessed 18.12.18).

[12] At the time of writing, the most recent of these yearly resolutions was United Nations General Assembly (UNGA) Sustainable Fisheries Resolution of 5 December 2017, A/RES/72/72.

[13] At the time of writing, the most recent occurrence of this procedure, which has so far taken place every four to six years, led to the report of the 2016 UNFSA Resumed Review Conference, A/CONF.210/2016/5, 1 August 2016.

[14] Voluntary Guidelines for flag State performance, FAO, Rome, (flag state Guidelines) <http://www.fao.org/3/a-i4577t.pdf>, 2015 (accessed 18.12.18).

[15] Report of the Thirty-First Session of the Committee on Fisheries (Rome, 9 – 13 June 2014), *FAO Fisheries and Aquaculture Report No. 1101*, Rome, 2015.

[16] K. Erikstein, J. Swan, Voluntary Guidelines for Flag State Performance: A New Tool to Conquer IUU Fishing, Int. J. Mar. Coastal Law 29 (2014) 116–147.

[17] Regional Statistical Analysis of Responses by FAO Members to the 2018 Questionnaire on the Implementation of the Code of Conduct for Responsible Fisheries and Related Instruments, COFI/2018/SBD.1 <http://www.fao.org/3/CA0465EN/ca0465en.pdf>, 2018 (accessed 18.12.18).

[18] Report of the Secretary-General to the 2016 UNFSA Resumed Review Conference, A/CONF.210/2016/1, 1 March 2016.

[19] The mention of "measures as set out in the IPOA-IUU and the Code . . ." at paragraph 47 of the flag state Guidelines implies that market related measures, including trade sanctions could be envisaged (see International Plan of Action to Prevent, Deter, and Eliminate IUU Fishing (IPOA-IUU), Rome, 2011, paras. 65–76).

[20] R.D. Eckert, for example was of the opinion that the EEZ can reduce unsustainable practices by creating an economic rent for the resource owner. However, he acknowledged that inefficiencies for straddling stocks and potential difficulties with enforcement could be expected (R.D. Eckert, The Enclosure of Ocean Resources: Economics and the Law of the Sea, Hoover Institution Press, Stanford, CA, 1979, 129–131).

[21] Shared stocks appear to be more likely to be overexploited (S.F. McWhinnie, The tragedy of the commons in international fisheries: An empirical examination, J. Environ. Econ. Manage. 57 (2009) 321–333.

[22] The state of exploitation of highly migratory and straddling stocks is considered particularly worrying (J.J. Maguire, M. Sissenwine, J. Csirke and S. Garcia, The state of world highly migratory, straddling and other high seas fishery resources and associated species, *FAO Fisheries Technical Paper No. 495*, Rome, 2006), even in comparison to the already dire status of marine fisheries in general (FAO, *The State of the World Fisheries and Aquaculture*, Rome, 2014; G. Ortuño Crespo and D.C. Dunn, A review of the impacts of fisheries on open-ocean ecosystems, ICES J. Mar. Sci. 74 (9) (2017) 2283–2297).

[23] F. Meere and C. Delpeuch, The challenge of combating illegal, unreported and unregulated (IUU) fishing, in: FAO and OECD, *Fishing for Development, FAO Fisheries and Aquaculture Proceedings No. 36*, Rome, 2015, pp. 31–52.

[24] FAO, *The State of the World Fisheries and Aquaculture*, Rome, 2016.

[25] S. Cullis-Suzuki, D. Pauly, Failing the high seas: A global evaluation of regional fisheries management organizations, Mar. Policy 34 (2010) 1036–1042.

[26] M.W. Lodge, D. Anderson, T. Lobach, G. Munro, K. Sainsburg, A. Willock, *Recommended Best Practices for Regional Fisheries Organizations*, Chatham House, London, 2007.

[27] M.J. Juan-Jordá, H. Murua, H. Arrizabalaga, N.K. Dulvy, V. Restrepo, Report card on ecosystem-based fisheries management in tuna regional fisheries management organizations, Fish Fish. 19 (2) (2018) 321–339.

[28] These questions were examined by the author in an earlier publication, see S. Guggisberg, *The Use of CITES for Commercially-exploited Fish Species*, Springer, Cham, 2016, pp. 43–49, 71.

[29] Report of the Twenty-Sixth Meeting of the Committee on Fisheries (Rome, 7 – 11 March 2005), *FAO Fisheries Report No. 780*, Rome, 2005.

[30] UNGA Sustainable Fisheries Resolution of 29 November 2005, A/RES/60/31.

[31] Report of the 2006 UNFSA Review Conference, A/CONF.210/2006/15, 5 July 2006.

[32] At the time of writing, one of the most recently established RFMOs, SPRFMO, was undertaking its first performance review [see SPRMO, Performance Review of RFMOs <https://www.sprfmo.int/about/the-convention/sprfmo-review-2018/>, 2018 (accessed 18.12.18)].

[33] M. Ceo, S. Fagnani, J. Swan, K. Tamada and H. Watanabe, Performance Reviews by Regional Fishery Bodies: Introduction, summaries, synthesis and best practices, Volume I: CCAMLR, CCSBT, ICCAT, IOTC, NAFO, NASCO, NEAFC, *FAO Fisheries and Aquaculture Circular No. 1072*, Rome, 2012.

[34] P.D. Szigeti and G. Lugten, The Implementation of Performance Review Reports by Regional Fishery Bodies, 2004–2014, *FAO Fisheries and Aquaculture Circular No. 1108*, Rome, 2015.

[35] Report of the Sixth Round of Informal Consultations of States Parties to the UNFSA, ICSP6/UNFSA/REP/INF.1, 29 May 2007.

[36] SPRFMO Convention, Auckland, 14 November 2009, in force 24 August 2012.

[37] Terms of Reference and criteria to conduct the Second Performance Review of the IOTC, in: *Report of the Eighteenth Session of the IOTC*, 2014, Appendix XVI.

[38] CCAMLR, Report of the Thirty-Fifth Meeting of the Commission (Hobart, Australia, 17–28 October 2016) Annex 8.

[39] Approach to a Second Review of ICCAT, Annex 1 <https://www.iccat.int/intermeetings/Performance_Rev/ENG/PER_FINAL_TOR_ENG.pdf>, (accessed 18.12.18).

[40] On the development of port state measures in the fisheries field, see J. Swan, Port State Measures—from Residual Port State Jurisdiction to Global Standards, Int. J. Mar. Coastal Law 31 (2016) 395–421.

[41] Report of the first meeting of the Parties to the Agreement on Port State Measures to Prevent, Deter and Eliminate Illegal, Unreported and Unregulated Fishing (Oslo, 29 – 31 May 2017), *Fisheries and Aquaculture Report No. 1211*, Rome, 2017.

[42] IMO Member State Audit Scheme & Implementation Support <http://www.imo.org/en/OurWork/MSAS/Pages/default.aspx>, (accessed 18.12.18).

[43] H. Jessen, L. Zhu, From a voluntary self-assessment to a mandatory audit scheme: monitoring the implementation of IMO instruments, Lloyd's Marit. Commer. Law Q. 3 (2016) 389–411.

[44] Council Regulation (EC) No 1005/2008 of 29 September 2008 establishing a Community system to prevent, deter and eliminate illegal, unreported and unregulated fishing (EU IUU Regulation), *OJ L 286, 29 October 2008*, pp. 1 – 32.

[45] Overview of existing procedures as regards third countries as of December 2018 <https://ec.europa.eu/fisheries/sites/fisheries/files/illegal-fishing-overview-of-existing-procedures-third-countries_en.pdf>, (accessed 18.12.18).

[46] The reactions to, and consistency with trade law of, the EU IUU Regulation were examined by the author in an earlier publication, see S. Guggisberg, Recent developments to improve compliance with international fisheries law, *L'Observateur des Nations Unies* 42 (2017) 139–169.

[47] On this, see for example A. Leroy, F. Galetti and C. Chaboud, The EU restrictive trade measures against IUU fishing, *Mar. Policy* 64 (2016) 82–90.

CHAPTER 44

Understanding potential impacts of subsidies disciplines and small-scale fisheries

U. Rashid Sumaila

Fisheries Economics Research Unit, Institute for the Oceans and Fisheries & UBC School of Public Policy and Global Affairs, The University of British Columbia, Vancouver, BC, Canada

Chapter Outline

44.1 Introduction

In the World Trade Organization (WTO) negotiations several proposals suggest that some kinds of subsidies that would be otherwise prohibited, such as those to capital or operating costs, might continue to be provided to small-scale fishing of different kinds. Some of these proposals suggest that flexibilities for small-scale fishing subsidies could be based on a single agreed definition. The EU proposal, for example, defines "subsistence" fishing with reference to the use to which catch is put, while the proposal by Indonesia defines "artisanal" fishing with reference to geographic location of the fishing activity, the type of gear used, and the purpose to which catch is put, and defines "small-scale" fishing primarily with reference to vessel length. Other proposals, including those by the ACP (African, Caribbean, and Pacific) and Least Developed Country groups, suggest relying on national definitions of small-scale fishing to establish the scope of the exception.

In the next few paragraphs, I provide a brief literature review of the subject of small-scale fisheries (SSF). In particular, I present recent approaches to defining the "degree of

Predicting Future Oceans.
DOI: https://doi.org/10.1016/B978-0-12-817945-1.00048-4

small-scaleness" of fisheries. This is then followed with a presentation of data on the share of fisheries catch, revenues, and subsidies to SSF.

44.2 Distinguishing between small-scale and large-scale fisheries

It is widely agreed that SSF are a crucial part of the global fisheries sector that need to be protected and supported [1]. For instance, it is estimated that SSF contribute ~30% of the global landed value (revenues at the dock), and employ millions of people living in coastal communities, some of them in remote and rural regions. Hence, SSF provide crucial livelihood opportunities where such opportunities are limited. They therefore make vital contributions to some of the world's most vulnerable communities and people.

Agreeing on a clear universally accepted definition and distinction between SSF and large-scale fisheries (LSF) is truly difficult; a fact acknowledged by the FAO Advisory Committee on Fisheries Research in 2003. Since then, this conclusion of the FAO has been supported both in the literature and by key global documents such as the Voluntary Guidelines for Securing Sustainable Small-Scale Fisheries [1], where country-level definitions are applied. The main argument against trying to develop a universally accepted definition for SSF, as stated in the Voluntary Guidelines, is the fact that the sector is very diverse and dynamic plus the fact that SSF are generally place- and local community-based fisheries that are rich in customs, traditions, and values.

Even though there is no internationally agreed definition of small-scale fishing, the FAO use 24 m vessel length as a broad cutoff between SSF and LSF. As a result, academic studies have developed different methods for identifying it, four of which are described briefly in the paragraphs below.

Cumulative percent distribution: This approach categorizes fisheries as small- or large-scale on a relative rather than absolute scale. There are three steps in this approach: (1) categorize the fisheries in the political entity being analyzed by gear types and vessel sizes; (2) list and rank gear/vessel combinations in ascending order according to annual catch or landed value per vessel; and (3) construct a cumulative percentage distribution of catch or landed value with the ranked fisheries, and the group of fisheries that provides the first 50% of landed value is then classified as "small-scale" and the remainder as "large-scale." The approach, first proposed by Ruttan et al. [2] has been applied in a number of countries/regions, including the North Atlantic [3], New England [4], Northeast of Brazil [5], and British Columbia, Canada [6].

Vessel length split: This approach simply chooses a vessel length that serves as the cutoff for determining what is small-scale and what is large-scale. In Atlantic Canada, for example, a vessel of 67 m length has been used as a cutoff with any vessel that is less than 67 m length classified as small-scale and vessels with >67 m length considered large-scale. This is generally a very high cutoff point for SSF. As stated earlier, the FAO uses 24 m length as the cutoff. Also, the European Maritime and Fisheries Fund [Regulation (CE) No 508/2014] considers vessels to be small-scale if their length is <12 m length and not using towed gear. Here SSF are typically "artisanal" and coastal, using small vessels, targeting multiple resource species using traditional gears [7].

Point-based framework: This method uses a numerical descriptors approach (NDA), which was originally proposed for the segmentation of European fishing fleets [8]. It is a score-based approach that uses several structural and functional descriptors. The approach identifies a group of relevant technical, biological, and economic descriptors that are used to categorize marine fishing fleets (Table 44.1). Each descriptor is awarded a score of between 1 and 5 according to predetermined ranges for each descriptor (Table 44.1). The scores for each fishery are then summed-up and if the total score is above a certain threshold then the fishery is considered to be artisanal or small-scale [7]. It should be noted that the NDA can be used at any geographical scale by adapting the final score and/or numerical ranges of each descriptor.

The degree of *small-scaleness*: A more recent attempt at providing a lens through which SSF can be distinguished from LSF, pioneered by our group at UBC, asked the question,

Table 44.1: Point-based framework for separating the pacific fleet. ITQ is Individual Transferable Quota.

	Points				
Features	**1**	**2**	**3**	**4**	**5**
Overall vessel length (m)	<35	35–45	45–65	65–100	>100
Type of gear		Passive		Seine/active	
Catch/vessel (t)	<100,000	100,000–300,000	300,000–700,000	700,000–1,000,000	>1,000,000
Crew numbers	<2	3	4	5	>6
Gross revenue per harvester ($)	<50,000	50,000–250,000	250,000–1,000,000	1,000,000–2,000,000	>2,000,000
License value ($)	<150,000	150,000–250,000	250,000–500,000	500,000–1,000,000	>1,000,000
Vessel replacement cost ($)	<137,000	137,000–218,000	218,000–500,000	500,000–750,000	>750,000
ITQ fishery	No		Partial		Full

A fishery will be assigned a point based on where it falls on the scale for each feature. The totals will range from 9 to 39. Everything scoring 20 and below points is considered SSF. Note that blanks appear in the gear and ITQ features as they are categorical and nonnumerical categories. *SSF*, Small-scale fisheries.

Data from D. Gibson, U.R. Sumaila, Socio-economic contribution of small-scale versus large-scale fisheries in British Columbia, in: OceanCanada Working Paper #2017-05, 2017 [9].

what is the degree of "small-scaleness" of a given fishery? [6]. The approach consist of the following steps: (1) identify features or characteristics associated with SSF that are widely accepted by academics and practitioners (Table 44.2); (2) determine how many fisheries are active in the country or region to be studied; (3) run all the fisheries identified through the features and characteristics of SSF to determine whether or not they have a given feature or not and give a score of, for example, 1 if it has the feature or 0 otherwise; and (4) add up the scores for each fishery to obtain the total score, which is then an indicator of "small-scaleness." Achieving a total maximum score implies the fishery in question is as small-scale as possible while a score of 0 implies the opposite. Scores between 0 and the maximum depicts the degree of "small-scaleness." An application of this approach found that commonly identified features of SSF are present in British Columbia's fishing fleets to a varying degree [6]. Aboriginal Food, Social, and Ceremonial fisheries and all commercial fisheries in British Columbia are analyzed to determine the presence or absence of each of these SSF features. The results of this research create a gradient of fisheries from smallest to largest scale, indicating that many fisheries in British Columbia can be classified as small-scale [6].

An advantage of the above approaches is that they are applied at the scale of political entities (e.g., countries or regions within countries), and therefore allows for the fact that gear that is large-scale in one political entity may be categorized as small-scale in another. Still, no approach has universal acceptance. For instance, the 2012 Hidden Harvest study used context-specific definitions for SSF (e.g., dated or low levels of technology;

Table 44.2: List of common features of small-scale fisheries identified by Gibson and Sumaila [6].

Vessel features	Economic features	Social features
Vessel under 12 m (39.3 ft.)	Low fuel consumption (e.g., <$10,000)	Fish for food and community use
Nonmotorized vessel	Relatively little capital and energy input (e.g., <$250,000)	Support social and cultural values
Passive gear	Relatively low yield and income	Regulated through customary rules with some government involvement
Multigear	Part-time, seasonal, multioccupational	
Multispecies	Sold in local markets	
Dated or low levels of technology, labor intensive (Labour intensity in qualitative terms, not as quantitative measure of labour in proportion to capital.)	Sustain local or regional economies	
Inshore, limited range to fish, fishing pressure adjacent to community	Individual or community ownership	

labor intensive) [10]. In practice therefore country-based definitions are what is currently used in fisheries economics.

Notwithstanding the absence of an agreed international definition of small-scale fishing, there are some elements—referred to in the work above—that are widely accepted by practitioners to be features of small-scale fishing. These are set out in Table 44.2.

44.3 Share of fisheries catch, revenues, and subsidies to small-scale fisheries

Work by the *Sea Around Us* project has estimated the amount of catch, revenues, and subsidies provided to large- and small-scale fleets around the world.

For the analysis of catch proportion to SSF and LSF, we relied on the approach developed by the *Sea Around Us* [11], with SSF consisting of subsistence, artisanal, and recreational fisheries. On the other hand, LSF are made up of all the industrial fleets in a country. It should be noted that the definitions of each of these sectors are based on national categorizations sourced from national legislation reviewed by the researchers. The catch data used are 10-year averages for the period 2005–14.

In the case of landed value, we combined the catches from the *Sea Around Us* with the ex-vessel price database developed by the Fisheries Economics Research Unit in collaboration with the *Sea Around Us* [12–14]. Similar to the catch data, we used 10-year averages for the period 2005–14.

The starting point for splitting national subsidies into the proportion that goes to SSF is the country-level fisheries subsidies database reported in Sumaila et al. [15]. Of the 146 maritime countries that are included in the database, subsidies in 81 countries were analyzed by Schuhbauer et al. [16], selected based on data availability and the total amount of subsidies they provide globally. In all, these 81 countries gave 98% of the estimated US $35 billion annual global fisheries subsidies.

Details of the methodology used to split national subsidies estimates into the portion that goes to SSF are provided in Schuhbauer et al. [16]. For each subsidy subtype the collected information that was found in the literature was grouped into three data categories as illustrated in Fig. 44.1.

Group 1: Is there quantitative data available? If yes, then the indicated subsidy quantity to SSF is recorded (Fig. 44.1). Group 2: Is qualitative data available? If yes, use the qualitative information to estimate the amount of subsidies provided to SSF (Fig. 44.1) [16]. Qualitative data is often found in government documents or technical reports, in the form of bullet points and tables, which are broken down into objectives. If a subsidy

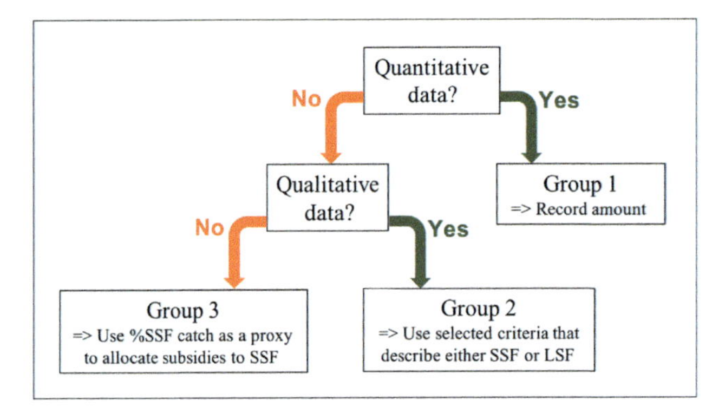

Figure 44.1

Illustrating the methodology used to divide 2009 subsidy amounts into SSF and LSF. *LSF*, Large-scale fisheries; *SSF*, small-scale fisheries. *Adapted from A. Schuhbauer, R. Chuenpagdee, W.W.L. Cheung, K. Greer, U.R. Sumaila, How subsidies affect the economic viability of small-scale fisheries, Mar. Policy 82 (2017) 114–121.*

amount was described by more than one objective/bullet point, we split the total subsidy equally between the stated objectives (see example in Fig. 44.1) [16]. To be consistent, the use of the following words describe SSF: artisanal, subsistence, small-scale, non-motorized, coastal, and community-based; and for LSF: industrial, large-scale, freezer trawlers, offshore, over sea, and deep sea [16].

Fig. 44.2 displays the proportions of catch, landed value, and subsidies that go to SSF versus those that go to their large-scale counterparts. The figure reveals that globally SSF caught 27.5 million tons, which was about 25% of the average total annual catch of 112 million tons from 2005 to 2014.

Of the estimated average annual landed value of US$164 billion generated in the period from 2005 to 2014, US$51 billion or 31% of the total was generated by the small-scale sector (Fig. 44.2).

Annually, fishing subsidies of US$35 billion (using 2009 estimate as an example) are given by governments worldwide to the fisheries sector. It is shown in Fig. 44.1 that only US$5.6 billion (i.e., 16% of the total) goes to SSF. The disparity between the small-scale and large-scale is even worse when one looks at the capacity-enhancing subsidies, where 90% of the nearly US$20 billion is estimated to go to LSF [13]. The largest subsidies are that for fuel, over 90% of which are estimated to be given to LSF through marine diesel subsidies, which are mostly out of the reach of small-scale fishers because of the high cost of purchasing and maintaining diesel motors [13]. These fuel subsidies promote fuel-inefficient technology and help large-scale fishers stay in business even when operating costs exceed total revenue gained from fishing. Subsidies for port development and boat construction, renewal, and

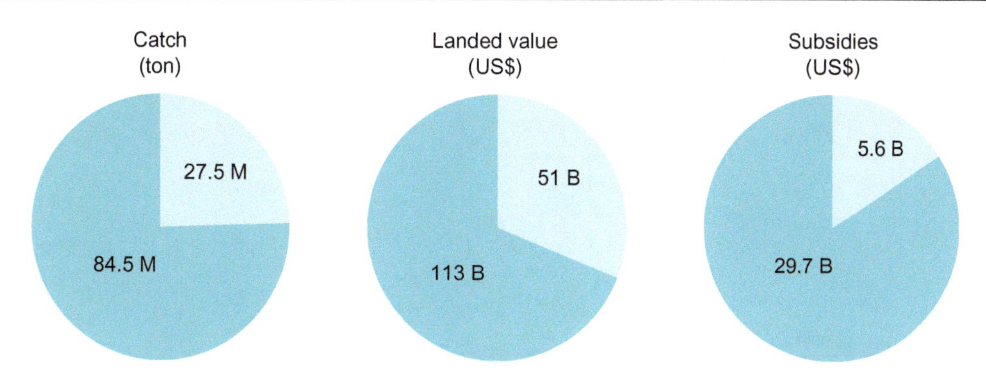

Figure 44.2
Share of annual average catch, landed value, and subsidies that goes to small-scale versus large-scale fisheries globally. Light blue is small-scale and dark blue is large-scale fisheries. "M" and "B" stand for "million" and "billion," respectively. *Adapted from data in A. Schuhbauer, R. Chuenpagdee, W.W.L. Cheung, K. Greer, U.R. Sumaila, How subsidies affect the economic viability of small-scale fisheries, Mar. Policy 82 (2017) 114–121*

modernization are also likely to give the LSF sector a huge advantage over their small-scale counterparts, who appear to receive only a small percentage of those subsidies.

Regionally, we see similar patterns to the global picture with the bias against SSF being stronger in some regions (Fig. 44.3A and B).

Interesting observations from Fig. 44.3 are (1) for all regions of the world, the share of landed values generated by SSF are larger than the share of their catch, implying that, on a per unit weight basis, their catch is more valuable than those of LSF on average; (2) the proportion of total subsidies given to SSF is lower than the proportion of the landed values they generate. This is highlighted more intensely in Oceania for reasons that are not obvious to the author; and (3) LSF receive approximately four times more subsidies than their small-scale counterparts, with up to 60% of those subsidies promoting overfishing [13].

44.4 Implications of using a common definition or relying on national definitions of small-scale fishing for the purpose of subsidy disciplines

In principle the scope and application of a multilateral agreement on fisheries subsidies including exceptions for subsidies to small-scale fishing would be clearer if the agreement contained a common universally accepted definition of SSF. There are, it seems, some descriptive elements that are frequently used to identify small-scale fishing in different contexts that could conceivably be used to generate a common definition in a subsidies

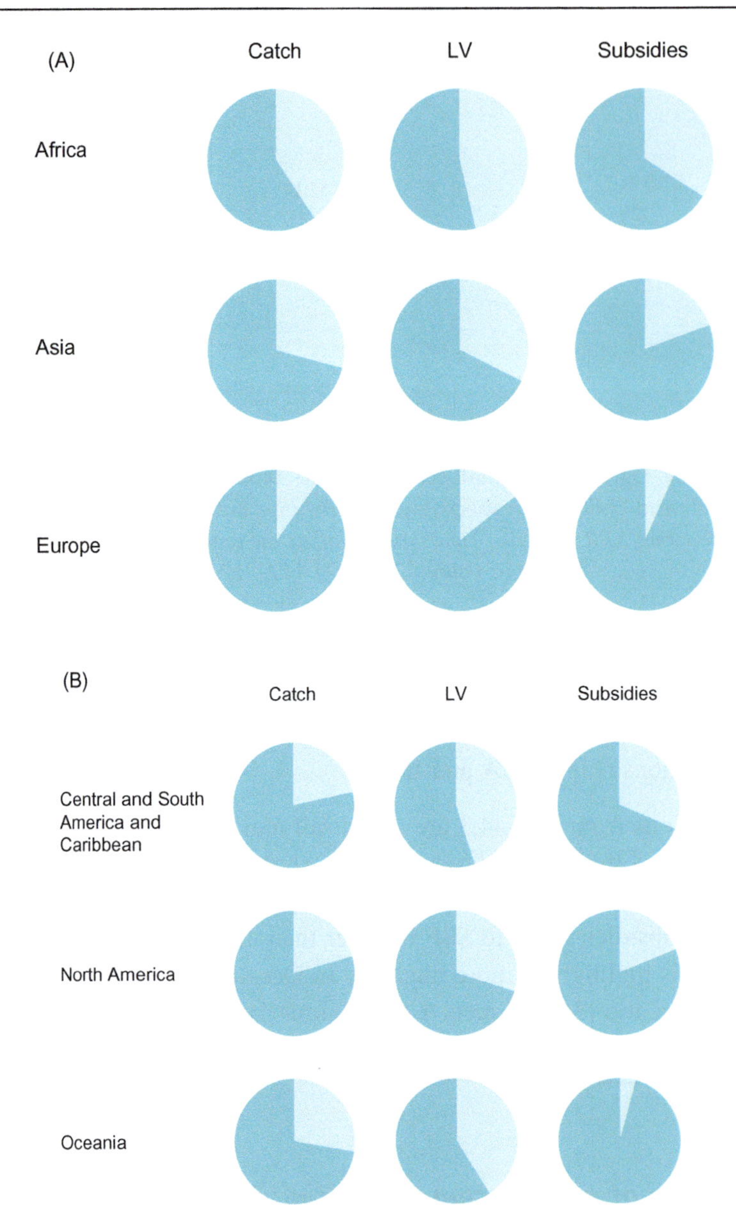

Figure 44.3

(A) Share of annual average catch, LV, and subsidies that go to small-scale versus large-scale fisheries in Africa, Asia, and Europe. Light blue is small-scale and dark blue is large-scale fisheries. (B) Share of annual average catch, LV, and subsidies that go to small-scale versus large-scale fisheries in Central and South America, North America, and Oceania. Light blue is small-scale and dark blue is large-scale fisheries. *LV*, Landed value. *Adapted from data in A. Schuhbauer, R. Chuenpagdee, W.W.L. Cheung, K. Greer and U.R. Sumaila, How subsidies affect the economic viability of small-scale fisheries, Mar. Policy 82, 2017, 114–121* agreement. However, the reality in the field is that SSF are very diverse even within a country, as well as globally, making it really difficult to develop and apply a single definition worldwide. Further, if consensus over a definition of small-scale fishing has eluded fisheries management experts, it is probably neither appropriate nor practical for governments to try to develop an agreed definition of small-scale fishing in the context of negotiations over subsidies in the WTO.

Using national definitions to define the scope of exceptions in a subsidy agreement might therefore seem more feasible compared to a global definition. It should be acknowledged, though, that this approach could introduce a lot of flexibility into the disciplines, as governments could (legitimately in that circumstance) choose to include activity of considerable scale in their national definitions of what is small-scale fishing and therefore, if a WTO agreement included exceptions for this fishing, could be eligible for subsidization. A wide degree of flexibility could potentially undermine the effectiveness of an agreement in supporting the reform of subsidy patterns.

Negotiators face a dilemma: a common definition would be difficult for countries to accept as it may not capture the place and community aspects of their fisheries. Adopt national definitions and you may end up of with potentially very wide exceptions that could reduce the effectiveness of subsidies disciplines.

If there were agreement that some flexibility should be provided for countries to continue to subsidize SSF, negotiators could consider establishing the scope of these exceptions by combining a degree of flexibility with reference to generally accepted concepts of small-scaleness. For example:

- Apply national definitions that can capture the differences in SSF found in different countries, but include reference in the agreement to an illustrative list of features that are commonly accepted features of SSF. Negotiators could also consider ensure there is a minimum level of transparency that would allow subsidy disciplines to be effectively monitored. This could include requiring notification of the subsidies provided to the small-scale sector, along with the national definition of what is small-scale, to the WTO Subsidies and Countervailing Measures (SCM) Committee. To help countries implement this obligation, the WTO's SCM Committee could seek the advice of a group of experts and practitioners who can work with countries to help define their SSF in a manner that captures key local aspects of the sector without being too flexible for tackling harmful subsidies.

44.5 Conclusion

Currently, national definitions of SSF are used in international guidelines, academic literature, and in fisheries management practice. This is because a universally accepted common global definition of SSF is not yet in place. It is probably neither realistic nor appropriate for the WTO to seek to agree on a new global definition of small-scale fishing. If there were agreement in the WTO to applying different rules to subsidies to small-scale fishing, rather than try to negotiate a new international definition, governments could consider agreeing to use national definitions of small-scale fishing for the purpose of subsidy rules, along with an illustrative list of features commonly accepted as describing small-scale fishing. Governments could also consider requiring that national definitions, and subsidies provided to fisheries meeting these definitions, were notified to the SCM Committee.

References

[1] FAO, Voluntary Guidelines for Securing Sustainable Small-Scale Fisheries in the Context of Food Security and Poverty Eradication, FAO, Rome, 2015.
[2] L.M. Ruttan, F.C.J. Gayanilo, U.R. Sumaila, D. Pauly, Small versus large-scale fisheries: a multi-species, multi-fleet model for evaluating their interactions and potential benefits, in: D. Pauly, T.J. Pitcher (Eds.), Methods for Evaluating the Impacts of Fisheries on North Atlantic Ecosystems, University of British Columbia, Vancouver, BC, 2000, pp. 64–78.
[3] U.R. Sumaila, Y. Liu, P. Tyedmers, Small versus large-scale fishing operations in the North Atlantic, in: T.J. Pitcher, U.R. Sumaila, D. Pauly (Eds.), Fisheries Impacts on North Atlantic Ecosystems: Evaluations and Policy Exploration, University of British Columbia, Vancouver, BC, 2001, pp. 28–35.
[4] N.O. Therkildsen, Small- versus large-scale fishing operations in New England, USA, Fish. Res. 83 (2) (2007) 285–296.
[5] L.D.M.A. Damasio, P.F.M. Lopes, M.G. Pennino, A.R. Carvalho, U.R. Sumaila, Size matters: fishing less and yielding more in smaller-scale fisheries, ICES J. Mar. Sci. 73 (2016) 1494–1502. Available from: https://doi.org/10.1093/icesjms/fsw016.
[6] D. Gibson, U.R. Sumaila, How small-scale are fisheries in British Columbia? in: OceanCanada Working Paper #2017-03, 2017.
[7] S. Villasante, et al., Fishers' perceptions about the EU discards policy and its economic impact on small-scale fisheries in Galicia (North West Spain), Ecol. Econ. 130 (2016) 130–138.
[8] L. García-Flórez, J. Morales, M.B. Gaspar, D. Castilla, E. Mugerza, P. Berthou, et al., A novel and simple approach to define artisanal fisheries in Europe, Mar. Policy 44 (2014) 152–159.
[9] D. Gibson, U.R. Sumaila, Socio-economic contribution of small-scale versus large-scale fisheries in British Columbia, in: OceanCanada Working Paper #2017-05, 2017.
[10] World Bank, FAO, and WorldFish Center, Hidden harvests: the global contribution of capture fisheries, in: Report No. 66469-GLB, World Bank, Washington, DC, 2012.
[11] D. Pauly, D. Zeller, Catch reconstructions reveal that global marine fisheries catches are higher than reported and declining, Nat. Commun. 7 (2016) 10244.
[12] U.R. Sumaila, A.D. Marsden, R. Watson, D. Pauly, A global ex-vessel fish price database: construction and applications, J. Bioecon. 9 (2007) 39–51.
[13] W. Swartz, U.R. Sumaila, R. Watson, Global ex-vessel fish price database revisited: a new approach for estimating 'missing' prices, Environ. Resour. Econ. 56 (2013) 467–480.
[14] T.C. Tai, T. Cashion, V.W. Lam, W. Swartz, U.R. Sumaila, Ex-vessel fish price database: disaggregating prices for low-priced species from reduction fisheries, Front. Mar. Sci. 4 (2017) 363.
[15] U.R. Sumaila, V. Lam, F. Le Manach, W. Swartz, D. Pauly, Global fisheries subsidies: an updated estimate, Mar. Policy 69 (2016) 189–193.
[16] A. Schuhbauer, R. Chuenpagdee, W.W.L. Cheung, K. Greer, U.R. Sumaila, How subsidies affect the economic viability of small-scale fisheries, Mar. Policy 82 (2017) 114–121.

CHAPTER 45

The trouble with tunas: international fisheries science and policy in an uncertain future

Andre Boustany[1,2]

[1]Principal Investigator, Fisheries, Monterey Bay Aquarium, CA, United Sates [2]Nicholas School of the Environment, Duke University, NC, United States

Chapter Outline

Management of fish populations has always been difficult due to the need to coordinate among various users, oftentimes with incomplete data or uncertain science on even the most basic biology of the fish species being managed. These problems are compounded when managing highly migratory fish species that cross international borders such as tuna, billfish, and many shark species. In these cases, it is necessary to harmonize management among a number of nations and fishing entities that may have competing views on the ultimate goals of fishery management and varying capacity or will to regulate their own fisheries or collect the necessary scientific data. Under these scenarios, there is a tendency to devolve toward the least common denominator in both the science and management of shared fish stocks. Bad or incomplete data on catch, fishing effort or basic fish biology by one fishing entity has the potential to make scientific models assessing the health of fish stocks unreliable, even if all the other fishing entities provide complete and accurate data. Likewise, with management and enforcement of fishing regulations, even one nation falling short will set up a situation where other nations feel less compelled to follow regulations, lest their conservation efforts be wasted and their foregone catch be harvested by fishers elsewhere. Added to that, international trade in the managed fish species means that seafood products from one nation compete in the same markets as product from other nations, oftentimes with an inability to distinguish those that were harvested responsibly from those that were not.

International management of highly migratory and shared fish populations is conducted through regional fisheries management organizations (RFMOs) [1]. These organizations comprise fishing entities, usually nations, that come together and agree to jointly manage the

Predicting Future Oceans.
DOI: https://doi.org/10.1016/B978-0-12-817945-1.00044-7

species and fisheries within the region of the RFMO. These nations are bound by the charter of the RFMO as well as by any other marine governance institutions, such as the United Nations Convention on the Law of the Sea (UNCLOS) and the UN Fish Stocks Agreement (UNFSA), to which the nations are signatories. All these agreements mandate that fisheries should strive toward sustainable, long-term yield, employing the precautionary approach while also protecting the environment [2,3]. However, several aspects of the structure and function of RFMOs limit their ability to effectively manage shared, highly migratory fish stocks. The fundamental characteristic of RFMOs that impacts their effectiveness is that membership in them is voluntary [4]. While the voluntary nature of membership in and of itself does not limit the ability of RFMOs to manage shared fish stocks, it creates a situation where regulations within the RFMO must be kept lax enough to encourage all fishing and market states to voluntary agree to participate and be regulated by the RFMO. No international fisheries management organization could be effective if significant portions of the fishing and market existed outside of its structure, and encouraging buy-in of all stakeholders is, therefore, paramount. The principal examples of these lenient regulations are in how RFMOs deal with enforcement of regulations and in the amount of leeway they give member states to voluntarily opt out of regulations. RFMOs have limited centralized ability to enforce agreed upon regulations, leaving it to member states to implement regulations within their own fleets and markets [5]. Additionally, RFMOs broadly allow member states to take a reservation (opt out) of any regulations to which they do not wish to abide [6]. Both these features are to encourage participation by states who may otherwise balk at signing up to an agreement that would compromise their sovereignty to manage their own industries.

One of the consequences of voluntary participation within RFMOs is they end up being composed only of nations that have a vested economic interest in the fish species being managed, either as fishing entities or as market states. As such, all decisions made by these nations through the RFMO carry direct economic consequences that may influence their decisions on harvest levels, fleet capacity, and other management measures. Contrast this to the Convention on International Trade in Endangered Species of Wild Fauna and Flora (CITES), which currently has 183 members [7]. These member states all have a say in regulating the trade of animal and plant species, regardless of whether or not they are a harvesting or end market state, which provides a forum for dispassionate voices whose decisions are not clouded by their own economic concerns.

Finally, the effectiveness of RFMOs is limited in that there exist few mechanisms for oversight of the decisions made by the RFMO. While RFMOs and member states are bound by the principles of the UNFSA and UNCLOS, there are few options for challenging RFMO decisions that are not in keeping with these measures. There is some ability to challenge the actions of individual member states through the International Tribunal for the Law of the Sea, but such actions are rarely used and implementation can drag on for years [8]. And, this avenue does not regulate the RFMO decisions directly, rather it has been applied to individual member states whose actions have been operating outside the bounds of the RFMO and/or other regulatory frameworks.

To examine how this lack of oversight affects the ability of outside interests to influence the adoption of resolutions at tuna RFMOs, I look at the 2018 Atlantic bigeye tuna assessment by the International Commission for the Conservation of Atlantic Tunas (ICCAT), the RFMO responsible for the management of highly migratory pelagic fish species in the Atlantic Ocean. The assessment for bigeye tuna used three different modeling frameworks, and the main assessment model was run 18 different ways with varying input parameters to account for uncertainty in any of these inputs [9]. All model types and all model runs came to the same conclusions; the population of bigeye tuna in the Atlantic Ocean was overfished, overfishing was still occurring, and catches would need to be curtailed considerably for the population to recover to sustainable levels. The ICCAT delegates were unable to agree to any reductions in catch, despite it being abundantly clear this course of inaction would be incongruent with obligations of ICCAT, UNCLOS, and UNFSA. This can be contrasted with the decisions at ICCAT on East Atlantic bluefin tuna quotas in 2017. In this case, due to lack of reporting by fishing nations and large changes to the operation of fisheries, the stock assessment models and individual model runs gave wildly different results for the state of the population, indicating estimates for the population were highly unreliable and precaution should be employed when setting catch limits. The Commission, however, decided to raise the total catch limits by >50% over the next 3 years [10]. Due to the lack of credible oversight of the Commission decisions, there is little to no ability to challenge either of these decisions outside of ICCAT. The contrasting paths ICCAT took with East Atlantic bluefin tuna and bigeye tuna highlight the ease with which quotas can be raised with the difficulty in their lowering. This has the potential to set up a dangerous ratchet effect that can create permanent headwinds to creating and maintaining sustainable fisheries.

These RFMO structural deficiencies can be compared to other, more effective regimes. As an example of a more effective fisheries management structure, I briefly identify the salient part of US fisheries management, primarily governed by the Magnuson–Stevens Act [11]. While far from perfect, the United States generally ranks high on effectiveness of fisheries management compared to other nations [12]. US fisheries are primarily managed by stakeholders of the fisheries through eight Regional Fisheries Management Councils (RFMCs). These Councils develop fisheries management plans (FMPs), which govern all aspects of the fisheries they manage, including catch quotas, fishing seasons, gear restrictions, etc. Similar to management at RFMOs, the Councils are made up of stakeholders in the fisheries they manage. In the case of RFMCs the stakeholders have tended to be heavily skewed toward extractive industry, with >80% of Council representatives coming from the fishing industry [13]. This sets up a situation similar to that for RFMO management, where the path of least resistance is to take on higher risk, increase (or not decrease) catch quotas, and maximize immediate profits for the industry. However, in the case of US management all council FMPs must ultimately be vetted and accepted by federal fisheries managers who can refuse any FMP that does not meet the legal standards. While there is always the possibility to put political pressure on federal managers to accept FMPs that allow higher risk or that fall short of some aspect of US fisheries law, the federal oversight provides a check on the decisions generated by the RFMCs.

In addition to the oversight duties provided by federal fisheries managers, parties can further challenge any accepted FMP through the US federal court system if they believe the plan falls short of legal requirements [14]. This second layer of review and oversight, through a process that is even more removed from the stakeholders that comprise the RFMC, allows for a greater voice by non-industry stakeholders, such as environmental organizations. In contrast, the RFMOs are generally lacking in either of these two levels of oversight. Also in contrast to the RFMO management structure, in the US membership within the regulatory framework is not voluntary and there is no ability for stakeholders to opt out of regulations.

Given that we know what structures allow for successful management of marine fisheries, what are the possibilities we can implement similar systems for international fisheries? While redesigning the entire make up of international fisheries governance seems highly unlikely, several cases exist where it has been attempted to implement these two effective principles (bringing nonfishing nations into the decision-making process and implementing additional layers of oversight). Attempts to include nations without a potential industry bias into the RFMO decision-making process have been made in the International Whaling Commission (IWC), an RFMO that was implemented to manage whale fisheries in 1946 [15]. The IWC was implemented and structured in a way similar to other RFMOs, providing countries with whale fisheries a venue to collectively manage harvest for long-term sustainability. Originally established by 15 countries, a three-fourths majority is necessary to implement regulatory changes. While this is a complex legal and sociological case that has unfolded over decades, at issue here is how non-whaling countries were encouraged by conservation minded entities to join the IWC and take a position in regulating whale fisheries for which they did not have a direct stake. In essence, this broadened the regulatory decision-making process to nations who did not have a direct economic interest in commercial whaling. This ultimately led to 89 countries joining the IWC and a moratorium on commercial whaling since 1982. While this course of action did have the effect of instituting greater conservation measures into the IWC, it ultimately resulted in a breakdown of the regulatory framework as nations that wished to harvest whales found the Commission unresponsive to their wishes, and they effectively removed themselves from the organizational restrictions on whaling [16]. Iceland and Norway accomplished this through taking reservations to the moratorium and Japan through the practice of scientific whaling and ultimately by announcing their withdrawal from the IWC completely [17].

Likewise, efforts have been made to implement additional levels of oversight of fisheries management and trade in fisheries product outside of the RFMO framework. For years the Eastern Atlantic stock of bluefin tuna was overfished, with ICCAT having little ability to either adopt sustainable fishing regulations or to enforce existing regulations. Again this was due mainly to the majority of ICCAT contracting parties having an economic stake in these fisheries and being unwilling to take the short-term economic hit to their domestic fisheries. With an inability to convince ICCAT nations to adopt more conservation-oriented practices, a move was made in 2010 by nations without fisheries for East Atlantic bluefin tuna to list the species under CITES [18]. While this would not have moved management out of the RFMO structure, it would have provided another level of oversight of the species by implementing additional requirements on the international trade of Atlantic bluefin tuna products. As CITES has 183 current members, this oversight would have effectively come from the global community as a whole. Since a two-thirds voting majority is needed to list a species under CITES and bluefin fishing and market states lobbied hard not to list, the proposal to list Atlantic bluefin tuna was not adopted [18]. Even if the proposal to list Atlantic bluefin had been adopted, CITES also contains an option to take reservations and it is unclear if fishing and market nations would have simply decided to opt out of any additional reporting and trade requirements.

The difficulties in managing internationally shared fish populations will likely be exacerbated as we enter into an era of climate change, which is expected to have a wide range of effects on the ocean environment and on fish. Oceans are predicted to become warmer [19], less productive [20], more acidic, less oxygenated [21], and more variable [22]. The first four of these are thought to have a negative effect on growth of fish and fish populations [23], necessitating decreases to catch rates by management bodies. The increase in oceanic variability will act to reduce the predictive power of stock assessment models, calling for an increased application of the precautionary approach as the scope of potential outcomes increases.

Given the stated deficiencies in RFMOs' ability or willingness to enact precautionary decreases in catch rate and the discrepancy in ease of raising versus lowering catches, climate change has the potential to exacerbate overfishing in internationally managed fish stocks. With limited ability to significantly change the current structure of RFMOs, what options remain for ensuring sustainable internationally shared fish populations? One option available within the RFMO framework is the use of target and limit reference points (see Chapter 41: BBNJ and the open ocean) [24]. These are conservation measures that are agreed to by fishing nations beforehand that are implemented automatically when given criteria are met. For example, if the population of the managed fish species falls below a given level, fishing quotas would automatically decrease by half. The benefit here is that conservation measures are easier to implement when they are viewed in the abstract than when RFMO delegates can see their direct economic implications.

An option to increase conservation measures outside of the RFMOs themselves is the creation of novel regulatory structures that have some ability to influence the regulation of fisheries. Any such structures would need to include fishing and nonfishing nations, and would need to be at a level superseding RFMOs, providing oversight to them. One novel framework that meets these requirements is the United Nations resolution on the conservation and sustainable use of marine biological diversity of areas beyond national jurisdiction (BBNJ), created through the auspices of the Convention on the Law of Sea and the UN General Assembly (UNGA) [25]. As a UNGA resolution this would include a broad array of nations, not only those who harvest fish from the high seas. And, as high seas fish species are by definition biodiversity beyond national jurisdiction and high seas fisheries have been shown to affect biodiversity in these species, it would appear as though this UN resolution would be able to influence the management of highly migratory, shared fish stocks. However, to date these efforts have explicitly excluded fisheries from the purview of the BBNJ process.

Another pathway for influencing the sustainability of highly migratory fish stocks lies not through regulatory oversight, but through market forces. Seafood certification, labeling, and ranking schemes have developed in recent decades as a way to inform fish buyers. Under such a system, independent organizations not affiliated with the seafood harvesting industry identify seafood products that have been harvested sustainably from healthy populations from those which have been harvested unsustainably. This identification allows ecologically minded buyers and consumers of seafood to refrain from purchasing products that have been identified as coming from non-sustainable sources. As purchases of sustainable seafood increase, products from these sources can charge a premium relative to that from unsustainably harvested products, setting up a market incentive for fisheries to follow responsible practices and harvest levels. The great advantage to this strategy is it is not dependent on any of the structures of the current RFMO-based management framework and all the limitations outlined above. The main constraint of the market-based system lies in the ability to differentiate sustainably harvested seafood from competing products in the market, a non-trivial task. While no solution promises to be a panacea for overfishing in internationally shared fish stocks, there are pathways toward sustainability within and outside the current management structure for highly migratory fish species.

References

[1] United Nations, United Nations conference on straddling fish stocks and highly migratory fish stocks, in: Agreement for the Implementation of the Provisions of the United Nations Convention on the Law of the Sea of 10 Dec 1982 Relating to the Conservation and Management of Straddling Fish Stocks and Highly Migratory Fish Stocks, UN Doc. A/Conf./164/37, 1995.

[2] S.M. Garcia, The precautionary approach to fisheries and its implications for fishery research, technology and management: an updated review, FAO Technical Paper, 1995, p. 350.

[3] United Nations, United Nations Convention on the law of the sea, in: UN Doc. A/Conf.62/122, 1982.

[4] R. Rayfuse, Regional fisheries management organizations, in: The Oxford Handbook of the Law of the Sea, 2015.
[5] Z. Tyler, Saving fisheries on the high seas: the use of trade sanctions to force compliance with multilateral fisheries agreements, Tulane Environ. Law J. 20 (1) (2006) 43–95.
[6] H.S. Schiffman, Reservations in marine environmental treaties: practical observations and legal limitations, Whittier L. Rev. 26 (2004) 1003.
[7] Convention on International Trade in Endangered Species of Wild Fauna and Flora (CITES), List of Parties to the Convention. Retrieved from: <https://www.cites.org/eng/disc/parties/index.php>.
[8] S. Marr, The southern bluefin tuna cases: the precautionary approach and conservation and management of fish resources, Eur. J. Int. Law 11 (4) (2000) 815–831.
[9] SCRS, Report of the Standing Committee on Research and Statistics (SCRS) (Madrid, Spain 1–5 October 2018), October 2018.
[10] ICCAT, Recommendation by ICCAT Amending the Recommendation 14-04 on Bluefin Tuna in the Eastern Atlantic and Mediterranean, BFT10-07, 2017.
[11] A.C.T.A., Magnuson-Stevens fishery conservation and management act, Public Law 94 (1996) 265.
[12] J. Alder, S. Cullis-Suzuki, V. Karpouzi, K. Kaschner, S. Mondoux, W. Swartz, et al., Aggregate performance in managing marine ecosystems of 53 maritime countries, Mar. Policy 34 (3) (2010) 468–476.
[13] T.A. Okey, Membership of the eight Regional Fishery Management Councils in the United States: are special interests over-represented? Mar. Policy 27 (3) (2003) 193–206.
[14] National Research Council, Evaluating the Effectiveness of Fish Stock Rebuilding Plans in the United States., The National Academies Press, Washington, DC, 2014. Available from: https://doi.org/10.17226/18488.
[15] International Convention for the Regulation of Whaling, Dec. 2, 1946, 62Stat. 1716, 161 U.N.T.S. 72.
[16] D.D. Caron, The international whaling commission and the North Atlantic Marine Mammal commission: The Institutional Risks of Coercion in Consensual Structures, Am. J. Int. Law, 89, 1995, pp. 154–174.
[17] The Guardian, Japan to Resume Commercial Whaling After Leaving IWC, 19 Dec. 2018. Retrieved from: <https://www.theguardian.com/environment/2018/dec/20/japan-to-resume-commercial-whaling-after-leaving-iwc-report>.
[18] A. Nayar, Bad news for tuna is bad news for CITES, Nat. News Mar. 23 (2010).
[19] S. Levitus, J.I. Antonov, T.P. Boyer, C. Stephens, Warming of the world ocean, Science 287 (5461) (2000) 2225–2229.
[20] J.J. Polovina, E.A. Howell, M. Abecassis, Ocean's least productive waters are expanding, Geophys. Res. Lett. 35 (3) (2008) 1–5.
[21] O. Hoegh-Guldberg, J.F. Bruno, The impact of climate change on the world's marine ecosystems, Science 328 (5985) (2010) 1523–1528.
[22] J.A. McGowan, D.R. Cayan, L.M. Dorman, Climate-ocean variability and ecosystem response in the Northeast Pacific, Science 281 (5374) (1998) 210–217.
[23] W.W. Cheung, J. Dunne, J.L. Sarmiento, D. Pauly, Integrating ecophysiology and plankton dynamics into projected maximum fisheries catch potential under climate change in the Northeast Atlantic, ICES J. Mar. Sci. 68 (6) (2011) 1008–1018.
[24] D.C. Dunn, G. Ortuño Crespo, P.N. Halpin, Incorporating the dynamic and connected nature of the open ocean into governance of marine biodiversity beyond national jurisdiction, in: W.W.L. Cheung, A.M. Cisneros-Montemayor, Y. Ota (Eds.), Predicting Future Oceans, Elsevier, 2019.
[25] J. Caddy, Checks and balances in the management of marine fish stocks: organizational requirements for a limit reference point approach, Fish. Res. 30 (1–2) (1997) 1–5.

CHAPTER 46

The road to implementing an ecosystem-based approach to high seas fisheries management

Guillermo Ortuño Crespo
Nicholas School of the Environment, Duke University, Beaufort, NC, United States

Chapter Outline

> Animals living in the waters, especially the sea waters, are protected from the destruction of their species by man. Their multiplication is so rapid and their means of evading pursuit or traps is so great, that there is no likelihood of his being able to destroy the entire species of any of these animals.
>
> —***Jean-Baptiste Lamarck (1809).***

> When everything is connected to everything else, for better or worse, everything matters.
>
> —***Bruce Mau.***

46.1 Introduction

Over the last century, the ecological footprint of *Homo sapiens* has expanded and intensified faster than ever before, transforming from a terrestrial and coastal hunter-gatherer species, to a both a terrestrial and marine "top predator" [1,2]. In oceanic systems, the extractive industries for biotic and abiotic resources (e.g., deep-sea mining or fishing) have ventured deeper and farther from shore than ever before [3,4]. While coastal and

Predicting Future Oceans.
DOI: https://doi.org/10.1016/B978-0-12-817945-1.00045-9

archipelagic states have the legal obligation under international law [5] to protect and preserve the marine environment within their 200 nautical mile exclusive economic zone (EEZ), no single nation has the legal responsibility of managing activities in the areas beyond national jurisdiction (ABNJ; or high seas); an area where marine fisheries have expanded significantly over the last 70 years. The socioeconomic relevance of fisheries for job security, food security, and other ecosystem services worldwide makes their sustainability and persistence imperative for achieving several of the Sustainable Development Goals set by the United Nations in 2015.

As we continue learning about the ecological ramifications of these emerging open ocean uses, the question remains as to whether we are managing and monitoring these activities with the necessary scrutiny to ensure the long-term health and stability of the global ocean and the ecosystem services that it provides. The call for a more ecologically holistic approach to management, particularly that of fisheries, has been growing over the last few decades [6]. This call stems from our increased understanding of the complexity and connectivity of marine biological communities and ecosystems and has paved the way for a series of policy and management measures that seek to understand and abate preventable or irreversible impacts on marine ecosystems, particularly those that are caused by anthropogenic forcings. Marine ecosystems can be viewed as "ecological engines," where you run the risk of engine failure if too many elements from said engine are removed or damaged; thus impeding the delivery of—ecosystem—services (e.g., seafood provisioning, carbon sequestration, protection against coastal erosion, or recreational and spiritual benefits) that would benefit society. Humanity now faces the ecological, social, and political challenges to accommodate a growing demand and catch of oceanic species [7] without permanently removing too many elements from said marine "ecological engine"; this is a particular challenge for oceanic ecosystems, which are large, dynamic, and poorly understood.

This chapter will explore oceanic fisheries through the lens of international governance, fisheries management, and pelagic ecology, in order to assess *if* and *how* the international community is implementing a more holistic ecosystem approach to fisheries (EAF) [also known as ecosystem-based fisheries management (EBFM)] in international waters. To this end this chapter commences by (1) briefly elaborating on the current definitions of EAF and EBFM; (2) explaining the expansion of oceanic fisheries over the last 70 years; (3) outlining the known ecological impacts of marine fisheries in the open ocean; (4) identifying the main conventions and international governance bodies tasked with managing not only resources of economic interest, but the entirety of the marine environment; (5) examining the existing monitoring and management initiatives to abate these impacts, as well as identifying existing data/knowledge gaps; and, finally, (6) envisaging the road ahead for the international community as it works toward an EAF in the high seas.

46.1.1 Expansion of marine fisheries

Since the mid-20th century, increases in seafood demand and advancements in fleet technology and capacity have catalyzed the range and capabilities of commercial fisheries, expanding from shallow coastal waters into the pelagic realm [4]. While the extractive capacity of high seas fisheries has increased almost uninterruptedly since the 1950s (Fig. 46.1), the relative contribution of high seas seafood toward global food security is still low [8]. Much of the fisheries landings from ABNJ are destined to fuel the sushi, sashimi, and canned tuna industries, which boomed in the 1960s and 1970s, but at what cost to the other components of oceanic ecosystems? As evidence mounts on the potential negative ecological impacts of high seas fisheries [9,10] and as the success of the current management strategies is put into question [11,12], some have gone as far as suggesting that fisheries in the high seas should be prohibited [13].

Besides landing estimates, there are various other ways of assimilating the magnitude of this expansion. For instance, according to the *Sea Around Us*, in 1950 a total 45 fishing states or territories fished in the high seas; by 2010, the number of states and territories had climbed to 132, although the ownership of these vessels is hard to assign given the widespread use of flags of convenience. In terms of kilowatts, Merrie et al. show there is an almost exponential increase in fishing effort from 1950 to 2006 [3]. Other ways in which the expansion of extractive power in the open ocean can be conceptualized is by looking at

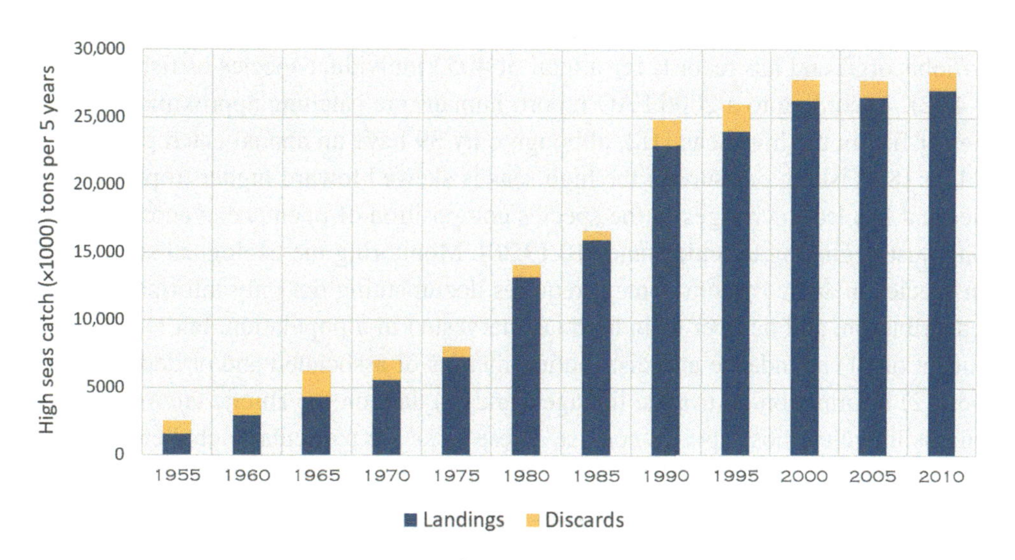

Figure 46.1
The trajectory of high seas fisheries landings and discards from 1950 to 2010 aggregated every 5 years. *Data from Sea Around Us Project (www.seaaroundus.org).*

the changes in the biomass of tuna, billfish, or sharks over time. For instance, the adult biomass of all monitored tuna populations declined by 52.5% from 1954 to 2006, while that of the tuna or tuna-like populations declined by 59.9% in the same time period [14]. As for sharks, there is similar evidence of severe declines in the abundance of populations caught by commercial fisheries [15]. The next section focuses on the impacts that the changes in extractive power have had on open ocean systems.

46.2 The ecological impacts of fisheries in the open ocean

The estimates of high seas fisheries landings reported by the *Sea Around Us* Project suggest that approximately 180 million tons of biomass have been extracted from ABNJ from 1950 to 2010, of which 36% were tuna or billfish species. For centuries, the immense size of the open ocean led many to believe that it was too vast for fisheries to impact the integrity of its ecological fabric, yet the evidence that has been accumulating over the last two decades demonstrates otherwise [10]. Historically, fisheries exploitation has been, in many senses of the word, asymmetrical. Many fisheries have targeted and extirpated large mature individuals, leading to age-truncated populations [16], as well as deletions of parts of the gene pool [10]. Similarly the spatio-temporal distribution of most fisheries is highly heterogeneous, meaning that target populations have not been harvested equally throughout their range [17]. These asymmetries also extend to the way biological communities have been exploited. The Ocean Biogeographic Information System (OBIS) is the most comprehensive repository of spatially explicit marine biological diversity data in the world (http://iobis.org/) and has records for a total of 4052 individual species of fish in ABNJ (Fig. 46.2). According to a 2006 FAO report, humans are catching approximately 200 species of fish in the high seas [18], although only 39 have an annual catch greater than 1000 tons [8]. Fishing pressure in the high seas is skewed toward higher trophic level species and has led to changes in the species composition of open ocean ecosystems that we are only just beginning to understand [10,19,20]. Monitoring the ecological status of open ocean species in an ecosystem context requires documenting not only information about the size, distribution, and number of individuals harvested in a population, but is also dependent on (1) abundance and distribution metrics of associated and/or dependent species, (2) information on trophic linkages, and (3) data on the abiotic environmental conditions in which those species coexist. These tasks are particularly challenging for oceanic species, which, unlike those in coastal or benthic environments, may cross multiple national jurisdictions, thus making the fisheries data aggregation and analysis process much more complicated.

Some scientists believe that the weak trophic linkages in the open ocean, the low density of species per unit volume and the ephemeral nature of most oceanographic features results in a lack of top-down trophic control [21]. Over the last 20 years there has been an increase in

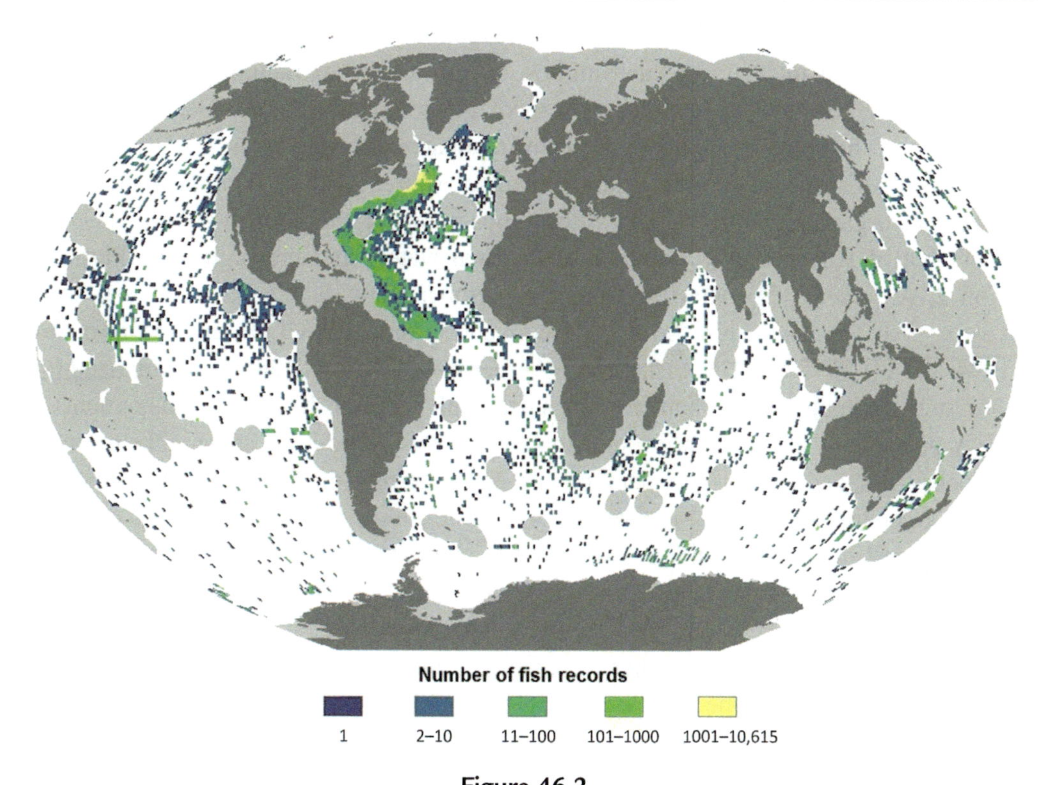

Figure 46.2
Spatial distribution of all the fish records for the 4052 species which comprise the known high seas fish biodiversity in OBIS. *OBIS*, Ocean Biogeographic Information System.

the evidence of fisheries-induced top-down forcings (such as trophic cascades) in oceanic biological communities [20,22]. While there are currently no examples of regime shifts in open ocean ecosystems, there is abundant evidence for this type of ecological disruptive event in pelagic and continental shelf systems [23].

46.3 Jurisdictional frameworks underpinning high seas ecosystem-based fisheries management

By establishing EEZs the 1982 United Nations Convention on the Law of the Sea (UNCLOS) placed 64% of the global ocean outside the jurisdiction of any one coastal or island state. Still, all states remained responsible for the conservation of the marine environment in the high seas. However, almost four decades later, the dimensions (and resulting impacts) of today's industries and activities in the marine realm are far greater than the ones that existed when UNCLOS was drafted. The degree of governance and management in the high seas varies across sectoral activities. For example, while the governance framework for seabed mining (the International Seabed Authority) manages the distribution and intensity of each mining operation, the extraction of marine genetic resources in ABNJ currently lacks any form of governance structure. Oceanic fisheries are somewhere in between. Among its many mandates, the parties to UNCLOS reached an agreement to cooperate in the establishment of subregional or regional fisheries management organizations (RFMOs) with competency for the conservation and management of fish resources within jurisdictional waters and the high seas (Part VII, Section 2, Article 11 [5]). RFMOs comprise member states which have interests in harvesting fish species within the RFMO convention area. UNCLOS entered into force in 1994 and by 1995 the UN Fish Socks Agreement (UNFSA; [24]) was finalized, further strengthening the mechanisms for sustainable oceanic fisheries management. The UNFSA promotes the conservation and management of straddling and highly migratory fish stocks—displaying transboundary movement patterns—through an ecosystem-based approach (General Principles—Article 5 [24]), which must be exercised both within and beyond the jurisdictional boundaries of coastal states. The UNFSA mandate includes requirements for monitoring and managing impacts not just to target species, but to "species belonging to the same ecosystem or associated with or dependent upon the target stocks" [24]. As pelagic fisheries expanded and diversified, so did the range of their ecological impacts. The increases in intensity [25], spatial range [4], and average depth [3] of pelagic fisheries led to an intensification of their overlap with both target and nontarget oceanic biodiversity, which has resulted in new population, community, and ecosystems-level ecological impacts, such as reduced ranges, trophic cascades, or reductions in biodiversity, respectively [11]. Although the mandates of many RFMOs are strongly aligned with the conservation and management principles laid out by the UNFSA, our current understanding of the ecological impacts of fisheries on open ocean ecosystems, unlike those in coastal or deep-sea systems, is quite limited and hinders our ability to abate them; which is dependent on the contribution of high-resolution catch and bycatch data by RFMO member states. The UN is engaged in a 2-year process (2018–20) to create an international legally binding instrument for the conservation and sustainable use of marine biodiversity in ABNJ. While the new instrument "should not undermine existing legal instruments and frameworks and relevant global, regional, and sectoral bodies,"[1] such as RFMOs, such an instrument may play a supporting role for sectoral fisheries bodies to fulfill their EBFM mandates by providing crucial scientific and logistical support to generate knowledge on the biological and oceanographic conditions in high seas ecosystems.

[1] UNGA Resolution 69/292 of July 6, 2015, at para 2; see also UNGA Resolution of December 24, 2017, at para 3.

46.4 Existing ecosystem-based fisheries management tools and gaps

While a comprehensive assessment of the spectrum of management actions and knowledge gaps affecting the implementation of an EBFM strategy in the high seas is beyond the scope of this section, I will provide some thoughts regarding the general trajectories of RFMOs and some of the main gaps in knowledge and monitoring of high seas species.

Various studies have explored the performance of RFMOs across a series of topics, including the way that target stocks are managed [15], existing RFMO bycatch management strategies [26], and existing strategies and data collection processes to implement an EAF across RFMOs [27]. As far as tuna RFMOs go, a recent study provides a comprehensive assessment of the degree to which these bodies are delivering across different EBFM criteria [26]: texts and main structures, management of target species, management of bycatch species, ecosystem properties and trophic relationships, and habitats. The authors conclude that, while most tuna RFMOs have adequate mandates and organizational structures in place, as well as adequate management strategies of target stocks, the knowledge and management strategies surrounding bycatch species, trophic interactions, or habitats is poor across the board [26]. The conservation and management status of pelagic/oceanic sharks exemplifies these deficiencies. A report by the UN Environment Programme (UNEP) and the Convention on Migratory Species on the ecological status of migratory or possibly migratory chondrichthyan species conclude that, of the 1093 species in the class Chondrichthyes, 153 are migratory or possibly migratory [28]. The range and migratory pathways of these species extend across jurisdictions and straddle into the high seas. The study determined that up to 46% of the migratory or possibly migratory cartilaginous fish are classified as Vulnerable, Endangered, or Critically Endangered under the Red List by the International Union for Conservation of Nature (IUCN) and those that are currently classified as Data Deficient comprise up to 23% of the species. Only 5.8% (9/153) species have stock assessments available in the RAM Legacy Database, one of the most comprehensive repositories of fisheries stock assessments in the world, and of these, six are classified as Vulnerable or Endangered by the IUCN Red List. There are stock assessments for at least two other oceanic shark species which are not in the RAM Legacy Database; ICCAT conducted them on the blue (*Prionace glauca*) and porbeagle sharks (*Lamna nasus*) (www.iccat.int). While not as robust as stock assessments, other forms of population abundance estimates exist. Productivity and susceptibility analysis have been performed for 16 species of migratory sharks and rays, most of which did not have a stock assessment [29]. If we combine the two, the proportion of migratory Chondrichthyes with some form of stock abundance or productivity estimate still remains low (16.3%).

The current knowledge deficiencies about the ecological status of species in the open ocean extend far beyond the main bycatch species. As previously mentioned, OBIS has records for just over 4000 species of fish in the high seas (Fig. 46.2), however, 1992 of these species only have one record in the system. It is important to note that the spatial distribution of the fish records on OBIS is heavily biased toward the Central and North Atlantic. The number of fish species that inhabit ABNJ is therefore likely to be significantly larger. This gap in knowledge is mirrored across the full breadth of high seas species recorded under OBIS: 20,355 species. Fisheries are in a unique position to help collect biological and ecological data on oceanic biodiversity.

The expansion of commercial fisheries into the open ocean has not been accompanied by monitoring and data collection structures that would allow for the implementation EBFM in the high seas. Severe knowledge gaps on the composition and dynamics of open ocean ecosystems still remain and, while progress has been made by RFMOs to improve their management of associated and dependent species, efforts are still biased toward a small proportion of high seas biodiversity. The gaps not only affect rare species, Collette et al. highlighted how 11 of the 61 species of scombrid or billfish caught by the global fishing fleet lack adequate data and are classified as Data Deficient by IUCN [30].

At the community level, an example of these differences in knowledge is apparent in the spatial distribution of ecosystem-level trophic models such as Ecopath with Ecosim (EwE) models [31]. According to the Ecobase repository (http://ecobase.ecopath.org/#discoverytools) there are 457 EwE models with metadata that reconstruct the trophic dynamics of marine ecosystems worldwide. The distribution of these models is severely biased toward coastal or continental shelf ecosystems, leaving large knowledge gaps about these interactions in the open ocean. Preserving community composition and trophic dynamics is a key objective of EBFM, which cannot be accomplished without accurate information about species composition and trophic relationships.

46.5 The future of an ecosystem approach to fisheries in the high seas

As uses of the open ocean continue to diversify and intensify, the international community faces the complex challenge of not only managing sectoral industries, but accounting for cumulative and synergistic impacts in a changing ocean. Our ability to learn from the past to predict the state of the ocean in the future will determine the success of fisheries management and processes such as the UN BBNJ efforts. In the case of marine fisheries, new technologies are not only allowing us to track the spatiotemporal footprint of fisheries [32,33], but also to use this information to predict their potential distribution [34]. While its use and accessibility is not widespread, this type of information could help empower governance bodies in the high seas, increasing their ability to trace and monitor ocean uses [35].

Much of this paradigm shift from single-species management to an EBFM approach will hinge on the availability of ecological, biological, oceanographic, and anthropogenic data;

our ability to understand the mechanisms and interaction pathways which interconnect ocean species, uses, and ecosystems; and our capacity to develop comprehensive management schemes which can adapt as climatological and ecological conditions change. For instance, a recent study explored the impacts on climate change on the effectiveness of area-based management tools in the North Atlantic Ocean [36]. The authors concluded that within the next 20–50 years, the efficacy of most of the spatial management measures in the region, many of which are important for fisheries management, would be compromised. This highlights the need for managers to consider not only current, but also future distributions of species, trophic relationships, or interaction pathways. Besides changes in species distribution at the population or stock level [37–39], there is mounting evidence that climate change is and will continue altering the biophysical and chemical properties of the ocean, which in turn are affecting the productivity [40] of marine systems and the phenology [41] and body size [42] of marine organisms. Tracking these changes will require RFMOs' Parties to reinforce their data collection mechanisms in a way that accounts for spatial and taxonomic biases, as well as research and management efforts which account for the dynamics of these systems and which can generate and respond to actionable knowledge. For instance, RFMOs must continue to expand the taxonomic scope of stock assessments or productivity analyses to encompass a larger number of those associated or dependent species that may be impacted directly or indirectly by fishing activities. At the state level, fishing nations should continue expending and strengthening their fisheries observer and data collection programs. Further, both states and RFMOs could benefit by increasing their engagement with members of civil society, such as academia, who are actively engaged in research that may help in the implementation of an ecosystem-based approach to fisheries management in the high seas.

References

[1] D.J. McCauley, et al., Marine defaunation: animal loss in the global ocean, Science 347 (2015) 1255641.

[2] C.T. Darimont, C.H. Fox, H.M. Bryan, T.E. Reimchen, The unique ecology of human predators, Science 349 (2015) 858–860.

[3] A. Merrie, et al., An ocean of surprises—trends in human use, unexpected dynamics and governance challenges in areas beyond national jurisdiction, Global Environ. Change 27 (2014) 19–31.

[4] W. Swartz, E. Sala, S. Tracey, R. Watson, D. Pauly, The spatial expansion and ecological footprint of fisheries (1950 to present), PLoS One 5 (2010) e15143.

[5] UNCLOS, United Nations Convention on the Law of the Sea, Opened for Signature 10 December 1982, 1833 UNTS397 (entered into force 10 November 1994), 1982.

[6] S.M. Garcia, The *Ecosystem Approach* to *Fisheries: Issues, Terminology, Principles, Institutional Foundations, Implementation* and *Outlook*, Food & Agriculture Org, 2003.

[7] V. Restrepo, M.J. Juan-Jordá, B.B. Collette, F.L. Frédou, A. Rosenberg, Tunas and Billfishes, First Global Marine Assessment (World Ocean Assessment), 2016.

[8] L. Schiller, M. Bailey, J. Jacquet, E. Sala, High seas fisheries play a negligible role in addressing global food security, Sci. Adv. 4 (2018) eaat8351.

[9] M.R. Clark, et al., The impacts of deep-sea fisheries on benthic communities: a review, ICES J. Mar. Sci. 73 (2015) i51–i69.
[10] G. Ortuño Crespo, D.C. Dunn, A review of the impacts of fisheries on open-ocean ecosystems, ICES J. Mar. Sci. 74 (9) (2017) 2283–2297.
[11] S. Cullis-Suzuki, D. Pauly, Failing the high seas: a global evaluation of regional fisheries management organizations, Mar. Policy 34 (2010) 1036–1042.
[12] M.J. Juan-Jordá, H. Murua, H. Arrizabalaga, N.K. Dulvy, V. Restrepo, Report card on ecosystem-based fisheries management in tuna regional fisheries management organizations, Fish Fish. 19 (2018) 321–339.
[13] C. White, C. Costello, Close the high seas to fishing? PLoS Biol. 12 (2014) e1001826.
[14] M.J. Juan-Jordá, I. Mosqueira, A.B. Cooper, J. Freire, N.K. Dulvy, Global population trajectories of tunas and their relatives, Proc. Natl. Acad. Sci. U.S.A. 108 (2011) 20650–20655.
[15] J.K. Baum, et al., Collapse and conservation of shark populations in the Northwest Atlantic, Science 299 (2003) 389–392.
[16] ISC Pacific Bluefin Tuna Working Group. Executive summary of the 2016 Pacific bluefin tuna stock assessment, in: 16th Meeting of the ISC Plenary, July 2016 (ISC16), Japan, 2016.
[17] B. Worm, D.P. Tittensor, Range contraction in large pelagic predators, Proc. Natl. Acad. Sci. U.S.A. 108 (2011) 11942–11947.
[18] J.-J. Maguire, The *State* of *World Highly Migratory, Straddling* and *Other High Seas Fishery Resources* and *Associated Species*, Food & Agriculture Org, 2006.
[19] S.P. Cox, et al., Reconstructing ecosystem dynamics in the central Pacific Ocean, 1952 1998. II. A preliminary assessment of the trophic impacts of fishing and effects on tuna dynamics, Can. J. Fish. Aquat. Sci. 59 (2002) 1736–1747.
[20] J. Hinke, I. Kaplan, K. Aydin, G. Watters, R. Olson, J.F. Kitchell, Visualizing the food-web effects of fishing for tunas in the Pacific Ocean, Ecol. Soc. 9 (1) (2004).
[21] J.H. Steele, From carbon flux to regime shift, Fish. Oceanogr. 7 (1998) 176–181.
[22] P. Ward, R.A. Myers, Shifts in open-ocean fish communities coinciding with the commencement of commercial fishing, Ecology 86 (2005) 835–847.
[23] C. Möllmann, R. Diekmann, 4 Marine ecosystem regime shifts induced by climate and overfishing: a review for the Northern Hemisphere, Adv. Ecol. Res 47 (2012) 303.
[24] UNFSA, United Nations Conference on Straddling Fish Stocks and Highly Migratory Fish Stocks, July 24–August 4, 1995, Agreement for the Implementation of the Provisions of the United Nations Convention on the Law of the Sea of 10 December Relating to the Conservation and Management of Straddling Fish Stocks and Highly Migratory Fish Stocks, U.N. DOCA/Conf. 164/37, 1995.
[25] J.A. Anticamara, R. Watson, A. Gelchu, D. Pauly, Global fishing effort (1950–2010): trends, gaps, and implications, Fish. Res 107 (2011) 131–136.
[26] E.L. Gilman, Bycatch governance and best practice mitigation technology in global tuna fisheries, Mar. Policy 35 (2011) 590–609.
[27] E. Gilman, K. Passfield, K. Nakamura, Performance of regional fisheries management organizations: ecosystem-based governance of bycatch and discards, Fish Fish. 15 (2014) 327–351.
[28] S. Fowler, The conservation status of migratory sharks, UNEPCMS Secr. Bonn Ger (2014).
[29] E. Cortés, et al., Expanded ecological risk assessment of pelagic sharks caught in Atlantic pelagic longline fisheries, Collect. Vol. Sci. Pap. ICCAT 71 (2015) 2637–2688.
[30] B.B. Collette, et al., High value and long life—double jeopardy for tunas and billfishes, Science 1208730 (2011).
[31] M. Colléter, et al., Global overview of the applications of the Ecopath with Ecosim modeling approach using the EcoBase models repository, Ecol. Model. 302 (2015) 42–53.
[32] F. Mazzarella, M. Vespe, D. Damalas, G. Osio, Discovering vessel activities at sea using AIS data: mapping of fishing footprints, in: Information Fusion (FUSION), 2014 17th International Conference on 1–7, IEEE, 2014.

[33] D.A. Kroodsma, et al., Tracking the global footprint of fisheries, Science 359 (2018) 904–908.
[34] G.O. Crespo, et al., The environmental niche of the global high seas pelagic longline fleet, Sci. Adv. 4 (2018) eaat3681.
[35] D.C. Dunn, et al., Empowering high seas governance with satellite vessel tracking data, Fish Fish. (2018).
[36] D. Johnson, M. Adelaide Ferreira, E. Kenchington, Climate change is likely to severely limit the effectiveness of deep-sea ABMTs in the North Atlantic, Mar. Policy 87 (2018) 111–122.
[37] A.L. Perry, P.J. Low, J.R. Ellis, J.D. Reynolds, Climate change and distribution shifts in marine fishes, Science 308 (2005) 1912–1915.
[38] N.K. Dulvy, et al., Climate change and deepening of the North Sea fish assemblage: a biotic indicator of warming seas, J. Appl. Ecol. 45 (2008) 1029–1039.
[39] W.W. Cheung, et al., Projecting global marine biodiversity impacts under climate change scenarios, Fish Fish. 10 (2009) 235–251.
[40] W.W. Cheung, et al., Large-scale redistribution of maximum fisheries catch potential in the global ocean under climate change, Global Change Biol. 16 (2010) 24–35.
[41] M.J. Genner, et al., Temperature-driven phenological changes within a marine larval fish assemblage, J. Plankton Res. 32 (2009) 699–708.
[42] W.W. Cheung, et al., Shrinking of fishes exacerbates impacts of global ocean changes on marine ecosystems, Nat. Clim. Change 3 (2013) 254.

CHAPTER 47

Ocean pollution and warming oceans: toward ocean solutions and natural marine bioremediation

Juan José Alava[1,2]

[1]Institute for the Oceans and Fisheries, University of British Columbia, Vancouver, BC, Canada
[2]Fundación Ecuatoriana para el Estudio de Mamíferos Marinos (FEMM), Guayaquil, Ecuador

Chapter Outline

47.1 Introduction

Marine pollution and anthropogenic climate change have been impacting and reshaping the chemistry of the global ocean at different scales since the onset of the anthropocene, that is, the epoch emerging around the late 1700s and early 1800s when humans and societies became a global geological and ecological driving forcing [1,2]. Increasing human populations and our associated demands for food, industrialization, and global transport with concomitant emissions of chemical contaminants are critically affecting the forces of nature [2]. Global temperatures continue to increase, currently at 1.0°C relative to preindustrial levels due to increasing CO_2 emissions [3]. Since the last century, legacy and emerging pollutants are working against the health of the global ocean and survival of marine species. Moreover, the bioaccumulation and effects of

Predicting Future Oceans.
DOI: https://doi.org/10.1016/B978-0-12-817945-1.00046-0

environmental–anthropogenic pollutants in combination with climate change are being amplified and exacerbated in marine food webs and ecosystems [4,5].

Ocean pollution by chemical contaminants has emerged since the 1500s (16th century) due to releases of anthropogenic mercury from mining [6,7], and recently by both persistent organic pollutants (POPs) and anthropogenic radionuclides before the mid-1900s (e.g., ~1930–45). Although some POPs, including polychlorinated biphenyls (PCBs) and dichloro-diphenyl-trichloroethanes (DDTs), were banned in developed and industrialized countries during the 1970s, some organochlorine pesticides (OCPs) are still used in developing countries to control malaria vectors and crop pests (i.e., DDT) and recent emerging POPs, including polybrominated diphenyl ethers (PBDE) flame retardants and perfluorinated chemicals, are contaminating food webs. The highest concentrations of PCBs and DDTs still tend to be reported from locations in temperate countries where usage was very intense or POPs were manufactured, stored, or underwent ocean disposal (e.g., Palos Verdes in the Southern California Bight) [8,9].

At the peak of the "Organochlorine era" (i.e., the period of OCP use from the late 1940s to early 1980s) [9], Rachel Carson's Silent Spring published in 1962 drew global attention to the potential effects of man-made chemicals, in particular pesticides, on wildlife populations (e.g., raptors and songbirds) and human health [10], which fundamentally influenced public and political opinion. Yet our blue planet is still contaminated with the legacy of that era of uncontrolled and misused agricultural pesticides and industrial chemicals. Legacy and emerging pollutants, including POPs, current-use pesticides, heavy metals, marine debris and microplastics, personal care products and pharmaceuticals (PCPPs), polycyclic aromatic hydrocarbons (PAHs), and radionuclides, are ubiquitous in the global ocean. For example, the bioaccumulative and toxic nature of POPs and organic mercury is a particular stressor affecting organisms at the top of marine food webs (i.e., seabirds, marine mammals, large pelagic fish, and humans) because of the inherent toxicity and health effects of these chemicals [11–17].

Likewise, recent studies and modeling work showed that a wide range of POPs deposited in sinks such as oceans and ice were unexpectedly remobilized and revolatilized into the atmosphere from repositories in Arctic regions over the past two decades as a result of climate change [18–21]. Both climate change and environmental pollution as the major human-induced drivers in the 21st century are affecting and transforming marine ecosystems and ocean health, threatening the viability of coastal resources, food security, and sustainable resource development [5,22]. Increasing global emissions of greenhouse gases will exacerbate the effects of climate change and environmental pollutants in marine food webs [4,5,23].

New emerging anthropogenic pollutants, such as microplastics, PCPPs, and radionuclides (e.g., cesium 137 emissions during the Fukushima aftermath in March 2011), pose new protracted risks for marine organisms and food webs. For instance, with the arrival and rapid expansion of the "Plastic era" since the last half century and the first report of plastics in the marine environment in the early 1970s [24], plastic pollution and microplastics are probably the number one contaminant and among the top threats impacting the health of the oceans (for impacts in the ocean and marine life, see Section 47.3.2) in what has now become the "Age of Plastics" [24–27].

The conservation and sustainable management of biodiversity and the marine environment, including transient estuarine, marine coastal, and offshore habitats, are the foremost priorities in the global environmental agenda of the United Nations Sustainable Development Goals (SDGs) as part of the UN 2030—Agenda for Sustainable Development, and Aichi Biodiversity Strategic Goals of the Convention on Biological Diversity aimed to significantly minimize habitat degradation and the existing rate of marine species loss to conserve the oceans. Achieving sustainable oceans is of paramount importance for coastal communities and marine industries, as stated within the United Nations SDG 14 (Life Below Water) for the sustainability, prosperity, and state of future oceans [28]. In doing so, the prevention and reduction of marine pollution, the sustainable management and restoration of marine ecosystems, the mitigation of ocean acidification, and the protection of marine areas are part of the oceans targets that were envisioned as co-benefits by the Nereus Program from achieving SDG 14 [28]. This obviously possesses tremendous implications for marine biodiversity and commercially important species and the sustainability of fisheries, mainly for less attended small-scale fisheries (i.e., artisanal, subsistence, recreational, and ceremonial) from developing nations, in the naissance of the "Blue Economy" [29].

In this context, marine pollution and warming oceans are among the major anthropogenic factors affecting marine biodiversity and fisheries in the global ocean, triggering the challenge to pursue new solution-oriented research for ecotoxicologists, marine biologists, oceanographers, ocean engineers, and environmental scientists. Thus, ecotoxicological research and science-based information are crucial to support the risk assessment and management of anthropogenic pollutants in the face of global environmental change. The interplay of climate change (i.e., ocean warming and ocean acidification) and marine pollution with associated impacts on marine organisms and food webs represents a new looming threat to be assessed and addressed at present. This chapter offers an appraisal of the weight of evidence of marine pollution and climate change interactions and the search for ocean pollution solutions focusing on natural marine bioremediation to hamper the exposure to, bioaccumulation of, and risk of pollutants to marine fauna, seafoods, and humans in our changing oceans.

To accomplish this goal, an overview is first presented on the exposure and accumulation of POPs and metals, with special emphasis on organic mercury in marine organisms and ecosystems under the influence of climate change. Ultimately, this work expects to bring new baseline information, tools, and potential management strategies with a particular emphasis on marine ecosystem services and nature-based solutions to provide knowledge and management tools to decision makers to better understand how to reduce and address both marine pollution and climate change with the aim of recovering and protecting marine ecosystems and the well-being of our future oceans.

47.2 Ocean pollutants and climate change

The production, use, emissions, and disposal of >100,000 chemical substances [30,31] have direct implications for the health of the world's oceans. The global ocean is the ultimate "sink" for anthropogenic pollutants and a reservoir uptaking about 28% of the atmospheric CO_2 emissions, causing ocean acidification [32–35]. The interaction of climate change and ocean pollution is readily affecting and projected to impact the chemistry and quality of our oceans, threatening the survival of many marine species, fisheries, and well-being for humans [4,5,22,36]. The interaction of climate change—warmer temperatures, ocean acidification—and pollutants is also predicted to increase the level of exposure and toxic effects of many environmental contaminants due to changes in the patterns, transport, and fate of chemicals [5,37–40], as shown in Fig. 47.1.

Of particular attention are the cumulative effects of these anthropogenic stressors in marine ecosystems and food webs because climate change can affect pollutant bioaccumulation by amplifying POPs (i.e., PCBs) and methylmercury (CH_3Hg) in marine food webs [4], as illustrated in Fig. 47.1. This interaction may then affect the viability of coastal resources, livelihood and culture, and food security [5,41]. The impacts may be particularly high for vulnerable communities such as coastal aboriginal people who rely heavily on living marine resources as a source of traditional food and culture [41–43].

POPs bioaccumulate in aquatic organisms and can readily biomagnify in marine food webs [44–48], in which the increase and amplification of chemical concentrations occur at each trophic level in the food web, from primary producers to top predators. Because POPs can be transported across abiotic compartments they can contaminate the multimedia environment of the ocean through cycling, including oceanic and atmospheric circulation, sinking of contaminants bound to particulate matter in the water column, and sediment exchange [18,19]. In the Arctic, for instance, the iconic polar bear (*Ursus maritimus*) and seabirds contain increasing concentrations of some POPs (e.g., PCBs, PBDEs, and β-hexachlorocyclohexane) and mercury because of changes in their diet composition and food web structure, and alteration in pollutants exposure and pathways due to the reduction and melting of ice cover, driven by climate change [23,49,50].

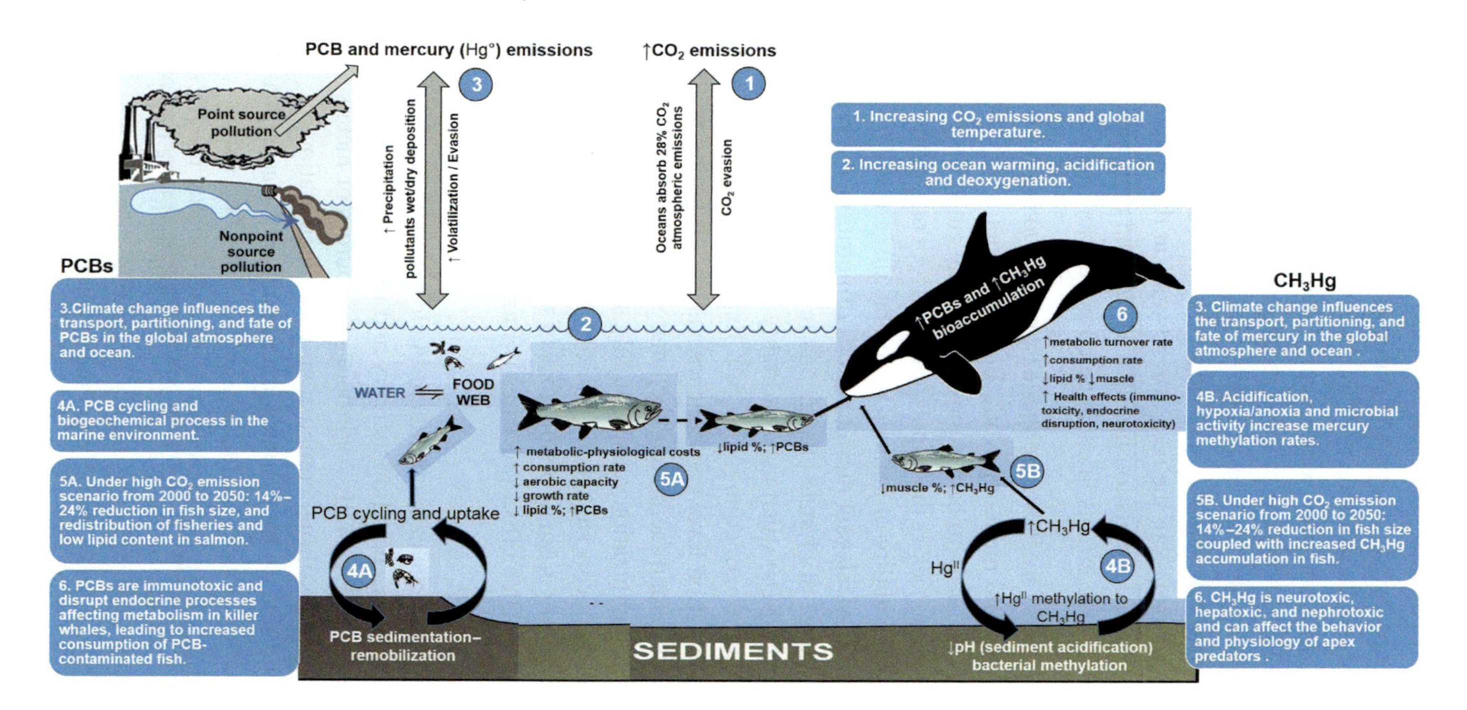

Figure 47.1

Conceptual illustration showing the interaction pathways and combined impact of pollutants (PCBs and methyl mercury—CH_3Hg) and climate change in the ocean, and alteration of exposure pathways and bioaccumulation in a piscivorous-marine mammalian food web (fish-eating killer whale), used here as an example. These interactions include phenomena deemed to be either climate change dominant (i.e., climate change—induced contaminant susceptibility) and/or contaminant dominant (i.e., contamination leads to an increase in climate change sensitivity). For more details on climate change—pollutant interactions, see Alava et al. [4,5]. *PCBs*, Polychlorinated biphenyls. *Based on J.J. Alava, W.W.L. Cheung, P.S. Ross, R.U. Sumaila, Climate change-contaminant interactions in marine food webs: towards a conceptual framework, Global Change Biol. 23 (2017) 3984–4001. doi:10.1111/gcb.13667.*

Considering marine mammals as sentinels of ocean pollution (i.e., *canaries in the coal mine*), a number of toxicological effects have been attributed to PCBs in this functional group of species. The weight of evidence based on field and semicaptive studies, lab research, and modeling work highlights the health effects of PCBs at the genetic, cellular, organism, and population levels, including immunotoxicity, reproduction impairments, and endocrine disruption toxic effects [16,17,51–56]. At present, recent model projections by Desforges et al. [17], for example, indicate that PCB-mediated effects on reproduction and immune function may threaten the long-term viability and lead to the collapse of >50% of the killer whale (*Orcinus orca*) populations within the next 30–50 years at the global level. The impact of climate change on PCB bioaccumulation in killer whales and their main prey may further exacerbate the PCB health toxic effects in endangered populations [4].

Indirect effects in organisms and food webs by POPs are the least understood impacts from both field and empirical research studies and modeling work. Both immunotoxicity and endocrine disruption effects by POPs are likely to be the most deleterious impairments in organisms with potential indirect consequences at the population level. A compromised immune and endocrine system affects the ability of animals to combat disease and to successfully reproduce, which is especially relevant during periods of nutritional stress and other types of stressors under the influence of climate change and in highly variable environments (e.g., El Niño events or ENSOs) when mass mortality occurs and populations often approach the critical tipping point of extinction [5,57–59]. Exposure to immunotoxic contaminants may have significant population level consequences as a contributing factor to increasing anthropogenic stress in wildlife and facilitating the emergence of infectious disease outbreaks [16,17,60]. Potential feminization causing highly skewed sex ratios (i.e., >females vs <males) by estrogenic chemicals such as DDTs [61–63] can have implications at the population level, mainly in small populations of threatened and endangered species facing bottleneck or Allee effect processes within their population dynamics.

Additional global pollutants impacting the ocean environment in close combination with ocean warming and acidification are metals, including mercury and its methylated form, methylmercury (i.e., CH_3Hg), as shown in Fig. 47.1. Anthropogenic sources of metals include mines, metal refineries/smelters, fossil fuel combustion, waste incineration, pesticides and wood preservatives, and release from domestic and industrial waste. While climate change can change the natural cycle of other metals such as zinc (Zn) and cadmium (Cd) altering exposure and sensitivity to these metals in Arctic biota, the Arctic is likely to become a more effective trap for these metals because precipitation is thought to increase in this polar region [18]. Methylmercury is highly neurotoxic and bioaccumulates throughout the food web almost entirely via dietary uptake, reaching the highest concentrations in fish and organisms at the top of the food web [13,64,65]. However, within the climate change context and given the global circulation and complexity of mercury cycling in the ocean, this is a metal of urgent concern that warrants special attention in different oceans basins.

An increase in exposure to mercury concentrations in biota, including apex predators due to differences in feeding preferences has been associated with changes in primary production and species composition of aquatic communities in Arctic ice-free waters throughout both bottom-up and top-down processes [66–68]. Changing oceans and shrinking of polar ice (i.e., Arctic and Antarctic) is affecting habitats and food webs with important consequences of mercury exposure to primary producers and top predators. For instance, mercury exposure and biomagnification has been linked to habitat selection, diet plasticity, and feeding behavior of beluga whales [69,70]. These climatic-induced changes in behavior can affect the accumulation of mercury in several other top predators of the Arctic, such as seals, polar bears, and killer whales, and Inuit people. For instance, the implications of a changing Arctic climate on mercury fate in beluga food webs and the consequences for the health of beluga whales remain pressing research needs [71]. Increasing risk of mercury toxicity due to bioaccumulation exacerbated by climate change has been projected for the food web and fisheries of the Faroe Islands, where fishing communities rely strongly on Atlantic cod (*Gadus morhua*) and pilot whales (*Globicephala melas*) [65]. The impact of increasing ocean surface temperature and acidification on mercury and methylmercury bioavailability and bioaccumulation in food webs is an ongoing research front raising red flags [4,5,65,72].

Contrasting with our current understanding on climate change interactions with POPs and mercury in marine ecosystem and food webs, questions remain concerning the impact of increasing sea surface temperature and ocean acidification on the bioavailability and food web bioaccumulation of microplastics, PCPPs, PAHs, antifouling paints, and radionuclides.

47.3 Ocean solutions and natural marine bioremediation

47.3.1 Ocean solutions for chemical pollution

Ecotoxicological risk assessments, environmental monitoring and scientific data provide decision makers in national governments with basic chemical, physical, and toxicological information in support of national regulations to implement command and control and conventional abatement methods to address water pollution such as wastewater treatment plants—WWTP (e.g., tertiary wastewater treatment plants)—and remediation approaches for organic pollutant- and metal-contaminated sediments (i.e., sediment capping, dry excavation, dredging and ocean disposal, geotextiles mats, human-made bioremediation, and bioaugmentation technologies) to recover and remediate marine environments from chemical contamination [73,74]. In the same way, a feasible and classical case to decrease both anthropogenic chemicals (e.g., mercury) and greenhouse gases is the technological abatement of atmospheric emissions from fossil fuel/coal combustion and electric power generating industries. While abatement is a reactive measure (i.e., command and control approach) to counteract emissions "at the end of the pipe" prior to chemical releases, more proactive and precautionary actions are fully required to target reduction goals of chemical emissions. For the case of POPs and mercury, cutting and preventing emissions locally and/or regionally has obviously varying degrees of effectiveness for local concentrations, depending on the situation, when compared to the international efforts by nations to foster and put in place international instruments (i.e., Stockholm Convention on POPs and Minamata Convention on Mercury, respectively) to curtail emissions at the global level.

Recently, precautionary actions have also been considered to address ocean pollution. According to a multistakeholder workshop "Bring your Solutions to Ocean Pollution—Recovering BC's marine mammals by tackling pollution" held in May 2015 at the Vancouver Aquarium (Ocean Wise) to identify priority pollutants of concern and sources of contaminants (i.e., by sector, site, or process) in marine mammals listed in the Canada's Species at Risk Act (SARA) in British Columbia, the designing of novel, effective, solution-oriented actions that could be implemented by private, government, academic, and/or conservation sectors was recommended [75,76]. At that workshop, five operational sectors that were evaluated as sources of pollution included urban, shipping and harbors, agroforestry, industry, and home and garden. As a result five candidate ocean pollution solution ideas were ultimately conceived [75,76], which, adapted for the purpose of this chapter, are shown in Table 47.1.

Particularly, the blue fueling initiative can be considered in close conjunction with cleaner, and more sustainable energy using renewable resources while governments follow the transition to a low carbon emissions economy with strong climate change mitigation. Alternative approaches to mitigate chemical pollution through climate change geoengineering can also be considered in the future.

47.3.2 Ocean solutions for plastic pollution

The oceans and marine life in general are greatly impacted and threatened by marine debris, particularly large plastics and microplastics. All oceans are polluted by marine debris, from which about 73% is plastic, and more than 1300 species are affected through entanglement and ingestion [77]. The volume of plastic that entered the oceans in 2010 from 192 coastal countries (93% of the global population) ranged from 4.8 to 12.7 million metric tonnes [26]. In addition to deleterious entanglements and ingestion, marine debris and plastics serve as vectors for organic pollutants [e.g., PAHs, phthalates, and nonylphenols, as well as POPs such as PCBs and dichlorodiphenyldichloroethylene (DDE), which is a byproduct of DDT] and invasive species, and leach toxic chemicals [78–81]. A recent study estimated that a minimum of 5.25 trillion particles weighing 268,940 tons are floating in the world's ocean [25].

Table 47.1: Proposed ocean solutions to proactively prevent, mitigate, and eliminate chemical pollution in the ocean.

Ocean solution pollution	Rationale
Pharmaceutical and personal care products' ecocertification (i.e., drug-free oceans)	Knowing that hundreds of PPCPs are being discharged through sewage effluent into freshwater systems and coastal waters, this solution envisions new, standardized product labels that enable consumers to target ocean-friendly choices when purchasing such products
Blue fueling initiative	Understanding that thousands of small-scale spills (e.g., gas, diesel, oil) and hydrocarbon leaks assault the marine environment every year, the adoption of new technologies and best management practices by the private and public sectors to prevent and reduce the risk of spills should be fostered and implemented when fueling maritime transportation and vessels
Blue furniture (i.e., ocean-friendly furnishing)	Knowing that furniture typically has several chemicals, including persistent flame retardants (e.g., PBDEs), which are found in sediments, fish, and apex predators of the ocean, the private or industrial sector should consider ocean pollution in the designing, manufacturing, and use of home and office furniture
Global and local ocean networks	Acknowledging that new multiparty forums are needed to provide leadership on ocean pollution issues and provide effective guidance on updating pollution regulations and determining action levels
Ocean pollution solution curriculum for schools	Recognizing outreach and education curricula as the foremost logical approaches and much needed steps to help to promote and sustain the health and sustainability of our oceans for future generations

PBDE, Polybrominated diphenyl ethers.
Source: *These solutions are summarized based on and adapted from information available in Vancouver Aquarium & Georgia Strait Alliance, Bring your Solutions to Ocean Pollution—Recovering BC's marine mammals by tackling pollution, Report of a workshop hosted by the Ocean Pollution Research Program, Vancouver Aquarium, May 21–22, 2015, Vancouver, BC, 2015; J.J. Alava, K. Gordon, C. Morales-Caselles, W. Gao, C.A. Wilhelmson, P.S. Ross, Toward ocean pollution solutions for marine mammals at risk in British Columbia, Canada, in: Snapshot Presentation at the 2016 Salish Sea Ecosystem Conference (SSEC), April 13–15, 2016, Vancouver, BC, Canada, 2016. <https://cedar.wwu.edu/ssec/2016ssec/salish_sea_snapshots/38/>.*

Of great concern are microplastics, which are defined as particles <5 mm and can be deliberately manufactured (plastic resin pellets and powder) or generated as breakdown by-products of larger debris and macroplastic (e.g., clothing, ropes, bags, bottles) [78–80]. Both the origin and final fate of most plastics is derived from land-based sources, including household and industrial water, aquaculture, shipping, and tourism. Microplastics can also carry on their surfaces epiplastic communities, including pathogenic bacteria such as *Vibrio*, *Aeromonas*, *Enterobacter*, *Halomonas*, *Mycobacterium*, *Photobacterium*, *Pseudomonas*, *Rhodococcus*, and *Shigella* [82,83]. In view of the accelerated rate of pervasive ocean pollution by plastics and microplastics, concerted management actions and proactive solutions are needed to hamper and eliminate plastics pollution.

Solutions for macroplastics and microplastics pollution are complex and entail several fronts of solution-oriented actions to combat the life cycle of plastics and restoring marine and coastal environments from the local level to global ocean governance [27,84,85]. To reduce the global ecological footprint by plastic pollution, scientific expertise, community participation, and market-based strategies are required [27]. While there are successful regulations to ban microbeads from the oceans [84], more endeavors to foster policies to phase out the production and alter the market of plastics and microplastics are needed. A key tactic is to search for loopholes in the vicious socioeconomic cycle of plastics to reduce and avert the consumption and production of plastic, as humans and associated plastic demand are the causes of plastic production and pollution. Although recycling is still a primary idea and effort to minimize plastic pollution [86], this waste management strategy is not sufficient as only 5% of plastic packing material is retained at the end of the recycling process for subsequent use, that is, 95% of plastic packaging material value is lost to the economy following a short first use or one use cycle [87]. The classification of plastic waste as "hazardous" was proposed as a policy approach for countries to managing plastic debris and to move to a closed loop system of plastic use and reuse (i.e., all plastics are reused and recycled) with the aim to restore affected habitats and prevent more dangerous debris from accumulating in the marine environment [88].

The London Convention/Protocol on the Prevention of Marine Pollution by Dumping of Wastes and Other Matter (The London Protocol) is the international instrument to promote policies for the effective control of all sources of marine pollution, including plastic pollution, to protect the marine environment from human activities [89]. Proposing microplastics as POPs under the criteria of the Stockholm Convention on POPs (i.e., persistence, bioaccumulation and toxicity or PBT chemicals) may also offer an international venue and opportunity for nations to further regulate, control and eliminate these pervasive micropollutants. However, our oceans continue to be polluted with plastics, highlighting the need of ocean pollution solutions, mitigation actions, and outreach strategies with capacity building to eliminate and reduce the influx of plastics and microplastics to the ocean. Changes in human behavior preferences and ocean-friendly choices, and market-based instruments for product consumption in close conjunction with the creation of proactive policies and industry-specific regulations can be framed within the solutions presented in Table 47.2.

47.3.3 Climate change ocean solutions

An outstanding set of ocean solutions and actions to mitigate and adapt to the impact of climate change on the marine ecosystem and the recognition of the uncertainties and limitations of currently available climate and ocean management options have been assessed recently [90–92]. A remarkable example of these ocean solutions is the natural uptake of CO_2 emissions (i.e., blue carbon: carbon stored in coastal or marine ecosystems)

Table 47.2: Proposed solutions to proactively prevent, mitigate, and eliminate plastic pollution in the ocean by targeting specific sectors from the local level to global ocean governance.

Target sector		Solutions/actions
Global, national, and local governments: authorities and politicians	←	Enact plastic reduction policies for phasing out of plastic bags, other microbead sources, and disposable plastics
↕		
Private sector and industries	←	Target industries with policies to minimize plastic production and packaging (e.g., eco-fee) and redesign plastic-free products by means of incentives, policy, and regulations
↕		
General public/nonprivate sector	←	Promote social mobilization and offer incentives and taxes (e.g., plastic-tax) and market-based instruments (e.g., plastic bottle deposit/cash back for containers) to encourage and call on consumers to alter consumption, littering, and throw-away habits
↕		
Fishing communities	←	Offer and implement good subsidies to collect plastic and marine debris instead of fishing fish as a new way of life. Fishers receive compensation and/or a salary for fishing plastic from the ocean. Alternatively, fishing for both fish and plastic
↕		
Schools' curricula	←	Democratize a network of knowledge on plastic pollution and education, integrating best individual/community practices and habits through environmental awareness programs and outreach to eliminate plastic pollution framed within a Pollution Solution Curriculum

The vertical bidirectional arrows (↕) indicate that the policy process can take either a bottom-up or top-down approach or both depending on the political framework or public policy philosophy of a given socioeconomic system or nation.

performed by marine vegetation and coastal plan communities such as mangrove forests, seagrass beds, and salt marshes [91,93–96]. Despite marine vegetation habitats covering $<2\%$ of the coastal ocean, these habitats have burial rates 40 times higher than tropical rainforests and account for more than 50% of the carbon buried in marine sediments [93]. Particularly mangrove forest ecosystems have been identified as efficient natural carbon sinks [95]; however, mangroves can deliver back stored carbon via methane (CH_4) emissions, partially offsetting blue carbon burial [95,97].

According to Gattuso et al. [90], the most effective approaches across all ecosystems to reduce the impacts of ocean warming, ocean acidification, and sea level rise are as follows (for a review see Gattuso et al. [90]):

- renewable energy (i.e., substituting fossil fuel energy by ocean energy);
- alkalinization (i.e., adding natural or human-made alkalinity to enhance CO_2 removal and carbon storage);

- hybrid methods (e.g., marine-biomass-fueled energy with carbon capture on land);
- local vegetation (i.e., restoring and conserving of coastal vegetation to enhance CO_2 uptake and prevent further emissions); and
- albedo enhancement (i.e., increasing surface ocean albedo by producing long-lived ocean foam).

The protection of marine habitats and ecosystems by spatial measures including marine protected areas, restoration of hydrology (i.e., maintaining and restoring coastal hydrological regimes), and eliminating overexploitation of living resources and overextraction of nonliving resources are also relatively effective to reduce impacts on seagrass habitats, mangroves, and salt marshes, serving as blue carbon ecosystems [90,91]. At this level, geoengineering and renewable (green) energy in combination with the conservation of natural sinks for blue carbon have emerged as state of the art technologies and human interventions in the planetary environment to prevent and minimize the accelerated rate of climate change.

47.3.4 On the pursuit of natural marine bioremediation

Contrasting the important number of human-propelled actions and alternatives for solutions to overcome marine pollution and the impact of climate change in the oceans, very little effort and a lack of solutions-oriented research exist to solve marine pollution and reduce contaminants by capitalizing the use of natural bioremediation performed by the intrinsic ecological services of marine ecosystems and the specific functioning role of marine species, that is, natural phytoremediation and zooremediation [98].

While the oceans provide ecological and physical services for the planetary system, including food for marine life and human beings, absorption of heat and energy from the climate system and atmospheric warming, uptake of anthropogenic CO_2 from the atmosphere, and the storage and global distribution of excess water from melting ice pack and glaciers [99], the oceans also harbor thousands of species with an enormous genetic diversity pool that can serve not only for sustaining the function of marine ecosystems, for food and for medicine and biomedical research purposes [100–103], but to naturally bioremediate and ecologically restore marine ecosystems and coastal areas from pollution [98,101]. This underscores the distinct need for identifying and conserving marine ecosystems with species able to act as marine ecosystem engineers to naturally uptake, filter, store, and buffer and reduce the exposure, accumulation, and effects of chemicals; and thus enhance natural remediation in the marine environment. Within this premise, a theoretical scheme to conceptualize the central idea of combined ecological functions of key marine species and interactive services of oceanic and coastal–marine ecosystems envisioned to benefit the natural recovery from marine pollution and climate change is presented in Fig. 47.2.

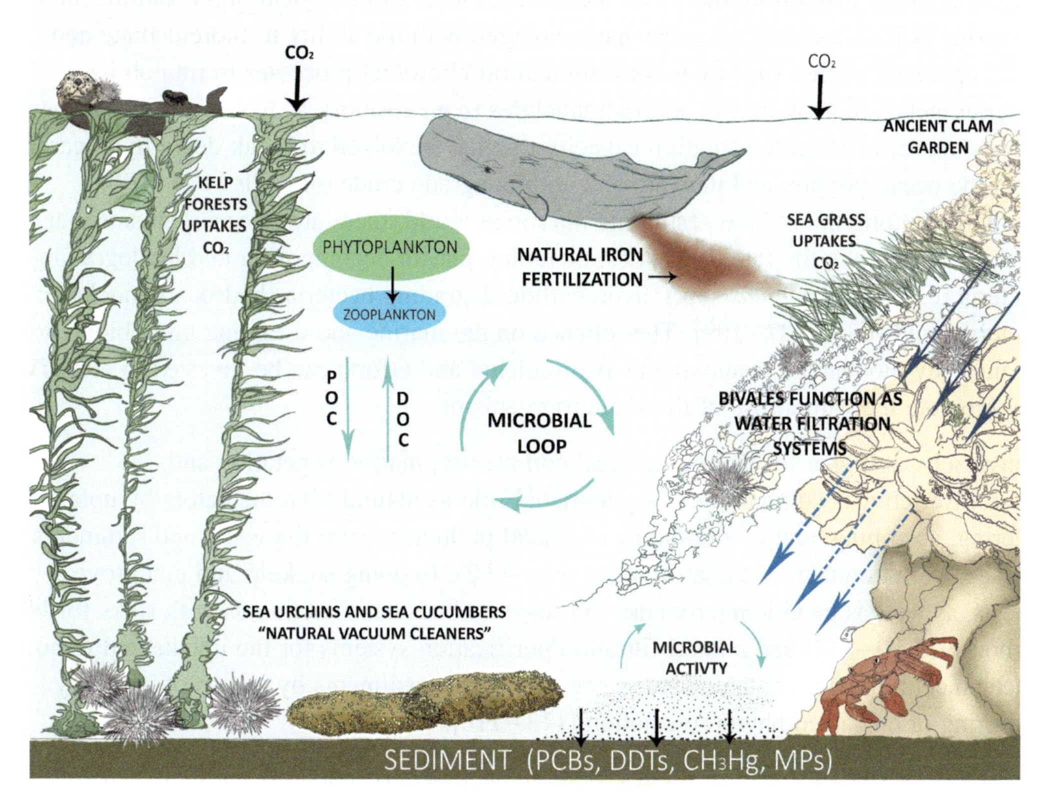

Figure 47.2

Conceptual model showing an example of the ecological function of marine species from a temperate marine ecosystem for pollutant depuration, natural bioremediation, and mitigation of climate change. The intrinsic capacity of marine vegetation acting as blue carbon sinks, including phytoplankton, kelp forests, microbenthic algae, and seaweeds, to uptake and store CO_2, functioning as "marine scrubbers" abating carbon injection into the ocean. At the bottom of the food web, plankton (i.e., phytoplankton and zooplankton) and microbiota (bacterioplankton) are key for the microbial loop in carbon and detritus cycling in marine biogeochemical processes. Marine bacteria also play a key role as bioremediators to break down and immobilize chemical pollutants (e.g., hydrocarbons, metals) in sediments contaminated by pollutants (e.g., PCBs, DDTs, CH_3Hg, microplastics—MPs). Low trophic level macroinvertebrates such as grazers (e.g., sea urchins) and suspension/deposit-feeders (e.g., sea cucumbers) serve as "living vacuum cleaners," while filter-feeders (e.g., bivalves) function as a "natural filtration system" to depurate and remediate the marine environment from pollution (e.g., filtering metals and pharmaceuticals/drugs). At the top of the food web, apex predators such as marine mammals (e.g., sperm whales) also contribute as carbon sinks by means of natural iron fertilization from their fecal matter by enhancing primary production and thus the absorption of CO_2 in the ocean (i.e., "whale pump"). Restoration of ancient clam gardens can potentially help to mitigate the impact of ocean acidification by enhancing the deposition of calcium carbonate ($CaCO_3$) and serve as an adaptation strategy against sea level rise for seafood security. *DDTs*, Dichloro-diphenyl-trichloroethanes; *DOC*, dissolved organic carbon; *PCBs*, polychlorinated biphenyls; *POC*, particulate organic carbon. *Artwork created by Nastenka Alava.*

Fig. 47.2 shows that within the lower hierarchical level of biological organization, naturally occurring pollution-eating microbes have emerged with the ability to bioremediate and clean up contaminated sites by harnessing natural microbial processes to immobilize harmful metals (e.g., mercury) and radionuclides (e.g., cesium), as well as natural bacteria possessing intrinsic bioremediation capacity that have evolved to break down hydrocarbons from the ocean bottom, and thus are able to biodegrade crude oil spills [104–106]. Similarly, among the marine–estuarine microbial world, there are microorganisms that could play a role in the fate of plastics, including plastic-decomposing and biodegrading bacteria (e.g., *Pseudomonas* spp.), hydrocarbon-degrading bacteria (hydrocarbonoclastic bacteria), and fungi [107–109]. The reliance on the marine and estuarine microbiome to intrinsically biodegrade contaminants is promising and efforts can be invested to conserve and promote the natural use of these microorganisms.

Conversely, at higher levels of biological complexity, marine vegetation and macroinvertebrates communities play a crucial role as natural bioremediators by uptaking, filtering, absorbing, and accumulating chemical pollutants from the water and sediments in the ocean environment and coastal areas (Fig. 47.2). In doing so, kelp and mangrove forests, as well as benthic microalgae and seagrass beds can function as both sinks for blue carbon [93,110–112] and natural filtration/purification systems for the uptake and removal of chemical pollutant-contaminated water and marine sediments by means of naturally occurring or enhanced phytoremediation [113–116].

Moreover, marine macroinvertebrates including echinoderms and bivalves can function as natural vacuum cleaners and water filtration systems, respectively. While sea urchins are benthic grazers controlling, reshaping, and feeding on macroalgae or kelp forests (e.g., *Macrocystis*, *Nereocystis*, *Laminaria*), reefs, and rocky intertidal algal communities [117–119], sea cucumbers are deposit- and suspension-feeders that consume organic matter and detritus from the water and sediments, reducing the organic load and redistributing surface sediments, making them bioremediators and playing a key role for bioturbation [120]. These keystone species are not only capable of contributing to bottom-up processes and the cycling of carbon and nutrients, but also have the potential to clean up and remove contaminants from living organisms (algal communities) and sediments to account for biogeochemical cycling and contribute to restoring the ecosystem to benefit the health of benthic communities (Fig. 47.2).

In fact, sea urchins (e.g., *Strongylocentrotus* spp.) appeared to be abundant bioremediators in barren polluted sites in the absence of giant kelp (*Macrocystis pyrifera*) near sewage discharges/outfalls by scraping on rock surfaces [121]. The bioremediation potential of deposit-feeding sea cucumbers (i.e., *Apostichopus japonicus*) has also been investigated in suspension aquaculture of bivalves, demonstrating a great bioremediation capacity in this aquaculture system [122]. More recently, sea cucumbers have been considered as nutrient recyclers and processors of particulate waste in polyculture or integrated multitrophic aquaculture systems given their capabilities to process and remediate biodeposits and impacted sediments [123]. Also, four species of deposit-feeding and suspension-feeding sea cucumbers were found to ingest microplastics and small plastic particles (0.25 to <15 mm) of nylon and polyvinyl chloride along with sediment and other plastic fragments [124]. These findings suggest the potential use of free-ranging populations of sea urchins and sea cucumbers as living "vacuum cleaners" of contaminated marine sediments, as illustrated in Fig. 47.2.

Equally important are the natural filtration treatment and restoration systems provided by filter feeders such as mussels, clams, and oysters, which are important bioremediators due to their high bioaccumulation capacity to bioconcentrate metals and organic pollutants from the water column [98,114,125,126], increased denitrification rates and enhanced nutrient sequestration [127], and the removal of suspended or/and resuspended sediments (e.g., total suspended and dissolved solids), by removing up to 30%–45% of local particle concentrations [128]. In the Northeastern Pacific, ancient clam gardens harboring intertidal benthic communities (e.g., sea urchins, sea cucumbers, and algal communities) are ecological, socioeconomic, and cultural production systems, enhancing ecosystem productivity to support traditional First Nation People's seafood production while sustaining resilient ecosystems [129,130].

In addition to their function as a feasible climate change adaptation measure for seafood security and to overcome sea level rise, these traditional bivalve systems can also potentially serve as natural "self-filtration treatment and bioreactor systems" that can be proactively managed by coastal communities to filter out pollutants and remediate coastal environments in remote areas (Fig. 47.2). Thus the natural enhancement of bivalve biotreatment systems and ancient clam gardens as ecological services offered by bivalves (e.g., mussel and oyster reefs) should be further fostered and applied as reliable approaches for marine zooremediation.

Finally, the role of top predators such as marine mammals, including large whales (i.e., sperm whales, *Physeter macrocephalus*; and blue whales, *Balaenoptera musculus*) in the ocean is of paramount importance to control and help to mitigate climate change by taking up CO_2 through iron fertilization released from whale fecal matter (e.g., the sperm whale populations in the Southern Ocean alone act as a carbon sink by removing 2×10^5 tonnes of carbon from the atmosphere), driving the ocean biological pump and productivity [131,132]. Similarly, several other species of baleen whales, toothed cetaceans, and seals perform a crucial task by contributing to the nitrogen recycling in coastal waters and enhancing primary production [133]. Seabirds are also critical for enhancing coral reef productivity and functioning, highlighting the need for restoring seabird-derived nutrient subsidies from large areas of ocean [134].

Holistic and integrated environmental management approaches in concert with natural bioremediation in the oceans should protect and enhance not only low trophic level marine biota and humans, but also upper trophic level marine fauna, especially apex predators as indicators of ocean health. This is of critical importance for top predators already facing and trying to adapt to climate change as they play a key role to control top-down processes over food webs [135,136], and protect and maintain carbon stocks and sinks in blue carbon ecosystems [137].

47.4 Discussion

Since fisheries, marine wildlife, and ecosystems provide life support, sustenance, ecological services and functions as well as economic opportunities through harvesting, marine fisheries catches, and ecotourism in many developed and developing countries, the health of the ocean cannot be viewed separately from human health. Coastal waters that are contaminated with persistent chemicals and exacerbated by climate change impacts can lead to human illness and adverse health, reduced fisheries quality and quantity, and impacts on the health of marine wildlife. This has obvious social and economic consequences.

Risk management and communication processes that balance the risks and benefits of a diet of traditional food have been successful through input from a diverse group of regional experts and communities aiming to incorporate many sociocultural and economic factors to arrive at a risk management decision that will be the most beneficial [138,139]. Coastal–marine environments and communities fostering enhancement of natural bioremediation, mitigation, and adaptation strategies as proactive solutions to face climate change and protect from chemical pollutants can provide for an abundance of clean fisheries products and healthy marine fauna for tourism, essential foundations for the well-being of marine biodiversity, human residents, and the ecotourism sector. A balance needs to be struck between public health and ecosystem health risks versus the use and release of chemical pollutants potentially exacerbated by climate change. While the toxicological paradigm '*the dose makes the poison*' provides a theoretical foundation for an approach that minimizes ecological damage by chemicals while optimizing human health benefits (e.g., DDT use to control the Malaria mosquito, *Anopheles*, in developing countries [58]), the outdated pollution rule '*the solution to pollution is dilution*' has proven to be erroneous since the 1950s [140] because of the persistent, bioaccumulative and toxic nature of many pollutants (e.g., PCBs, dioxins, DDTs, OCPs, CH_3Hg) discharged into the oceans.

Likewise, climate change policy and international regulatory instruments should be aimed at minimizing and adapting the socioeconomic and ecological challenges by looking at the resilience and intrinsic remediation capacity of natural marine ecosystems in the face of ocean warming and acidification in the oceans [99]. Simultaneously, government plans implementing green energy alternatives, carbon taxes and market-based instruments to phase out pollution, and limiting chemical emissions from coal-fired electricity generation and oil industries, would help to lessen the negative impacts by both anthropogenic stressors.

Finally, when climate change policies and pollution management actions are not sufficient to overcome impacts and reach reduction targets in a reasonable time, social mobilization is an important ally to engage and motivate the public and stakeholders to implement climate and pollution solutions through social learning, behavioral change, and community-based conservation actions.

Acknowledgments

The author thanks Nastenka Calle (Pacific Institute for Climate Solutions, Simon Fraser University) for providing meaningful comments to this chapter, and Nastenka Alava for drawing the artwork presented in Fig. 47.2. Special thanks to the editors A. Cisneros-Montemayor, Y. Ota, and W.W.L. Cheung for the invitation to contribute to this book, and Colin Thackray for providing edits and insights to improve this chapter. The author expresses his gratitude and appreciation to the Nippon Foundation Nereus Program at the University of British Columbia.

References

[1] P.J. Crutzen, Geology of mankind: the anthropocene, Nature 415 (2002) 23.
[2] W. Steffen, P.J. Crutzen, J.R. McNeill, The anthropocene: are humans now overwhelming the great forces of nature, AMBIO 36 (8) (2007) 614–621.
[3] IPCC, Global Warming of 1.5°C, Summary for Policy Makers, Intergovernmental Panel on Climate Change, WMO, UNEP, October 6, 2018. <http://www.ipcc.ch/report/sr15/>.
[4] J.J. Alava, A.M. Cisneros-Montemayor, R. Sumaila, W.W.L. Cheung, Projected amplification of food web bioaccumulation of MeHg and PCBs under climate change in the Northeastern Pacific, Sci. Rep. 8 (2018) 13460. Available from: https://doi.org/10.1038/s41598-018-31824-5.
[5] J.J. Alava, W.W.L. Cheung, P.S. Ross, R.U. Sumaila, Climate change-contaminant interactions in marine food webs: towards a conceptual framework, Global Change Biol. 23 (2017) 3984–4001. Available from: https://doi.org/10.1111/gcb.13667.
[6] D.G. Streets, M.K. Devane, Z. Lu, T.C. Bond, E.M. Sunderland, D.J. Jacob, All-time releases of mercury to the atmosphere from human activities, Environ. Sci. Technol. 45 (2011) 10485–11049.
[7] C.H. Lamborg, C.R. Hammerschmidt, K.L. Bowman, G.J. Swarr, K.M. Munson, D.C. Ohnemus, et al., A global ocean inventory of anthropogenic mercury based on water column measurements, Nature 512 (2014) 65–68. Available from: https://doi.org/10.1038/nature13563.
[8] M.E. Blasius, G.D. Goodmanlowe, Contaminants still high in top-level carnivores in the Southern California Bight: levels of DDT and PCBs in resident and transient pinnipeds, Mar. Pollut. Bull. 56 (2008) 1973–1982.
[9] L.J. Blus, Organochlorine pesticides, in: D.J. Hoffman, B.A. Rattner, G.A. Burton, J. Cairns (Eds.), Handbook of Ecotoxicology, CRC Press and Taylor & Francis, Boca Raton, FL, 2003, pp. 313–339.
[10] R. Carson, Silent Spring., Houghton Mifflin Company, Boston, MA, 1962.
[11] T. Colborn, F.S. Vom Saal, A.M. Soto, Developmental effects of endocrine-disrupting chemicals in wildlife and humans, Environ. Health Persp. 101 (1993) 378–383.
[12] J.E. Elliott, K.H. Elliott, Tracking marine pollution, Science 340 (6132) (2013) 556–558.

[13] J. Wiener, R. Bodaly, S. Brown, M. Lucotte, M. Newman, D. Porcella, et al., Monitoring and evaluating trends in methylmercury accumulation in aquatic biota, in: R. Harris, D.P. Krabbemhoft, R. Mason, M.W. Murray, R.J. Reash, T. Saltman (Eds.), Ecosystem Responses to Mercury Contamination, CRC Press and Taylor & Francis Group, Boca Raton, FL, 2007, pp. 87–122.
[14] A.M. Scheuhammer, M.W. Meyer, M.B. Sandheinrich, M.W. Murray, Effects of environmental methylmercury on the health of wild birds, mammals, and fish, AMBIO 36 (1) (2007) 12–19.
[15] A. Scheuhammer, B. Braune, H.M. Chan, H. Frouin, A. Krey, R. Letcher, et al., Recent progress on our understanding of the biological effects of mercury in fish and wildlife in the Canadian Arctic, Sci. Total Environ. 509–510 (2015) 91–103.
[16] J.P.W. Desforges, C. Sonne, M. Levin, U. Siebert, S. De Guise, R. Dietz, Immunotoxic effects of environmental pollutants in marine mammals, Environ. Int. 86 (2016) 126–139.
[17] J.P. Desforges, A. Hall, B. McConnell, A. Rosing-Asvid, J.L. Barber, A. Brownlow, et al., Predicting global killer whale population collapse from PCB pollution, Science 361 (6409) (2018) 1373–1376.
[18] R.W. Macdonald, T. Harner, J. Fyfe, Recent climate change in the Arctic and its impact on contaminant pathways and interpretation of temporal trend data, Sci. Total Environ. 342 (2005) 5–86.
[19] J. Ma, Z. Cao, Quantifying the perturbations of persistent organic pollutants induced by climate change, Environ. Sci. Technol. 44 (2010) 8567–8573.
[20] J. Ma, H. Hung, C. Tian, R. Kallenborn, Revolatilization of persistent organic pollutants in the Arctic induced by climate change, Nat. Clim. Change 1 (2011) 255–260.
[21] J. Ma, H. Hung, Correspondence: arctic contaminants and climate change, Nat. Clim. Change 2 (2011) 829–830.
[22] M. Barange, G. Merino, J.L. Blanchard, J. Scholtens, J. Harle, E.H. Allison, et al., Impacts of climate change on marine ecosystem production in societies dependent on fisheries, Nat. Clim. Change 4 (2014) 211–216. Available from: https://doi.org/10.1038/nclimate2119.
[23] M.A. McKinney, S. Pedro, R. Dietz, C. Sonne, A.T. Fisk, D. Roy, et al., A review of ecological impacts of global climate change on persistent organic pollutant and mercury pathways and exposures in arctic marine ecosystems, Curr. Zool. 61 (4) (2015) 617–628.
[24] C.J. Moore, How much plastic is in the ocean? You tell me, Mar. Pollut. Bull. 92 (1–2) (2015) 1–3.
[25] M. Eriksen, L.C. Lebreton, H.S. Carson, M. Thiel, C.J. Moore, J.C. Borerro, et al., Plastic pollution in the world's oceans: more than 5 trillion plastic pieces weighing over 250,000 tons afloat at sea, PLoS One 9 (12) (2014) e111913.
[26] J.R. Jambeck, R. Geyer, C. Wilcox, T.R. Siegler, M. Perryman, A. Andrady, et al., Plastic waste inputs from land into the ocean, Science 347 (6223) (2015) 768–771.
[27] J. Vince, B.D. Hardesty, Plastic pollution challenges in marine and coastal environments: from local to global governance, Restor. Ecol. 25 (1) (2017) 123–128.
[28] Nippon Foundation-Nereus Program, Oceans and Sustainable Development Goals: Co-benefit, Climate Change and Social Equity, Vancouver, Canada, 2017, 28 pp. <www.nereusprogram.org>.
[29] D. Pauly, A vision for marine fisheries in a global blue economy, Mar. Policy 87 (2018) 371–374.
[30] A.J. Hendriks, How to deal with 100,000 + substances, sites, and species: overarching principles in environmental risk assessment, Environ. Sci. Technol. 47 (8) (2013) 3546–3547.
[31] D.C. Muir, P.H. Howard, Are there other persistent organic pollutants? A challenge for environmental chemists, Environ. Sci. Technol. 40 (23) (2006) 7157–7166.
[32] S.C. Doney, The growing human footprint on coastal and open-ocean biogeochemistry, Science 328 (2010) 1512–1516.
[33] J.P. Gattuso, L. Hansson, Ocean acidification: background and history., in: J.P. Gattuso, L. Hansson (Eds.), Ocean Acidification, Oxford University Press, Inc., New York, 2011, pp. 1–20. 352 pp.
[34] S.C. Doney, M. Ruckelshaus, J.E. Duffy, J.P. Barry, F. Chan, C.A. English, et al., Climate change impacts on marine ecosystems, Annu. Rev. Mar. Sci. 4 (2012) 11–37.
[35] C. Le Quéré, R.M. Andrew, P. Friedlingstein, S. Sitch, J. Pongratz, A.C. Manning, Global Carbon Budget 2017, Earth Syst. Sci. Data Discuss. 10 (1) (2017) 405–448. Available from: https://doi.org/10.5194/essd-2017-123.

[36] D.P. Krabbenhoft, E.M. Sunderland, Global change and mercury, Science 341 (6153) (2013) 1457–1458.

[37] D. Schiedek, B. Sundelin, J.W. Readman, R.W. Macdonald, Interactions between climate change and contaminants, Mar. Pollut. Bull. 54 (12) (2007) 1845–1856.

[38] P.D. Noyes, M.K. McElwee, H.D. Miller, B.W. Clark, L.A. Van Tiem, K.C. Walcott, et al., The toxicology of climate change: environmental contaminants in a warming world, Environ. Int. 35 (6) (2009) 971–986.

[39] R.J. Wenning, S.E. Finger, L. Guilhermino, R.C. Helm, M.J. Hooper, W.G. Landis, et al., Global climate change and environmental contaminants: a SETAC call for research, Integr. Environ. Assess. Manage. 6 (2010) 197–198.

[40] J.M. Balbus, A.B.A. Boxall, R.A. Fenske, T.E. Mckone, L. Zeise, Implications of global climate change for the assessment and management of human health risks of chemical in the natural environment, Environ. Toxicol. Chem. 32 (1) (2013) 62–78.

[41] M.C. Tirado, R. Clarke, L.A. Jaykus, A. McQuatters-Gollop, J.M. Frank, Climate change and food safety: a review, Food Res. Int. 43 (7) (2010) 1745–1765.

[42] A.M. Cisneros-Montemayor, D. Pauly, L.V. Weatherdon, Y. Ota, A global estimate of seafood consumption by coastal Indigenous peoples, PLoS One 11 (12) (2016) e0166681. Available from: https://doi.org/10.1371/journal.pone.0166681.

[43] L.V. Weatherdon, Y. Ota, M.C. Jones, D.A. Close, W.W.L. Cheung, Projected scenarios for coastal First Nations' fisheries catch potential under climate change: management challenges and opportunities, PLoS One 11 (1) (2016) e0145285. Available from: https://doi.org/10.1371/journal.pone.0145285.

[44] B.C. Kelly, M.G. Ikonomou, J.D. Blair, A.E. Morin, F.A.P.C. Gobas, Food web specific biomagnification of persistent organic pollutants, Science 317 (2007) 236–239.

[45] B.C. Kelly, M.G. Ikonomou, J.D. Blair, B. Surridge, D. Hoover, L. Haviland, et al., Perfluoroalkyl contaminants in an Arctic marine food web: trophic magnification and wildlife exposure, Environ. Sci. Technol. 43 (2009) 4037–4043.

[46] F.A.P.C. Gobas, J. Arnot, Food web bioaccumulation model for polychlorinated biphenyls in San Francisco Bay, California, USA, Environ. Toxicol. Chem. 29 (2010) 1385–1395.

[47] J.J. Alava, F.A.P.C. Gobas, Assessing biomagnification and trophic transport of persistent organic pollutants in the food chain of the Galapagos sea lion (*Zalophus wollebaeki*): conservation and management implications, in: A. Romero, E.O. Keith (Eds.), New Approaches to the Study of Marine Mammals, InTechOpen, Croatia, 2012, pp. 77–108. Available from: https://doi.org/10.5772/51725.

[48] D.L. Cullon, M.B. Yunker, J.R. Christensen, R.W. Macdonald, M.J. Whiticar, N. Dangerfield, et al., Biomagnification of polychlorinated biphenyls in a harbor seal (*Phoca vitulina*) food web from the Strait of Georgia, British Columbia, Canada, Environ. Toxicol. Chem. 31 (2012) 2445–2455.

[49] B.M. Braune, A.J. Gaston, K.A. Hobson, H.G. Gilchrist, M.L. Mallory, Changes in food web structure alter trends of mercury uptake at two seabird colonies in the Canadian Arctic, Environ. Sci. Technol. 48 (2014) 13246–13252.

[50] B.M. Jenssen, G.D. Villanger, K.M. Gabrielsen, J. Bytingsvik, T. Bechshoft, T.M. Ciesielski, et al., Anthropogenic flank attack on polar bears: interacting consequences of climate warming and pollutant exposure, Front. Ecol. Evol. 3 (2015) 16. Available from: https://doi.org/10.3389/fevo.2015.00016.

[51] A.H. Buckman, N. Veldhoen, G. Ellis, J.K.B. Ford, C. Helbing, P.S. Ross, PCB-associated changes in mRNA expression in killer whales (*Orcinus orca*) from the NE Pacific Ocean, Environ. Sci. Technol. 45 (2011) 10194–10202.

[52] S. De Guise, D. Martineau, P. Beland, M. Fournier, Effects of in vitro exposure of beluga whale leukocytes to selected organochlorines, J. Toxicol. Environ. Health, A. 55 (1998) 479–493.

[53] A.J. Hall, O.I. Kalantzi, G.O. Thomas, Polybrominated diphenyl ethers (PBDEs) in grey seals during their first year of life—are they thyroid hormone endocrine disruptors? Environ. Pollut. 126 (2003) 29–37.

[54] B.E. Hickie, P.S. Ross, R.W. Macdonald, J.K.B. Ford, Killer whales (*Orcinus orca*) face protracted health risks associated with lifetime exposure to PCBs, Environ. Sci. Technol. 41 (2007) 6613–6619.

[55] L. Mos, M. Cameron, S.J. Jeffries, B.F. Koop, P.S. Ross, Risk-based analysis of PCB toxicity in harbor seals, Integr. Environ. Assess. Manage. 6 (2010) 631–640.

[56] I. Peñín, M. Levin, K. Acevedo-Whitehouse, L. Jasperse, E. Gebhard, F.M.D. Gulland, et al., Effects of polychlorinated biphenyls (PCB) on California sea lion (*Zalophus californianus*) lymphocyte functions upon in vitro exposure, Environ. Res. 167 (2018) 708–717.

[57] J.J. Alava, P.S. Ross, M.G. Ikonomou, M. Cruz, G. Jimenez-Uzcategui, S. Salazar, et al., Gobas FAPC DDT in endangered Galapagos Sea Lions (*Zalophus wollebaeki*), Mar. Pollut. Bull. 62 (2011) 660–671.

[58] J.J. Alava, S. Salazar, M. Cruz, G. Jimenez-Uzcategui, S. Villegas-Amtmann, D. Paez-Rosas, et al., DDT Strikes Back: Galapagos sea lions face increasing health risks, AMBIO 40 (2011) 425–430.

[59] J.J. Alava, P.S. Ross, Pollutants in tropical marine mammals of the Galapagos Islands, Ecuador: an ecotoxicological quest to the last Eden (Chapter 8), in: M.C. Fossi, C. Panti (Eds.), Marine Mammal Ecotoxicology: Impacts of Multiple Stressors on Population Health, Elsevier/Academic Press, London, UK, 2018, pp. 213–234.

[60] P.S. Ross, The role of immunotoxic environmental contaminants in facilitating the emergence of infectious diseases in marine mammals, Hum. Ecol. Risk Assess. 8 (2002) 277–292.

[61] D.M. Fry, C.K. Toone, DDT-induced feminization of gull embryos, Science 213 (4510) (1981) 922–924.

[62] L.J. Guillette Jr., T.S. Gross, G.R. Masson, J.M. Matter, H.F. Percival, A.R. Woodward, Developmental abnormalities of the gonad and abnormal sex hormone concentrations in juvenile alligators from contaminated and control lakes in Florida, Environ. Health Persp. 102 (1994) 680–688.

[63] W.R. Kelce, C.R. Stone, S.C. Laws, L.E. Gray, J.A. Kemppainen, E.M. Wilson, et al., Metabolite *p*, *p*-DDE is a potent androgen receptor antagonist, Nature 375 (1995) 581–585.

[64] M.B. Sandheinrich, J.G. Wiener, Methylmercury in fish: recent advances in assessing toxicity of environmentally relevant exposures, in: W.N. Beyer, J.P. Meador (Eds.), Environmental Contaminants in Biota: Interpreting Tissue Concentrations, 2, CRC Press, Boca Raton, FL, 2011, pp. 169–190.

[65] S. Booth, D. Zeller, Mercury, food webs, and marine mammals: implications of diet and climate change for human health, Environ. Health Persp. 113 (5) (2005) 521–526.

[66] T. Douglas, M. Amyot, T. Barkay, T. Berg, J. Chételat, What is the fate of mercury entering the arctic environment (Chapter 3)? in: P. Outridge, R. Dietz, S. Wilson (Eds.), AMAP Assessment 2011: Mercury in the Arctic, Arctic Monitoring and Assessment Programme (AMAP), Oslo, Norway, 2011, pp. 45–65.

[67] R.W. Macdonald, L.L. Loseto, Are Arctic Ocean ecosystems exceptionally vulnerable to global emissions of mercury? A call for emphasised research on methylation and the consequences of climate change, Environ. Chem. 7 (2010) 133–138. Available from: https://doi.org/10.1071/EN09127.

[68] G.A. Stern, R.W. Macdonald, P.M. Outridge, S. Wilson, J. Chetelat, A. Cole, et al., How does climate change influence arctic mercury? Sci. Total Environ. 414 (2012) 22–42.

[69] L.L. Loseto, G.A. Stern, D. Deibel, T.L. Connelly, A. Prokopowicz, L. Fortier, et al., Linking mercury exposure to habitat and feeding behaviour in Beaufort Sea beluga whales, J. Mar. Syst. 74 (2008) 1012–1024. Available from: https://doi.org/10.1016/J.JMARSYS.2007.10.004.

[70] L.L. Loseto, G.A. Stern, S.H. Ferguson, Size and biomagnification: how habitat selection explains beluga mercury levels, Environ. Sci. Technol. 42 (2008) 3982–3988. Available from: https://doi.org/10.1021/ES7024388.

[71] H. Frouin, L.L. Loseto, G.A. Stern, M. Haulena, P.S. Ross, Mercury toxicity in beluga whale lymphocytes: limited effects of selenium protection, Aquat. Toxicol. 109 (2012) 185–193.

[72] S. Jonsson, A. Andersson, M.B. Nilsson, U. Skyllberg, E. Lundberg, J.K. Schaefer, et al., Terrestrial discharges mediate trophic shifts and enhance methylmercury accumulation in estuarine biota, Sci. Adv. 3 (1) (2017) e1601239.

[73] L.W. Perelo, In situ and bioremediation of organic pollutants in aquatic sediments, J. Hazard. Mater. 177 (1–3) (2010) 81–89.

[74] C. Zeller, B. Cushing, Panel discussion: remedy effectiveness: what works, what doesn't? Integr. Environ. Assess. Manage. 2 (1) (2006) 75–79.

[75] Vancouver Aquarium & Georgia Strait Alliance, Bring your Solutions to Ocean Pollution—Recovering BC's marine mammals by tackling pollution, Report of a workshop hosted by the Ocean Pollution Research Program, Vancouver Aquarium, May 21–22, 2015, Vancouver, BC, 2015.

[76] J.J. Alava, K. Gordon, C. Morales-Caselles, W. Gao, C.A. Wilhelmson, P.S. Ross, Toward ocean pollution solutions for marine mammals at risk in British Columbia, Canada, in: Snapshot Presentation at the 2016 Salish Sea Ecosystem Conference (SSEC), April 13–15, 2016, Vancouver, BC, Canada, 2016. <https://cedar.wwu.edu/ssec/2016ssec/salish_sea_snapshots/38/>.

[77] M. Bergmann, M.B. Tekman, L. Gutow, Marine litter: sea change for plastic pollution, Nature 544 (7650) (2017) 297.

[78] GESAMP, IMO/FAO/UNESCO-IOC/UNIDO/WMO/IAEA/UN/UNEP Joint Group of Experts on the Scientific Aspects of Marine Environmental Protection, in: T. Bowmer, P.J. Kershaw (Eds.), Proceedings of the GESAMP International Workshop on Plastic Particles as a Vector in Transporting Persistent, Bio-Accumulating and Toxic Substances in the Oceans. GESAMP Rep. Stud. No. 82, 2010, 68 pp.

[79] M.R. Gregory, Environmental implications of plastic debris in marine settings—entanglement, ingestion, smothering, hangers-on, hitch-hiking and alien invasions, Philos. Trans. R. Soc. Lond. B: Biol. Sci. 364 (1526) (2009) 2013–2025.

[80] C.J. Moore, Synthetic polymers in the marine environment: a rapidly increasing, long-term threat, Environ. Res. 108 (2008) 131–139.

[81] M. Sigler, The effects of plastic pollution on aquatic wildlife: current situations and future solutions, Water Air Soil Pollut. 225 (11) (2014) 2184. Available from: https://doi.org/10.1007/s11270-014-2184-6.

[82] I.V. Kirstein, S. Kirmizi, A. Wichels, A. Garin-Fernandez, R. Erler, M. Löder, et al., Dangerous hitchhikers? Evidence for potentially pathogenic *Vibrio* spp. on microplastic particles, Mar. Environ. Res. 120 (2016) 1–8.

[83] S. Oberbeckmann, M.G.J. Löder, M. Labrenz, Marine microplastic-associated biofilms—a review, Environ. Chem. 12 (5) (2015) 551–562.

[84] P. Dauvergne, The power of environmental norms: marine plastic pollution and the politics of microbeads, Environ. Polit. 27 (4) (2018) 579–597.

[85] M. Haward, Plastic pollution of the world's seas and oceans as a contemporary challenge in ocean governance, Nat. Commun. 9 (1) (2018) 667.

[86] J. Hopewell, R. Dvorak, E. Kosior, Plastics recycling: challenges and opportunities, Philos. Trans. R. Soc. Lond. B: Biol. Sci. 364 (1526) (2009) 2115–2126.

[87] World Economic Forum, Ellen MacArthur Foundation and McKinsey & Company. The New Plastics Economy—Rethinking the Future of Plastics, 2016, 117 pp. <http://www.ellenmacarthurfoundation.org/publications>.

[88] C.M. Rochman, M.A. Browne, B.S. Halpern, B.T. Hentschel, E. Hoh, H.K. Karapanagioti, et al., Policy: classify plastic waste as hazardous, Nature 494 (7436) (2013) 169–171.

[89] IMO, Convention on the Prevention of Marine Pollution by Dumping of Wastes and Other Matter, International Maritime Organization (IMO), 2017. <http://www.imo.org/en/OurWork/Environment/LCLP/Pages/default.aspx>.

[90] J.-P. Gattuso, A.K. Magnan, L. Bopp, W.W.L. Cheung, C.M. Duarte, J. Hinkel, et al., Ocean solutions to address climate change and its effects on marine ecosystems, Front. Mar. Sci. 5 (2018) 337. Available from: https://doi.org/10.3389/fmars.2018.00337.

[91] D. Herr, G.R. Galland, The Ocean and Climate Change. Tools and Guidelines for Action, IUCN, Gland, Switzerland, 2009. 72 pp.

[92] Magnan, A.K. *et al.* Ocean-Based Measures for Climate Action, IDDRI, Policy Brief N°06/18. 2018.

[93] C.M. Duarte, J.J. Middelburg, N. Caraco, Major role of marine vegetation on the oceanic carbon cycle, Biogeosciences 1 (2005) 173–180.

[94] C.M. Duarte, I.J. Losada, I.E. Hendriks, I. Mazarrasa, N. Marbà, The role of coastal plant communities for climate change mitigation and adaptation, Nat. Clim. Change 3 (11) (2013) 961.

[95] E. Mcleod, G.L. Chmura, S. Bouillon, R. Salm, M. Björk, C.M. Duarte, et al., A blueprint for blue carbon: toward an improved understanding of the role of vegetated coastal habitats in sequestering CO_2, Front. Ecol. Environ. 9 (10) (2011) 552–560.

[96] L. Pendleton, D.C. Donato, B.C. Murray, S. Crooks, W.A. Jenkins, S. Sifleet, et al., Estimating global "blue carbon" emissions from conversion and degradation of vegetated coastal ecosystems, PLoS One 7 (9) (2012) e43542.
[97] J.A. Rosentreter, D.T. Maher, D.V. Erler, R.H. Murray, B.D. Eyre, Methane emissions partially offset "blue carbon" burial in mangroves, Sci. Adv. 4 (6) (2018) eaao4985. Available from: https://doi.org/10.1126/sciadv.aao4985.
[98] S. Gifford, R.H. Dunstan, W. O'Connor, C.E. Koller, G.R. MacFarlane, Aquatic zooremediation: deploying animals to remediate contaminated aquatic environments, Trends Biotechnol. 25 (2007) 60–65. Available from: https://doi.org/10.1016/j.tibtech.2006.12.002.
[99] T.F. Stocker, The silent services of the world ocean, Science 350 (6262) (2015) 764–765.
[100] E. Kenchington, M. Heino, E.E. Nielsen, Managing marine genetic diversity: time for action? ICES J. Mar. Sci. 60 (6) (2003) 1172–1176.
[101] E.A. Norse, Global Marine Biological Diversity: A Strategy for Building Conservation into Decision Making, Island Press, Washington, DC, 1993. 383 pp.
[102] C. Pedrós-Alió, Marine microbial diversity: can it be determined? Trends Microbiol. 14 (6) (2006) 257–263.
[103] D.P. Tittensor, C. Mora, W. Jetz, H.K. Lotze, D. Ricard, E.V. Berghe, et al., Global patterns and predictors of marine biodiversity across taxa, Nature 466 (7310) (2010) 1098–1101.
[104] T.C. Hazen, E.A. Dubinsky, T.Z. DeSantis, G.L. Andersen, Y.M. Piceno, N. Singh, et al., Deep-sea oil plume enriches indigenous oil-degrading bacteria, Science 330 (6001) (2010) 204–208.
[105] E. Kintisch, Can Microbes Save the Gulf Beaches? The Challenges Are Myriad, Science News, April 30, 2010.
[106] T. Zwillich, A tentative comeback for bioremediation, Science 289 (5488) (2000) 2266–2267.
[107] J.P. Harrison, M. Schratzberger, M. Sapp, A. Osborn, Rapid bacterial colonization of low-density polyethylene microplastics in coastal sediment microcosms, BMC Microbiol. 14 (2014) 1–15.
[108] A. McCormick, T.J. Hoellein, S.A. Mason, J. Schluep, J.J. Kelly, Microplastic is an abundant and distinct microbial habitat in an urban river, Environ. Sci. Technol. 48 (2014) 11863–11871.
[109] E.R. Zettler, T.J. Mincer, L.A. Amaral-Zettler, Life in the "plastisphere": microbial communities on plastic marine debris, Environ. Sci. Technol. 47 (13) (2013) 7137–7146.
[110] I.K. Chung, J.H. Oak, J.A. Lee, J.A. Shin, J.G. Kim, K.S. Park, Installing kelp forests/seaweed beds for mitigation and adaptation against global warming: Korean Project Overview, ICES J. Mar. Sci. 70 (5) (2013) 1038–1044.
[111] C.D. Hepburn, D.W. Pritchard, C.E. Cornwall, R.J. McLeod, J. Beardall, J.A. Raven, et al., Diversity of carbon use strategies in a kelp forest community: implications for a high CO_2 ocean, Global Change Biol. 17 (7) (2011) 2488–2497.
[112] Dd'A. Laffoley, G. Grimsditch, The Management of Natural Coastal Carbon Sinks, IUCN, Gland, Switzerland, 2009. 53 pp.
[113] E.B. Barbier, S.D. Hacker, C. Kennedy, E.W. Koch, A.C. Stier, B.R. Silliman, The value of estuarine and coastal ecosystem services, Ecol. Monogr. 81 (2) (2011) 169–193.
[114] S. Gifford, R.H. Dunstan, W. O'Connor, T. Roberts, R. Toia, Pearl aquaculture: profitable environmental remediation? Sci. Total Environ. 319 (2004) 27–37. Available from: https://doi.org/10.1016/S0048-9697(03)00437-6.
[115] G.R. MacFarlane, A. Pulkownik, M.D. Burchett, Accumulation and distribution of heavy metals in the grey mangrove, *Avicennia marina* (Forsk.) Vierh.: biological indication potential, Environ. Pollut. 123 (2003) 139–151.
[116] T. Yamamoto, I. Goto, O. Kawaguchi, K. Minagawa, E. Ariyoshi, O. Matsuda, Phytoremediation of shallow organically enriched marine sediments using benthic microalgae, Mar. Pollut. Bull. 57 (2008) 108–115.
[117] B. Konar, J.A. Estes, The stability of boundary regions between kelp beds and deforested areas, Ecology 84 (1) (2003) 174–185.

[118] N. Kriegisch, S. Reeves, C.R. Johnson, S.D. Ling, Phase-shift dynamics of sea urchin overgrazing on nutrified reefs, PLoS One 11 (2016) 1–15. Available from: https://doi.org/10.1371/journal.pone.0168333.

[119] J. Watson, J.A. Estes, Stability, resilience, and phase shifts in rocky subtidal communities along the west coast of Vancouver Island, Canada, Ecol. Monogr. 81 (2011) 215–239. Available from: https://doi.org/10.1890/10-0262.1.

[120] S.W. Purcell, C. Conand, S. Uthicke, M. Byrne, Ecological roles of exploited sea cucumbers, in: R.N. Hughes, D.J. Hughes, I.P. Smith, A.C. Dale (Eds.), Oceanography and Marine Biology: An Annual Review, Taylor & Francis Group and CRC Press, Boca Raton, FL, 2016, pp. 375–394. 510 pp.

[121] D.L. Leighton, L.G. Jones, W.J. North, Ecological relationships between giant kelp and sea urchins in southern California, in: E. Gordon Young, J.L. McLachlan (Eds.), Proceedings of the Fifth International Seaweed Symposium, August 25–28, 1965, Halifax, Canada, 1966, pp. 141–153. <https://doi.org/10.1016/B978-0-08-011841-3.50023-9>.

[122] X.T. Yuan, H.S. Yang, Y. Zhou, Y.Z. Mao, Q. Xu, L.L. Wang, Bioremediation potential of *Apostichopus japonicus* (Selenka) in coastal bivalve suspension aquaculture system, J. Appl. Ecol. 19 (4) (2008) 866–872.

[123] L.N. Zamora, X. Yuan, A.G. Carton, M.J. Slater, Role of deposit-feeding sea cucumbers in integrated multitrophic aquaculture: progress, problems, potential and future challenges, Rev. Aquacult. 10 (1) (2018) 57–74.

[124] E.R. Graham, J.T. Thompson, Deposit-and suspension-feeding sea cucumbers (Echinodermata) ingest plastic fragments, J. Exp. Mar. Biol. Ecol. 368 (1) (2009) 22–29.

[125] J. Gomes, A. Matos, R.M. Quinta-Ferreira, R.C. Martins, Environmentally applications of invasive bivalves for water and wastewater decontamination, Sci. Total Environ. 630 (2018) 1016–1027.

[126] I.C. Rosa, R. Costa, F. Gonçalves, J.L. Pereira, Bioremediation of metal-rich effluents: could the invasive bivalve *Corbicula fluminea* work as a biofilter? J. Environ. Q. 43 (5) (2014) 1536–1545.

[127] M.L. Kellogg, J.C. Cornwell, M.S. Owens, K.T. Paynter, Denitrification and nutrient assimilation on a restored oyster reef, Mar. Ecol. Progr. Ser. 480 (2013) 1–19.

[128] R.H. Carmichael, W. Walton, H. Clark, Bivalve-enhanced nitrogen removal from coastal estuaries, Can. J. Fish. Aquat. Sci. 69 (7) (2012) 1131–1149.

[129] A.S. Groesbeck, K. Rowell, D. Lepofsky, A.K. Salomon, Ancient Clam Gardens increased shellfish production: adaptive strategies from the past can inform food security today, PLoS One 9 (3) (2014) e91235.

[130] D. Lepofsky, N.F. Smith, N. Cardinal, J. Harper, R. Bouchard, D.I.D. Kennedy, et al., Ancient shellfish mariculture on the Northwest coast of North America, Am. Antiquity 80 (2) (2015) 236–259.

[131] T.J. Lavery, B. Roudnew, J. Seymour, J.G. Mitchell, V. Smetacek, S. Nicol, Whales sustain fisheries: blue whales stimulate primary production in the Southern Ocean, Mar. Mammal Sci. 30 (3) (2014) 888–904.

[132] T.J. Lavery, B. Roudnew, P. Gill, J. Seymour, L. Seuront, G. Johnson, et al., Iron defecation by sperm whales stimulates carbon export in the Southern Ocean, Proc. R. Soc. Lond. B: Biol. Sci. (2010) 1–5. Available from: https://doi.org/10.1098/rspb.2010.0863. rspb20100863.

[133] J. Roman, J.J. McCarthy, The whale pump: marine mammals enhance primary productivity in a coastal basin, PLoS One 10 (2010) e13255. Available from: https://doi.org/10.1371/journal.pone.0013255.

[134] N.A. Graham, S.K. Wilson, P. Carr, A.S. Hoey, S. Jennings, M.A. MacNeil, Seabirds enhance coral reef productivity and functioning in the absence of invasive rats, Nature 559 (7713) (2018) 250–253.

[135] J.C. Pistevos, I. Nagelkerken, T. Rossi, M. Olmos, S.D. Connell, Ocean acidification and global warming impair shark hunting behaviour and growth, Sci. Rep. 5 (2015) 16293. Available from: https://doi.org/10.1038/srep16293.

[136] W.J. Sydeman, E. Poloczanska, T.E. Reed, S.A. Thompson, Climate change and marine vertebrates, Science 350 (6262) (2015) 772–777. Available from: https://doi.org/10.1126/science.aac9874.

[137] T.B. Atwood, R.M. Connolly, E.G. Ritchie, C.E. Lovelock, M.R. Heithaus, G.C. Hays, et al., Predators help protect carbon stocks in blue carbon ecosystems, Nat. Clim. Change 5 (12) (2015) 1038–1045. Available from: https://doi.org/10.1038/nclimate2763.

[138] S.G. Donaldson, J. Van Oostdam, C. Tikhonov, M. Feeley, B. Armstrong, et al., Environmental contaminants and human health in the Canadian Arctic, Sci. Total Environ. 408 (22) (2010) 5165–5234.

[139] J.L. Kirk, I. Lehnherr, M. Andersson, B.M. Braune, L. Chan, A.P. Dastoor, et al., Mercury in Arctic marine ecosystems: sources, pathways and exposure, Environ. Res. 119 (2012) 64–87.

[140] M.D. Hollis, G.E. McCallum, Dilution is no longer the solution for pollution, Wastes Eng. 30 (1959) 578–581.

CHAPTER 48

Beyond prediction—radical ocean futures—a science fiction prototyping approach to imagining the future oceans

Andrew Merrie

Stockholm Resilience Centre, Stockholm University, Stockholm, Sweden

Chapter Outline

48.1 Introduction

This chapter[1] first introduces the concept of scenarios and salient critiques of prevailing quantitative/analytical approaches to scientific scenario development. Then, building on the scientific findings that make up this book, the chapter argues for why our changing oceans require a more radical, imaginative approach to complement existing quantitative models and analytical scenarios to paint a full picture of the future ocean. The specific approach of Science Fiction prototyping is described in detail as it was applied in the radical ocean futures project and the chapter concludes with key learnings and takeaways.

Stories are the glue that build and hold human civilization together. Humans use stories as important teaching and survival tools. Narratives, in short, can make us care about one another and our world, but they are double-edged swords. Stories can also deceive us, tear

[1] Please note that this chapter is adapted from and builds upon; A. Merrie, P. Keys, M. Metian, H. Österblom, radical ocean futures —scenario development using science fiction prototyping, Futures 95 (2018) 22–32. DOI: https://doi.org/10.1016/j.futures.2017.09.005.

Predicting Future Oceans.
DOI: https://doi.org/10.1016/B978-0-12-817945-1.00047-2

us apart, lull us into a false sense of security, unquestioningly accept the status quo. Just like science, narratives can be a tremendously powerful tool, but they are always normative and they can hold terrifying amounts of tangible, cultural, social, and political power [3]. They create our reality as much as they explain it. The futures we envision, be they positive or futures of collapse, may make us more likely to respond to events in the world in a way that moves us closer toward creating that envisioned future [4]. Our imaginations are key to many of the social, cultural, and technological developments that have contributed to and in some cases created the challenges we face today [5]. For example, the vision of those who were able to imagine shifting an entire society toward industrialization as occurred during the industrial revolution has transformed many societies and yet led to the creation of previously unimaginable problems that then had to be overcome, such as the use of child labor in factories during the 19th century. Our imagination has led us to envision the kind of societies in which we currently live, willingly or unwillingly. Critically, however, our capacity to imagine can also help us find our way through the planetary scale challenges humanity faces in creating an equitable and sustainable future on planet earth. Traditionally, the practice of imagination within science has been restricted to a very particular approach to developing scenarios within the field.

Scenarios can help organizations, individuals, communities, corporations, and nations develop the capability to deal with the unknown and unpredictable, or the unlikely, but possible. A scenario, defined as "a coherent, internally consistent and plausible description of a potential future trajectory of a system" [6], represents an important tool for proactively thinking about and anticipating things to come. Scenarios have a long history in military and corporate strategy [7], and are increasingly used by governments as a tool to support policy making in the field of natural resource management [8–10]. This is most evident in the field of climate change, where targets that were agreed as part of the United Nations Framework Convention on Climate Change "Paris Agreement" substantially rely on estimated future trajectories given different pathways for greenhouse gas emissions [11].

Diverse, multiple, cumulative human activities influence marine ecosystems [12], with dramatic increases in fishing pressure over time as one of the clearest pieces of evidence for such impacts. However, scenarios for the future of marine fish stocks [13,14] have been criticized for the assumptions they make about linear dynamics operating in marine ecosystems or the limited adaptive capacity of societies [15,16]. They have also been critiqued for providing limited empirical support for suggested management tools [17] and a lack of recognition of dynamics operating at the global scale (such as fishing industry consolidation), which can be of critical importance for both fish stocks and the changing state of the oceans more broadly [18]. Together these criticisms highlight the challenge of incorporating nonlinear change, radical uncertainty, unpredictability, and surprise in systems such as the global oceans, which are closely intertwined with multiple human activities operating across geographic scales [19]. What is more, scientific scenarios, such as those highlighted, tend to focus on the use of analytical tools and quantitative techniques, almost always at the expense of storytelling, imagination, metaphor, and creativity. Indeed, guidance around scientific writing and practice often actively discourages the incorporation of such elements, as they are suggested to undermine scientific rigor and the credibility of the scientists applying them [20]. Adding these components is not only about outreach and connecting scenarios to a wider audience, but when integrated into the scenario creation process itself, these elements can reveal issues that are hidden, highlight the importance of trends in new ways, and potentially lead to the development of new or augmented ocean science that is able to investigate novel questions that are posed in the process of creating scenarios [21].

48.2 Science fiction—practical, structured blending of science with imagination, storytelling, and speculation

Science fiction, part of a broader genre of speculative fiction, and dealing principally with the impact of actual or imagined science on society and/or individuals, represents one way to imagine alternative futures that can provide a useful contribution to "predicting the future oceans". Francis Bell describes classic works of science fiction by Arthur C. Clarke, Ursula K. LeGuin, and others as applied fictions. Their books and short stories provoke discussions and dialogues about the future of technological evolution, robots, and the future of gender norms in society, as only a few examples. Science fiction may help us connect to better, more healthy, and sustainable oceans. To head toward different, perhaps even better, futures, we have to be able to imagine what those futures might look, sound, smell, and feel like, and what the impacts and implications of different types of change might be [22].

The narratives that describe the Earth's oceans have shifted from describing the oceans as infinite, wild, and thriving...to finite, fragile and full of garbage—oily, hot, acidic, plastic, and nearly empty. Human impact on marine ecosystems and the scale and pace at which change happens today, means we must be better at anticipation and we need to be fast learners, despite history showing how challenging this is in practice. The discipline of anticipation distinguishes between planning for the future, preparing for the future, and using novel futures to discover new ways to make sense of the emergent present and take advantage of the unknowable future [23]. Stories about our future oceans fall into the latter category, and can operate as narrative simulations to help us address the complex challenges we face today as well as act as a bridge between our complex reality in the present and our radically uncertain future.

48.3 The method of science fiction prototyping

Science fiction prototyping is a method [24,25] developed by Brian David Johnson in his role as a futurist at the Intel Corporation (Fig. 48.1). The approach was designed to assist Intel engineers in thinking "humanistically" about technologies they were developing [25].

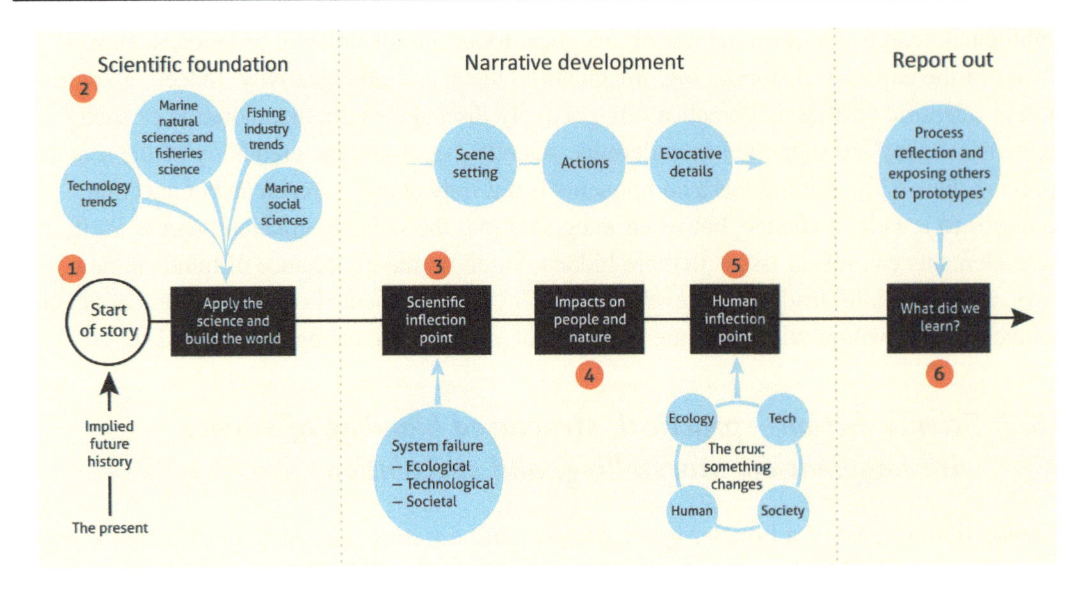

Figure 48.1

This figure illustrates the process of science fiction prototyping. The process here is shown along a line, but can be recursive and iterative. Reproduced with permission from the Authors and Elsevier. *Figure Design: Azote/Jerker Lokrantz.*

Science fiction prototypes are "short works of fiction, grounded in scientific fact and crafted for the purpose of starting a conversation about the implications, effects, or ramifications of technology and the future" [26]. Because of the purpose they serve, it is important to point out that although science fiction prototypes draw inspiration from science fiction, the authors are not making the argument that these prototypes approach the richness, thematic sophistication, or societal visioning that is achieved in high-quality literary science fiction [26]. However, science fiction prototypes allow for a focused, tailored, and creative way to think about possible futures around a particular issue—in this case the future of marine fisheries and the global oceans, building on the latest marine science. The process for science fiction prototyping begins with the present, and an acknowledgment of an implied future based on a robust scientific foundation. To generate a high degree of content credibility [27], the "radical ocean futures" were developed on four pillars that together formed the scientific base for the scenarios and when combined, helped account for the ability to apply "tacit knowledge while minimizing tacit biases" [28]. The four pillars of the scientific foundation draw on a diverse set of data sources and include: (1) technology trends linked to key story elements; (2) marine natural sciences and fisheries science; (3) fishing industry trends; and (4) marine governance trends. Building a believable world that pushes the boundaries of current technological and social/institutional frameworks is critical. Revealing too much of the world, however, can draw power from the narrative. As important as it is to describe the future world, it is equally important to not describe too much, in order to allow for the reader to explore and imagine; a process that has been called "mystery boxing" by Hollywood director J.J. Abrams in a March 2007 TED talk [29]. Once the scientific foundation has been applied and the world is "built," the following stage involves developing a clear turning point in the narrative, where the world changes in some significant way. During the moment of the "scientific inflection point," ecological, technological, and/or societal problems introduce tensions to the story that the characters must navigate. The impacts on people and nature around this inflection point are explored, via the plot and action within the world of the narrative. Interacting with the plot and action is the more personal, "human inflection point," which illustrates how individuals do not simply react to changes, but rather interact with them.

The process and method is both recursive and iterative. During the final step of the process, the science fiction prototyping method advises finishing each scenario with some form of reflection, reporting-out, and learning.

48.4 Radical ocean futures—scientifically valid scenarios

How then do we put this concept into practice? How can one weave together science and storytelling in a way that is both scientifically credible and narratively compelling? radical ocean futures, a science fiction protoyping approach to imagining our future oceans [30,31], provides one such example. The heart of the project consists of four short "radical ocean futures" created following the method of science fiction prototyping, and representing scientifically grounded narratives of potential future oceans. The four scenarios take place between 2050 and 2070—far enough in the future for more radical changes to occur, while still close enough to the present to be relatable to the reader and relevant for anticipatory governance interventions. While all four scenarios include similar information based on outputs from existing analytical scenarios on likely future dynamics (sea level rise and increasing temperatures, see for example [32]), each interprets more uncertain predictions differently. Each is narrated from a distinct perspective, a naïve scientist, an intrepid journalist, a lone fisher, and an ambitious CEO. The scenarios play with these archetypes, while emphasizing the human character in each of them, to evoke engagement and interest. Oceans Back from the Brink, Fish Inc., Rime of the Last Fisherman, and Rising Tide are considered along both an ecological (sustainable–collapse) and a social (connected–fragmented) dimension, thereby creating a "scenario space" [33] for some of the many possible futures that could emerge depending on the emphasis society decides to place on the implementation of different ecological, sociopolitical, technological, and economic changes (see Fig. 48.1). In "Fish Inc." we imagine Astrid Amundsen, the Norwegian Seafood CEO who successfully privatized global food security. "Rime of the Last Fisherman" gives us a peek at the journals of Alejandro Balmaceda, reportedly the world's last ocean-going fisherman. In "Rising Tide" a journalist tells the story of Tarawa Station and how small

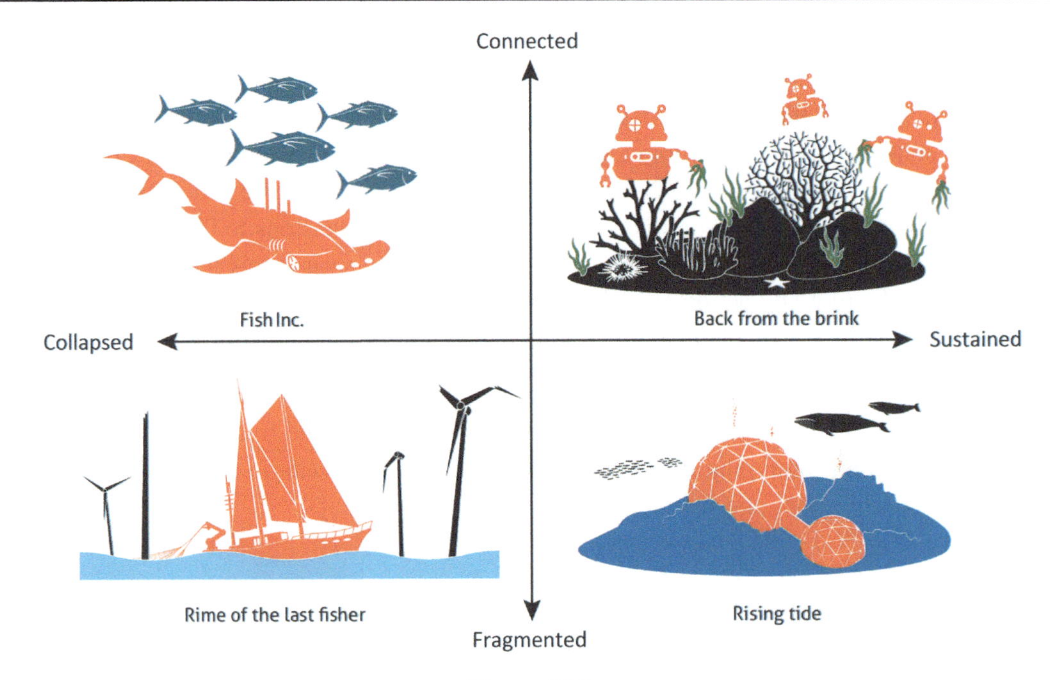

Figure 48.2
The scenario space. Narrative scenarios are represented across a two-dimensional space and each illustration is a representation of a key defining element for each of the four "radical ocean futures." These four "radical ocean futures" do not represent "the" future but four possible futures in a space of many possible, interacting futures. The "collapsed to sustained" on the horizontal axis refers to the ecological dimension and the "fragmented to connected" on the vertical access refers to the societal dimension. Reproduced with permission from the Authors and Elsevier. *Figure Design: Azote/Jerker Lokrantz.*

island states in the Pacific ocean banded together to survive and thrive, despite several meters of sea level rise. In "Oceans Back From the Brink" Prof. Michelle Ching, a scientist descended from a great pirate queen, tells us how, through artificial intelligence and collective action, the oceans were brought back from the brink of collapse (Fig. 48.2).

The characters in the crafted scenarios are experiencing both desirable and undesirable existences. Likewise in the real world, desirability is relative. For example, a fishing conglomerate that is aiming for large-scale harvest of skipjack tuna (*Katsuwonus pelamis*) in the Western Pacific is likely to have very different ideas about what is "desirable" (or even what is "sustainable") compared to a group of small-scale fishers in Palau. As such it is important to represent a spectrum of perspectives to account for the diverse lived experiences of groups and individuals across a set of scenarios crafted using the science fiction protoyping approach.

The scenarios also underline the importance of creating unusual and challenging collaborations between those engaged with managing marine natural resources and technology, military, governance, business, and local and Indigenous communities to expand the thinking space about policy development in marine fisheries and the future oceans. Furthermore, these scenarios help illustrate the need to think actively about preserving and restoring the global commons, including actively considering the costs and benefits of protecting the high seas, but also ways to invest in future transformative efforts. The different scenarios do not only represent interesting stories, but also illustrate a method for synthesizing information from diverse disciplines and knowledge sources, while accounting for nonlinear change and coevolutionary processes. These aspects are difficult to include and model with other scientific methods and tools (e.g., more quantitative scenarios) and may therefore be omitted.

48.5 Conclusion

The value of these imaginative approaches is not in attempting to supplant the critical scientific work of "predicting the future ocean," but rather to provide a powerful complementary set of tools that enables scientists, policy makers, and others to engage with plausible and complex futures. The act of imagining and envisaging diverse futures can elucidate unseen pathways for mobilizing human ingenuity while also highlighting the limitations of human agency in a complex world. This approach to scenarios can contribute to building anticipation and response capacities in marine science, governance, and management organizations and, more normatively, for those seeking to change the world for the better. Where anticipation of nonlinear change is impossible, such imaginative approaches can contribute—through the creation of alternate possibility spaces (which can be used as a type of "narrative simulation")—a way to experiment with and learn from the many potential surprising situations and shifts that might occur in the future and are hinted at in the wide range of research insights collected in this very volume. Such a radical approach to imagining the future can help us to see the possibilities that exist for unusual and challenging collaborations to expand the thinking space around the health and sustainable use of the future oceans for a diverse humanity on a crowded planet.

References

[1] D.J. McCauley, M.L. Pinsky, S.R. Palumbi, J.A. Estes, F.H. Joyce, R.R. Warner, Marine defaunation: animal loss in the global ocean, Science 347 (6219) (2015) 1255641.

[2] T.S. Galloway, M. Cole, C. Lewis, Interactions of microplastic debris throughout the marine ecosystem, Nat. Ecol. Evol. 1 (5) (2017) 0116.

[3] M. Foucault, Power/Knowledge: Selected Interviews and Other Writings, Pantheon, 1980, pp. 1972–1977.

[4] A. Evans, The Myth Gap: What Happens When Evidence and Arguments Aren't Enough, Transworld Publishers, Ealing, 2017.

[5] N.Y. Harari, Sapiens: A Brief History of Humankind., Vintage, 2018.

[6] E. Oteros-Rozas, B. Martín-López, T.M. Daw, E.L. Bohensky, J.R.A. Butler, R. Hill, et al., Participatory scenario planning in place-based social-ecological research: Insights and experiences from 23 case studies, Ecol. Soc. 20 (2015). Available from: https://doi.org/10.5751/ES-07985-200432.

[7] P. Schwartz, The Art of the Long View: Planning for the Future in an Uncertain World, John Wiley & Sons, Chichester, West Sussex, 1996.

[8] G.D. Peterson, G.S. Cumming, S.R. Carpenter, Scenario planning: a tool for conservation in an uncertain world, Conserv. Biol. 17 (2) (2003) 358–366. Available from: https://doi.org/10.1046/j.1523-1739.2003.01491.x.

[9] L.S. Evans, C.C. Hicks, P. Fidelman, R.C. Tobin, A.L. Perry, Future scenarios as a research tool: investigating climate change impacts, adaptation options and outcomes for the Great Barrier Reef, Australia, Hum. Ecol.: Interdisc. J. 41 (2013) 841–857. Available from: https://doi.org/10.1007/s10745-013-9601-0.

[10] I.M. Rosa, H.M. Pereira, S. Ferrier, R. Alkemade, L.A. Acosta, H.R. Akcakaya, et al., Multiscale scenarios for nature futures, Nat. Ecol. Evol. 1 (10) (2017) 1416.

[11] IPCC, in: C.B. Field, V.R. Barros, D.J. Dokken, K.J. Mach, M.D. Mastrandrea, T.E. Bilir, et al. (Eds.), Climate Change 2014: Impacts, Adaptation, and Vulnerability. Part A: Global and Sectoral Aspects. Contribution of Working Group II to the Fifth Assessment Report of the Intergovernmental Panel on Climate Change, Cambridge University Press, Cambridge, United Kingdom and New York, 2014, p. 1132.

[12] J.B.C. Jackson, M.X. Kirby, W.H. Berger, K.A. Bjorndal, L.W. Botsford, B.J. Bourque, et al., Historical overfishing and the recent collapse of coastal ecosystems, Science 629 (2001) 629–637. Available from: https://doi.org/10.1126/science.1059199.

[13] B. Worm, E.B. Barbier, N. Beaumont, J.E. Duffy, C. Folke, B.S. Halpern, et al., Impacts of biodiversity loss on ocean ecosystem services, Science 314 (2006) 787–790. Available from: https://doi.org/10.1126/science.1132294.

[14] C. Costello, D. Ovando, T. Clavelle, C.K. Strauss, R. Hilborn, M.C. Melnychuk, Global fishery prospects under contrasting management regimes, Proc. Natl. Acad. Sci. U.S.A. 113 (18) (2016) 1–5. Available from: https://doi.org/10.1073/pnas.1520420113.

[15] T. Branch, Not all fisheries will be collapsed in 2048, Mar. Policy 32 (2008) 38–39.

[16] R. Hilborn, Reinterpreting the state of fisheries and their management, Ecosystems 10 (2007) 1362–1369. Available from: https://doi.org/10.1007/s10021-007-9100-5.

[17] B. Worm, Averting a global fisheries disaster, Proc. Natl. Acad. Sci. U.S.A. 113 (18) (2016) 4895–4897. Available from: https://doi.org/10.1073/pnas.1604008113.

[18] H. Österblom, J. Jouffray, J. Spijkers, Where and how to prioritize fishery reform? Proc. Natl. Acad. Sci. U.S.A. 113 (25) (2016) 124–126. Available from: https://doi.org/10.1073/pnas.1605723113.

[19] A. Merrie, D. Dunn, M. Metian, A. Boustany, Y. Takei, A. Oude Elferink, et al., An ocean of surprises—trends in human use, unexpected dynamics and governance challenges in areas beyond national jurisdiction, Global Environ. Change Hum. Policy Dimensions 27 (2014) 19–31.

[20] A.H. Hofmann, Scientific Writing and Communication: Papers, Proposals, and Presentations., Oxford University Press, 2014.

[21] H. Österblom, B. Crona, C. Folke, M. Nyström, M. Troell, Marine ecosystem science on an intertwined planet, Ecosystems 20 (2017) 54–61. Available from: https://doi.org/10.1007/s10021-016-9998-6.

[22] E. Bennett, M. Solan, R. Biggs, T. McPhearson, A. Nörstrom, P. Olsson, et al., Bright Spots: seeds of a good anthropocene, Front. Ecol. Environ. 14 (8) (2016) 441–448. Available from: https://doi.org/10.1002/fee.1309.

[23] R. Miller, R. Poli, P. Rossel, The discipline of anticipation: exploring key issues, in: Bellagio Document 4: Working Paper 1, 2013. Accessed Online October 23, 2018. <http://filer.fumee.dk/5/The%20Discipline%20of%20Anticipation%20-%20Miller,%20Poli,%20Rossel.pdf>.

[24] B.D. Johnson, Science fiction prototypes or: how I learned to stop worrying about the future and love science fiction, in: V. Callaghan, A. Kameas (Eds.), Intelligent Environments 2009: Proceedings of the 5th International Conference on Intelligent Environments, Barcelona 2009, Amsterdam, IOS Press BV, 2009, pp. 3−19. https://doi.org/10.3233/978-1-60750-034-6-3.

[25] B.D. Johnson, Science Fiction Prototyping: Designing the Future With Science Fiction, Morgan & Claypool, San Rafael, 2011.

[26] M. Burnam-Fink, Creating narrative scenarios: science fiction prototyping at emerge, Futures 70 (2015) 48−55. Available from: https://doi.org/10.1016/j.futures.2014.12.005.

[27] C. Selin, Trust and the illusive force of scenarios, Futures 38 (1) (2006) 1−14. Available from: https://doi.org/10.1016/j.futures.2005.04.001.

[28] G. Graham, A. Greenhill, V. Callaghan, Technological forecasting and social change special section: creative prototyping, Technol. Forecasting Soc. Change 84 (2014) 1−4. Available from: https://doi.org/10.1016/j.techfore.2013.11.007.

[29] J.J. Abrams, The Mystery Box. TED—Ideas Worth Spreading March 2007. <https://www.ted.com/talks/j_j_abrams_mystery_box>, 2007 (accessed 13.02.17).

[30] Radical Ocean Futures. <https://radicaloceanfutures.earth/>, 2017 (Last Updated: 30.06.17).

[31] A. Merrie, P. Keys, M. Metian, H. Österblom, Radical ocean futures—scenario development using science fiction prototyping, Futures 95 (2018) 22−32. Available from: https://doi.org/10.1016/j.futures.2017.09.005.

[32] J.P. Gattuso, A. Magnan, R. Bille, W.W.L. Cheung, E.L. Howes, F. Joos, et al., Contrasting futures for ocean and society from different anthropogenic CO_2 emissions scenarios, Science 349 (6243) (2015) aac4722-1−aac4722-10. Available from: https://doi.org/10.1126/science.aac4722.

[33] D. Staley, M.P. Dias, V. Tzankova, T. Schiphorst, K.E. Behar, Imagining possible futures with a scenario space, Parsons J. Inform. Mapping 1 (4) (2009) 1−8.

SECTION 8

Conclusion

CHAPTER 49

In conclusion: Sustainable and equitable relationships between ocean and society

Yoshitaka Ota

Nippon Foundation Nereus Program, School of Marine and Environmental Affairs, University of Washington, Seattle, WA, United States

Chapter Outline

49.1 Introduction

Predicting future oceans requires integrated interdisciplinary research and examining mechanisms of socio-ecological linkages at multiple spatial scales. In the preceding chapters we attempted to answer a simple but far-reaching question: what will future oceans be like? We investigated how oceans will be affected by climate change, predicting that ocean systems will be driven into unprecedented levels of biophysical changes. We explored various theories of causal linkages between global environmental changes, states of fisheries, sociocultural conditions of coastal communities, and ocean governance. By merging both natural and social science perspectives, we also proposed visions for sustainable and equitable relationships between human societies and marine and coastal ecosystems.

As sea temperature increases and ocean acidification intensifies, the patterns and magnitudes of marine primary production will shift and lead to ecological changes, from the geographical redistribution of species and ecosystem compositions to biodiversity

Predicting Future Oceans.
DOI: https://doi.org/10.1016/B978-0-12-817945-1.00058-7

losses. These changes in marine ecosystems will affect fisheries productivity and thus our human societies. While some gains in both catch volume and value are anticipated and observed regionally, our research suggests that the overall impact will be negative, and management inefficiency and inability to respond appropriately will exacerbate inequalities for economically disadvantaged and politically marginalized populations. This outcome is highly likely under current trends, but it is not inevitable.

Our outlooks are founded upon a framework of socioecological systems, integrating mechanisms of natural systems and societal aspects without being constrained to a single model or by theories from a specific discipline. While we strived to offer a diversity of perspectives and expertise on the relationships between oceans and human societies, this is by no means comprehensive. Yet the limitations in our understanding do not diminish our confidence in the overall conclusions that can be drawn from these chapters. As we endeavor to address our guiding question on future oceans, this concluding chapter explores the idea of changing oceans through two lenses: first, that the future will be a convergence of multiscaled, complex interactions, some of which are compounding while others will be conflicting; and second, our relationships with oceans are heterogeneous, and without explicit recognition of such heterogeneity, sustainability and equity across our relationships may be in jeopardy. We will thus conclude by proposing further and deeper interdisciplinarity in the way in which we study ocean and human relationships.

49.2 Climate change impacts

Global scale models provide long-term projections of climate change impacts and multiscale interactions between marine and coastal ecosystem. Projections generally extend to around 2050–2100 and these models present strategic policy options as alternative outcomes. They identify a chain of causal mechanisms, potential tradeoffs, and attempt to minimize model uncertainties by integrating layers of risks and outliers (e.g., simulation model ensemble, see Chapter 7: Building confidence in projections of future ocean capacity). The model of cascading impacts from oceanography to ecology is the initial framework to assess the changes and the principle drivers of future ocean conditions, the impact of climate change.

As Stock et al. (Chapter 2: Changing ocean systems: from understanding to prediction) argue, the four aspects of biophysical changes—sea temperature rise, acidification, deoxygenation, and shifting primary productivity—are "in a state of constant flux, interacting with a broad range of natural and unprecedented anthropogenic drivers." Frölicher (Chapter 5: Extreme climatic events in the ocean) shows that frequency, intensity, and duration of marine heat waves, for example, will increase under anthropogenic climate change even if the emissions targets of the Paris accord are reached. Changes in oceanographic conditions are projected to have effects on marine life and generally expected to shift their distribution poleward toward cooler waters. Reygondeau (Chapter 9: Current and future biogeography of exploited marine exploited groups under climate change) projects that the loss of species richness will be amplified in the equatorial regions, while du Pontavice (Chapter 12: Changing biomass flows in marine ecosystems: from the past to the future) finds a 20% overall decline in global fish biomass under the current climate change scenario. Greater occurrences of harmful algal blooms, mass strandings of mammals, and mass mortalities due to marine heat waves are also projected by Frölicher (Chapter 5: Extreme climatic events in the ocean).

With regard to the downstream effects of climate change, the prevalence of chemical contaminants is a good example of how compounding and conflicting interactions in ocean systems lead to a high degree of prediction uncertainties. Thackray and Sunderland (Chapter 6: Seafood methylmercury in a changing ocean) posit that increasing seawater temperature should intensify bioaccumulation in fish species as their higher metabolism in warmer water increases overall food consumption; yet warmer, more productive ecosystems will be dominated by shorter food webs with high turnover, thus dampening the biomagnification effect throughout food webs. Other microlevel effects such as changing size of phytoplankton communities and migration and geographical redistributions of nekton species will likely have various, often competing effects on contamination of fish species and seafood.

The projected ecological responses to climate change are therefore not always geographically uniform, and the reactions of marine ecosystems and organisms vary with local biophysical and ecological factors including cyclical climate oscillations [see Roberts (Chapter 13: The role of cyclical climate oscillations in species distribution shifts under climate change)] and life histories of local biota. Asch (Chapter 4: Changing seasonality of the sea: past, present, and future) notes, for example, that seasonal responses of marine animals to climate change differ across trophic levels. Phenological plasticity of a given population is dictated by its genetic variance, while on the individual scale, factors such as size and age can influence how they cope with environmental variabilities. The impacts of fishing across marine food webs can also influence ecosystem responses, not only because of direct tolerance of fish species to changes in ocean warming but also through their links to other marine organisms through feeding and other interspecies relations. As highlighted by Henschke (Chapter 14: Jellyfishes in a changing ocean), even efficient adapters to extreme environments such as jellyfish react to climate change not in relation to sea temperature but through indirect effects of increase in their zooplankton prey biomass.

Thus beyond projections of global trends, the focus of research on climate and oceans has shifted to our management adaptations, including how fisheries models can best incorporate projected population changes [see Tanaka (Chapter 19: Integrating environmental information into stock assessment models for fisheries management)] to implications on short-term stock assessments and ecosystem-based management [see Selden and Pinsky (Chapter 20: Climate change adaptations and spatial fisheries management)]. As the toolbox of ecological and biological management expands, science will help quantify our uncertainties, not only for species responses but in how fisheries themselves adapt [see Gonzalez Taboada (Chapter 15: Understanding variability in marine fisheries: importance of environmental forcing)].

49.3 Links to human society

As the impacts of climate change transcend ocean systems, there is increasing emphasis on identifying and addressing linkages with emerging and ongoing human development issues, such as socioeconomic inequity and vulnerabilities at the intersect between environmental stresses and the politically marginalized. The loss of revenue in marine fisheries is projected to be higher amongst developing countries, and the global market may exacerbate such trends [see Lam (Chapter 18: Projecting economics of fishing and fishing effort dynamics in the 21st century under climate change) and Chen (Chapter 23: The big picture: future global seafood market)]. Meanwhile, the promotion of aquaculture to meet growing seafood demand also carries risks such as increased pollution and economic barriers to production for food security [see Oyinlola (Chapter 22: Mariculture: perception and prospect under climate change)], and the development of alternative industries such as ecotourism will require more resources and investments to benefit development [see Wabnitz (Chapter 21: Adapting tourist seafood consumption practices in Pacific Islands to climate change)]. Complexities embedded in these multiscale interactions demand that the development of ocean governance be a constant negotiation process across scales and priorities.

Stakeholder engagement has been proposed as a key factor in responding to complex impacts beyond changes in ecosystems, namely, through socioecological interactions. Both Blasiak (Chapter 34: Climate change vulnerability and ocean governance) and Petersson (Chapter 36: New actors, new possibilities, new challenges—nonstate actor participation in global fisheries governance) call for engagement and involvement of civil society partners in ocean policies using existing and potential new platforms, and Boustany (Chapter 45: The trouble with tunas: international fisheries science and policy in an uncertain future) argues for greater emphasis on scientific legitimacy and oversight among international institutions and regional fisheries management organizations. The gaps identified in these contributions are not limited to the issue of knowledge transfer and integration, however, but also include the lack of states' capacity and willingness to abide by compliance mechanisms [see Guggisberg (Chapter 43: Verifying and improving states' compliance with their international fisheries law obligations)], which are in some cases embedded within neoliberal ideology [see Seto (Chapter 35: The last commons: (re)constructing an ocean future)]. The challenge in advancing more equitable and engaged ocean governance is therefore dependent on both efforts to promote governance in policy design, and to review and implement what state, intergovernmental, and regional governing bodies have agreed upon without compromising legitimacy, accountability and responsibility. This is crucial given that the stakes of livelihood or food security risks are much more urgent than a future decades away, and we thus argue science can provide more than strategic perspectives on management but can help incorporate and support local knowledge and trust [see contributions from Mason (Chapter 28: Ocean policy on the water—incorporating fishermen's perspectives and values) and Vierros and Ota (Chapter 29: Integration of traditional knowledge in policy for climate adaptation, displacement and migration in the Pacific)].

The collage that is ocean management requires an ambitious commitment of resources and public engagement, and an increasing reliance on innovation to meet monitoring needs as the use of ocean spaces intensifies from fisheries, mining, energy, bioprospecting, and other sectors. The concern over the development of multiple uses of ocean spaces is that they would synergistically damage marine ecosystems, for which we know much less than any other part of global landscapes. Recent developments and debates on high seas governance shows that we need to increase our capacity to comprehend the roles of connectivities and transient nature across marine systems in order to successfully manage social and ecological systems spanning within and beyond national jurisdictions [see Dunn et al. (Chapter 41: Incorporating the dynamic and connected nature of the open ocean into governance of marine biodiversity beyond national jurisdiction)]. As noted by Ortuno Crespo (Chapter 46: The road to implementing an ecosystem-based approach to high seas fisheries management), modern technology can aid in tracking and monitoring fish and fisheries footprints across the oceans.

49.4 Values and relationships of oceans

The recognition of global climate change as the main driver of environmental changes in future oceans implies a need for capacity building that can accommodate the complexities arising across both natural and social systems. Studies on climate change impacts address the gaps in scientific and management capacity, balancing between responses and tradeoffs among different socioecological links. Though we identify these gaps from available data and models and call for further innovation, we also recognize that incapacity is not simply due to the lack of knowledge, but also to the ways in which global projections are set within worldviews that are inherently confined within limited parameters, namely, the quality of the environment and quantities of resources.

The importance of recognizing heterogeneity manifests in how we respond to multiscale interactions and impacts of environmental changes between different communities, social groups, and individuals. As noted by Oestreich et al. (Chapter 26: The impact of environmental change on small-scale fishing communities: Moving beyond adaptive capacity to community response), two artisanal fishing villages, both on the same lagoon in India, reacted differently to their environmental challenges. These differences in the ways communities respond—either reactively or proactively—are not limited to the availability of alternative resources or financial capacity. Also in Southeast Asia, the future of mangrove fishing is driven by demographic changes of the fishing communities, but does not follow the prevailing Malthusian view that directly links population increases and overexploitation. Seary (Chapter 27: The future of mangrove fishing communities) found that fishers' response to the change is not the intensification of fishing, but finding alternative uses for the ecosystem, such as switching to different mangrove-associated species, and only if their social needs are accommodated by the response strategy. Based on her field-based observations, it is possible that an increase in household members and diversification of household income can allow fishers to accept reductions in catch.

In the case of indigenous peoples, community concerns related to climate-mediated environmental changes are not restricted to impacts on the biophysical state of the environment but also include impacts on livelihoods and disruptions to people's knowledge systems, and relationships (with each other, with local species, and with local landscapes) [see Cisneros-Montemayor and Ota (Chapter 30: Coastal indigenous peoples in global ocean governance)]. Still, biophysical changes in the environment may reinforce existing socioeconomic, and political imbalances—particularly those related to resource management, conservation, and self-determination. Despite considerable advances in the comanagement of Arctic species, decision-making process do not always include values of importance for coastal indigenous peoples. Kenny (Chapter 24: Climate change, contaminants, and country food: collaborating with communities to promote food security in the Arctic) points out that many of the impacts that matter most to local people are enveloped within values that are not strictly ecological, but that articulate relationships with "nature" as part of life and identity.

In the current system of ocean economy and development, however, the scope of values and relationships scarcely reflect the diverse positions of those engaged with the oceans. Current trajectories of existing relationships limit our efforts to imagine transformative pathways and, rather, simply increase the number of uses for ocean space. Consequent future scenarios might then be dominated by collaborations among powerful existing industries rather than creating a new policy space for more diverse and equitable partnerships [see Merrie (Chapter 48: Beyond prediction—radical ocean futures—a science fiction prototyping approach to imagining the future oceans)]. Given the limited availability of resources required for such development, the objectives of a "Blue Economy" may be attainable only if governments prioritize the transformative nature of economic development through proactive integration of various ecological and social values and especially by creating opportunities to advance social equity [see Cisneros-Montemayor (Chapter 38: A Blue Economy: equitable, sustainable, and viable development in the world's oceans)]. Currently, financial gains in ocean-based economic activities are dominated by extractive industries [see Cisneros-Montemayor (Chapter 38: A Blue Economy: equitable, sustainable, and viable development in the world's oceans)] while subsidies continue to distort the economic playing field in favor of industrial fleets [see Sumaila (Chapter 44: Understanding potential impacts of subsidies disciplines and small-scale fisheries)].

Beyond economic gains, ensuring that human security is considered is an urgent goal in our response to global environmental changes. Climate change impacts on both individual wellbeing and political stability are already discussed within the science and policy interface, yet business sectors are still incrementally engaging with the issues, including labor abuses in seafood supply chains. There have been calls for socially responsible seafood production [see Teh et al. (Chapter 31: The relevance of human rights to socially responsible seafood)], that urges industry to avoid ongoing risks for food, human, and cultural security in seafood production systems, caused by consolidation of access rights and expansion of international market operations, as opposed to a focus on livelihoods. Typically, the responses of the private sector, primarily through their Corporate Social Responsibility strategies, are still limited and lack accountability. Thus interventions by the public sector remain critical [see Swartz (Chapter 32: The emergence of corporate social responsibility in the global seafood industry: potentials and limitations)]. Regarding conflicts at the individual and national level, we know little about the mechanisms underlying these negative impacts, but can begin to draw inferences based on analyses of historical and more recent data [see Spijkers (Chapter 37: Exploring the knowns and unknowns of international fishery conflicts)]. More precisely, we have less confidence in predicting the impacts of social and political changes as opposed to those driven by climate, even though these human dimensions, in the Anthropocene, are more rapidly controllable through our efforts.

49.5 Predicting our future oceans

In this book, we have focused on what we have become and who is at risk now rather than what we can prepare for, to advance our discussion for future oceans. What is certain is that the multidimensional dynamics of our relationships with oceans will evolve in the future. Some changes may be gradual while others could be more abrupt, and we have already seen such disturbances in various marine ecosystems around the world. Through these evolving interactions, the dynamics between society and the environment, or more precisely various components of both, will likely be affected. Such impacts should not be viewed simply in terms of gains and losses or winners and losers, but must be examined within a much broader context of political powers and economic capacities. As human societies adapt to the new contexts of ocean systems, we must reassess how we value our relationships with our oceans and how we wish to secure our livelihoods. The common causal thread linking through climate, environment, economy, society, and a policy response may not be the appropriate model for understanding the intricate dynamics of current and future oceans, and solutions based on this domino theory may fail as an adaptation policy and furthermore cause imbalanced burdens and inequalities in our relationships with the oceans.

Faced with unprecedented changes to the oceans, we must demand changes of a similar magnitude in the way we govern oceans, including increased accountability, transparency, and equity in the distribution of policy costs and resource benefits. We know that conferring control and access to ocean resources to a restricted number of stakeholders would inevitably delay responses with wider benefits. Our historical paths suggest that environmental quality and human wellbeing are not as tightly coupled as often understood; tradeoffs between environmental protection and societal burdens are skewed, especially against people who are politically marginalized. Creating safe operating spaces *for all* must be emphasized in our responses, building and promoting economic and social foundations for environmental action with a strategic focus on diverse values and needs [see Singh (Chapter 39: Can aspirations lead us to the oceans we want?)].

Future oceans will demand that scholars reframe our research so that social and ecological links are examined through the dual lenses of multidimensional interactions and of the heterogeneity of values and relationships embedded in social, cultural, and political contexts. As Harvey [1] suggests in his discussion of environment and ecology, therefore we must recognize that "all critical examinations of the relation to nature are simultaneously critical examinations of society." We must recognize that our perspective in studying oceans cannot be solely dominated by propositions for causal linkages from environmental changes to ecological impacts to economic effects to societal inequalities to insecurity for individuals. As we have argued throughout, links between the oceans and society are not a set of uniform drivers and impacts, but are interchangeable; impacts can turn into drivers through our actions, and our actions are prompted by choices in our values and relationships.

Our desires for the oceans are not future concerns, but are a matter of urgency. However, if we only accommodate the immediate needs of people, all values will not be represented; the future oceans will be inequitable. Our response to ocean changes is critical given that we need to be responsible for both current and future, local and global impacts, and both ecological and societal outcomes. Those connections are hard to make, but our desire to understand the oceans (and being humble upon facing their vastness) and our human values provide a nexus and the pathways to achieve sustainable and equitable relationships between oceans and peoples.

Reference

[1] D. Harvey, The nature of environment: dialectics of social and environmental change, Soc. Regist. 29 (29) (1993) 1–51.

Index

Note: Page numbers followed by "*f*" and "*t*" refer to figures and tables, respectively.

F

G

J

K

L

N

O

P

Q

R

T

U

V

W

Y

Z